PIPELINE RISK ASSESSMENT

The Definitive Approach and Its Role in Risk Management

PIPELINE RISK ASSESSMENT

The Definitive Approach and Its Role in Risk Management

W. Kent Muhlbauer

PIPELINE RISK ASSESSMENT: The Definitive Approach and Its Role in Risk Management
Publisher: Expert Publishing, LLC
Author: W. Kent Muhlbauer
Layout and Artwork: Meredith Foster, Total Communications, Inc
Additional Artwork: Chelsea Ilyse Scott

ISBN 978-0-9906700-0-1

Contents

I.1 ACRONYMS

ACVG AC (alternating current) Voltage Gradient
AGA American Gas Association
ANSI American National Standards Institute
API American Petroleum Institute
APWA American Public Works Association
ASME American Society of Mechanical Engineers
AST Above ground Storage Tank
CGA Common Ground Alliance
CIS Close Interval Survey
CLSM Controlled Low-Strength Material
CoF Consequence of Failure
CPM Computational Pipeline Monitoring
CP Cathodic Protection
CSA Canadian Standards Association
D/t Diameter to wall thickness ratio
DAMQAT Damage Prevention Quality Action Team
DCS Distributed Control Systems
DCVG DC (direct current) Voltage Gradient
DIN Deutsches Institut fur Normung (the German Institute for Standardization)
DIRT Damage Information Reporting Tool
DOT (U.S.) Department of Transportation
DSAW Double Submerged Arc Welding
Dt Ratio Diameter-to-Thickness Ratio
EAC Environmentally Assisted Corrosion
ECDA External Corrosion Direct Assessment
EE Essential Elements
EGIG European Gas Pipeline Incident Group
EL Expected Loss
ERW Electric Resistance Welding
EMAT Electromagnetic Acoustic Transducers
EPA Environmental Protection Agency
EPRG European Pipeline Research Group
ERCB Energy Resources Conservation Board (formerly Alberta Energy and Utilities)
ERW Electric Resistance Weld
ESR Epoxy Sleeve Repair
ERF Estimated Repair Factor
EUB Alberta Energy and Utility Board
FBE Fusion Bonded Epoxy
FFS Fitness For Service
FEA Finite Element Analysis
FMEA Failure Modes and Effects Analysis
FRC Fiber-Reinforced Concrete
GIS Geographic Information System
GMAW Gas Metal Arc Welding
GPR Ground-Penetrating Radar
GPS Global Positioning System
GRI Gas Research Institute
GTAW Gas Tungsten Arc Welding
HAZ Heat Affected Zone
HAZOPS Hazard and Operability Study
HCA High-Consequence Area
HDPE High Density Polyethylene
HF High Frequency
HIC Hydrogen Induced Cracking
HSE Health and Safety Executive (UK)
HUD Housing and Urban Development
HVA High value area
ICS Industrial Control System
ILI In-Line Inspection
IPL Independent Protection Layers
Km Kilometer
LDPE Low Density Polyethylene
Limit states 'ultimate' (ULS), 'leakage' (LLS), and 'serviceability' (SLS)

LOPA Level Of Protection Analysis
LUT Look Up Table
MAOP Maximum Allowable Operating Pressure
MAWP Maximum Allowable Working Pressure
MFL Magnetic Flux Leakage
mi Mile
MOP maximum operating pressure
MPI Magnetic Particle Inspection
MPY mils per year
NAPSR National Association of Pipeline Safety Representatives
NDE Non-Destructive Examination
NDT Non-Destructive Testing
NEB National Energy Board (Canada)
NOP Normal Operating Pressure
NPS Nominal Pipe Size
NRA Nuclear Regulatory Agency
NTSB National Transportation Safety Board
OD Outer/Outside Diameter
OPS Office of Pipeline Safety
OSHA Occupational Safety and Health Administration
PCS Process Control System
PE Polyethylene
PGD Permanent Ground Deformation
PHA Process Hazard Analysis
PHMSA Pipeline and Hazardous Materials Safety Administration
PIPES Pipeline Inspection, Protection, Enforcement, and Safety Act
PLC Programmable Logic Controller
PLRMM *Pipeline Risk Management Manual, 3rd edition*
PoD Probability of Damage
PoF Probability of Failure
PP Polypropylene
PPTS Pipeline Performance Tracking System
PRA Probabilistic Risk Assessment
PRCI Pipeline Research Council International, Inc.
PFD Probability of Failure on Demand
PL Protection Layer
PSA Petroleum Safety Authority (Norway)
psi Pounds Per Square Inch
PVC Poly Vinyl Chloride
PXX abbreviation for conservatism level: P50, P99.9, etc
QA/QC Quality Assurance/Quality Control
QRA Quantitative Risk Assessment
RBD Reliability Based Design
ROV Remotely Operated Vehicle
ROW Right Of Way
RPR Rupture Repair Ratio
SCADA Supervisory Control And Data Acquisition
SCC Stress Corrosion Cracking
SLOD Significant Likelihood Of Death
SIL Safety Integrity Layer
SME Subject Matter Expert
SMYS Specified Minimum Yield Strength
SSC Sulphide Stress Corrosion
TSB Transportation Safety Board of Canada
TTF Time To Failure
UAV Unmanned Airborne Vehicle
UKOPA UK Onshore Pipeline Operators Association
ULCC Utility Location and Coordinating Council
UST Underground Storage Tank
UT Ultrasonic Testing
UTS Ultimate Tensile Strength
Yr Year

CAUTION

This text describes an approach to comprehensive pipeline risk assessment. While the underlying methodology has been proven over years of practice, not every nuance of application is documented here. The user must understand that, as with all technical approaches, a qualified person must oversee its use and accepts sole responsibility for any and all results of applying methodologies described herein and their subsequent uses.

PREFACE

Formal risk management has become an essential part of pipelining. As an engineered structure placed in a constantly changing natural environment, a pipeline can be a complex thing. Good risk assessment is an investigation into that complexity; providing an approachable, understandable, manageable incorporation of the physical processes potentially acting on a pipeline: external forces, corrosion, cracking, human errors, material changes, etc.

Recent work in the field of pipeline risk assessment has resulted in the development of methodologies that overcome limitations of the previous techniques while also reducing the cost of the analyses. Alternative approaches simply no longer compete. This more-defensible, more-efficient, more-useful, i.e., definitive, approach is detailed here.

This text recommends the abandonment of some previous risk assessment methodologies. Our reasons for building and using certain older models are no longer valid. We no longer have to take short-cuts to work around computer processing limitations or to approximate underlying scientific/engineering principles. We don't need extensive component failure histories to produce absolute estimates of risks, as once believed, nor do we have to use data that is so generalized that it does not fairly represent the specific assets being studied. We now have strong, reliable, and easily applied methods to estimate actual risks, and no longer must accept the compromises generated by intermediate scoring schemes or statistics-centric approaches.

A goal of this book is to provide an intuitive, transparent, and robust approach to help a reader put together an efficient risk assessment tool and, with that, optimize the management of pipeline risks.

Therefore, this book is also about risk management—not just risk assessment. Risk is a fuzzy topic, and managing risk involves numerous social and psychological issues. It is by no means a strictly technical endeavor. This book advocates a single, very efficient risk assessment methodology, developed and tuned over years of applications, as the starting point of risk management. The practice of risk assessment can now be fairly standardized.

However, it is a disservice to the reader to imply that there is only one correct risk management approach. Those embarking on a formal pipeline risk management process should realize that, once an improved risk understanding is obtained, they have many options with which to react to that risk. This should not be viewed as negative feature, in my opinion. The choices in technical, business, and social problem-solving surrounding risk management makes the process challenging and exciting.

So, my advice to the reader is simple: arm yourself with this 'next generation' knowledge of how to measure risk, adopt an investigative mind set—good risk management requires sleuthing!—and then, enjoy the journey!

I Introduction

Pipeline risk management is a complex and fascinating practice, bringing together aspects of science (including physics, chemistry, biology, geology, and more), engineering, history, probability theory, human psychology, and even philosophy.

It begins with assessing the risks. Here is the typical challenge: decades ago, someone designed a multi-component engineered structure using pressurized pipe, valves, fittings, compressors, pumps, tanks, etc. It was installed in a highly variable natural/man-made environment across deserts, jungles, farms, rivers, lakes, mountains, urban centers—often with changing soils, temperature extremes, micro-organism activity, magnetic field effects, etc. Now, years and years later, we are trying to determine where weaknesses and more consequential failure locations exist. A myriad of scientific phenomena—both natural and man-made—are interacting to complicate our ability to understand and creating a puzzle with thousands of pieces to fit together. What an interesting confluence of engineering coexisting with Mother Nature!

Next comes the practical applications of having 'solved' this puzzle: armed with an understanding of the risks, what can and should now be done? This is where we must leave the realm of pure science and engineering and enter into aspects of the human behavioral sciences.

This text endeavors to examine more completely the solving of the puzzle—the risk assessment—and then lightly step into the issues of managing risk.

The intention is to equip the risk manager with the tools to understand the risk and the ability to efficiently apply this knowledge when making decisions.

I.1 THE PUZZLE

Today, we have an unprecedented amount of data available to solve this pipeline risk puzzle. Let's say we want to understand internal corrosion potential on a natural gas pipeline. We examine some recent ILI results, looking for internal corrosion metal loss indications. We find some. Are they occurring at bottom o'clock positions of the pipe circumference? If so, that is a clue. We plot the ILI anomalies in GIS, add aerial photography, add topography, and look for more clues. Do we see clusters of metal loss at possible low spots—where the pipe is crossing creeks, valleys, etc? Let's overlay elevation data—are there steep inclinations here where liquids/solids could accumulate and persist? Are we close to gas inputs, where historical liquid excursions (carryovers) might have accumulated and might first impact piping?

Next, we examine gas quality records and the performance record of the input gas streams that might have put contaminants into the gas stream. Given this, we need to understand the chemistry—what combinations of chemicals and environmental factors could be generating corrosion and at what rates? Then we can study fluid flows, thermodynamics, and hydraulics to understand how contaminants might behave inside the product stream. For those who like engineering detective work—isn't such sleuthing compelling?

This is essentially what good risk assessment is doing. But it is far more efficient than what we would-be detectives can do individually. The risk assessment can broadcast our detective work over tens of thousands of miles of pipelines almost instantly. This effectively replaces thousands of man-hours of investigation and instantly puts key information into the hands of decision-makers.

It really is exciting to see large quantities of data drawn into a model and immediately see meaningful, actionable information come out. Turning data into information ensures that the right decisions can be made.

The risk assessment should add clarity. Some risk assessments add complexity. The real world is sufficiently complex that no unnecessary additional complexity should be tolerated. In a good risk assessment, if complexity appears, it should only be because the underlying science is complex.

Assessment is of course, just the beginning of risk management. Even with complete understanding of risk—via the risk assessment—we still have the challenges of how to manage this risk. Again, a host of factors comes into play: how much risk reduction is warranted? How quickly should risk reduction occur? Which is better—much risk reduction at a specific location or more modest risk reduction but over many miles of pipeline? All strive to answer the key underlying question: how safe is 'safe enough'?

I.2 HOW RISK ASSESSMENT HELPS

Achieving safety while undertaking a potentially dangerous activity means identifying and managing risks. Although they seem simple in concept, pipelines are actually complex, dynamic systems, operating in often-challenging environments and subject to a vast and varying array of integrity threats.

While risk has always been an interesting topic to many, it is also often clouded by preconceptions. Many equate risk analyses with requirements of huge databases, complex statistical analyses, and obscure probabilistic techniques. In reality, good risk assessments can be done with only moderate effort and even in a data-scarce environment. This was the major premise of the earlier PRMM[1].

PRMM has a certain sense of being a risk assessment cookbook—"Here are the ingredients and how to combine them." Feedback from readers indicates that this was

1 *Pipeline Risk Management Manual, 3rd Edition*, hereinafter referred to as PRMM

useful to them. That aspect is reflected in this book, even as the new methodologies shown here are far superior to our past practices.

Beyond the desire for a straightforward approach, there also seems to be an increasing desire for more sophistication in risk modeling. This is no doubt the result of an unprecedented number of practitioners pushing the boundaries as well as more widespread availability of data and more powerful computing environments. Today, it is easy and cost-effective to consider many more details in a risk model. Initiatives are currently under way to generate more widespread, complete, and useful databases to further our knowledge and to better support the detailed risk modeling efforts.

The desire for 'more'—more accuracy, more knowledge, more decision-support—is also fueled by the knowledge that potential consequences of incorrect risk management are higher now than in the past and will likely continue to increase. Aging infrastructure, system expansions, and encroaching populations are primary drivers of this change. Regulatory initiatives reflect this concern in many parts of the world.

I.3 ROBUSTNESS THROUGH REDUCTIONISM

The best practice in risk assessment is to assess major risk variables by evaluating and combining many lesser variables, generally available from the operator's records or public domain databases. This is sometimes called a reductionist approach, reducing the problem to its subparts for examination. This allows assessments to benefit from direct use of measurements or evaluations of multiple smaller variables, rather than a single, high-level variable, thereby reducing subjectivity. If the subparts—the details—are not yet available, then higher level inputs must suffice.

The reductionist approach also applies to the physical dimensions of the system. The risk for a pipeline is assessed as the sum of the risk of its components, where the components are the pipe, fittings, valves, tanks, pumps, compressors, meters, etc.

A critical belief underlying this book is that all pertinent information should be used in a risk assessment. There are very few pieces of collected pipeline information that are not useful to the risk assessment. The risk evaluator should expect any piece of information to be useful until he absolutely cannot see any way that it can be relevant to risk or decides its inclusion is not cost-effective.

Any and all experts' opinions and thought processes can and should be codified, thereby demystifying the experts' personal assessment processes. The experts' analysis steps and logic processes can be replicated to a large extent in a risk assessment model. A detailed model should ultimately be 'smarter' than any single individual or group of individuals operating or maintaining the pipeline—including that retired guy who 'knew everything'. It is often useful to think of the assessment process as 'teaching the model'. We 'tell' the model what we know and what it means to know various things. We are training the model to 'think' like the best experts and giving it the benefit of the collective knowledge of the entire organization and all the years of record-keeping.

I.4 CHANGES FROM PREVIOUS APPROACHES

Previous risk assessment approaches served us well in the past. They helped support decision making by crystallizing thinking, removing subjectivity, and helping to ensure consistency. But the era of many older approaches has passed, due to increased expectations as well as the now superior analyses techniques and availability of powerful and inexpensive computer tools.

Our regulators, attorneys, neighbors, and other stake holders are no longer satisfied that we can successfully manage risk using tools that are not modern and robust. We now have strong, reliable, and easily applied methods to estimate actual risks, and no longer must accept the compromises generated by intermediate scoring schemes or statistics-centric approaches. The modern approach to pipeline risk assessment is presented here. It is superior—in accuracy, defensibility, and cost of analyses—to all alternative approaches since it incorporates the best and eliminates the weaknesses from others. The migration from the older approaches is described in the following sections.

A substantial improvement in risk assessment methodology should not be a surprise. Changes to risk algorithms have always been anticipated, and every risk model—even the most advanced—should be regularly reviewed in light of its ability to incorporate new knowledge and the latest information.

This book presents the newer risk assessment methodologies for evaluating all aspects of pipeline risk. This approach reflects the advances in risk assessment technology from research & development efforts as well as years of input of pipeline operators, pipeline experts, and risk assessors.

A migration from both relative risk assessment and 'classical' QRA is central to better understanding risk. There is no longer any valid reason to use a relative, scoring type risk assessment approach. There is also no reason to adopt the statistics-centric 'classical' QRA approaches. We now have updated techniques and a powerful, but simple framework to capture and more efficiently use all available information. When much more useful results are available with no additional cost or effort, why use lesser solutions?

I.4.1 Key Changes

Early chapters of this book offer foundational and background information. The experienced, practicing risk manager may wish to move directly to the how-to chapters. It is advisable to quickly become familiar with the most essential elements of the newer methodology presented in this book. Central to this much-improved methodology are several key features:

1. The abandonment of all scoring (point assignment systems) which is now replaced by measurements.
2. The PoF triad—exposure, mitigation, and resistance—the essential ingredients to understand PoF.
3. The use of OR and AND gate math.

4. The use of both measurements and estimates to replicate an SME's decision processes.
5. The calculation of hazard zones to drive CoF estimates.

Many other aspects of risk assessment remain similar to previous approaches. Pipeline risk factors are generally well understood. It is only the better capturing of their role in risk that changes. The estimation of consequences has generally been more grounded in physics and engineering principles already. Fewer changes in those methodologies are warranted.

Armed with these key changes in methodology, the more experienced reader can scan Chapter 2 for basic definitions and application nuances and then move to Chapters 5-11 to efficiently begin assessing risks.

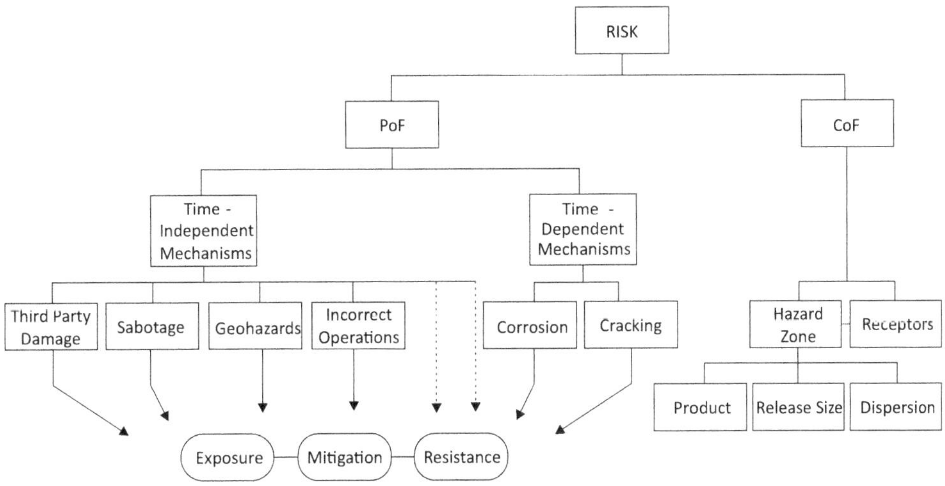

Figure I.1 Modeling of Pipeline Risk

I.4.2 Migration from previous methodologists

If you have a complete risk assessment system based on older methods, that system can usually be readily migrated to a modern platform. The previous work is preserved and can be more efficiently employed while measured data from today's modern inspection and integrity-evaluation tools is also integrated.

Table I.1 below shows an example of converting input data from an older, scoring type risk assessment approach into a modern risk assessment. The first step is to identify what aspect of risk is impacted by the previously-collected data. All inputs should inform estimates of either: PoF-exposure, PoF-mitigation, PoF-resistance, or CoF. Then, the previously assigned scores or point values can be linked to measurement values. This allows rapid conversion of even the largest scoring type risk databases.

Table I.1
Example Conversion of Scores to Measurements

Risk Issue	Old Index/Score	New PoF Element	Measurement/Estimate
depth cover	shallow = 8 pts	mitigation	15%
wrinkle bend	yes = 6 pts	resistance	-0.07″ pipe wall
coating condition	fair = 3 pts	mitigation	0.01 gaps/ft2
soil	moderate = 4 pts	exposure	4 mpy

Some calibrations and handling of special cases will usually be needed, and documentation will need to be updated, but the whole conversion/migration effort should consume only dozens of man-hours, not hundreds.

In re-using previous data, there should be some similarities in results when comparing old versus new. But there should also be new and important insights emerging, as the modern approaches provides superior results that more accurately represent real-world risks.

Sidebar

The Outlook for Pipeline Risk Assessment: An Interview

US regulators have recently expressed criticism regarding how Integrity Management Plan (IMP) risk assessment (RA) for pipelines is being conducted. Do you also see problems?

There is a wide range of practice among pipeline operators right now. Some RA is admittedly in need of improvement—not yet meeting the intent of the IMP regulation. However, I believe that is not due to lack of good intention but rather incomplete understanding of risk. Risk is a relatively new concept and not easy to fully grasp. To address PHMSA's concerns, we as an industry need to improve our understanding of risk and how to measure it.

What's new in the world of pipeline risk assessment?

In the last few years, the emergence of the US IMP regulations has prompted the development of more robust RA methodologies specifically designed for pipelines. Even though PHMSA and others have identified weaknesses among some practitioners, much progress has been made. Previous methodologies fell into two categories: 1) scoring systems designed for simple ranking of pipeline segments, and 2) statistics-based quantitative risk assessments (QRA's) used in more robust applications, often for industrial sites and for certain regulatory and legal needs. The first were popular among the pre-IMP voluntary practitioners but were limited in their ability to accurately measure risk and to meet IMP regulatory requirements. The second category was costly and ill-suited for long linear assets, like pipelines.

You note two categories of previous risk assessment methodologies. What about others, like 'scenario-based' or 'subject matter experts', that are listed in some standards?

I think that listing is confusing tools with risk assessment methodologies. The two examples you mention are important ingredients in any good risk assessment but they are certainly not complete risk assessments themselves.

What are the newest pipeline risk assessment methodologies like?

They're powerful, intuitive, easy to set up, less costly, and vastly more informative than either of the previous approaches. By independent examination of key aspects of risk and the use of verifiable measurement units, the whole landscape of the risks becomes apparent. That leads to much improved decision-making.

How can they be both easy and more informative?

More informative since they produce the same output as the classic QRA but are more accurate. Easy because they directly capture our understanding of pipelines and what can cause them to fail. The word 'directly' is key here. Previous methods relied on inferential data and/or scoring schemes that tended to interfere with our understanding.

If they do the same thing as QRA, why not just use classical QRA?

Several reasons: classic QRA is expensive and awkward to apply to a long, linear asset in a constantly changing natural environment—can you imagine developing and maintaining event trees/fault trees along every foot of every pipeline? Classical QRA was created by statisticians and relies heavily on historical failure frequencies. Ask a statistician how often something will happen in the future and he will ask how often it has happened in the past. I often hear something like "we can't do QRA because we don't have data." I think what they mean is that they believe that databases full of incident frequencies—how often each pipeline component has failed by each failure mechanism—are needed before they can produce the QRA type risk estimates. That's simply not correct. It's a carryover from the notion of a purely statistics-driven approach. While such historical failure data is helpful, it is by no means essential to RA. We should take an engineering- and physics-based approach rather than rely on questionable or inadequate statistical data.

But if I need to estimate ('quantify)'how often a pipeline segment will fail from a certain threat, don't I need to have numbers telling me how often similar pipelines have failed in the past from that threat?

No, it's not essential. It's *helpful* to have such numbers, but not necessary and sometimes even counterproductive. Note that the historical numbers are often not very relevant to the future—how often do conditions and reactions to previous incidents remain so static that history can accurately predict the future? Sometimes, perhaps, but caution is warranted. With or without historical comparable data, the best way to predict future events is to understand and properly model the mechanisms that lead to the events.

Why do we need more robust results? Why not just use scores?

Even though they were developed to help simplify an analysis, scoring and indexing systems add an unnecessary level of complexity and obscurity to a risk assessment. Numerical estimates of risk—a measure of some consequence over time and space, like 'failures per mile-year'—are the most meaningful measures of risk we can create. Anything less is a compromise. Compromises lead to inaccuracies; inaccuracies lead to diminished decision-making, leading to mis-allocation of resources; leading to more risk than is necessary. Good risk estimates are gold. If you can get the most meaningful numbers at the same cost as compromise measures, why would you settle for less?

Are you advocating exclusively a quantitative or probabilistic RA?

Terminology has been getting in the way of understanding in the field of RA. Terms like quantitative, semi-quantitative, qualitative, probabilistic, etc. mean different things to different people. I do believe that for true understanding of risk and for the vast majority of regulatory, legal, and technical uses of pipeline risk assessments, numerical risk estimates in the form of consequence per length per time are essential. Anything less is an unnecessary compromise.

What about the concern that a more robust methodology suffers more from lack of any data? (i.e.," If I don't have much info on the pipeline, I may as well use a simple ranking approach".)

That is a myth. In the absence of recorded information, a robust RA methodology forces SME's to make careful and informed estimates based on their experience and judgment. From direct estimates of real-world phenomena, reasonable risk estimates emerge, pending the acquisition of better data. Therefore, I would respond that lack of information should drive you towards a more robust methodology. Using a lesser RA approach with a small amount of data just compounds the inaccuracies and does not improve understanding of risk—it is largely a waste of time.

It sounds like you have methods that very accurately predict failure potential. True?

Unfortunately, no. While the new modeling approaches are powerful and the best we've ever had, there is still significant uncertainty. We are unable to accurately predict failures on specific pipe segments except in extreme cases. With good underlying data, we can do a decent job of predicting the behavior of numerous pipe segments over longer periods of time—the behavior of a population of pipeline segments. That is of significant benefit when determining risk management strategies.

Nonetheless, it sounds like you're saying there are now pipeline RA approaches that are both better and cheaper than past practice... ?

True. RA that follows the *Essential Elements* guidelines avoids the pitfalls that befall many past practices. Yet, we can still apply all of the data that was collected for the previous approaches. Pitfall avoidance, full transparency, and re-use of data makes the approach more efficient than other practices. Plus, the recommended approaches now generate the most meaningful measurements of risk that we know of.

Sounds too good to be true. What's the catch?

One catch is that we have to overcome our resistance to the kinds of risk estimate values that are produced. When faced with a number such as 1.2E-4 failure/mile-year, many react with immediate negative reaction, far beyond a healthy skepticism. Perhaps it is the scientific notation, or the probabilistic implication, or the 'illusion of knowledge', or some other aspect that evokes such reactions. I find that such biases disappear very quickly however, once an audience sees the utility of the numbers and makes the connection — 'Hey, that's actually a close estimate to what the real-world risk is.'

Another 'catch' is the one we touched on previously. Rare events like pipeline failures have a large element of randomness, at least from our current technical perspective. That means that, no matter how good the modeling, some will still be disappointed by the high uncertainty that must often accompany predictions on specific pipeline segments.

How can industry as a whole improve RA, especially in the eyes of the public and regulators?

A degree of standardization that serves all stake holders is needed. A list of essential elements sets forth the minimum ingredients for acceptable pipeline risk assessment. Every risk assessment should have these elements. A specific methodology and detailed processes are intentionally NOT essential elements, so there is room for creativity and customized solutions. If regulators encounter too many substandard pipeline RA practices, then prescriptive mandates might be deemed necessary. Such mandates are usually less efficient than approaches that permit flexibility while prescribing only certain ingredients.

1 RISK ASSESSMENT AT A GLANCE

Highlights

The following is a summary of the risk evaluation framework described in subsequent chapters. The framework forms the foundation for risk assessment on any pipeline system or component or collection of components within the system.

1.1 RISK ASSESSMENT AT-A-GLANCE

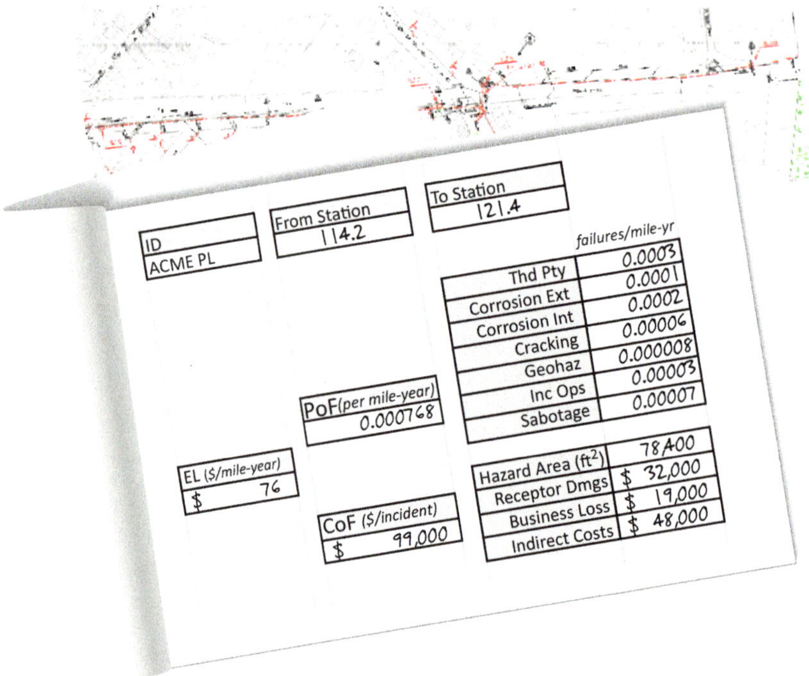

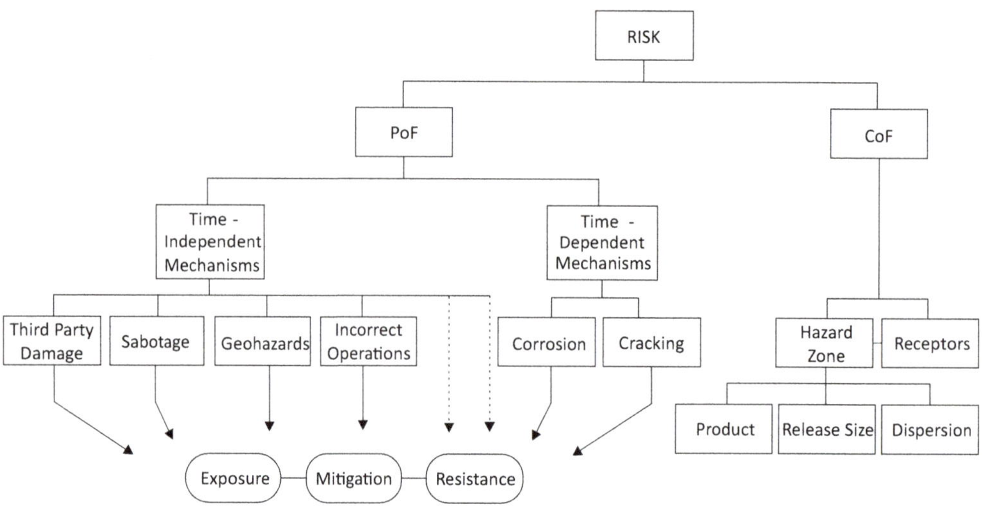

Figure 1.1 Modeling of Pipeline Risk

Risk assessment should be consistent—there is no reason for multiple types of risk assessment. The same framework applies to very robust as well as very simple assessments. An example of a rudimentary (high level, few details) application of this risk assessment strategy using this approach is shown in Chapter 1.3.3.3 Rudimentary Risk Assessment.

1.2 RISK: THEORY AND APPLICATION

1.2.1 The Need for Formality

Humans are poor estimators of risk without formality. We routinely overestimate and underestimate true risks due to influences of emotion, memory, or personal preference. Here is an insightful quote from a bestselling book on risk [10]:

> *"Nature is so varied and so complex that we have a hard time drawing valid generalizations from what we observe.*
>
> *We use shortcuts that lead us to erroneous perceptions, or we interpret small samples as representative of what larger samples would show.*
>
> *We display risk aversion when we are offered a choice in one setting and then turn into risk seekers when we are offered the same choice in a different setting.*
>
> *We have trouble recognizing how much info is enough and how much is too much.*
>
> *We pay excessive attention to low-probability events accompanied by high drama and overlook events that happen in routine fashion.*
>
> *We start out with a purely rational decision about how to manage our risks and then extrapolate from what may be only a run of good luck."*

There are also those who opine that attempts to quantify risk are generally flawed. So-called Black Swan (taleb[1]) events are so complex and rare as to be essentially unpredictable. These are events previously thought to be impossible, until they actually happen—for example, the first sighting of a black-colored swan. Even the specialized statistical theories and associated distributions (for example, extreme value, etc.) are thought by some to be vain attempts to know the unknowable. Most will agree that extensive and complex modeling can quickly become impractical for real-world risk management and that over-reliance on modeled values with high uncertainty can lead to misdirection of resources.

---

1 Taleb, N.N., 2010, *The Black Swan: the Impact of the Highly Improbable*, Random House Trade Paperbacks.

However, extreme positions against measuring risk underestimate the value of the attempt itself. Such critics miss the key point that the measurement effort itself yields great rewards, even when the measurement is imperfect: "anything that is measured, improves." Striving to assign some realistic value to any phenomena yields benefits far beyond the value produced. Even when results are imprecise and may not fully capture the unforeseeable, the knowledge gained by earnest attempts to include all possibilities and to assign meaningful values is a significant benefit of risk management.

Much has been written on the general topics of the scientific method and modeling in both science and engineering. See PRMM for a relevant discussion of these principles, and their nuanced application in engineering and risk assessment for pipelines.

The objective is to build a useful tool—one that is regularly used to aid in everyday business and operating decision making, one that is accepted and used throughout the organization, and one that is robust and defensible.

1.2.2 Complexity

In any modeling effort, complexity should exist only because the underlying real-world phenomenon is complex. **The RA should not add complexity**.

Ironically, a scoring type risk assessment, intended to simplify the modeling of real-world phenomena, actually adds complexity. By converting real-world phenomena into 'points' via an assignment protocol, an artificial layer of complexity has been introduced. This is unnecessary.

A robust risk assessment, covering complex scientific elements such as corrosion mechanisms and stress-strain relationships, may require a level of complexity in order to fully represent the associated risk issues. In this case, the complexity reflects the complexity of the science and is appropriate for certain kinds of risk assessment. In contrast, a risk assessment that requires the assignment of scores to various conditions—for example, soil corrosivity, CP effectiveness, etc—and then the assignment of weightings to each, and then the combination of the scores using non-intuitive algorithms, is adding complexity that probably adds no value to the analyses. As a matter of fact, such artificial complexity probably detracts from the accuracy and usability of the risk assessment.

1.2.3 Intelligent Simplification

The challenge when constructing a risk assessment model is to fully understand the mechanisms at work and then to identify the optimum number of detailed variables for the model's intended use. This follows the reductionist approach previously discussed—breaking the problem down into pieces for later reassembly into meaningful risk estimates.

We must understand and embrace the complexity in order to achieve the optimum amount of simplification—this is the process of 'intelligent simplification.' The best approach is to begin with the robust solution, including all details and all nuances that make up the real-world phenomena. Only then can a shortcut be contemplated. That way, what is sacrificed by the simplification is clear.

Furthermore, the robust solution will be immediately appropriate for many practitioners and eventually appropriate for many more (ie, a desired future level of detail in the risk assessment). Modeling complex phenomena such as AC induced corrosion, vapor cloud explosion potential, and many others, requires numerous inputs and interactions among inputs. Understanding what those inputs are and how they should be used to best model scenarios is the first step. With that understanding, simplifications without excessive loss of accuracy may be possible.

When simplifications are not appropriate, the robust solution should be employed, but perhaps in such a way that it does not interfere with the assessment's efficiency. Many processes, originating from sometimes complex scientific principles, are "behind the scenes" in a good risk assessment system. These must be well documented and available, but need not interfere with the casual users of the methodology (everyone does not need to understand the engine in order to benefit from use of the vehicle). Engineers will normally seek a rational basis underpinning a system before they will accept it. Therefore, the basis must be well documented.

Deciding not to include a detailed variable directly in the risk assessment does not necessarily mean it is ignored. The detail may already be part of an evaluation being conducted elsewhere. For instance, the corrosion department may have a very sophisticated analyses of AC induced corrosion potential. Rather than replicate this analysis in the risk assessment, perhaps only the results need to be migrated into the risk assessment.

Among all possible variables, choices are required that yield a balance between a comprehensive model and an unwieldy model—inclusion of every possible detail versus loss of important information. Users should be allowed to determine their own optimum level of complexity. Some will choose to capture much detailed information because they already have it available; others will want to get started with a high-level framework. However, by using the same overall risk assessment framework, results can still be compared: from very detailed approaches to overview approaches.

Figure 1.2 illustrates the use of a 'short circuit' pending availability of full soil corrosivity information. A 16 mpy soil corrosivity value is used pending information regarding soil moisture, pH, and contaminant levels which will lead to more accurate soil corrosivity values. Having the details shown, but not populated, in the risk assessment model has advantages. It documents that further analyses is possible, if not warranted, and that the entered value is thought to capture the sub-variables that are not yet known.

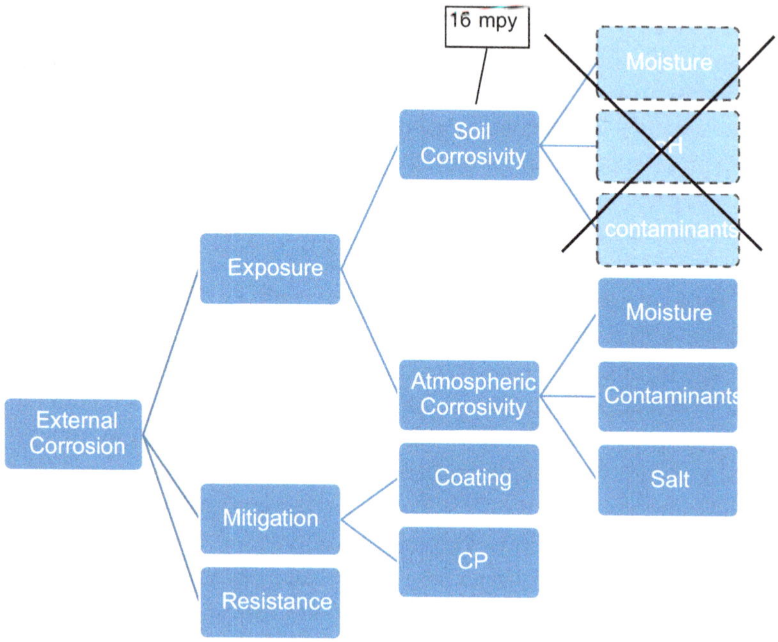

Figure 1.2 Using Short Circuit, Pending Full Data Availability

Having flexibility in the level of rigor of a risk assessment is a large advantage. While detailed, technically rigorous analyses will always strengthen the assessment, it will not always be warranted. By this we mean, the cost/benefit of the rigor does not always justify the effort. In some instances, this will be a guess—a perceived low-value analysis may actually turn out to be a critical consideration and its absence is lamented. For instance, discounting the potential for H_2 permeation through a steel component's wall seems reasonable until the rare phenomenon contributes to a failure and prompts regret that it wasn't previously a consideration.

See also the discussion of Chapter 3.7 Verification, Calibration, and Validation.

1.2.4 Classical QRA versus Physics-based Models

Most documented risk assessment approaches are based on statistical analyses. This is because the problem of risk assessment was initially given to statisticians to solve. Ask a statistician how often something will happen in the future, and their first question will be 'how often has it happened in the past.' This is reasonable for a methodology that deals exclusively with analyses of how numbers are 'behaving.' The ability of statistics to model the behavior of larger populations over longer periods of time is undisputed. But this does not provide a complete solution for practitioners of risk management.

Historical data should always influence our estimates of risk. However, it will rarely capture all the pertinent considerations. Even purists will usually agree that statistics can only fully describe very simple and rather uninteresting systems in the universe. Coin flips and games of chance (cards, roulette, etc) are examples. Real-world systems are complex and require many insights well beyond statistical analyses for understanding.

Scientists and engineers, rather than statisticians, have been more involved in certain portions of the risk assessment, notably consequence modeling. Historically, consequence assessments have made sound use of science and engineering where probability assessments often have not. Consequence evaluations have, for years, routinely used dispersion modeling, thermal effects predictions, heat transfer equations, kinetics and thermodynamics of fluid movements, and many others. On the other hand, probability was simply based on historical rates. Perhaps the historical rates were modified by some very subjective 'adjustment factors' to account for instances when the subject pipeline was thought to behave differently from the statistical population. But, too often, little science and engineering was applied to the problem of measuring failure potential in a formal but efficient manner.

Underlying most meanings of risk is the key issue of 'probability.' Statistics and probability are closely intertwined. But, as is detailed in this text, probability expresses a degree of belief beyond statistical analyses. 'Degree of belief' is the most compelling definition of probability because it encompasses statistical evidence as well as science, engineering, interpretations, and judgment. Our beliefs should be firmly rooted in fundamental science, engineering judgment, and reasoning. This does not mean ignoring statistics—proper analysis of historical data—for diagnosis, to test hypotheses, or to uncover new information. Statistics help us understand our world, but it certainly does not explain it.

The assumption of a predictable distribution of future leaks predicated on past leak history might be realistic in certain cases, especially when a database with enough events is available and conditions and activities are constant. However, one can easily envision scenarios where, in some segments, a single failure mode should dominate the risk assessment and result in a very high probability of failure rather than only some percentage of the total. Even if the assumed distribution is valid in the aggregate, there may be many locations along a pipeline where the pre-set distribution is not representative of the particular mechanisms at work there.

There is an important difference between using statistics to better understand numbers—inputs and results—versus basing a risk assessment predominantly on historical incident rates, using statistics to support the belief that the past is the best way to predict the future. This is admittedly an oversimplification and is debatable in several key ways, especially when considering that all techniques are strengthened by simultaneous understanding of both the underlying physics and the statistics. However, this distinction emphasizes a core premise of this recommended methodology in this book. That premise is that the understanding of the physical phenomena behind pipeline failure should be the dominant basis of a risk assessment. Statistics, in particular historical event frequencies, should be secondary inputs.

The exposure-mitigation-resistance analyses that is an essential element of PoF assessment, is a key aspect that differentiates a modern pipeline risk assessment from classical QRA. Classical QRA does not seek the exposure-mitigation-resistance differentiation. Without this insight, past failure rates typically used in such assessments have questionable relevance to future failure potential.

Failure to quantify the exposure-mitigation-resistance influences leads to incomplete understanding which makes risk management problematic. Ideally, historical event rate information will be coupled with the exposure-mitigation-resistance analyses to yield the best PoF estimates.

The exposure-mitigation-resistance analyses is an indispensable step towards full understanding of PoF, as is detailed in later chapters. Without it, understanding is incomplete. Full understanding leads to the best risk management practice—optimized resource allocation—which benefits all stakeholders.

More will be said about improvements over Classical QRA approaches in later sections.

1.2.5 Statistical Modeling

To be clear, the message here is NOT that statistical theory is to be avoided but rather that statistics should supplement rather than drive risk modeling. Science and physics provide the model basis but statistics is very useful in tuning or calibrating inputs and results. Failure to use statistical theory would be an error.

In fact, the risk assessment framework proposed in this text has been successfully deployed as a model making increased use of statistical techniques. In one such application, Bayesian networks were established to better incorporate probability distributions, rather than point estimates, and learning or feedback processes were included. The same essential elements as recommended here should be used in this application. This is especially important for the breakdown of PoF into separate, but connected, measurements of exposure, mitigation, and resistance.

In addition to the classical models of logic, new logic techniques are emerging that seek to better deal with uncertainty and incomplete knowledge. Methods of measuring "partial truths"—when a thing is neither completely true nor completely false—have been created based on fuzzy logic originating in the 1960s from the University of California at Berkley as techniques to model the uncertainty of natural language. Fuzzy logic or fuzzy set theory resembles human reasoning in the face of uncertainty and approximate information. Questions such as "To what degree is x safe?" can be addressed through these techniques. They have found engineering application in many control systems ranging from "smart" clothes dryers to automatic trains.

1.3 THE RISK ASSESSMENT PROCESS

1.3.1 Fix the Obvious

Where formal risk assessment is not yet in place, potential practitioners sometimes feel overwhelmed, hesitating to get started due to the apparent magnitude of the task ahead. How to assess the risks associated with hundreds or thousands of miles of pipeline system, especially when desired information is scarce?

It is important to recognize that, even in the absence of a formal risk assessment, risk assessment has always been occurring, usually successfully, in all pipeline operations since their inception. The formalization of risk understanding should not interfere with the practice of 'fix the obvious.' A formal risk assessment is not needed to show where large issues are already apparent. The risk assessment can refine and improve resource allocation and bring to light the less apparent or distant future risk issues. But when an indisputable risk issue is identified and mitigation actions are obvious and available, time should not be wasted in extensive study or other formalization.

So, the obvious advice is: While seeking to improve risk management processes, continue the practice of risk management.

1.3.2 Using this Manual

Robust pipeline risk assessment generates a risk profile, showing changes in risk along a pipeline route. Risk management uses that profile to identify ways to effectively minimize the risk. Chapters 1 - 12 discuss the risk assessment process as it can be applied to all types of facilities, handling any kind of product, and traversing any location. Chapter 13 describes the transition from risk assessment to risk management.

1.3.3 Quickly getting answers

FOCUS POINT
A good risk assessment process supports rapid, easy to obtain risk estimates as well as detailed, robust risk estimates.

Formal pipeline risk assessment does not have to be highly complex or expensive. A savvy risk manager can, in a relatively short time, have a fairly detailed pipeline risk assessment system set up, functioning and producing useful results. Simple computer tools such as a spreadsheet or desktop database can efficiently and completely support even the most robust of assessments. Then, by establishing some administrative protocols around the processes, the quick-start applicator now has a complete system to fully support risk management. The underlying ideas are straightforward, and rapid establishment of a very useful decision support system is certainly possible. Initial information and processes may not be of sufficient rigor for full decision-support, but the user will nonetheless immediately have a formal structure from which to better ensure decisions of consistency and completeness of information.

Both a rudimentary, quick assessment and a robust, detailed assessment will follow the same procedure. This provides for the assessment to grow—getting more accurate with the inclusion of more and more details. The difference between the simple assessment and the robust lies only in the depth of investigation. Before examining this in more detail, consider also that a risk conceptualization exercise is also available to 'get answers quick'.

1.3.3.1 Risk Conceptualization—Getting 'In the Ballpark'

There exists a type of risk analysis that is even more preliminary than the rudimentary assessment to be presented in a following section. This might be termed more of a risk *conceptualization* rather than assessment and is based solely on basic deductive reasoning. Illustrated by an example, an analyst may posit that a pipeline's future risks will mirror the losses shown by recent historical annual US gas transmission pipeline experience. He assumes that the subject pipeline 'behaves' as an average[2] US gas transmission pipeline. Under this assumption, he deduces that future risks on the subject pipeline are 1.2 significant leak/ruptures per 2,000 mile-years that generate $1,200/ mile-year of losses. He scales these values to the length of his subject pipeline and uses results in decision-making.

A similar approach is the use of historical leak/break rates to predict future behavior of sections of distribution pipeline systems. With larger counts of leak/break events, these produce more statistically valid summaries and are sometimes used to understand system deterioration rates.

These generalized, statistical approaches obviously are limited, especially when applied to a particular pipeline segment (see numerous discussions later in this text regarding pitfalls associated with use of general statistics in this way). They do, however offer useful risk context, providing insights into behaviors of populations of components over long periods of time. In the absence of any other information, this approach provides estimates that may often be a close approximation—perhaps within an order of magnitude or so—of average future performance of many pipelines.

2 Actually, more of a 'composite' performance since the vast majority of miles of pipeline have incident rates and losses much lower than implied by an average

1.3.3.2 Risk Assessment Steps

True risk assessment must consider the specifics of the asset being assessed and not be unduly influenced by historical data from other assets. The following minimum steps are required for assessment of pipeline risk, regardless of level of rigor:

1. Segmentation: Identify the components that comprise the segment being assessed
   a. A new component is needed for every significant change in the pipeline's current and historical construction/operating/maintenance practice and every significant[3] change in the pipeline's surroundings.
2. Exposure: Estimate each component's unmitigated exposure from each threat, recognizing the two types of exposure
   a. Degradation rate from time-dependent failure mechanisms
   b. Event rate from time-independent failure mechanisms.
3. Mitigation: Estimate effect of each mitigation measure for each component's threats
   a. Identify all mitigation measures
   b. Rate effectiveness of each
   c. Combine and apply estimates to appropriate exposures.
4. Resistance: Estimate each component's resistance to failure from each mitigate exposure
   a. Theorize amount of resistance available in the absence of defects
   b. Estimate the role of possible defects present in each component, considering rates of defect emergence and age and accuracy of all inspections and integrity assessments.
5. PoF: Calculate PoF from each threat
   a. Risk Triad: combine Exposure, Mitigation, Resistance
   b. Estimate TTF and then PoF for time-dependent failure mechanisms
   c. Estimate PoF for time-independent failure mechanisms
   d. Combine all PoF's.
6. Calculate CoF for each component, based on desired level of conservatism and
   a. Possible failure scenarios
   b. Possible damages from each scenario.
7. Combine PoF and CoF into a risk estimate for each component. Combine component risk estimates as needed.

These steps show the minimum amount of inputs and analyses necessary to produce plausible estimates of risk along a pipeline. Experience has shown that any of these can independently dominate the actual risk. Therefore, each warrants consider-

---

3 Significant from a risk standpoint; ie, anything that can impact the probability of failure or the consequences should a failure occur.

ation and should be documented in the assessment, even if only a cursory level of effort can be applied to generate initial estimates.

Implicit in these steps is the initial recognition that a pipeline (or pipeline station or any other portion of a pipeline system) is a collection of components. Each component will contribute to the risk associated with the whole collection. Each component is exposed to threats from its immediate surroundings. These normally include corrosion, external forces, and others. Each component also generates some amount of consequence potential to its surroundings. This is the reality that should be captured in any risk assessment.

Even the most rudimentary risk assessment needs to acknowledge the individual components that comprise the pipeline system and their individual surroundings. To do this, a list of components is needed. This can be very detailed or, at the other extreme, very generalized.

As described above, for each component, three inputs are needed to characterize each plausible threat. Each component also requires one input for consequence potential. These four component-specific inputs are best obtained by examination of all of the pertinent underlying features but can be simply assigned a preliminary general estimate, pending the deeper analyses. In a very rudimentary assessment, the four ingredients are directly input for each component based perhaps solely on SME judgment.

1.3.3.3 Rudimentary Risk Assessment

Beyond a much generalized risk conceptualization exercise as discussed previously, a rudimentary risk assessment for a specific pipeline or portion of a pipeline under specific operational and maintenance protocols, can be conducted with a minimum of inputs. With fewer inputs, it will suffer from reduced accuracy—there are always trade-offs between rigor and accuracy/defensibility.

A rudimentary initial risk assessment can be created by obtaining SME estimates for each of the inputs implied in the list above and in Figure 1.3.

With a bit of guidance, the SME's can provide the necessary PoF inputs. Then consequence potential needs to be estimated. This may require a different SME since the operator/maintainer, while hopefully being well schooled in incident response, may have little or no experience with consequence valuations. The kinds of damage scenarios potentially created are first identified by an appropriate SME. These will fall into one or more of the categories of thermal (fire and explosion), toxicity (including pollution damages), and mechanical (the non-explosion phenomena associated with pressurized components). Then, the receptors potentially exposed to these scenarios are characterized. Categories include people, property, environment, commercial activities, service interruption, and others, depending on the scope of the risk assessment.

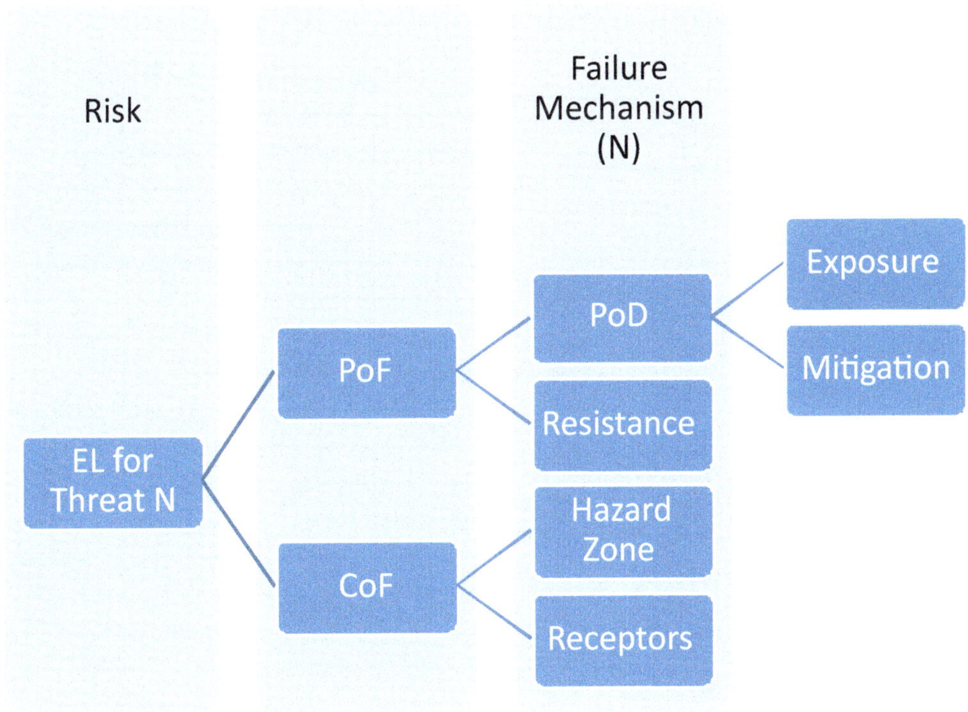

Figure 1.3 Risk Assessment Structure for Each Failure Mechanism on Each Component Assessed

Example: 1.1

For example, an assessor thinks that a portion of a pipeline system can be characterized by four different combinations of pipe characteristics, soil types, product corrosivities, potential excavation activities, and nearby population densities. He creates groupings using these parameters. The groups will serve as surrogates for the segments that actually exist. In other words, prior to the full solution of a dynamic segmented pipeline with risk estimates for each segment, he is employing a short cut by modeling the risks in terms of four general combinations of characteristics occurring along this pipeline.

He models each 'segment' as being exposed to 4 general types of failure mechanisms, requiring 12 inputs for each segment and: 4 segments x 4 failure mechanisms per segment x 3 PoF inputs per failure mechanism = 48 inputs as the minimum requirement for a PoF estimate representing the threats to all segments. He also needs an estimate of CoF for each segment, A through D, for a total input count of 52 inputs. He builds a framework to capture the needed inputs and calculations:

Table 1.1
Sample Rudimentary P90+ Risk Assessment, Part 1: Structure

			Units	**A**	**B**	**C**	**D**
			failures/year				
		Exposure	mpy				
Ext Corr	PoF	Mitigation	%				
		Resistance	%				
			failures/year				
		Exposure	mpy				
Int Corr	PoF	Mitigation	%				
		Resistance	%				
			failures/year				
		Exposure	events/year				
External Force	PoF	Mitigation	%				
		Resistance	%				
			failures/year				
		Exposure	events/year				
Human error	PoF	Mitigation	%				
		Resistance	%				
PoF total			failures/year				
CoF			$/failure				
Risk (EL)			$/year				
			$/year				

From a properly structured SME team meeting, the assessor now populates the inputs for each risk element based on the team's judgment and specific knowledge of each pipeline segment assessed. Their inputs have a targeted P90 level of conservatism—ie, they provide values that most likely overstate the actual risk.

Table 1.2
Sample Rudimentary P90+ Risk Assessment, Part 2: Inputs

			Units	A	B	C	D
			failures/year				
		Exposure	mpy	16	8	8	12
Ext Corr	PoF	Mitigation	%	0.9	0.9	0.9	0.9
		Resistance	%	0.25	0.375	0.375	0.25
			failures/year				
		Exposure	mpy	0.1	0.1	4	2
Int Corr	PoF	Mitigation	%	0.5	0.5	0.5	0.5
		Resistance	%	0.25	0.375	0.375	0.25
			failures/year				
		Exposure	events/year	2	5	0.2	0.5
External Force	PoF	Mitigation	%	0.95	0.95	0.95	0.95
		Resistance	%	0.9	0.95	0.95	0.9
			failures/year				
		Exposure	events/year	0.1	0.1	0.1	0.1
Human error	PoF	Mitigation	%	0.99	0.99	0.99	0.99
		Resistance	%	0.9	0.9	0.9	0.9
PoF total			failures/year				
CoF			$/failure	$50	$200	$50	$50
Risk (EL)			$/year				
			$/year				

Having obtained the needed inputs, the assessor then uses simple equations, discussed in this text, to arrive at preliminary risk estimates for each component. The simple equations used are summarized as follows:

Risk = Expected Loss (EL) = PoF x CoF

PoF_time-independent = exposure x (1 - mitigation) x (1 - resistance)

PoF_time-dependent = *f* (Time-to-Failure, TTF)

TTF = resistance / [exposure x (1 - mitigation)]

Sample Rudimentary P90+ Risk Assessment, Part 3: Results

			Units	**A**	**B**	**C**	**D**
			failures/year	0.006	0.002	0.002	0.005
		Exposure	mpy	16	8	8	12
Ext Corr	PoF	Mitigation	%	0.9	0.9	0.9	0.9
		Resistance	%	0.25	0.375	0.375	0.25
			failures/year	0.0002	0.0001	0.005	0.004
		Exposure	mpy	0.1	0.1	4	2
Int Corr	PoF	Mitigation	%	0.5	0.5	0.5	0.5
		Resistance	%	0.25	0.375	0.375	0.25
			failures/year	0.010	0.013	0.001	0.003
		Exposure	events/year	2	5	0.2	0.5
External Force	PoF	Mitigation	%	0.95	0.95	0.95	0.95
		Resistance	%	0.9	0.95	0.95	0.9
			failures/year	0.0001	0.0001	0.0001	0.0001
		Exposure	events/year	0.1	0.1	0.1	0.1
Human error	PoF	Mitigation	%	0.99	0.99	0.99	0.99
		Resistance	%	0.9	0.9	0.9	0.9
PoF total			failures/year	0.017	0.015	0.008	0.011
CoF			$/failure	$50	$200	$50	$50
Risk (EL)			$/year	$835	$2,973	$403	$570
			$/year	**$4,782**			

The various combinations of PoF and CoF yield differing risks for each segment. Armed with these estimates, the decision-makers now move into a risk management phase. This phase will often include improving upon the initial risk estimates, either with deeper analyses or with actual inspections, surveys, investigations, and tests.

In a short period of time, the assessor has produced a rudimentary estimate of risk, documenting key inputs and intermediate calculations associated with the estimate. Furthermore, he has established a framework from which the subsequent robust risk assessment can emerge.

Each of his preliminary inputs can now be reviewed and revised in light of appropriate additional inputs from measurements and investigations. For instance, he may use actual soil resistivity measurements to better estimate exposure rates, mpy, to external corrosion, creating additional components (segments) as the new data provides more granularity since it captures changes along the route. Similarly, he can use depth of cover surveys to modify mitigation estimates for external forces. He can consult previous HAZOP studies to improve his human error inputs. He can use ILI results for improved resistance estimates. He may choose to put additional analyses focus where it is warranted, for example, rare, but sometimes critical phenomena such as AC induced corrosion, landslide potential, SCC, etc. There are countless ways to continuously make the assessment better without changing any aspect of the underlying methodology.

1.4 PIPELINE RISK ASSESSMENT: EXAMPLE 2

The previous section described a simple risk assessment application that employed a short-cut solution—avoiding the need to dynamically segment a pipeline. While it illustrates the framework of good risk assessment, this short cut compromises the risk assessment and should only be used for limited applications and under special circumstances.

Risk assessment on any facility is most efficiently done by first dividing the facility into components with unchanging risk characteristics. For a cross-country pipeline, this involves collecting data on all portions of the pipeline and its surroundings and then using this data to 'dynamically segment' the pipeline into segments of varying length. Risk algorithms are applied to each of the segments, producing risk estimates that truly reflect changing risks along the pipeline.

The risk estimating algorithms are conceptually very straightforward. However, as with any assessment of a complex mechanical system installed in a varying, natural environment, there are many details to consider. This is illustrated by an example risk assessment on a hypothetical pipeline.

Varying levels of analyses rigor are available to risk assessors. For example, a resistance estimate might be modeled as simply being related to stress level and pipe characteristics or, for more robust analyses, could include sophisticated finite element analyses. In the following example, details are omitted in order to better demonstrate the higher level principles.

To illustrate key concepts, one time-independent failure mechanism (third party damage) and one time-dependent failure mechanism (external corrosion) are assessed. All other failure mechanisms will follow one of these two forms. Estimates from all failure mechanisms can be combined in various ways to meet the needs of the subsequent risk management processes.

Example: 1.2

A 120 mile pipeline is to have a risk assessment performed. For the assessment, failure is defined as loss of integrity leading to loss of pipeline product. Consequences are measured as potential harm to public health, property, and the environment and are expressed in units of dollars loss; for example, all consequences are monetized.

Verifiable measurement units for the assessment are as follows:

MEASUREMENT	UNITS
Risk	$/year
Probability of Failure (PoF)	failures/mile-year
Consequence of Failure (CoF)	$/failure
Time to Failure (TTF)	years
Exposure	events/mile-year
Mitigation	%
Resistance	%

Data is collected and includes Subject Matter Expert (SME) estimates where actual data is unavailable. The integrated data shows changes in risk along the pipeline route—6,530 segments are created by the changing data with an average length of 87 ft. This relatively short average length shows that a risk profile with adequate discrimination has been generated.

A level of conservatism is defined as P90 for all inputs that are not based on actual measurements. This is conservative—a bias towards overestimation of actual risks. P90 means that risk is underestimated once out of every 10 inputs, ie, there will be a negative surprise only 10% of the time. The risk assessors have chosen this level of conservatism to account for plausible (albeit extreme) conditions and to ensure that risks are not underestimated.

For assessing PoF from time-independent failure mechanisms—those that do not worsen over time, such as third party damage and human error—the summary equation is as follows:

PoF_time-independent = exposure x (1 - mitigation) x (1 - resistance)

As an example for applying this to PoF due to time-independent third-party damage, the following inputs are identified (by SME's) for a certain portion of the subject pipeline.

- **Exposure** (unmitigated 'attack') is estimated to be three (3) third-party damage events per mile-year. This means that, over this mile of pipeline, excavators will be operating 3 times per year and, in the absence of mitigation, will cause damage to the pipeline three times per year
- Using a **mitigation** (defense) effectiveness analysis, SME's estimate that 1 in 50 of these exposures will not be successfully prevented by existing mitigation measures. This results in an overall mitigation effectiveness estimate of 98% mitigated.

- SME's perform a **resistance** analysis to estimate that, of the exposures that are not mitigated, 1 in 4 will cause immediate failure, not just damage. This estimate includes the possible presence of weaknesses due to threat interaction and/or manufacturing and construction issues. So, the pipeline in this area is judged to have a 75% resistance to failure (survivability) from this mechanism, given the failure of mitigations.

Assuming that frequencies and probabilities are practically interchangeable, these inputs result in the following assessment:

PoF_third-party damage

= (3 damage events per mile-year) x (1 - 98% mitigated) x (1 - 75% resistive)
= 1.5% (0.015) per mile-year
(a failure every 67 years along this mile of pipeline)

Note that a useful intermediate calculation, 'probability of damage' (but not failure), emerges from this assessment and can be verified by future inspections.

(3 damage events per mile-year) x (1 - 98% mitigated)
= 0.06 damage events/mile-year
(damage occurring about once every 17 years).

This same approach is used for other time-independent failure mechanisms and for all portions of the pipeline.

In assessing PoF due to **<u>time-dependent</u>** failure mechanisms—corrosion and cracking, the previous algorithms are slightly modified:

PoF_time-dependent = f (Time-to-Failure, TTF)
TTF = resistance / [exposure x (1 - mitigation)]

To continue the example, SME's have determined that, at certain locations along the 120 mile pipeline, soil corrosivity leads to 5 mpy external corrosion exposure (if left unmitigated). Analyses of coating and CP effectiveness leads SME's to assign a mitigation effectiveness of 90%.

Recent inspections, adjusted for uncertainty and considering possible era-of-manufacture weaknesses, result in an effective pipe wall thickness estimate of 0.220" (remaining resistance). Use of these inputs in the PoF assessment for the next year is shown below:

TTF = 220 mils / [5 mpy x (1 - 90%)] = 440 years
PoF = 1 / TTF = [5 mpy x (1 - 90%)] / 220 mils = 0.11% PoF

So, the combined PoF from these two threats—third party excavators and external corrosion—is estimated to be 0.015 + 0.0011 = 0.016 failures/mile-year. This 1.6% failure probability can now be used with estimates of consequence potential to arrive at overall risk estimates generated by these two threats.

SME's have analyzed potential scenarios and determined the range of possible consequences generated by a failure. After assignment of probabilities to each scenario, a point estimate representing the distribution of all future scenarios yields the value of $18,500 per failure. This can be thought of as a probability-adjusted 'average' consequence per failure.

Risk assessors similarly calculate all risk elements for each of the 6,530 segments. To estimate PoF for any portion of the 120 mile pipeline, a probabilistic summation is used to ensure that length effects and the probabilistic nature of estimates are appropriately considered. To estimate total risk, an expected loss calculation for the full 120 miles yields $25,200 of risk exposure from this pipeline per year of operation. The average is $210/mile-year.

Risk Management

The risk estimates generated in this way are extremely useful to decision makers. Such estimates can become part of the budget setting and valuation processes. In this example, the company first uses these values to compare to, among other benchmarks, a US national average for similar pipelines of $350/mile-year. The comparison needs to consider the P90 level of conservatism employed. Often, a P90 or higher level of conservatism is appropriate for determining risk management on specific pipeline segments, but will not compare favorably to historical incident data since those generally reflect P50 estimates.

Understanding how each pipeline segment contributes to the overall risk sets the stage for efficient risk management.

For risk management at specific locations, cost / benefits of various risk mitigation measures can be compared by running 'what if' scenarios using the same equations with anticipated mitigation effectiveness arising from the proposed action(s).

These estimates can also be used to establish 'safe enough' limits by comparing to pre-determined risk acceptability criteria such as those proposed in ref [95, 9988].

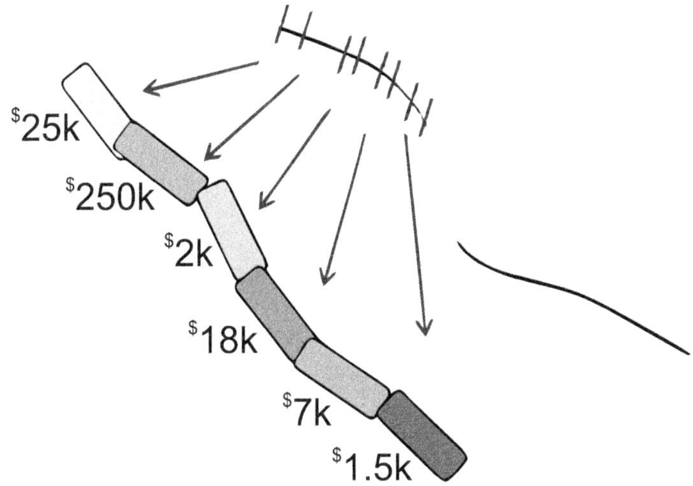

Figure 1.4 Changing Risk Along a Segmented Pipeline EL in $/yr

1.5 VALUES SHOWN ARE SAMPLES ONLY

To help with understanding and preparations of both preliminary and complete risk assessments, this text offers sample valuations for many pipeline risk factors. As with any engineered system (the risk assessment system described herein employs many engineering principles), a degree of caution and due diligence is warranted. The experienced pipeline operator should challenge the example value assignments offered: Do they match your operating experience in general? Are they appropriate for the subject component being assessed? Read the reasoning behind all valuations: Do you agree with that reasoning? Invite (or require) input from employees at all levels. Is there more definitive or more recent data suggesting alternative valuations?

2 DEFINITIONS AND CONCEPTS

Highlights

The following discussions show how certain terms and concepts are used in this text. They may differ from definitions/interpretations when used elsewhere. Additional definitions related to the mechanics of performing data collection and risk assessment are shown in a following chapter.

Definitions and Concepts

As a reference chapter, the following discussions present terms and concepts used throughout this text.

2.1 PIPE, PIPELINE, COMPONENT, FACILITY

As used in this book, a pipeline segment can be any length of pipe, not necessarily a 'joint' length. A component is a part of a pipeline that is other than a pipe segment and can be a flange, valve, fitting, tank, pump, compressor, separator, filter, regulator, or any of many other portions of a typical pipeline. A pipeline is a collection of pipe segments and components. A facility is a collection of components. A system is one or more pipelines and associated facilities. See also the discussion of segmentation for purposes of assessing risk under Chapter 4.5 Segmentation.

Risk concepts covered in this book are meant to apply to any segment of pipe, component, entire pipeline, facility, or system. While pipe is often used to illustrate a concept, the concept also applies to any other component.

As a convenience, the terms component and segment will be used most often in this book.

The basic risk concepts also apply to all component material types. While steel is often the focus of discussion, risks associated with all other materials of construction such as plastic, cast iron, concrete, and others, can be efficiently assessed using these same methods.

Owner/Operator references are used interchangeably here, both referring to the decision-makers who control choices in pipeline design, operations, and maintenance.

2.1.1 Types

Pipeline systems are often categorized into types such as transmission, distribution, gathering, offshore, and others, as discussed in Chapter 3.8 Types of Pipeline Systems. All types are appropriately assessed using the same methodology.

2.1.2 Facility

Facility, station, etc refers to one or more occurrences of, and often a collection of, equipment, piping, instrumentation, and/or appurtenances at a single location, typically where at least some portion is situated above-ground (unburied). Facilities and their subparts are efficiently assessed using the same methodology.

2.1.3 System

The word 'system' has many uses in this text. It is used in context such as safety system, control system, management system, procedure system, training system, to indicate a collection of parts or sub-systems. While no set definition exists, a pipeline *system* normally refers to a large collection of pipeline segments and related stations/ facilities.

2.2 HAZARDS AND RISK

As detailed in many references, risk is most commonly defined as the probability of an event that causes a loss and the potential magnitude of that loss. Risk changes with changes in either the probability of the event or when the magnitude of the potential loss (the consequences of the event). In common use, the term 'hazard' generally refers more to the consequence. It has commonly been said that the hazard associated with a thing or an action is unchangeable but the risk is changeable. Transportation of products by pipeline entails the hazards of the pipeline failing, releasing its contents, and causing damage (in addition to the potential loss of the product itself). The risk associated with this transportation is highly changeable by numerous means.

The most commonly accepted definition of risk is often expressed as a mathematical relationship:

Risk = (event likelihood or probability) × (event consequence)

Risk is best expressed as a measurable quantity such as the expected frequency of certain types of incidents, human fatalities or injuries, or economic loss.

A complete understanding of the risk requires that three questions be answered:

1. What can go wrong?
2. How likely is it?
3. What are the consequences?

The risk assessment approach recommended here measures risk in terms of expected loss (EL).

2.3 EXPECTED LOSS

A powerful approach to measuring and reporting risk is to combine the range of possible consequence scenarios, and their respective probabilities of occurrence, into a single value representing all potential losses over time. Risk expressed in this fashion is called "expected loss" (EL). It encompasses the classical definition of risk: probability x consequences, but expresses risk as a probability of various potential consequences over time. While the concept of expected loss is not a new concept in risk, especially in financial matters, it is perhaps unfamiliar to many practicing pipeline risk assessment.

EL measurement units present the risk as a loss over time, often based on average expected behavior—dollars per year, for instance, for a particular pipeline system. The value is intended to embody all possible consequences (losses) with their respective likelihoods. This value can be viewed as the amount of potential future loss that has been created by the presence of the facility. Costs are a convenient common denominator for all types of losses, and monetized losses are used in the examples presented here.

An EL analysis captures the high-consequence-extremely-improbable scenarios; the low-consequence-higher-probability scenarios, and all variations between. It does this without overstating the influence of either end of the range of possibilities. The use of probabilities ensures that the influences of certain scenarios are not over- or under-impacting the results. All scenarios are considered with appropriate 'weight' for more objective decision support.

Each point on a pipeline produces its own unique set of potential probability-consequence pairings and hence its own expected loss. Theoretically, each possible dollar consequence scenario is multiplied by a probability of occurrence to arrive at a probability-adjusted consequence value (dollars) for each possible consequence scenario. Each point on the pipeline therefore has a distribution of possible failure and consequence scenarios. For practical reasons, a subset of all possible scenarios is used to approximate the distribution of all possible scenarios. This distribution can be expressed as a single point estimate—the expected loss at that location.

The individual expected values for all scenarios at all points along the pipeline can then be combined to produce an expected loss for the entire pipeline (or any portion of any pipeline). Multiple pipelines can have their EL's combined for a measure of the risk of an entire operation. These values show decision-makers the overall risks and suggest levels of appropriate risk management actions, as will be discussed later.

Annualizing all potential consequences into an EL is a modeling convenience. A $100,000 loss event that occurs once every 10 years is mathematically equivalent to an expected loss of $10,000 per year. However, a uniform loss rate—X dollars of loss each period—is really not the expectation. Only the long-term expected losses over time—the behavior of the population—are thought to be fairly represented by the average annual expectation. This presents some financial planning challenges when one considers that while the expected loss on an annualized basis might be acceptable to an organization, that cost might actually occur in a tremendous one-year event and then no other losses occur for decades—no doubt a much less acceptable situation. Similarly, from a risk-tolerance perspective, a once-every-10-years $100,000 event is usually quite different from an annual $10,000 event. While the mathematical equivalence is valid, other considerations challenge the notion of equivalency.

The phrase 'expected loss' carries some emotionalism. It implies that a loss—including injuries, property damages, and perhaps even fatalities—is being forecast as inevitable. This often leads to the question: 'why not avoid this loss?' Most can understand that there is no escaping the fact that risks are present. Society embraces risk and even specifies tolerable risk levels through its regulatory and spending habits. EL is just a measure of that risk. Nonetheless, such terms should be used very carefully, if at all, in risk communications to less-technical audiences. This is more fully discussed elsewhere.

In summary, the EL, as it is proposed here, will represent an average rate of loss from the combination of all loss scenarios at a specific location along a pipeline. An $11K/year EL may represent a $100K loss every ten years and an annual $1K loss ($100K / 10 yrs + $1K/yr = $11K/yr). It is therefore a point estimate representing a sometimes wide range of potential consequences. The EL sets the stage for cost/benefit analyses of possible projects and courses of action as is discussed under Chapter 2.19 Valuations (cost/benefit analyses).

2.4 OTHER RISK UNITS

The most compelling presentation of risk is perhaps in EL values—that is, in monetary terms. They are easily recognizable and provide context that most will understand.

However, they are not without controversy, as discussed later. When alternate, non-monetary presentations of risk is required, options are available.

As an example, consider a table of risk estimates presented in PRMM, based on a real evaluation of a 700 mile gasoline pipeline. That table presents risk as the expected frequencies of certain consequences. Careful examination of this presentation shows many different aspects of risk being considered:

- Leak count as consequence—2.6 leaks over the project life is, itself, an expression of risk
- Receptor damages as consequence—the frequencies of specific damages are shown, recognizing that not all receptors are exposed to all miles and that only some of the 2.6 leaks will result in measurable damage (ie, some will be too small, be rapidly contained, or otherwise not really cause damage)
- Risk presented by the entire project, ie 700 miles of pipeline operating for 50 years. While accurate as a summary value, this will compare unfavorably to most other facilities (non-pipelines), operating within fenced boundaries and having very limited geographical impact potential.
- Length effects. While 700 miles of pipeline actually exists and does expose receptors along its route, a risk value based on total length can be misleading. Each potential receptor is only exposed to a certain length. In this example, usually 2,500 ft of pipe is conservatively assumed to expose a certain point location. So, a 100 ft creek crossing would be exposed to leak/rupture potential from 2,500 + 100 = 2,600 ft of pipeline.
- Annual risks versus lifetime risk (50 years, in this example) is another presentation choice that is potentially misleading.

Many measures of acceptable risk are linked to fatality, specifically annual individual fatality risk. See related discussions under value of human life and risk acceptability criteria under Chapter 11.8.2.4 Value of statistical life and injury.

2.5 FAILURE

As detailed in PRMM, answering the question of "what can go wrong?" begins with defining a pipeline failure. A failure implies a loss or consequence.

The *unintentional release of pipeline contents* is one common definition of a failure. *Loss of integrity* is a type of pipeline failure also implying leak/rupture. The difference between the two may lie in some scenarios such as tank overfill that may include the first but not the latter.

A more general definition of failure is *'no longer able to perform its intended function'*. The risk of service interruption, includes failure from all scenarios resulting in the pipeline not meeting its delivery requirements (its intended purpose).

The concept of limit state can be useful here. In structural engineering, a limit state is a threshold beyond which a design requirement is no longer satisfied (CSA Z662 Annex O). The structure is said to have failed when it fails to meet its design intent which in turn is an exceedance of a limit state. Typical limit states include 'ultimate'—corresponding to a rupture or large leak—'leakage', and 'serviceability'.

Complicating the quest for a universal definition of failure in the pipeline industry is the fact that municipal pipeline distribution systems (water, wastewater, natural gas) tolerate some amount of leakage. Failure may be defined as 'excessive' leakage in contrast to pipelines where any amount of leakage is considered 'failure'.

The most used definition of failure in this book will be leak/rupture. The term *leak* implies that the release of pipeline contents is unintentional, distinguishing a failure from a venting, de-pressuring, blow down, flaring, or other deliberate product release.

2.6 FAILURE MECHANISM, FAILURE MODE, THREAT

Digging deeper, we often need a definition of 'failure' from a material science point of view. Loss of load carrying capacity is a good working definition of material failure. 'Load carrying capacity' is also an appropriate definition for resistance, as measured in a risk assessment. In this text, a failure mechanism is the driving force that can cause a failure.

The failure *mode* is the manner in which the material fails. Common failure mode categories are ductile (yield), brittle (fracture) or a combination, with sub categories of tensile, compressive, and shear. The failure mode is the end state.

The failure *mechanism* is the process that leads to the failure mode. Failure mechanisms include corrosion, impact, buckling, and cracking.

A failure *scenario* is the complete sequence of events that, when combined, result in the failure.

A failure manifesting as a leak is included in the 'load carrying capacity' definition for most pipeline components, since the load of internal pressure is no longer completely carried once a leak of any size forms.

As detailed in PRMM, the ways in which a pipeline can fail can be categorized according to the behavior of the failure mechanisms relative to the passage of time. When the failure rate tends to vary only with a changing environment, the underlying mechanism is considered *time-independent* and should exhibit a constant failure rate as long as the environment stays constant. When the failure rate tends to increase with time and is logically linked with an aging effect, the underlying mechanism is *time-dependent*.

Pipelines tend to avoid early-life leak/rupture failures by commonly used techniques such as manufacture/construction quality control (for example, pipe mill pressure testing, weld inspection) and post-installation pressure test.

Pipelines are often constructed of materials such as steel that has no known degradation mechanism other than corrosion and cracking. By controlling these, a steel pipeline is thought to have an indefinite life-span. See discussion under 'design life'.

Estimates of pipe strength are essential in risk assessment. This is discussed in Chapter 10 Resistance Modeling.

2.7 PROBABILITY

PRMM provides a compelling discussion of probability as it applies to pipeline risk management. The most useful definition of probability is a *degree of belief*. Probability of anything more than 'systems' such as a simple game of chance (coin flip, poker, roulette, dice, etc) requires analysis beyond simple examination of historical event rates and their accompanying statistics. It includes engineering judgment, expert opinion, and an understanding of the underlying physical phenomena of the 'event' whose probability is being assessed.

2.8 PROBABILITY OF FAILURE

When we speak of the probability of a pipeline failure, we are expressing our belief regarding the likelihood of an event occurring in a specified future period. Probability is most often expressed as a decimal ≤ 1.0 or a percentage $\leq 100\%$. Historical data, usually in the form of summary statistics, often partially establishes our degree of belief about future events. Such data is not, however, the only source of our probability estimates.

Probability is often expressed as a forecast of future events. In this application, the expression has the same units as a measured event frequency, i.e. events per time period. When event frequencies are very small, they are, for practical purposes, interchangeable with probabilities: 0.01 failures per year is essentially the same as a 1% probability of one or more failures per year, for purposes here. When event frequencies are larger, a mathematical relationship—reflecting an assumed underlying distribution—is used to convert them into probabilities, ensuring that probabilities are always between 0 and 100%.

The pipeline risk assessment model described here is designed to incorporate all conceivable failure mechanisms that can contribute to probability of failure. Emerging or yet to be identified failure causes are readily added to this framework, once they are understood. The risk assessment is calibrated using appropriate historical incident rates, tempered by knowledge of changing conditions. This results in estimates of failure probabilities that are realistic, utilize all available information appropriately, and match the judgments and intuition of those most knowledgeable about the pipelines.

2.8.1 PoF Triad

> FOCUS POINT
> The critically important fundamentals of measuring PoF are examined here.

In risk assessment, there is the need for a very specific approach to measuring failure probability (PoF). Three factors must be independently measured/estimated in order to fully understand PoF. The reasoning here is that the PoF is being examined in distinct pieces—a reductionist approach—prior to their aggregation into a PoF estimate.

Regardless of the definition of 'failure' being used, failure ***only*** occurs when there is a failure mechanism ***and*** preventive measures are insufficient ***and*** there is insufficient resistance to the failure mechanism. All three must occur before failure occurs. This is the genesis of the proper way to measure PoF.

We also recognize that there is more than one potential failure mechanism that can lead to failure. These two basic concepts lead to one of the most important of the essential elements of pipeline risk assessment:

All plausible failure mechanisms must be included in the assessment of PoF. Each failure mechanism must have each of the following three aspects measured or estimated in verifiable and commonly used measurement units:

> **Exposure (attack)**— an exposure[1] is defined as an event which, in the absence of any mitigation, can result in failure, if insufficient resistance exists. The type and unmitigated aggressiveness of every force or process that may precipitate failure is an exposure.
>
> **Mitigation (defense)**—the type and effectiveness of every mitigation measure designed to block or reduce an exposure.

1 This can be confusing to some since 'exposure' is a term also commonly applied to a location on an originally buried pipeline that has experienced a depletion of cover, rather than as one of the essential elements of a PoF measurement.

Resistance—a measure or estimate of the ability of the component to absorb the exposure force without failure, once if the exposure reaches the component.

For each time-dependent failure mechanism, a theoretical remaining life estimate must be produced and expressed in a time unit.

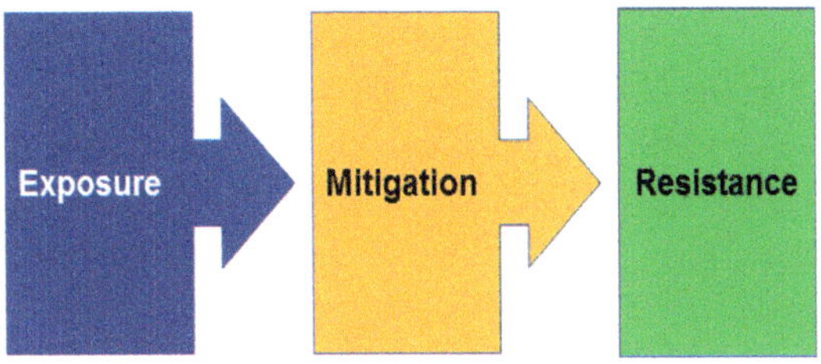

Figure 2.1 Exposure, Mitigation, Resistance

An analogous naming convention is 'attack', 'defense', and 'survivability', respectively, for these three terms. The evaluation of these three elements for each threat to each pipeline component within a segment results in a PoF estimate for that segment.

Measuring exposure—attack—independently generates knowledge of the 'area of opportunity' or the aggressiveness of the attacking mechanism. Then, the separate estimate of mitigation—defense—effectiveness shows how much of that exposure should be prevented from reaching the component being assessed. Finally, the resistance estimate shows how often the component will failure—it's survivability, if the exposure actually reaches the component.

This three-part assessment also helps with model validation and most importantly, with risk management. Fully understanding the exposure level, independent of the mitigation and system's ability to resist the failure mechanism, puts the whole risk picture into clearer perspective. Then, the roles of mitigation and system vulnerability are both known independently and also in regards to how they interact with the exposure. Armed with these three aspects of risk, the manager is better able to direct resources appropriately.

In risk management, where decision-makers contemplate possible additional mitigation measures, additional resistance, or even a re-location of the component (often the only way to change the exposure), this knowledge of the three key factors will be critical.

The simple equation for PoF shows two ways to reduce PoF—either increase mitigation—blocking the failure mechanism—or increase resistance—making the structure stronger to absorb more forces. This independent evaluation of exposure and mitigation also captures the idea that "no exposure" will inherently have less risk than "mitigated exposure," regardless of the robustness of the mitigation measures. As well, the notion that a very stout component is intrinsically safer, is captured.

In estimating future exposures, it is important to first list all potentially damaging mechanisms that could occur at the subject location. Then, numerical exposure values should be assigned to each.

Pre-dismissal of exposures should be avoided—the risk assessment will show, via low PoF values, where threats are insignificant. It will also serve as documentation that all threats are considered.

For example, falling trees, walls, utility poles, etc are often overlooked in a pipeline risk assessment. This is an understandable result of discounting such threats via an assumption that a buried component is often virtually immune from such damage. While this is normally an appropriate assumption, the risk assessment errs when such threat dismissal occurs without due process. Pre-screening of threats as insignificant weakens the assessment. The independent evaluation of exposure and mitigation ensures that, should depth of cover condition change, ie, the component is relocated above grade; or a particular falling object indeed can penetrate to the buried pipeline; are not lost to the assessment.

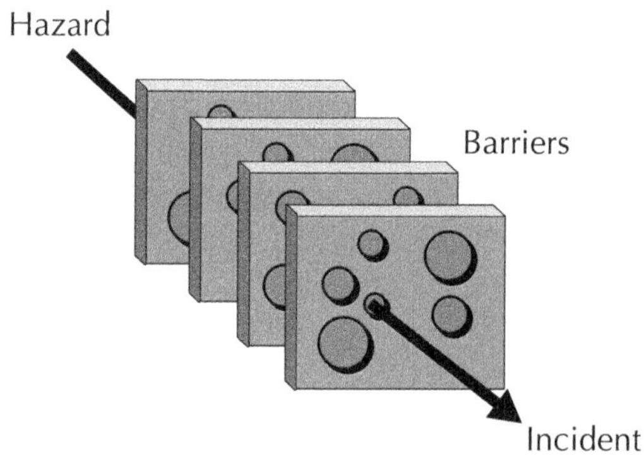

Figure 2.2 Swiss Cheese Analogy: More Slices and/or Fewer Holes Reduces Event Probability

2.8.2 Units of Measurement

Units of measurement should always be transparent and intuitive. In one common application of the exposure, mitigation, resistance triad, units are as follows. Each exposure is measured in one of two ways—either in units of 'events per time and distance', ie events/mile-year, events/km-year, etc, or in units of degradation—metal loss or crack growth rates, ie mpy, mm per year, etc. An 'event' is an occurrence that, in the absence of mitigation and resistance, will result in a failure. To estimate exposure, we envision the component completely unprotected and highly vulnerable to failure (think 'tin can' wall thickness). So, an excavator working over a buried pipeline is an event. This is counted as an event regardless of depth of burial, use of one-call, signs/markers, patrol, etc.

Units of measure, beginning with exposure estimates and carried through until final risk estimates, include time and distance. As time periods and distances increase, so too does risk. This is intuitive—more miles and more years of operation logically suggests that more things can go wrong—a greater area of opportunity. The probability (future frequency) of a corrosion leak at any location may only be 0.001 leaks per mile-year, but with hundreds of miles and/or decades of operation, the probability grows to almost 100% of at least one corrosion leak somewhere along the route within the time period.

Mitigation and Resistance are each measured in units of % representing 'fraction of damage or failure scenarios avoided'. A mitigation effectiveness of 90% means that 9 out of the next 10 exposures will not result in damage. Resistance of 60% means that 40% of the next damage scenarios will result in failure, 60% will not.

For assessing PoF from time-independent failure mechanisms—those that appear random and do not worsen over time—the top level equation can be as simple as:

$$\text{PoF\_time-independent} = \text{exposure} \times (1-\text{mitigation}) \times (1-\text{resistance})$$

With the above example units of measurement, PoF values emerge in intuitive and common units of 'events per time and distance', ie events/mile-year, events/km-year, etc.

A risk assessment measures the aggressiveness of potential failure mechanisms and effectiveness of offsetting mitigation measures and design features. The interplay between aggressiveness of failure mechanisms and mitigation/resistance effectiveness yields failure potential estimates.

2.8.3 Damage Versus Failure

Another benefit emerges from the exposure/mitigation/resistance triad. Probability of Damage—damage without immediate failure—can be measured independently from PoF. Using the first two terms without the third—exposure and mitigation, but not resistance—yields the probability of damage.

$$\text{Probability of Damage (PoD)} = f\ (\text{exposure, mitigation})$$

$$\text{Probability of Failure (PoF)} = f\ (\text{PoD, resistance})$$

Damage results from an exposure that reaches the component but does not cause failure.

Damage that does not result in immediate failure may cause reduced resistance against future failure mechanisms. Some damage may also trigger or accelerate a time-dependent failure mechanism. Calculation of both PoD and PoF values creates better understanding of their respective risk contributions and provides the ability to better respond with risk management strategies.

2.8.4 From TTF to PoF

Estimation of PoF for time-dependent failure mechanisms requires an intermediate calculation of time-to-failure (TTF).

PoF_time-dependent = f(TTF_time-dependent)

TTF_time-dependent = resistance / [exposure x (1–mitigation)]

The relationship between an estimated TTF and the probability of failure can be complex and warrants special discussion. The PoF is normally calculated as the chance of ***one or more*** failures in a given time period. In the case of time-dependent failure mechanisms, TTF estimates are first produced. The associated failure probability assumes that at least one point in the segment is experiencing the estimated degradation rate and no point is experiencing a more aggressive degradation rate.

The TTF estimate is expressed in time units and is calculated by using the estimated pipe wall degradation rate and the theoretical pipe wall thickness and strength, as was shown above. In order to combine the TTF with PoF from all other failure mechanisms, it is necessary to express the time-dependent failure potential as PoF. This requires a conversion of TTF to PoF. It is initially tempting to use the reciprocal of this time-to-failure number as a leak rate—failures per time period. For instance, 20 years to failure implies a failure rate of once every twenty years perhaps leading to the assumption of 0.05 failures per year. However, a logical examination of the TTF estimate shows that it is not really predicting a uniform failure rate. The estimate is actually predicting a failure rate of ~0 for 19+ years and then a nearly 100% chance of failure in the final year. Nonetheless, use of a uniform failure rate is conservative and helps overcome potential difficulties in expressing degradation rate in probabilistic terms. This is discussed later.

An exponential relationship can be used to show the relationship between PoF in year one and failure rate. Using the conservative relationship of [failure frequency] = 1/TTF, a possible relationship to use at least in the early stages of the risk assessment is:

PoF = 1-EXP(-1/ TTF)

Where
PoF = probability of failure in year one
TTF = time to failure

This relationship ensures that PoF never exceeds 1.0 (100%). As noted, this does not really reflect the belief that PoF's are very low in the first years and reach high levels only in the very last years of the TTF period. The use of a factor in the denominator will shift the curve so that PoF values are more representative of this belief. A

Poisson relationship or Weibull function can also better show this, as can a relationship of the form PoF = 1 / (fctr x TTF^2) with a logic trap to prevent PoF from exceeding 100%. The relationship that best reflects real-world PoF for a particular assessment is difficult, if not impossible to determine. Therefore, the recommendation is to choose a relationship that seems to best represent the peculiarities of the particular assessment, chiefly the uncertainty surrounding key variables and confidence of results. The relationship can then be modified as the model is tuned or calibrated towards what is believed to be a representative failure distribution.

The relationship between TTF and PoF includes segment length as a consideration. PoF logically increases as segment length increases since a longer length logically means more opportunity for active failure mechanisms, more uncertainty about variables, and more opportunities for deviation from estimated degradation rates. This is discussed more fully in a later section. See also Chapter 2.8.8 Probabilistic Degradation Rates and Chapter 2.8.9 Capturing "Early Years' Immunity" for a continuation of the TTF to PoF discussion.

2.8.5 Age as a Risk Variable

Age-based or historical leak-rate based estimates are readily generated when data is available and can be useful for quick or initial risk estimates. Statistical examination of historical leak and break data provides insights into behaviors of populations of components over long periods of time. When such populations are similar in characteristics and environment to a collection of components being assessed, such statistical analyses have some predictive capability. This is often an approach for general predicting of leaks in larger distribution systems.

While age is often used as a gross indicator of leak/break likelihood, especially on distribution systems where some amount of leakage is tolerable and is tracked over time, neither age nor historical leak rates indicate the presence of degradation mechanisms at any specific location.

Age is rarely a direct indicator of risk. It does, however, suggest indirect risk indications related to issues such as era of manufacture/construction and extent of degradation where time-dependent mechanisms are active. Location-specific failure probability is best estimated by assessment of relevant exposure, mitigation, and resistance characteristics at that location and system-wide deterioration is best estimated by accumulating all location-specific damage potentials. The more useful risk assessment will evaluate the actual mechanisms possibly at work at any location and then supplement this with population statistical data.

2.8.6 The Test of Time Estimation of Exposure

In the absence of more compelling evidence, an appropriate starting point for the exposure estimation may be the fact that a component or collection of components has not failed after x years in service. This involves the notion of having 'withstood the test of time'. A component having survived a threat, especially for many years, is evidence of the exposure level. This is best illustrated by example. If 10 miles of pipe, across an area with landslide potential, has been in place for 30 years without experiencing any landslide effects, then a failure tomorrow perhaps suggests an event rate of 1/(10 miles x 30 years) = 1/300 mile years.

This simple estimate will not address the conservatism level. The estimator will still need to determine if this value represents more of a P50 estimate or perhaps a more conservative P90+ value.

In some cases, the evidence is actually of the mitigated exposure level. That is, the component has survived the threat, but perhaps at least partially due to the presence of effective mitigation. This makes the separation of exposure more challenging.

Despite the lack of complete clarity, this 'test of time' rationale can be a legitimate part of an exposure estimate.

2.8.7 Time-dependent vs independent

Risk assessment begins with understanding potential failure scenarios. While both types of failure mechanism— time-dependent vs time independent—can be involved in a failure scenario, it might not be immediately obvious how to treat the combined effect scenario in a risk assessment. Fortunately, in a good risk assessment methodology, the contributions from each type are automatically and intuitively considered. All exposures should be included and any degradation effects should be factored into the ability to resist all corresponding stress levels.

There should not be any confusion regarding when a time-dependent mechanism is involved in a failure scenario. Consider an investigation of a failed component. The dominant failure mechanism type can normally be determined by simply answering the question; 'why did it fail today and not yesterday?'. If the component performed without failure for some previous period of time, and was not subjected to new stresses, then logically, some degradation occurred to cause the failure 'today' and not 'yesterday'. Degradation indicates a time-dependent failure mechanism at work.

In other words, unless the component has never before been subjected to the failure stress, the fact that it fails 'today' versus 'yesterday' implies a time factor—ie, some time-dependent mechanism was active and weakened the pipe since the previous application of that stress level. If the stress level is simply 'recently new', ie, hasn't been experienced lately, then degradation is still likely the dominant mechanism. Reductions in resistance (effective wall thickness, as detailed in Chapter 10.4.3 Effective Wall Thickness Concept) hasten time to failure and increase failure potential upon application of stress.

Even if the failure scenario does not involve a typical degradation process, but a time element is nonetheless inferred, the assessment can efficiently include it as a time-dependent failure mechanisms. Consider a leak at a threaded connection, where no corrosion or cracking is found. If the connection was leak free at one time and no new stresses were applied, the loosening of the connection can still be efficiently modeled as a degradation mechanism (see discussion in Chapter 6.8.4.3 Vibrations/ Oscillations).

2.8.8 Probabilistic Degradation Rates

Degradation rates are among the most difficult aspects of risk assessment to accurately estimate. Rates are highly variable over time and even in very localized areas. For instance, an aggressive pitting corrosion rate of 50 mpy can commonly exist within millimeter fractions where virtually zero degradation is occurring. It can also reach 50 mpy for some period of time and then become inactive for long periods. Our understanding of many of the even more common mechanisms, requires us to model a degree of randomness in the occurrence locations and possible rates. We use probabilities to recognize this randomness.

It would be convenient to model a 10% chance of a 50 mpy degradation as a 5 mpy degradation. But these two values have different implications. If it takes 50 mils of wall loss to cause a leak in a component, then a 10% chance of 50 mpy suggests that there is a 10% chance of a leak every year. However, a 5mpy would not result in a leak until 10 years have passed.

Both scenarios can be accommodated in the assessment by appropriate treatment of the conversion from TTF to PoF.

2.8.9 Capturing "Early Years' Immunity"

Using the basic relationship employing some form of PoF as a function of 1/TTF described above can result in excessive conservatism. Consider a very new, thick-walled component whose early years are virtually unthreatened by any plausible degradation rate. Even a 100 mpy degradation rate should not threaten the year one (or even year three) integrity of a 0.400" pipe. New components, those with heavy wall thicknesses, those in very benign environments, those with very accurate and recent inspections, etc, all have some amount of immunity to failure from slow-acting degradations, at least in the early years of exposure.

However, this immunity is uncertain and temporary for most. Using a relationship such as lognormal or Weibull to show failures only in later years of the TTF estimates risks missing the often small but real chance of very aggressive degradations or unexpectedly thin component wall thickness. Recall the example where a 10% chance of 50 mpy can suggest a real chance of leak in the next year.

A two-part relationship between PoF and TTF solves this issue and is often warranted. By adding an extreme value analysis to the basic TTF analysis, early year TTF's can be dismissed in certain scenarios.

The extreme value analysis requires the creation of a variable called TTF99. TTF99 is the minimum plausible TTF—a value that is lower than any actual value will be, 99-99.9% of the time—for example, the subject matter expert (SME) is 99+% confident that the TTF cannot be worse than this value, even considering a highly improbable coincidence of very unlikely factors. Establishing this extreme value can be done by taking the best pipe wall thickness estimate and degrading that by the highest plausible unmitigated corrosion/cracking rate. Alternatively, statistical methods can be used to establish the 99% confidence level, when data is available.

Using both TTF and TTF99 creates four scenarios, each with its own relationship to PoF. These scenarios involving TTF (best estimate of current time to failure) and TTF99 (lowest plausible TTF) are examined to arrive at an estimate of PoF:

The scenarios are summarized as follows, assuming the time of interest is 1 year[2]—a year one PoF is sought (what is the probability of failure in the next 12 months?) Note that TTF is the best estimate—ie, thought to be the most likely value—and TTF99 is the very conservative estimate:

> If TTF99 less than 1 year AND TTF less than 1 year, then PoF = 99+%
>
> If TTF99 less than 1 year AND TTF greater than 1 year, then use constant failure rate, basically the reciprocal of the TTF, to estimate PoF
>
> If TTF99 greater than 1 year AND TTF greater than 1 year, then 'use more optimistic relationship' (such as lognormal(TTF99)) to estimate PoF from TTF

Scenario 1. If it is plausible to have a year one failure AND the best estimate of TTF is also less than one year.

> If TTF is less than 1 year, then failure during year one is likely and PoF is assigned 99%. Pipeline segments are conservatively assigned this value when little information is available and a very short TTF cannot be ruled out.

Scenario 2. If it is plausible to have a year one failure AND the best estimate of TTF is greater than one year.

> If TTF > 1 year but TTF99 is < 1 year, then we believe year one failure is unlikely but cannot be ruled out. PoF needs to reflect the probabilistic mpy embedded in the TTF estimate. Probabilistic mpy means that, for instance, a 10 mpy includes a scenario of '10% chance of a 100mpy degradation rate'. To ensure that the PoF estimate captures the small chance of a 100mpy rate actually occurring next year, a constant and conservative failure probability—PoF = 1/TTF—is associated with the 10mpy. Pipeline segments will fall into this analysis category when very short TTF is possible but the most probable TTF values exceed the year for which PoF is being assessed.

2 Any future time can be used; producing risk estimates for the following year is common and used as an example here.

Scenario 3. If it is not plausible to have a year one failure, even using extreme values

If TTF99 > 1 year then we believe that, even under worst case scenarios, failure in year one will not happen. TTF99, rather than the actual TTF governs PoF. The relationship between TTF99 and PoF can be assumed to be lognormal or Weibull or some other distribution, with parameters selected from actual data or from judgments as to distribution shapes that are reflective of the degradation mechanism being modeled. Very low year one PoF's will emerge. A new pipeline even with a high plausible degradation rate, will have a PoF governed by this analysis (for example, a 0.250" thick wall will not experience a through-wall leak in year one even with a 100 mpy pitting corrosion rate).

Scenario 4. TTF is very high

Consider yet one more scenario: when TTF is very high, it may override TTF99 for PoF. This is again logical. Even if TTF99 is close to one—PoF approaching 100%—TTF might indicate that the segment's actual TTF (best estimate) is so far from this low probability event, that it should govern the final PoF estimate. A pipeline segment with very high confidence in both current pipe wall and a low possible degradation rate will have a high TTF. Even if a short TTF is theoretically possible—as shown by TTF99—a sufficiently high confidence in the estimated TTF can govern. Such high confidence is often obtained via repeated, robust inspections and when the degradation rate required for early failure would be an extreme aberration.

Scenarios 3 & 4 are appropriate only when TTF99 > 1 year or can be dismissed as implausible—virtually no chance of failure in year one. Then the worst case between scenario 3 and scenario 4 governs.

See the figure below showing the two-part curve, where PoF is on the vertical axis and time is on the horizontal axis.

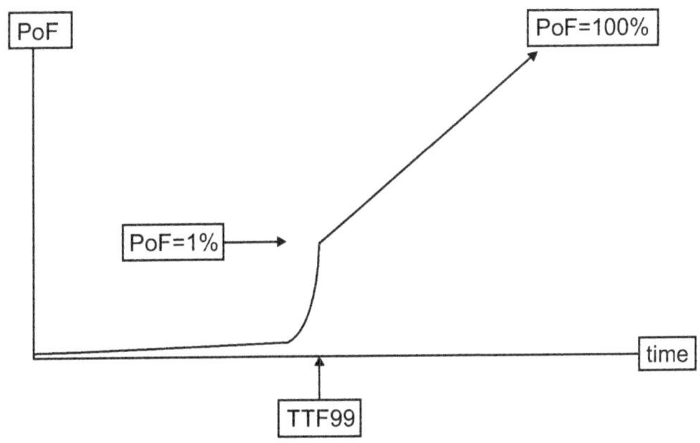

Figure 2.3 TTF to PoF

Again, the rationale for use of a two-part curve is intuitive. A new pipeline has little chance of corrosion leak in the early years, even when aggressive corrosion rates are possible. Therefore, even if a worst case TTF is 5 years, the new pipeline enjoys a very low PoF in year one. Use of the simple PoF = 1/TTF does not show this. It yields a 20% chance of failure in year one, requiring an extreme value analysis to demonstrate that this is over-conservative.

Alternatively, when conditions or uncertainty suggest a plausible near-term failure due to degradation, the use of TTF as a direct mean-time-to-failure link to PoF is more appropriate.

2.8.10 Example Application of PoF Triad

As an example (part of full example shown in Chap 1.4) of applying the PoF triad to a time-independent and a time-dependent failure mechanism, consider the following. For failure potential from third party excavations, the following inputs are identified for a hypothetical pipeline segment:

- Exposure (unmitigated) is estimated to be 3 excavation events per mile-year. A previous column discusses how these estimates can be made.
- Using a mitigation effectiveness analysis, SME's estimate that 1 in 50 of these exposures will not be successfully kept away from the pipeline by the existing mitigation measures. This results in an overall mitigation effectiveness estimate of 98%.
- Of the exposures that result in contact with the pipe, despite mitigations, SME's perform an analysis to estimate that 1 in 4 will result in failure, not just damage. This estimate includes the possible presence of weaknesses due to threat interaction and/or manufacturing and construction issues. So, the pipeline in this area is judged to be 75% resistive to failure from these excavation events, once contact occurs.

These inputs result in the following assessment:

(3 excavation events per mile-year) x (1–98% mitigated) x (1–75% resistive)
= 0.015 failures per mile-year

This suggests an excavation-related failure about every 67 years along this mile of pipeline.

This is a very important estimate. It provides context for decision-makers. When subsequently coupled with consequence potential, it paints a valuable picture of this aspect of risk.

Note that a useful intermediate calculation, probability of damage (but not failure) also emerges from this assessment:

$$(3 \text{ excavation events per mile-year}) \times (1\text{–}98\% \text{ mitigated}) = 0.06 \text{ damage events/mile-year}$$

This suggests excavation-related damage occurring about once every 17 years.

This damage estimate can be verified by future inspections. The frequency of new top-side dents or gouges, as detected by an ILI, may yield an actual damage rate from excavation activity. Differences between the actual and the estimate can be explored: for example, if the estimate was too high, was the exposure overestimated, mitigation underestimated, or both? This is a valuable learning opportunity.

This same approach is used for other time-independent failure mechanisms and for all portions of the pipeline.

For assessment of PoF for time-dependent failure mechanisms—those involving degradation of materials—the previous algorithms are slightly modified to yield a time-to-failure (TTF) value as an intermediate calculation in route to PoF.

$$\text{PoF\_time-dependent} = f(\text{TTF\_time-dependent})$$

$$\text{TTF\_time-dependent} = \text{resistance} / [\text{exposure} \times (1\text{–mitigation})]$$

As an example, SME's have determined that, at certain locations along a pipeline, soil corrosivity creates a 5 mpy external corrosion exposure (unmitigated). Examination of coating and cathodic protection effectiveness leads SME's to assign a mitigation effectiveness of 90% . Recent inspections, adjusted for uncertainty, result in a pipe wall thickness estimate of 0.220" (resistance). This includes allowances for possible weaknesses or susceptibilities, modeled as equivalent to a thinning of the component's wall thickness.

Use of these inputs in the PoF assessment is shown below:

$$\text{TTF} = 220 \text{ mils} / [5 \text{ mpy} \times (1\text{–}90\%)] = 440 \text{ years.}$$

Next, a relationship between TTF and PoF for the future period of interest, is chosen. For example, a simple and conservative relationship yields the following.

$$\text{PoF} = 1 / \text{TTF} = [5 \text{ mpy} \times (1\text{–}90\%)] / 220 \text{ mils} = 0.11\% \text{ PoF.}$$

In this example, an estimate for PoF from the two failure mechanisms examined—excavator damage and external corrosion—can be approximated by 1.5% + 0.1% = 1.6% per mile-year. If risk management processes deem this to be an actionable level of risk, then the exposure-mitigation-resistance details lead the way to risk reduction opportunities.

2.8.11 AND gates OR gates

Combining variables often involves the choice of multiplication versus addition. Each has advantages. Multiplication allows variables to independently have a great impact on a result. Addition better illustrates the layering of adverse conditions or mitigations. In formal probability calculations, multiplication usually represents the *and* operation.

Probabilistic math is used to combine variables to represent real-world phenomena. This means capturing various relationships among variables using "OR" & "AND" "gates." This OR/AND terminology is borrowed from flowchart techniques. The use of OR/AND math in pipeline risk assessment modeling represents a dramatic improvement over most older methods that used simple summations, averages, maximums, and other summary mathematics or statistics that often masked critical information.

2.8.11.1 OR Gates

OR gates imply independent events that can be added. The OR function calculates the probability that ***any*** of the input events will occur. If there are *i* input events each assigned with a probability of occurrence, P_i, then the probability that any of the *i* events occurring is:

$$P = 1 - [(1-P_1) * (1-P_2) * (1-P_3) *...*(1-P_i)]$$

This is the same as 1 – (the probability that none of the *i* events occur)

OR gates are extremely useful in that they capture, in a 'real-world' way, both the effects of single, large contributors as well as the accumulation of lesser contributors. With an OR gate, there is no 'averaging away' effect. In a pipeline risk assessment, this type of math better reflects reality since it uses probability theory of accumulating impacts to:

- Avoid masking some influences;
- Captures single, large impacts as well as accumulation of lesser effects;
- Shows diminishing returns;
- Avoids the need to have pre-set, pre-balanced list of variables;
- Provides an easy way to add new variables; and
- Avoids the need for re-balancing when new info arrives.

When summarizing the PoF of a component, the central question of 'what is the PoF?' is actually asking 'what is the PoF from either PoF1 or PoF2 or PoF3 or...?' where 1, 2, 3, etc represent all the ways in which the component can fail, ie, external corrosion, outside forces, human error, etc. The overall PoF can therefore be relatively high if any of the sub-PoF's are high or if the accumulation of small sub-PoF's adds up to something relatively high.

This is consistent with real-world risk. The question of overall PoF does NOT presume that all PoF's must 'fire' before the overall PoF is realized—it only takes one. A segment survives only if failure does not occur via ***any*** of the failure mechanisms. So, the probability of surviving is (third-party damage survival) AND (corrosion survival) AND (design survival) AND (incorrect operations survival). Replacing the ANDs with multiplication signs provides the relationship for probability of survival. Subtracting this resulting product of multiplication from one (1.0) gives the probability of failure.

OR Gate Example:

To estimate the overall probability of failure based on the individual probabilities of failure for stress corrosion cracking (SCC), external corrosion (EC) and internal corrosion (IC), the following formula can be used.

$P_{failure}$ = OR[P_{SCC}, P_{EC}, P_{IC}] = P_{SCC} OR P_{EC} OR P_{IC}
= OR [1.05E-06, 7.99E-05, 3.08E-08] (*using some sample values*)
= 1- [(1-1.05E-06)*(1-7.99E-05)*(1-3.08E-08)]
= 8.10E-05

The OR gate is also used for calculating the overall mitigation effectiveness from several independent mitigation measures. This function captures the idea that probability (or mitigation effectiveness) rises due to the effect of either a single factor with a high influence or the accumulation of factors with lesser influences (or any combination).

Mitigation % = M_1 OR M_2 OR M_3.....
= 1–[(1-M_1) * (1-M_2) * (1-M_3) *...*(1-M_i)]
= 1 – [(1-0.40) * (1-0.10) * (1-0.05)]
= 49%

or examining this from a different perspective,

Mitigation % = 1 – [remaining threat]
Where
[remaining threat] = [(remnant from M_1) AND (remnant from M_2) AND (remnant from M_3)] ...

The OR gate math assumes independence among the values being combined. While not always precisely correct, the advantages of assuming independence as a modeling convenience will generally outweigh any loss in accuracy.

The independence is often difficult to visualize, especially when assigning effectiveness values to mitigation. For instance, the effectiveness of a line locating program (see Chapter 5 Third-Party Damage) should be judged by estimating the fraction of future damaging events that are avoided by the line locating program ONLY—ie, imagining no depth of cover (but still out of sight), no signs, no markers, no public education, no patrol, etc.

2.8.11.2 AND Gates

AND gates imply "dependent" measures that should be combined by multiplication. With an AND gate, any sub-variable can alone have a dramatic influence. This is captured by multiplying all sub-variables together. In measuring mitigation, when all things have to happen in concert in order to achieve the mitigation benefit, a multiplication is used—an AND gate instead of OR gate. This implies a dependent relationship rather than the independent relationship that is implied by the OR gate.

AND Gate Example[3]:
The modeler is assessing a variable called "CP Effectiveness" (cathodic protection effectiveness) where confidence in ***all*** sub-variables is necessary in order to be confident of the CP Effectiveness—[good pipe-to-soil voltage readings] AND [readings close to segment of interest] AND [readings are recent] AND [proper consideration of IR was done] AND [low chance of interference] AND [low chance of shielding]... etc. If any sub-variable is not satisfactory, then overall confidence in CP effectiveness is dramatically reduced. This is captured by multiplying the sub-variables.

When the modeler wishes the contribution from each variable to be slight, the range for each contributor is kept fairly tight. Note that four things done pretty well, say 80% effective each, result in a combined effectiveness of only ~40% (0.8 x 0.8 x 0.8 x 0.8) using straight multiplication.

2.8.12 Nuances of Exposure, Mitigation, Resistance

In most instances, the categorization of each piece of information into one of these three is obvious—most variables are clearly telling us more about either the exposure, the mitigation, or the resistance. To some, the surrogate terms of 'attack', 'defense', and 'survivability' add clarity. Focusing on PoF only, here are some examples to help solidify the categorization:

2.8.12.1 The obvious

Variables can inform multiple aspects of a risk assessment, but usually, one category is more directly influenced by the variable. Soil corrosivity, excavator activity, vehicle traffic, seismic activity, flood potential, surge potential, landslides, are examples of phenomena that obviously inform exposure estimates. They tell us about the frequency and severity of 'attack'.

3 This example assumes some basic knowledge of protection of buried steel pipeline by cathodic protection. See chapter 6 and PRMM if more background is needed.

Coatings, depth of cover, training, procedures, maintenance pigging, are examples that, to most, are clearly defenses against damage. They are best modeled as mitigation measures. When the same mitigation measure protects against multiple exposures, it is valid to include their benefits in all relevant threats. For instance, depth of cover protects against impacts, excavations, and some types of geohazards.

Metal loss, cracks, lack of toughness, SMYS, wall thickness are examples of variables that inform resistance estimates.

2.8.12.2 The less obvious

Casings: a casing (see full discussion later) sometimes causes confusion when one focuses on corrosion problems potentially caused by their presence and loses sight of the original intent. Casings are usually installed as mitigation to external forces. They also serve other purposes such as consequence reduction, but they are mostly intended to protect a carrier pipe. Their role in a risk assessment should show their benefit in preventing excavation damages, traffic loads, and others. However, a casing's role as a corrosion issue should also be acknowledged. A casing changes the external corrosivity exposure (electrolyte in the annular space and possible electrical connections) and the ability to apply CP. Both should appear in the risk assessment. So, the presence of a casing is captured as a mitigation against external forces, an influencing factor for external corrosion exposure and mitigation (shielding of CP), and perhaps also in CoF.

ILI: some may initially think protection occurs with the activity of performing an ILI. Actually, as with other inspections and tests, neither the exposure nor the mitigation nor the resistance has changed because of the ILI. What has changed is the evidence—knowledge of resistance has increased, often dramatically, and uncertainty regarding exposure and mitigation is different because of the ILI. For instance, at every identified location of external metal loss on a buried pipeline, we know that both coating and CP have failed, so mitigation is reduced, perhaps to zero, pending repairs. We usually do not know when mitigation failed, so might not be able to directly modify exposure (mpy rate of corrosion) estimates without more information. So, the role of ILI is first in resistance estimates and secondarily in exposure and mitigation estimates. Of course, action prompted by the ILI will often change exposure and mitigation.

Laminations, wrinkle bends, and arc burns are resistance issues. They are not 'attacking' the pipe, nor do they contribute to or impair mitigation. They represent potential weaknesses, sometimes only under the influence of exacerbating factors such as certain loadings (for example, causing stress concentrations) or environment (for example, sources of H_2 that aggravate laminations and facilitate blistering or HIC). They are best modeled as potential losses of strength—ie, as resistance issues.

2.8.12.3 Additional Gray Areas

When information can logically be categorized in more than one place, the choice is usually a matter of preference and does not weaken the assessment. Choices of the role of the information usually leads to the same mathematical result. So, the choice is often not critical to the PoF estimate. Some examples of such choices are discussed below.

Note that while several 'gray area' examples are discussed here, the vast majority of information is very easily and intuitively categorized into its appropriate place in the risk assessment. The reader should not leave this section believing that any more than a very few scenarios have some ambiguity regarding modeling choices.

2.8.12.4 What Constitutes 'Exposure'? Normalizing Exposure and Resistance

Since PoF measures 'failure', the definition of exposure is linked to that of failure. An exposure must be able to cause a 'failure' if it is truly an exposure. If failure is defined, for instance, as 'permanent deformation', then exposures that could cause that to a pipe component, are counted. If failure is defined as 'loss of integrity', events causing immediate leaks/ruptures are obviously needed, but so are damage-only events. In fact, most assessments will appropriately include all events that can at least cause damage. Even when immediate failure from the event is not possible, the damage may contribute to a subsequent failure and is therefore of interest to the measurement of PoF.

Should excavation by hand shovel be considered an exposure for a steel pipeline? Yes, if any structural damage at all is possible—even a scratch. This scratch may directly reduce resistance to some future failure mechanism, although it is often an immeasurably small reduction. The scratch can also theoretically occur exactly at some point of pre-existing weakness, resulting in immediate failure.

Excavation by a plastic shovel probably cannot cause even minor scratch damage to a steel pipeline and need not be counted as an exposure. However, the indirect role of a 'hand shovel contact' event must be considered. Both the metal and plastic shovel should be counted as causes of damage to corrosion coating systems. Since coating is a mitigation measure, damage to a coating reduces mitigation effectiveness. This is different from an exposure. If concrete coating or rock shielding is present, it provides mitigation against coating damage.

Vandalism can be considered a type of sabotage. However, defacing (for example, spray painting) or minor theft of materials are actions that are readily resisted by most pipeline components. If the sabotage exposure count includes vandalism events, then resistance estimates must consider the fraction of exposure events that are vandalism spray-paint-type events and therefore 100% resisted by the component.

Exposure and resistance estimates for risk assessments of failure = 'service interruption' similarly revolve around the definition of failure. Just as with leak/rupture assessments, a probability of damage also emerges from the service interruption assessment. See full discussion in Chapter 12 Service Interruption Risk.

This nuance—what constitutes an 'exposure'—revolves around failure definition and also the choice of baseline resistance, which warrants further discussion.

2.8.12.5 Continuous Exposure

Unlike the discreet events measured in other time-independent failure mechanisms, some aspects of failure potential involve continuous exposure—ie, there is a constant force present that can fail the component, rather than an intermittent threatening force. A common example is a component connected to a pressure source that can create pressure in excess of the component's capability to withstand. This is not an uncommon scenario for pipelines since they are routinely connected to wells, pumps, compressors, foreign pipelines, and other pressure sources that, at times, can be too high for the connected components. The potentially damaging pressure source does not cause damage because control and safety systems protect downstream components.

Even desirable or normal loads can be viewed as continuous exposures. Any amount of internal pressure becomes a damage potential as resistance decreases; any span can be too long for a pipe with no resistance to gravity forces (weight). Pressure as a constant exposure is generally only mitigated when excessive, since some pressure is a desirable part of operability. Intended pressure does not lead to failure only because resistance prevents it. Gravity as a constant exposure is mitigated by having uniform support and, if mitigation fails, is resisted by the bending and shear capacities of the 'structure'.

Measuring this type of exposure appropriately in a risk assessment model requires the correct coupling of the continuous exposure with a corresponding mitigation effectiveness. A high-demand or continuous exposure requires mitigation with very high reliability. The modeling issue with continuous exposure is the choice of time units in which to express the rate of exposure. How do you express 'continuous'? In units of events per year? Or per day? Or even per minute? Since 'continuous' means an infinite number of occurrences per unit time, it is difficult to capture numerically.

For purposes of modeling, any unit can be chosen, so long as the mitigation is calibrated to the same unit. The continuous exposure can be counted as one event per day, once per hour, once per minute, once per second, or even less. Any of these is appropriate as long as the corresponding mitigation—for example, the regulator or relief valve effectiveness—is measured in the same per day, per hour, per minute, etc units of reliability. For instance, choosing units of one event per second to represent continuous exposure from a high pressure connecting pipeline requires that the pressure regulating valve's reliability be expressed in the same units—ie, failure rate for each second in service. If one exposure event per day is chosen to represent continuous, then the regulator's reliability must also be expressed in the context of how many days between failures of such regulators to prevent overpressure.

In some estimates, the use of 'probability of failure on demand' estimate for a safety or control device will automatically make the exposure and mitigation units of measure equivalent. However, in the above example, the regulator's 'demand' is continuous—its function is being continuously demanded—requiring attention to the units of each. Therefore, the regulator's mitigation effectiveness—its reliability—must be expressed in similar units—failures per day, per hour, per minute, per second, or even smaller. Then, when exposure is multiplied by (1 – mitigation), the resulting PoD is appropriate.

2.8.12.6 Spans

An interesting nuance arises in a risk assessment involving spans. A span makes the component susceptible to the effects of gravity. While the exposure of 'gravity' has always been present, its role goes unnoticed in a fully supported pipeline segment. If an event can result in loss of support, but not failure, how is it to be modeled? Has the span created an exposure—ie, a new attack? Or is it causing the loss of some resistance to an attack (gravity) that has always been there?

This warrants some discussion. The frequencies of exposures should include all events that can damage the theoretical component. Technically, only events that cause excessive stress cause damage. So, only spans of a certain length, given pipe and contents weight, buoyancy, lateral forces, vibration potential, other stresses, etc, are events that potentially result in damage. Rigid pipe and mechanical couplers generally have less resistance to spans compared to flexible, welded systems.

The full solution is to discriminate among events that cause varying amounts of span length to the component. This involves an initial measurement of the PoF of the supported pipe in terms of continuous exposure to gravity which is fully mitigated by the uniform support, with resistance available but uninvolved so long as the mitigation is in place. PoF from gravity effects would logically be nearly 0% as long as the support remains. If any portion of the length becomes unsupported, then the mitigation against the force of gravity is zero and damage is theoretically possible at that location. Realistically, only spans of a certain minimum length can result in damage for most

pipeline components. Minor spans will typically have no effect on either damage or failure potential. A few inches of span rarely causes damage to any component.

As span length increases, damages become possible and then eventually failure occurs. Assigning probability estimates to each possible span length will be challenging in many real-world applications. Furthermore, determining minimum span lengths for various damage and failure scenarios involves structural calculations that are redundant to the resistance estimations.

Therefore, a modeling choice emerges. An exposure count may include all span-producing events or only those events generating potentially damaging span lengths. The former results in an over-estimation of damage producing events, since even the insignificant spans are counted. The latter requires a pre-determination of damaging span lengths. This is not a trivial exercise since the following considerations are important: material characteristics, dimensions, contents, internal pressure, lateral forces, etc.

A simplification may be appropriate for some risk assessments. From a modeling perspective, it may be simpler to count any span-producing event as an exposure rather than pre-determine what span length is critical for each set of conditions. With a conservative assumption that any span length can cause damage, the inaccuracy that is generated is the production of a PoD that is conservatively overstated. Components that are unharmed by loss of support will show low PoF after resistance estimates are applied. However, they may show inappropriate PoD levels due to the over-counting of exposures (ie, including exposure events that can't cause damage). Perhaps this is tolerable in exchange for modeling convenience.

As an example of this simplification: consider a soil erosion event creating a one foot span as compared to a continuously supported 12" steel pipeline. If the erosion event is counted as an exposure (an 'attack') with a frequency of 0.1/year and no mitigation is provided, the model reports a 0.1 frequency of damaging events, even though damage is realistically not occurring with only a one foot span. The PoF will not be impacted by this inaccuracy in the intermediate PoD estimate. In the absence of severe weakness, the resistance prevents failure virtually 100% of the time. The resistance of the 12" steel pipe shows that essentially none of the 0.1 spans per year will result in failure.

Longer span lengths would generally require more resistance. Since some resistance is now being used to resist gravity, some load carrying capacity may no longer be available to resist other loads. So, a third modeling approach may state a definition of exposure as only events that can produce at least, say a 20 ft span (or whatever the calculation determines is a potentially damaging span, under a set of assumed component characteristics). A related solution is to create categories of span-producing events based on the length of span potentially produced. Each is assigned an exposure frequency. Some will exceed the point where damage is possible and some will be insignificant, from a structural damage perspective. A version of this approach is to begin with an exposure frequency that captures all span-creating events and then assign fractions to create categories of longer-span events. For example, 0.3 span-creating events per mile-year are expected; 55% of those will produce spans less than 3 ft in width; 40% produce spans greater than 3 ft but less than 10 ft; and 5% produce spans greater than 10 ft in width.

2.8.12.7 Mitigation vs Resistance

Some methods of protection from mechanical damage present a rare case where mitigation and resistance become a bit blurred. A concrete coating or casing reduces the frequency of contact with the pipe steel. That is a reduction in PoD and therefore can be thought of as a mitigation. This requires that the protection be viewed as independent from the component—it is something added to the component as a protective measure. That is clear for slabs and even casings, but a coating, even concrete coating, is often viewed as part of the component, especially when used as a buoyancy control. In that case, contacting the coating counts as contacting the component. This is also influenced by the definition of 'damage' implicit in the PoD. Does damage to a concrete coating constitute damage to the component? This is a matter of perspective and definition. The loss of a buoyancy control feature is analogous to the challenge of modeling spans, as previously discussed.

For consistency, the sample assessments offered here consider slabs, casings, and concrete coatings to be distinct from the component and therefore best treated as mitigation measures. Under this view, the component is not damaged when only the protection is damaged. Alternative views may be more appropriate for certain risk assessment situations.

2.8.12.8 Mitigation-by-others

Because mitigations can originate at facilities not under the control of the pipeline operator, there may be both foreign (owner of the origination point of the exposure) mitigations and operator (of the segment being assessed) mitigations. For instance, the highway department and law enforcement agencies will mitigate some of the threat of vehicle impact to nearby pipelines via barriers, speed limits, road configuration, etc. An operator of nearby facilities will mitigate the potential for rupture or explosion of their facilities, reducing the exposure to the assessed component.

From the perspective of the pipeline operator, the protective measures employed by others reduce the exposure to the pipeline. These actions taken by others are additive to the protective measures installed and maintained by the pipeline operator. Since these mitigations-by-others effectively change the rate of pipeline exposures, and since it will often be difficult to assess and track changes in mitigations of non-owned facilities, it is usually more efficient to include foreign mitigations in the exposure rate estimate assigned to the non-owned facility. Otherwise, the risk assessment tends to expand into an assessment of non-owned systems. The mitigations done by others are often still important to understand and perhaps quantify, but keeping them separate from mitigations applied by the assessed component owner is a modeling convenience.

Other examples include natural mitigation measures and indirect actions taken by others. Consider traffic impact potential where trees, berms, ditches, fences, etc are de facto barriers (mitigations) to vehicle impacts. Treatment of these features as mitigation-by-others, and including their role as exposure-reducers, is the simplest approach. However, should the trees be removed or die, the ground leveled, or the fence be removed, having the rates of 'vehicle leaves roadway' separate from the benefits of the features would be useful.

Similarly, when water depth is sufficient to preclude anchoring, dredging, fishing, and other third-party activities as possible damage sources, damage probability to offshore components is reduced. Just as with other natural barriers, the water depth can be treated as a mitigation in the risk assessment. The fact that the water depth may also preclude certain other activities can be factored into the exposure estimate without triggering an inappropriate 'double-counting' effect in the risk assessment.

A general rule of thumb may be to include all features and actions not under the control of the component owner as influences to exposure rates. Actions and features that are controlled by the component operator are treated as mitigation measures. That is, if foreign, then exposure, otherwise mitigation. An exception may be cases where it is desirable to develop an argument, via cost/benefit analyses, for a change in mitigation activities, even if performed by others.

2.8.12.9 Resistance Baseline

SECTION THUMBNAIL
A nuance of resistance modeling: should it start with 'zero' strength? Or 'normal' strength?

There is an interesting interplay between exposure and resistance since both are sensitive to the exact definition of 'failure'. Exposure measurement implicitly involves a theoretical baseline for resistance since an exposure is defined as an event that causes 'failure' and resistance is a measure of invulnerability to 'failure'. So, the definition of 'failure' is a component of resistance, just as it is for exposure. This is again best illustrated by examples. If failure = 'permanent deformation', then resistance measures the invulnerability to permanent deformation, given the presence of a force (an exposure) that can cause permanent deformation if there is insufficient resistance. If failure = 'leak/rupture', then resistance measures the invulnerability to leak/rupture, given the presence of a force (an exposure) that can cause leak/rupture if there is insufficient resistance.

If resistance is to be measured in simple terms of percentage or fraction of mitigated exposure events that do not result in failure, there is a need to define a starting point or baseline. That baseline must be consistent with the definition of the exposure event. If the baseline is to be 'zero resistance' then exposure involves imagining that there is no resistance at all. A thin-walled aluminum can or cardboard tube, egg-shell vessel, etc, crushable between two fingers—is the right mental image for almost complete lack of resistance. So, the image of an unprotected beverage can or cardboard tube sitting atop the ground, is the correct image to estimate exposure event frequencies when a 'zero resistance' baseline is chosen. If such a can may be broken /crushed/deformed by the event, then it should be counted as an exposure.

There are obviously many more exposure events that could break an aluminum beverage can compared to a steel pipeline. So exposure counts are dramatically increased when zero resistance is assumed. As a matter of fact, the number of potentially damaging events always increases when the threshold for damage is lowered.

If the risk assessment designer feels that zero resistance results in excessive exposure counts, he can define the resistance baseline as something other than zero. For instance, he may set the resistance baseline as the fraction of exposures above 'normal', which do not result in failure. Then resistance is the amount of 'extra' stress carrying capacity once 'normal' loads have been accommodated. This can theoretically lead to negative values. Perhaps failure has not yet occurred in a weakened component only because the upper limits of 'normal' have not recently occurred. If there is not only no 'extra' resistance, but not even 'sufficient' resistance, then a negative value is warranted.

This is a modeling choice. A changing resistance baseline—potentially different for each component under varying 'normal' loads—may be confusing to some. On the other hand, the imagineering of a no-resistance component and the associated need to count many seemingly minor exposures might be more troublesome for others.

Exposure Influenced by Resistance

When a resistance baseline other than 'zero resistance' is used, exposure varies, as was suggested in the previous section. Exposure rates are sensitive to changing resistance. When material characteristics degrade or are changed, a greater number of exposure scenarios can cause failure. Examples of such material changes include:

- creation of a HAZ,
- extreme temperatures effects reducing material stress-carrying capabilities,
- UV degradation,
- Hydrogen embrittlement.

Other examples of changing resistance include metal loss by corrosion, crack progression through a component wall, unanticipated or intermittent external loadings such as debris impingement in flowing water or gravity effects when support is lost, and others.

The most robust assessment can provide for a continuous updating of exposure estimates based on changing resistance. That is, if a resistance baseline other than zero has been chosen, then the count of exposures—events that can cause failure—will increase as resistance decreases.

Similarly, when modeling time-dependent failure mechanisms like cracking, the TTF shortens when either the modeled rate of cracking increases or the effective wall thickness is reduced. If material degradation or change (for example, creation of a HAZ) causes the material toughness/brittleness to change, is that better modeled as increased crack propagation rate (ie, more exposure)? or rather as reduced effective wall thickness (ie, less resistance)? Fortunately, the suggested mathematics ensures the same result regardless of chosen approach. While either will work in the proposed PoF model, it may be more intuitive to model this as a change in effective wall thickness. That way, this potential change in a material's property is readily seen alongside any other potential change in component strength.

As another example of the modeling choices for exposure-resistance interaction, consider the role of an expansion loop in a pipeline. If the expansion loop is present to reduce thermal stresses and fatigue, most would agree that resistance has been improved rather than exposure reduced or mitigation improved. After all, the changes in temperature still occur and the pipe is not protected from those resulting forces. Only the pipe's reaction, its ability to absorb the forces without damage, have changed.

However, a counter could be that each temperature cycle is now no longer imparting the same stresses and, hence, exposure estimates should be reduced. Again, either choice yields the same PoF under the suggested modeling approach.

Aspects such as inclusion of suspect weaknesses will always be necessary in the risk assessment. Other aspects will be discretionary. The risk assessor can decide, in the context of desired PXX and trade-offs between complexity and robustness, the optimum way to handle resistance and resistance-exposure issues such as:

- Yield vs ultimate stress levels.
- Inclusion of intermittent loadings.
- The extent of simultaneous consideration of changing resistance with loadings potentially causing exceedance of stress-carrying capability. See discussions of unanticipated spans and loss of buoyancy control features in Chapter 2.8.12.6 Spans.

2.9 FREQUENCY, STATISTICS, AND PROBABILITY

There is a difference between 'frequency' and 'probability' even though in some uses, they are mostly interchangeable. As used in this book, *frequency* refers to a count of events while *probability* refers to the likelihood of one or more events over some future time period. Either frequency or probability are suitable metrics in a risk assessment. If values are small, the two are numerically equivalent, ie, at very low frequencies of occurrence, the probability of failure will be numerically equal to the frequency of failure.

The actual relationship between failure frequency and failure probability is often modeled by assuming an underlying distribution from which probabilities can be determined. For example, the Poisson equation relating spill probability and frequency is

$$P(X)SPILL = [(f^{*}t)X/X!\,]^{*} \exp(-f^{*}t)$$

Where

P(X)SPILL	=	probability of exactly X spills
f	=	the average spill frequency for a segment of interest (spills/year)
t	=	the time period for which the probability is sought (years)
X	=	the number of spills for which the probability is sought, in the pipeline segment of interest.

The probability for one or more spills is evaluated as follows:

$$P(\text{probability of one or more})SPILL = 1 - P(X)SPILL$$

Where X = 0

Frequency may be more useful when conservative risk assessments produce high probabilities. For example, a P99 risk assessment will often be more useful if estimates are expressed as frequencies versus probabilities, due to the high number of 90+% probability estimates that commonly emerge in initial, conservative assessments. Frequencies are able to discriminate between, say 10 events/year and 20 events/year, while a per-year probability estimate (of one or more events per year) based on 10 and 20 events/year yields high and virtually indistinguishable values (99+%, dependent upon the relationship between frequency and probability used). Large probability numbers typically also emerge in a pipeline risk assessment for high exposure rates—for example, 8 excavations per mile-year—and in other values generated from very conservative assumptions. These too are better captured as frequencies.

A statistic is a value calculated from a set of numbers—it is not a probability. *Statistics* refers to the analyses of data; and the most compelling definition of *probability* is "degree of belief," which normally utilizes statistics but is rarely based entirely on them.

Statistics are methods of analyzing numbers or the numbers emerging from the analyses. While they are usually an important ingredient in predictions, statistics are based on past observations—past events. Statistics from historical incidents do not imply anything about future events until inductive reasoning is employed. As discussed in PRMM, historical failure frequencies—and the associated statistical values—are normally used in a risk assessment but must be used carefully. Extrapolating future failure probabilities from historical information alone can lead to significant under—or over—estimations of risk.

2.10 FAILURE RATES

A failure rate is simply a count of failures over time, by some definition of 'failure'.

Pipeline failure rates have historically been starting points for determining absolute risk values. Past failures on the pipeline being assessed are often pertinent to future performance. Beyond that, representative data from other pipelines are sought. Failure rates are commonly derived from historical failure rates of similar pipelines in similar environments. That derivation is by no means a straightforward exercise. In most cases, the evaluator must first find a general pipeline failure database and then make assumptions regarding the best "slice" of data to use. This involves attempts to extract from an existing database of pipeline failures a subset that approximates the characteristics of the pipeline being evaluating. Ideally, the evaluator desires a subset of pipelines with similar products, pressures, diameters, wall thicknesses, environments, age, operations and maintenances protocols, etc. It is very rare to find enough historical data on pipelines with enough similarities to provide data that can lead to confident estimates of future performance for a particular pipeline type. Even if such data are found, estimating the performance of the individual from the performance of the group presents another difficulty. In many cases, the results of the historical data analysis will only provide starting points or comparison points for detailed estimates of future failure frequency.

As a common damage state of interest, fatality rates are a subset of pipeline failure rates. Very few pipeline failures result in a fatality. A rudimentary frequency-based assessment will simply identify the number of fatalities or injuries per incident and use this ratio to predict future human effects. For example, even in a database with much missing detail (as is typically the case in pipeline failure databases), one can extract an overall failure rate and the number of fatalities per length-time (i.e., mile-year or km-year). From this, a "fatalities per failure" ratio can be calculated. These values can then be scaled to the length and design life of the subject pipeline to obtain some very high-level risk estimates on that pipeline. Samples of high-level data that are useful in frequency estimates for failure and fatality rates is given in PRMM Tables 14.1 through 14.4.

Several sources of failure data are cited and their data presented in this book. In most instances, details of the assumptions employed and the calculation procedures used to generate these data are not provided. Therefore, it is imperative that data tables not be used for specific applications unless the user has determined that such data appropriately reflect that application. The user must decide what information may be appropriate to use in any particular risk assessment. The evaluator will usually need to make adjustments to the historical failure frequencies in order to more appropriately capture a specific situation.

The recommendation is to make use of historical failure rates as calibration or benchmark tools. A risk assessment of a collection of components can be compared to relevant historical failure data as a validation tool. This is further discussed in later chapters.

2.10.1 Additional failure data

Historical failure rate data are also sometimes used to suggest statistical distinctions for pipeline characteristics such as wall thickness, diameter, depth of cover, and potential failure hole size. Such distinctions drawn from historical incident data are useful but can be misleading. The general warning regarding cause-and-effect before using any results of statistical analyses is germane: if a suggested correlation does not make sense, perhaps it does not really exist.

Several studies estimate the benefits of particular mitigation measures or design characteristics. These estimates are often based on statistical analyses of historical incidents. A study may often rely solely on the historical failure rate of a pipeline with a particular characteristic, such as a particular wall thickness or diameter or depth of cover. To be useful, this type of analyses must isolate the factor from other confounding factors and should also produce a rationale for the observation. For example, if data suggest that a larger diameter pipe ruptures less often on a per-length, per-year basis, is there a plausible explanation? In that particular case, higher strength due to geometrical factors, better quality control, and higher level of attention by operators are plausible explanations, so the premise could be tentatively accepted even though the diameter does not cause all of the effect seen. In other cases, the benefit from a mitigation is derived from engineering models or simply from logical analysis with assumptions. Observations from various studies are sometimes available and useful in assigning mitigation values.

Inferences drawn from statistical examinations of large populations of pipelines must be used carefully. They will not reflect conditions at specific locations of certain pipelines. Since risk management is ultimately interested in the specific locations, generalized data can be misleading.

Note discussions of individual behavior versus population behavior in many places in this text. If a certain population of pipeline segments does indeed 'behave' differently, that is useful insight, especially when a segment to be assessed can be assigned to that population.

Potential risk reduction benefits from several mitigation measures, as suggested by various references, have been compiled in PRMM Table 14.11. These are often based on statistical examinations of large populations of pipelines and may not reflect conditions at specific locations.

Other examples of statistical relationships include the possible mitigative effects of depth of cover and resistive benefits of pipe diameter, as discussed in ref [1043].

2.11 CONSEQUENCES

Implicit in any risk assessment is the potential for consequences. This is the last of the three risk-defining questions: If something goes wrong, what are the consequences?

Consequence implies a loss of some kind. The loss or damage state of interest must be pre-determined for a risk assessment.

Consequences that are commonly measured in a risk assessment include:

- Leaks and ruptures.
- Leaks and ruptures beyond a pre-specified threshold of loss.
- Results of leaks and ruptures:
    - o Fatalities and injuries,
    - o Property loss,
    - o Environmental harm,
    - o Monetary losses, including service interruption costs.

Some losses are more readily quantified than others. Both direct and indirect costs are often included in a modern risk assessment. See Chapter 11.8.9 Indirect costs and PRMM for further discussion.

2.12 RISK ASSESSMENT

Risk assessment is a measuring process capturing both the probability and consequences of the potential events of interest. The most useful risk assessment results are expressed in verifiable measurement units such as incidents per year, dollars per mile-year, and many others.

Risk is not a static quantity. Along the length of a pipeline, conditions are changing. As they change, the risk is also changing in terms of what can go wrong, the likelihood of something going wrong, and/or the potential consequences. Because conditions also change with time, risk is not constant even at a fixed location. When we perform a risk evaluation, we are actually taking a snapshot of the risk picture at a moment in time.

It is important to recognize what a risk assessment can and cannot do, regardless of the methodology employed. The ability to predict pipeline failures—when and where they will occur—would obviously be a great advantage in reducing risk. Unfortunately, this cannot be done except in extreme cases. Pipeline accidents are relatively rare and often involve the simultaneous failure of several safety provisions. This makes accurate failure predictions almost impossible. So, modern risk assessment methodologies provide a surrogate for such predictions. Assessment efforts by pipeline operating companies are normally *not* attempts to predict how many failures will occur or where the next failure will occur. Rather, efforts are designed to systematically and objectively capture everything that can be known about the pipeline and its environment, to put this information into a risk context, and then to use it to make better decisions.

A common incompleteness in risk assessment is to characterize a risk solely in terms of an average from a population distribution thought to generally represent the pipeline being assessed. While it is appropriate to seek an understanding of the distribution from which this individual is likely a part, the individual's position within that distribution must be characterized. The distribution of, for instance, event frequencies, will often include values that are orders of magnitude higher and lower than the average.

2.13 RISK ASSESSMENT VS RISK ANALYSES TOOLS

A risk assessment for a pipeline should meet a minimum set of requirements before it is labeled an assessment rather than a more limited analysis of risk. There are many risk analysis techniques that are better labeled as tools—ingredients or supplements to a risk assessment. See full discussion in Chapter 3 Assessing Risk.

2.14 MEASUREMENTS AND ESTIMATES

SECTION THUMBNAIL
We will often have both actual measurements and inferred values. Whichever is more informative should be used in the risk assessment.

Proper risk assessment uses all available information. Information used in a risk assessment takes two general forms—measurements and estimates. An often-used aspect of this assessment methodology is the simultaneous use of both actual measurements and inferential estimates. For purposes here, a measurement is a reading or value obtained using an instrument on a specific component while an inferred estimate emerges from secondary or indirect information, often produced based on the underlying physics or even from engineering judgment.

Measurements include inspections for corrosion feature dimensions, corrosion rates from coupons, crack depths, metal toughness, soil resistivity, and many others. Inferential or indirect information is often based on material science—for example, possible corrosion rates associated with metals in certain soils, potential crack growth rates in certain materials exposed to certain loadings, etc. For example, obtaining a pipe wall thickness by UT instrument is a measurement, while estimating wall thickness based on the pipe date and possible degradations (for example, mpy by corrosion) is an estimate. Inferential estimates are normally applied to all components for which a measurement is not available.

The final value to be used in the risk assessment emerges from an examination of both, after adjustments for information age and accuracy have been made to each. The assessment chooses the best value based on the strength of evidence—newer and more accurate information is chosen over older, less accurate information.

It is common for information collection on long, linear assets such as pipelines to be non-uniform. The disparities in information availability along the route is accommodated by this simultaneous use of measurements and estimates.

Examples of measurements used in a typical risk assessment include:

- Visual and NDE inspections performed on accessible components
- CIS[4] inspections for CP effectiveness

4 Some may argue that overline surveys such as CIS, DCVG, etc are inferential, ie, inferring conditions on a buried pipe some distance from the actual measurement. For purposes here, these surveys are considered measurements, recording actual values that represent a condition, even if that condition is used to infer other characteristics. Error rates increase by influences such as proficiency of surveyor and surface conditions.

- DCVG/ACVG coating holiday surveys
- Coupon corrosion rates
- ILI anomalies, especially when UT is providing a direct pipe wall thickness measurement[5]
- Test lead readings of pipe-to-soil potential.

In addition to instrument inaccuracies and operator errors, additional nuances have to be considered in using measurements. For example, highly accurate measurements, but taken some distance from the point of interest (such as internal corrosion coupons and test lead readings). Where conditions are not consistent, extrapolations can be very inaccurate. The age of the measurement is also important. The pipe wall thickness measured at the pipe mill 20 years ago may have little relevance to the actual wall thickness of the buried pipe 20 years later.

Each measurement and each inferential estimate requires an adjustment for its age and accuracy. The superior value—for example, the value among all possible measurements and estimates with the best age/accuracy combination should determine the value used in the risk assessment. A process is used to adjust each of these to reflect the confidence in the current (for example, age-adjusted and accuracy-adjusted) validity of their values.

Once adjusted, selection of the more optimistic value ensures that better information overrides lesser information in a conservative risk assessment. This same technique compares and chooses better measurement data over lesser measurement data, when multiple measurements are available at the same location (for example, multiple ILI's or multiple overline surveys on the same segments).

Again, with a consistent application of conservatism in uncertainty estimates, the more optimistic value—the information suggesting the best wall thickness—will usually govern.

Inappropriate overrides of inspection/test information are avoided by the careful application of consistent confidence values. When the confidence value is based upon 'damages since inspection/test', it must be ensured that equivalent PXX levels are used everywhere. For instance, it would not be correct to adjust P99 estimates from a 5-year old inspection by using P50 estimates of what may have happened in the last 5 years (for example, the wall thickness could be as low as 0.200" but let's assume that only 0.1 mpy of corrosion could possibly have occurred.)

Nevertheless, even with carefully chosen uncertainty values, the more optimistic value will not always be the best value to use.

5 ILI by MFL can similarly be said to be an inferential measurement, but is also treated as a measurement for purposes here.

Example: 2.1 This is illustrated in a specific example:

One set of measurements/estimates shows a 10% certainty that no more than 20 mils of wall loss has occurred in a certain thick-walled component. This uncertain estimate implies a maximum wall loss rate of 20 mils / 10% = 200 mils could actually have occurred. Taken from an original 0.500" wall thickness, this suggests a current wall thickness of 0.300".

As another piece of evidence, a recent ILI shows, with 90% confidence (including general- and run-specific inaccuracies) a 300 mil wall loss feature, leading to a current wall estimate of 0.500" - 0.300" = 0.200".

In this case, the ILI value should obviously govern. The original estimate, while seemingly conservative (tending to overestimate actual risk) was actually not conservative enough, as demonstrated by the recent ILI. A real-world example of this occurred when an operator installed a new gathering system which, after only a few years in service, experienced internal corrosion leaks. Upon investigation, corrosion rates in excess of 200 mpy were discovered—far exceeding what was thought possible in the design phase.

Differences in measurements/estimates compared to actual values will be influenced by both uncertainty and conservatism. Uncertainty causes an unintentional, undesirable difference that must be tolerated. In conservatism, the difference is intentional, in acknowledgment of the natural (random) variability inherent in real-world phenomena. Both are discussed in following sections.

2.15 UNCERTAINTY

The role of uncertainty in risk management is multifaceted as noted in PRMM:

- Risk assessment measurement error and uncertainty arise as a result of the limitations of the measuring tool, and the processes of taking the measurement, including the skills of the person performing the measurement. Pipeline risk assessment also involves the compilation of many other measurements (pipe strength, component wall thickness, depth of cover, pipe-to-soil voltages, pressure, etc.) and hence absorbs all of those measurement uncertainties. Risk assessment also makes use of engineering and scientific models (corrosion rates, stress formulae, thermal effects and overpressure estimates, etc.) each with accompanying errors and uncertainties.
- Adding to the uncertainty is the fact that the thing being measured in pipeline risk assessment is undergoing continuous change due to changing surroundings, as well as sometimes changing service conditions and possible degradation.

- A risk assessment must identify the role of uncertainty in its use of assumptions including how the condition of "no information" is to be handled in the assessment. For many applications of risk assessment results, it is advantageous to incorporate a conservative underlying philosophy of:

Uncertainty = increased risks

This not only encourages the frequent acquisition of information, but it also enhances the risk assessment's credibility. Unless a conservative 'guilty until proven innocent' approach is used, there will be no incentive to regularly inspect and verify conditions that influence risk, Riskier conditions may only be discovered when incidents occur. Investigating the incident will inevitably find that the risk assessment had assumed favorable, low risk conditions, in the absence of confirmatory information. This often implicitly discredits all other results of the risk assessments.

2.16 CONSERVATISM (PXX)

Conservatism is generally taken to mean an intentional bias towards over-estimation of the true risk. Risk assessment incorporating a high level of conservatism will tend to overstate the risks, perhaps by several orders of magnitude. This occurs through the use of input values and calculations that are based on worst-case or at least 'higher risk' assumptions. Risk assessment conducted with no conservatism always assumes the most likely values and the calculations that produce results that are most often true to the most common actual conditions.

Conservatism is a useful characteristic in many applications of risk management. However, conservatism may also be excessive, leading to inefficient and costly choices when not properly acknowledged in decision-making.

A risk assessment should be performed with a target level of conservatism. As used here, the PXX designations indicate a level of confidence that actual experience will be no worse than estimated. For instance, P90 is the point where 90% of future performance should be at or below this value. It is the point where one would be negatively surprised 10% of the time—once out of every ten episodes.

A P90+ assessment intentionally contains layers of conservatism. This is often done to encourage future data collection as a means of risk reduction and, more importantly, to ensure that risks are not underestimated.

For simplicity, the PXX refers to the conservatism of inputs rather than to the resulting conservatism of the assessment. Each risk assessment is obtained via a collection of inputs, each with an estimated level of uncertainty equal to PXX. Actual conservatism of final risk estimates often increase dramatically due to layering of conservatism due to a bias towards conservatism for each input. This layering also produces increasingly unlikely scenarios since multiple low probability events are assumed to occur simultaneously. Therefore, the PXX refers to the intended level of uncertainty associated with each input rather than the risk estimates. The PXX level of the final risk assessment is identified and managed in the calibration/validation phase (see Chapter 3.7 Verification, Calibration, and Validation).

Less conservative assumptions are sometimes needed for practical reasons. For instance, a defect over 95% through a pipe wall could exist and survive a pressure test or be undetected in an inspection. It would be counter-productive to assume that such rare defects exist everywhere, even though such an assumption would be very conservative. Rather, the wall thickness implied by a Barlow stress calculation (perhaps adjusted by a factor showing some localized thinning could have occurred) can be used as the primary means to estimate the probable—and still conservative—wall thickness when no other confirmatory integrity information is available.

P84, as representing approximately one standard deviation of a one-sided normal distribution, might be appropriate as a target level of conservatism that may have some consistency with certain design practices.

Some practitioners also produce P10 or similar estimates, reflecting best case or at least more optimistic inputs. As with the more conservative layering, choosing multiple optimistic inputs produces a combination with even more optimism, as well as more rarity.

The user should determine the level of conservatism appropriate to his needs. Often a P99 level—negative surprises only 1% of the time—or higher is warranted for assessments supporting new projects or presentations in public forums. A P50 to P70 level of analysis might be more appropriate for budget setting or long range planning.

See also the discussion of calibration in Chapter 3.7 Verification, Calibration, and Validation.

2.17 RISK PROFILES

While a profile can mean different things—risk changes over time, types of events possible (see FN curve discussion), etc—the focus here is on risk changes over 'space'. Generation of a profile of changing risks along a pipeline is essential to the understanding of risk and the subsequent management of those risks.

The PoF profile is produced by the risk assessment and shows location-specific PoF values. PoF changes along a route in response to dozens of factors that might change, including pipe/component specification (age, wall thickness, diameter, etc), coating condition, soil corrosivity, pressure, road crossings, foreign pipeline crossings, AC electrical power lines, depth of cover, and many more. Per-incident consequence costs can vary dramatically along a pipeline, changing with differences in pressure, flowrate, topography, receptor proximities, and, to a lesser degree, differences in pipe characteristics (ie, age, coating, wall loss).

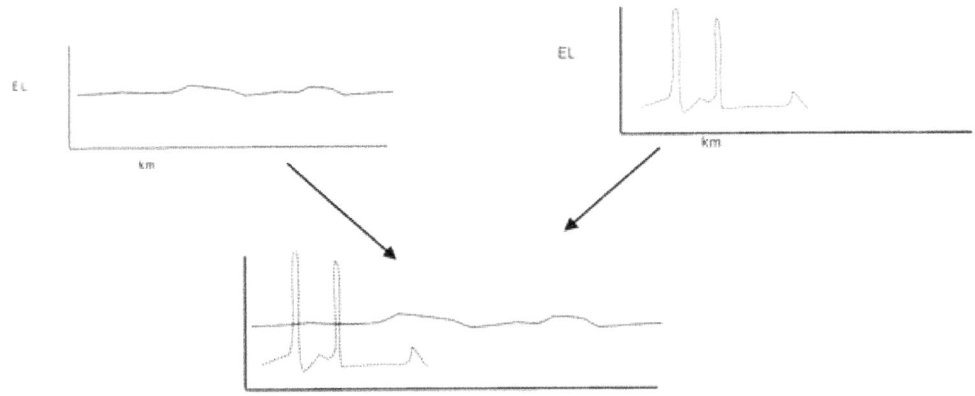

Figure 2.4 Two very different risk profiles, but perhaps with the same cumulative risk

Risk profiles are critical aspects of risk assessment, as discussed in Chapter 4.5 Segmentation and in risk management, as noted in Chapter 13.8.2 Profiling.

2.18 CUMULATIVE RISK

While the profile is an essential element in understanding, presenting, and managing risk, it is not an efficient tool for setting higher-level risk management strategies. Higher level risk management strategizing is distinct from foot-by-foot risk management. It involves risk summaries and comparisons of sometimes long segments.

Cumulative risk is a metric used to gauge the risk posed by any length of pipeline or any collection of components. Because risk values are very location specific along the pipeline, a method of 'rolling up' or aggregating all of the risks for any portion of a pipeline is important.

A typical pipeline risk assessment should show the level of risk that each point along the pipeline presents to its surroundings. Two pipeline segments, say 100 and 1,800ft, respectively, may have the same 'rate-of-risk', expressed in units such as $/mile or incidents/mile-year, for all portions. So each point along the 100-ft segment presents the same risk as does each point along the 1,800-ft length. Of course, the 1,800-ft length presents more overall risk than does the 100-ft length, because it has many more risk-producing locations. These two pipelines may also have exactly the same total risk, in which case, the shorter line has a much higher rate of risk than does the longer.

Longer pipeline lengths logically have higher risk values, since a longer line logically has a higher 'area of opportunity' for failure and generally exposes more receptors to consequences. Both the risk and the rate-of-risk are important to risk management.

In reality, both the 100 ft and 21,800 ft segments will be comprised of multiple components, each with its own length and contribution to risk. Many pipelines will have short lengths of relatively higher risk among long lengths of lower risk, as demonstrated in their risk profiles. In summarizing the risk for the entire pipeline, a simple average or median will hide the shorter, higher risk sections. A 'weak-link-in-the-chain' strategy—focusing on the maximum risk or rate-of-risk alone—will similarly not reflect the full risk. A cumulative risk—each portion with its respective length aggregated into a summary number—will produce the most meaningful measure. This is the area under the risk profile curve. As with any area-under-the-curve summarization, the shape of the curve—the profile, in this case—remains critical to the understanding.

The cumulative risk characteristic is also measured in order to track risk changes over time or compare widely different types of risk mitigation projects. Projects such as public education, ROW maintenance, and patrol are not usually assigned large mitigation benefits on a per-foot basis, but can impact many miles of pipe and hence have a large impact on risk. If, for example, we want to compare the risk benefit of clearing 20 miles of pipeline ROW and installing new signs to the value of lowering and re-coating 100 feet of pipeline. On one hand, the failure potential can be reduced significantly along a short stretch of pipeline. On the other hand, a more modest mitigation could be broadcast over a long length of pipeline. The comparison is not intuitive unless an accurate method of aggregation is established.

See Chapter 4.6 Results roll-ups for discussion of proper measurement of cumulative risk.

2.18.1 Changes over time

Note that the cumulative risk values can also demonstrate the natural risk increase over time. Recall the entropy analogies—risk will increase over time unless offsetting energies are applied. The risk assessment measures the risk at all points along a pipeline, at a specific point in time. The risk numbers are therefore a snapshot. They represent all conditions and activities at the time of the snapshot. If inspections and maintenance are not done, safety degrades. The most meaningful measure of changes in the risk situation will be how the risk for the length of interest changes over time.

Changes in risk are easily tracked by comparing risk snapshots. This can be done for a specific point on a pipeline, an entire pipeline, or any collection of components from any pipeline. It can also be done for any set of pipelines, such as "all pipelines in Texas," "all propane lines," "all mainlines > 12", "all lines older than 20 years," and so on.

The cumulative risk calculation also remedies the difficulties encountered in tracking risk changes when segment boundaries change after every assessment. The CR can be calculated for any length of pipe, regardless of segment boundaries.

2.19 VALUATIONS (COST/BENEFIT ANALYSES)

Note that a superior risk assessment can show the value of changes in practice by estimating the corresponding changes in failure potential and/or consequence. Many common practices are intuitively important and necessary, but rely on subjective choices regarding level of rigor. Examples of practice whose role is otherwise difficult to estimate include:

- Instrument maintenance/calibration—can be linked to outage rates
- Training—can be linked to human error rates
- Procedures—can be linked to human error rates
- Monitoring—can be linked to intervention opportunities
- Marking/labeling of critical equipment—can be linked to human error rates.

While the risk-assessment-generated estimates of benefits will also contain some subjectivity, the reductionist approach allows many more opportunities for concurrence among stakeholders regarding specifics of the role of the practice in risk reduction, thereby helping to ensure more objective results.

For example, incident investigations frequently cite the role of inadequate procedures as an aspect of the incident. Absent such incidents, the role of procedures, and an argument to improve the practice within a company, may generate widely differing beliefs regarding expected benefits. However, the risk assessment approach that dissects the specific aspects of procedures and their role in incident prevention, as discussed in Chapter 8.8.2 Procedures, allows all parties to identify specific points of divergence of opinion and opportunities to collect pertinent information or otherwise come to an agreement on appropriate valuations.

2.20 RISK MANAGEMENT

Risk management is the intentional changing of risk levels. As a reaction to perceived risk, it is the set of choices in action undertaken in support of a strategy towards a level of risk deemed acceptable. Like all management initiatives, risk management involves establishing priorities and making judgments about trade-offs such as cost vs. benefit. Even with very accurate risk assessment, risk management can be challenging, involving socio-economic and political decisions around acceptable or tolerable risk, urgency with which risks may need to be reduced, and many others.

Since risk is the product of probability of failure (PoF) and consequence of failure (CoF), either or both can be changed to change the risk. Typically, PoF offers more opportunities for controlling risk than CoF. For this reason, effective risk management programs generally concentrate more on PoF aspects.

Practically speaking, our objective is not the elimination of risk, but the management of it for an acceptable result. We cannot eliminate risk without sacrifices that would be unacceptable—like halting the benefits derived from the use of a pipeline.

See Chapter 13 Risk Management for a discussion of pipeline risk management.

3 ASSESSING RISK

Highlights

There is a real difference between identifying elements of risk and performing a risk assessment.

Assessing Risk

As far as the laws of mathematics refer to reality, they are not certain; and as far as they are certain, they do not refer to reality.

Albert Einstein

SECTION THUMBNAIL
- The mechanics of assessing pipeline risk.
- Evaluating a risk assessment.

The risk management process comprises five basic steps:
1. Risk modeling
2. Data collection and preparation
3. Segmentation
4. Assessing risks
5. Managing risks.

The first four are actually components of assessing risk while the last is the reaction to what the assessment has revealed. This section provides some background to the assessment of risk with a focus on applications to pipeline systems.

FOCUS POINT
Many techniques labeled 'risk assessment' are really risk analyses tools, not assessment methodologies.

3.1 RISK ASSESSMENT BUILDING BLOCKS

Risk assessment practitioners have varying ideas of how to understand and measure risk. Many tools and techniques are available to help. While almost all can improve understanding, few should be considered to be comprehensive risk assessment techniques. There is a real difference between identifying elements of risk and performing a risk assessment.

Ref [1052] provides a list and discussion of "risk assessment techniques":
- Brainstorming
- Structured or semi-structured interviews

- Delphi
- Checklists
- Primary hazard analysis
- Hazard and operability studies (HAZOPS)
- Hazard Analysis and Critical Control Points
- Environmental risk assessment
- What if? analysis
- Scenario analysis
- Business impact analysis
- Root cause analysis
- Failure mode effect analysis
- Fault tree analysis
- Event tree analysis
- Cause and consequence analysis
- Cause-and-effect analysis
- Layer of protection analysis (LOPA)
- Decision tree
- Human reliability analysis
- Bow tie analysis
- Reliability centered maintenance
- Sneak circuit analysis
- Markov analysis
- Monte Carlo simulation
- Bayesian statistics and Bayes
- FN Curves
- Risk indices
- Consequence/probability matrix
- Cost/benefit analysis
- Multi-criteria decision analysis.

Each are described in the reference along with a complexity rating and an opinion as to whether each can produce 'quantitative' results.

For improved clarity, these techniques should be categorized according to the role they play in risk assessment. Several ways to group them could be appropriate but for discussion purposes here, the following categories are suggested:

- Risk Assessment techniques—full risk assessment methodologies, meeting all requirements of an actual risk assessment.
- Risk Tools—ingredients or supplements to a risk assessment.

Where 'tools' can be further categorized into:

- Hazard/threat identification—techniques focused on presenting lists of or confirming hazards or threats to a system. Examples include HAZOPS, brainstorming, check lists.

- Scenario identification—techniques focused on the chain of events leading to a failure or unfolding once failure has occurred. Examples include event trees, fault trees, cause-effect analyses.
- Analyses support—usually statistically based, these techniques work with a risk assessment model to improve outputs. Techniques are applied both to risk assessment inputs and outputs (results). Examples include Monte Carlo simulation, Bayesian statistics, and Markov analyses.
- Visualization—techniques, usually with a strong graphical nature, used to support presentation or visualization of risk results or inputs. Examples include bowtie, matrix, FN curves.

Since many techniques can be used in differing ways, not all fit neatly into one of these categories. This does not detract from the central idea here that risk tools play various roles in a risk assessment, but are NOT complete risk assessment methodologies.

3.1.1 Tools vs Models

An important distinction has been drawn between risk assessments—meaning methodologies, techniques, etc that produce complete risk estimates; versus risk analyses tools that play a more limited role, such as hazard identification or analyses of specific cause-consequence pairings.

One of the simplest discriminators between a risk model and risk tool is the 'map point' test. This test simply means that, using a real risk assessment approach, one can pick any point on any pipeline and should have access to all pertinent risk information for that location. If the so-called risk assessment cannot support this straightforward and intuitive task, then it is probably a risk tool rather than a complete risk assessment model, at least for purposes of this discussion. This and other ways to identify a true risk assessment are presented in a later section. But first is an examination of some of the more popular risk tools.

3.1.1.1 Hazard Identification/Evaluation Techniques

In addition to the techniques from ref [1012], eleven hazard evaluation procedures used in the chemical industry have been identified [9]. Each of these tools has strengths and weaknesses, including amount of benefit derived from the application, costs of applying the tool, and appropriateness to a situation.

Some of the more formal risk tools in common use by the pipeline industry include:

- HAZOP
- Fault-tree/event-tree analysis.

See PRMM for details on these.

3.1.1.2 Analyses Support Tools

Some risk techniques, often noted as stand-alone risk assessments, are actually processes that can be applied to 'real' risk assessment models. Bayesian, Markov, Monte Carlo, and others are better viewed as processing techniques rather than risk assessments themselves. They supplement a risk assessment by providing better understanding of patterns and numerical 'behavior' of the data.

Other tools such as Layer of Protection Analysis (LOPA) typically focus on an aspect of risk such as control and safety equipment/instrumentation analyses.

3.1.1.3 Visualization Tools

Matrix

One of the simplest risk visualization structures is a matrix. It displays risks in terms of the likelihood and the potential consequences associated with an asset or process. The vertical and horizontal scales may be qualitative, using a simple scale, such as high, medium, or low, or using detailed descriptors guiding the assignments of matrix positions. The scales may also employ numbers; often relative—for example, from 1 to 5—or possibly using categories of absolute risk values. See Figure 3.1.

Events or collection of events are assigned to cells of the matrix based on perceived or estimated likelihood and consequence. Risks with both a high likelihood and a high consequence appear in one corner, usually the upper right part of the matrix. This approach may simply use expert opinion or a more complicated application might use quantitative information to rank risks. While this tool cannot incorporate all pertinent factors and their relationships, it may help to crystallize thinking by at least displaying the risk as two parts (probability and consequence) for separate examination.

Some may believe that risks with, say highest consequences but low probability, require different management from those with, say lower consequences but higher probability, even if both scenarios show equal risk (see further discussion in Chapter 13 Risk Management). A risk matrix therefore sometimes supports corporate decision-making or risk tolerance guidance, whereby response urgencies to manage risk emerge from the various combinations of probability and consequence.

While sometimes interesting presentation/visualization tools, matrices are not risk assessment models—they arguably fail all of the tests proposed to determine whether they can serve as an assessment technique. A matrix, even as only a presentation tool is also rather clumsy, since it cannot appropriately illustrate many important considerations. For example, maskings include differences in risk due solely to differences in pipeline length; whether risk is due to a handful of peaks or rather from a consistent, but high level; the range of consequence scenarios possible; etc. There is a certain disservice in presenting risk information in a way that is incomplete or potentially misleading.

Lowest <-----------------------> Highest

Scale of Probability of Occurrence

5	6	7	High Risk	Highest Risk
4	5	6	7	High Risk
3	4	5	6	7
Low Risk	3	4	5	6
Lowest Risk	Low Risk	3	4	5

Scale of Consequence

Figure 3.1 Example of qualitative risk criteria matrix

Others

Other visualization tools often found in risk assessment presentations include FN curves and bowtie. An FN curve is a variation on the matrix, showing both probability and consequence of various event scenarios. The bowtie combines an event tree—the event leading to the 'knot'—and the fault tree—event emerging from the 'knot', where the knot is the event or asset whose risk is being displayed.

Risks at specific locations are often shown on FN curves where the relationship between event frequency and severity is shown. FN curves display failure count or frequency (F) versus consequence, where consequences are often a number of fatalities (N). This type of risk presentation, often called a depiction of societal risk, is a usually a plot of the frequency, f, at which N or more persons are expected to be fatally injured.

Event and fault tree analyses also serve as visualization tools. The distinction is blurred when values are assigned to branches and nodes—it now becomes more than a simple visualization tool but is still unable to complete a risk assessment methodology for an entire pipeline.

Presentation graphics/charts are further discussed in Chapter 4 Data Management and Analyses. GIS also has a strong visualization aspect, as noted in later sections.

3.1.1.4 What is a risk assessment model?

Although we understand the underlying engineering concepts related to pipeline failure, predicting failures beyond a laboratory in a complex "real" environment can prove impossible. No one can definitively state where or when an accidental pipeline failure will occur. However, the more likely failure mechanisms and the more susceptible locations, can be identified in order to focus risk management efforts.

An assessment of failure probability requires the independent estimation of the three elements of PoF for each failure mechanism: exposure, mitigation, and resistance (see Chapter 2.8.1 PoF Triad)

The potential consequences must also be assessed. Risk assessments can incorporate dose–response and exposure analyses into the risk evaluation by considering the possible pathways, the intensity of exposures, and the amount of time a receptor could be vulnerable. The possible effects of these, overlaid with possible receptor types and quantities, leads to consequence estimations.

A full and complete risk assessment captures all of these aspects for every portion of the system being assessed. The risk estimates produced should provide understanding and insights far beyond what can be done informally or with lesser tools.

3.2 MODEL SCOPE AND RESOLUTION

Assessment scope and resolution issues complicated previous risk assessment techniques. In both relative risk models and classical QRA, choices in the ranges of certain risk variables were required. The assessments of relative risk characteristics were especially sensitive to the range of possible characteristics in the pipeline systems to be assessed. If only natural gas transmission pipelines were to be assessed, then the model was not set up to capture liquid pipeline variables such as surge potential and contamination potential. The model designer had a choice of either keeping such variables and scoring them as "no threat" or he had to redistribute the weighting points to other variables that do impact the risk.

As another example, earth movements often pose a very localized threat on a relatively few stretches of pipeline. When the vast majority of a pipeline system to be evaluated is not exposed to any land movement threats, relative risk points assigned to earth movements did not help to make risk distinctions among most pipeline segments. To some, it appeared beneficial to reassign them to other variables, such as those that warrant full consideration. However, without the direct consideration for this variable, comparisons with the small portions of the system that are exposed, or future acquisitions of systems that have the threat, became problematic. Classical QRA had similar limitations since the historical data often forced the land-movement threat to be kept low, even for short segments for which it was the dominant threat. This is further discussed elsewhere in this text.

In a relative risk assessment, the ability to discriminate differences in risk was also sensitive to the characteristics of the systems to be assessed. A model that was built for parameters ranging from, say, a 40-inch, 2000-psig propane pipeline to a 1-inch, 20-psig fuel oil pipeline was not able to make many risk distinctions between a 6-inch natural gas pipeline and an 8-inch natural gas pipeline. Similarly, a model that was sensitive to differences between a pipeline at 1100 psig and one at 1200 psig might have to treat all lines above a certain pressure/diameter threshold as the same [PRMM]. Classical QRA's had an analogous issue in determining the representative population of pipelines upon which to base the statistical future estimates.

Fortunately, such issues of model scope and resolution disappear with the advent of a physics-based approach to risk assessment. By mirroring real-world phenomena as closely as practical, the assessment automatically and appropriately responds to all changes in factors.

3.3 HISTORICAL APPROACHES

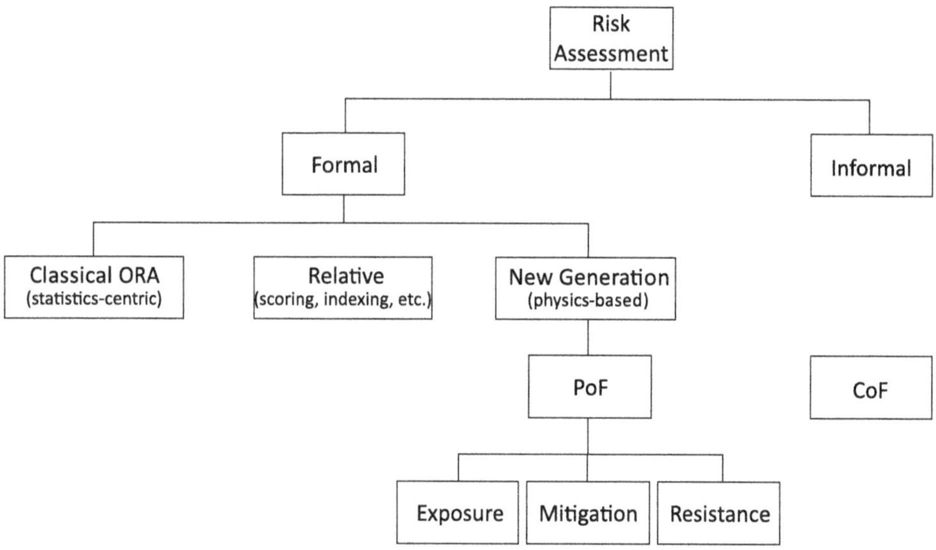

Figure 3.2 Pipeline Risk Modeling Options

Sidebar

Perspectives—Is Formal Risk Management Helping Me?

Ever consider that true risk management sometimes occurs only at the lower levels of some pipeline organizations? That is, personnel performing field activities are in effect setting risk levels for the company. Their choices of day-to-day activities are essentially driving risk management and thereby establishing corporate risk levels. This is not just theoretical—real choices are being made. While there are regulations and company-specific procedures to control certain actions, the on-the-ground team is often relied upon to prioritize, allocate, act, and request additional resources based solely on their perceptions.

Fortunately, we have a generally savvy work force that usually makes good choices. But why would top company executives choose to delegate company-wide risk management decision-making? In effect abdicating their own power to manage the risk of the organization?

In at least one sense, this delegation of risk management decision-making is a good thing. Those most knowledgeable in location-specific conditions/characteristics are often in the best position to make certain decisions. They are the subject matter experts in the pipeline's often-highly-variable immediate environment.

But such distributed control also has its weaknesses. In their risk 'assessments', the field team may not utilize all of the available information, for example, ILI details, operational data, learnings from other pipelines, etc. They also may not use a formal structure to find and manage the non-obvious risks. Even if they do use formal techniques, without a centralized view of risks across the entire organization, imbalances are certain to occur.

So, if the alternative is not superior, then why is centralized risk management not the standard? At least one explanation lies in the perceived accuracy and usefulness of risk assessments. Some risk issues are very apparent and no formal assessment is needed to understand them. Good inspection techniques take much subjectivity out of certain resource allocations—a list of identified critical anomalies is like a ringing telephone that must be answered. The 'fix-the-obvious' opportunities for risk management are hopefully fully addressed in inspection follow-ups and in the day-to-day O&M. A regional approach can be very efficient in managing obvious risk issues.

However, there are other risks and risk reduction opportunities that are not so obvious. Humans can judge a thing based on a subjective and simultaneous interpretation of a handful of factors—maybe 3-5. Real risk scenarios may involve a dozen or more factors. Remember, many modern pipeline incidents are of the 'perfect storm' type. Rare chains of events, often involving multiple improbable and non-apparent factors, lead to the incident. This is where formality is needed. The formal risk assessment, when done properly, finds those highly improbable scenarios, involving multiple, non-intuitive, overlapping issues that can generate the perfect storm event.

The previously unrecognized event is now revealed and quantified.

A Portfolio View

How can upper levels in the organization gain the risk understanding required to be fully engaged in risk management? By knowing the risk associated with every asset. The corporate-level decision-maker should seek a portfolio view of the company's assets, showing all costs of ownership. Just as with a portfolio of stocks and bonds, each asset ties up capital and has carrying costs. The revenue streams, capital cost, the O&M costs, tax liabilities, etc, have always been well understood. The risk cost?—perhaps not as much. Most know that risk is part of the cost of ownership but how many really use that knowledge in everyday decision-making? The key lies in reliable risk assessments whose results truly represent real-world cost of ownership risks. Then, and only then, is the top level decision-maker in a position to most efficiently allocate resources across the entire organization.

So, in a moment of self-evaluation, perhaps this question arises: is your risk assessment helping you? Some may answer "sure, I get a checkmark on my regulatory audit form." But most recognize that so much more is at stake. Beyond regulatory compliance, how much value emerges from the risk assessment effort? Some must admit that their assessments are mostly window-dressing—not really helping decision-making. Perhaps their risk assessment is only documenting what is already perceived. There is some value in such documentation. But there should also be some 'ah-ha' moments. After all, the whole point of a formal risk assessment is to provide the structure that can and does reveal the otherwise unknown.

FOCUS POINT
Pipeline risk assessment has matured. There are compelling reasons to migrate from previous approaches

3.3.1 Formal vs. informal risk management

PRMM discusses the transition from informal risk management to the formal processes. Some background on the maturation of the formal techniques is offered in the following section.

3.3.2 Scoring/Indexing models

Prior to the availability of an efficient physics-based approach to pipeline risk assessment, perhaps the most popular pipeline risk assessment technique in pipeline risk assessment efforts was the index or some similar scoring technique. Scoring systems are common in many applications, particularly where there is limited data or the information is subjective. Examples include sports and other competitive activities; finance and economics; and credit rating

In the pipeline risk assessment application of this approach, numerical values (scores) were assigned to conditions and activities on the pipeline system thought to contribute to the risk picture. This included both risk-reducing and risk-increasing factors, or variables. They were often a simple summation of numbers assigned to conditions and activities that are expected to influence risks. Whenever more risk-increasing conditions are present with fewer risk-reducing activities, risk was shown to be relatively higher. As risky conditions decrease or are offset by more risk-reduction measures, risk is relatively lower

Weightings were usually assigned to each risk variable or to groupings of factors. The relative weight reflected the importance of the item in the risk assessment and was based on statistics where available and on engineering judgment where data was not available. Pipeline sections were scored based on their attributes. The various pipe segments' scores were then available for uses such as ranking according to their relative risk scores in order to prioritize repairs, inspections, and other risk mitigating efforts. This technique ranged from a simple one- or two-factor model (where only factors such as leak history and population density are considered) to models with dozens of factors considering numerous aspects of risk influences.

The form of these pipeline assessments was normally some variation on:

CondA +CondB +... CondN = Relative Probability of Failure (or relative Consequence of Failure)

Or sometimes:

(CondA x WeightA) + (CondB x WeightB) + ... (CondN x WeightN) = Probability of Failure

Where

CondX represents some condition or factor believed to be related to risk, evaluated for a particular piece of pipeline.

WeightX represents the relative importance or weight placed on the corresponding condition or factor—more important variables have a greater impact on the perceived risk and are assigned a greater weight.

Even if the quantification of the risk factors was imperfect, the results were believed to give a reliable picture of places where risks are relatively lower (fewer "bad" factors present) and where they are relatively higher (more "bad" factors are present).

Early published works from the late 1980's and early 1990's in pipeline scoring type risk assessments are well documented.[1] Such scoring systems for specific pipeline operators can be traced back even further, notably in the earlier 1980's with gas distribution companies faced with repair-replace decisions involving problematic cast iron pipe.

Variations on this type of scoring assessment were in common use by pipeline operators for many years. The choices of categorization into failure mechanisms, scale direction (higher points = higher risk or vice versa) variables, and the math used to combine factors are some of the differences among these type models.

The scoring approach was often chosen for its intuitive nature, ease of application, and ability to incorporate a wide-variety of data types. Prior to the year 2000, such models were used primarily by operators seeking more formal methods for resource allocation—how to best spend limited funds on pipeline maintenance, repair, and replacement. Risk assessment was not generally mandated and model results were seldom used for purposes beyond this resource allocation. There are of course some notable exceptions where some pipeline operators incorporated very rigorous risk assessments into their business practices, notably in Europe where such risk assessments were an offshoot of applications in other industries or already mandated by regulators.

The use of indexing/scoring methodologies came into question in the US with new regulations focusing on pipeline integrity management. The role of risk assessment expanded significantly in the early 2000's when the DOT, OPS—now, Pipeline and Hazardous Materials Safety Administration (PHMSA)—began mandating risk analyses of all jurisdictional gas and hazardous liquid pipelines that could affect a High Consequence Area (HCA). Identified HCA segments were then scheduled for integrity assessment and application of preventative and mitigative measures depending on the integrity threats present. The entire integrity management process was intended to be risk-driven, with pipeline operators choosing risk assessment methodologies that could produce required integrity management decision-support.

The simple scoring assessments were generally not designed nor intended for use in applications where outside parties were requesting more rigorous risk assessments. Due in part to the US IMP regulations, risk assessment is now commonly used in project presentation and acceptance in public forums; legal disputes; setting design factors; addressing land use issues; etc, while previously, the assessment was typically used for internal decision support only.

Given their intended use, the earlier models did not really suffer from "limitations" since they met their design intent. They only now appear as limitations as the new uses are factored in. Those still using older scoring approaches recognized the limitations brought about by the original modeling compromises made.

1 Dr. John Kiefner's work for AGA, Dr. Mike Kirkwood from British Gas, W. Kent Muhlbauer's early editions of *The Pipeline Risk Management Manual*, and Mike Gloven's work at Bass Trigon.

In an attempt to simplify, these models actually introduced an extra and now unnecessary level of complexity. The real-world phenomena being modeled had to first be understood. Then a surrogate—the scoring process—for the actual phenomena was created and had to be maintained. The surrogate also had to keep up with a potentially evolving understanding of the underlying phenomenon.

Some of the more significant compromises arising from the use of the simple scoring type assessments included:

- Without an anchor to absolute risk estimates, the assessment results were useful only in a rather small analysis space. The results offered little information regarding risk-related costs or appropriate responses to certain risk levels. Results expressed in relative numbers were useful for prioritizing and ranking but were limited in their ability to forecast real failure rates or costs of failure. They could not be readily compared to other quantified risks to judge acceptability.
- Assessment inputs and results cannot be directly validated against actual occurrences of damages or other risk indicators. Even the passage of time and gaining of more experience, which normally improves past estimates, the scoring models' inputs generally were not tracked and improved.
- Results do not normally produce a time-to-failure, without which there is no technical defense for integrity assessments scheduling. Without additional analyses, the scores did not suggest appropriate timing of ILI, pressure testing, direct assessment, or other required integrity verification efforts.
- Potential for masking of effects when simple expressions could not simultaneously show influences of large single contributors and accumulation of lesser contributors. An unacceptably large threat—very high chance of failure from a certain failure mechanism—could be hidden in the overall failure potential if the contributions from other failure mechanisms were very low. This was because, in some scoring models, failure likelihood only approached the highest levels when all failure modes were coincident. A very high threat from only one or two mechanisms would only appear at levels up to their pre-set cap (weighting). In actuality, only one failure mode will often dominate the real probability of failure. Similarly, in the scoring systems, mitigation was generally deemed 'good' only when all available mitigations were simultaneously applied. The benefit of a single, very effective mitigation measure was often lost when the maximum benefit from that measure was artificially capped. See note 1.
- Some relative risk assessments were unclear as to whether they are assessing damage potential versus failure potential. For instance, the likelihood of corrosion occurring versus the likelihood of pipeline failure from corrosion is a subtle but important distinction since damage does not always result in failure.
- Some previous approaches had limited modeling of interaction of variables, a requirement in some regulations. Older risk models often did not adequately represent the contribution of a variable in the context of all other variables. Simple summations would not properly integrate the interactions of some variables.

- Some models forced results to parallel previous leak history—maintaining a certain percentage or weighting for corrosion leaks, third party leaks, etc.—even when such history might not be relevant for the pipeline being assessed.[1]
- Balancing or re-weighting was often required as models attempt to capture risk in terms that represent 100% of the threat or mitigation or other aspect. The appearance of new information or new mitigation techniques required re-balancing which in turn made comparison to previous risk assessments problematic.
- Some models could only use attribute values that are bracketed into a series of ranges. This created a step change relationship between the data and risk scores. This approximation for the real relationship was sometimes problematic.
- Some models allowed only mathematical addition, where other mathematical operations (multiply, divide, raise to a power, etc) would better parallel underlying engineering models and therefore better represent reality.
- Simpler math did not allow orders of magnitude scales and such scales better represent real-world risks. Important event frequencies can commonly range, for example, from many times per year to less than 1 in ten million chance per year. An underlying difficulty in the calibration of any scoring type risk assessment are the limitations inherent in such methodologies. Since the scoring approaches usually make limited use of distributions and equations that truly mirror reality (see previous discussion on limitations), they will not always closely track 'real-world' experience. For example, a minor 1 or 2% change in a risk score may actually represent an equivalent change in absolute estimates for one threat but a 100 fold change in another threat.
- Lack of transparency. a scoring system adds a layer of complexity and interferes with understanding of the basis of the risk assessment. Underlying assumptions and interactions are concealed from the casual observer and require an examination of the 'rules' by which inputs are made, consumed by the model, and results generated.

Note:

1. *See cautions against the use of weightings, Chapter 3.7.11 Weightings. The assumption of a predictable distribution of future leaks predicated on past leak history might be realistic in certain cases, especially when enough events are available and conditions and activities are constant. However, in some segments, a single failure mode will dominate the risk assessment and result in a very high probability of failure rather than only some small percentage of the total. Even if the assumed distribution is valid in the aggregate, there may be many locations along a pipeline where the pre-set distribution is not representative of the particular mechanisms at work there, leading to incorrect conclusions.*

Serious practitioners always recognized these "limitations" and worked around them when more definitive applications were needed.

3.3.3 Classical QRA Models

Numerical techniques are required in order to obtain estimates of absolute risk values, expressed in fatalities, injuries, property damages, etc., per specific time period. The more rigorous and complex risk assessment approaches in common use in many industries are typically referred to as probabilistic risk assessment (PRA), quantitative risk assessment (QRA), or numerical risk assessment (NRA). While some recognize differences among these labels, they are often used interchangeably.

Recall the earlier discussion on Classical QRA—the statistics-driven approach to risk assessment. For discussion purposes here, currently documented methodologies labeled PRA, QRA, NRA, including their common supporting processes such as Monte Carlo simulation, Markov analyses, Bayesian statistics, and other statistics-centric approaches are treated as variations on a single technique, which we will call Classical QRA for convenience. Classical QRA will compared to a physics-based approach—the preferred approach in pipeline risk assessment—in an upcoming discussion on 'myths'. Here, the discussion will examine this practice.

These techniques are assembled together under the premise that they all use statistics as the primary driver in understanding risk. The applicability of the oft-used supporting techniques further illustrates this point: Bayesian begins with statistics, sometimes modified by physics (a priori info). Markov links a future state with a current state, through initial state probabilities and probabilities of change. These are in contrast to an approach that begins with physics and then refines preliminary results using historical event frequencies. Both approaches benefit from the use of statistics but the primary focus is different.

Classical QRA is a technique used in the nuclear, chemical, and aerospace industries and, to some extent, in the petrochemical industry. The output of a classical QRA is usually in a form whereby its output can be directly compared to other risks such as motor vehicle fatalities or tornado damages. It can be thought of a statistical approach to the quantification of risks, emerging from numerical analyses applied to scenario structures such as event trees and fault trees (see discussion in Chapter 1 Risk Assessment at a Glance).

Classical QRA is a rigorous mathematical and statistical technique that relies heavily on historical failure data and event-tree/fault-tree analyses. Initiating events such as equipment failure and safety system malfunction are flow-charted forward to all possible concluding events, with probabilities being assigned to each branch along the way. Failures are backward flow-charted to all possible initiating events, again with probabilities assigned to all branches. All possible paths can then be quantified based on the branch probabilities along the way. Final accident probabilities are achieved by chaining the estimated probabilities of individual events.

This technique, when applied robustly, is usually very data intensive. It attempts to provide risk estimates of all possible future failure events based on historical experience. The more elaborate of these models are generally more costly than other risk assessments. They can be technologically more demanding to develop, require trained practitioners (statisticians), and need extensive data. A detailed classical QRA is usually the most expensive of the risk assessment techniques due to these issues.

The classical QRA methodology was first popularized through opposition to various controversial facilities, such as large chemical plants and nuclear reactors [88]. In addressing the concerns, the intent was to obtain objective assessments of risk that were grounded in indisputable rigorous analyses. The technique makes extensive use of failure statistics of components as foundations for estimates of future failure probabilities.

However, it was also recognized that statistics paints an incomplete picture at best, and many probabilities must still be based on expert judgment. In attempts to minimize subjectivity, applications of this technique became increasingly comprehensive and complex, requiring thousands of probability estimates and like numbers of pages to document. Nevertheless, variation in probability estimates remains, and the complexity and cost of this method does not seem to yield commensurate increases in accuracy or applicability. In addition to sometimes widely differing results from "duplicate" classical QRAs performed on the same system by different evaluators, another criticism includes the perception that underlying assumptions and input data can easily be adjusted to achieve some predetermined result [88]. Of course, this latter criticism can be applied to any process involving much uncertainty and the need for assumptions.

3.3.4 Myths

While the practice of formal pipeline risk assessment has been on-going for many years, the practice is by no means mature (as of this writing). There still exist some common misconceptions and myths. This is not unexpected, given the difficult nature of risk concepts themselves and the absence of detailed guidance documents (prior to this textbook).

3.3.4.1 Myth1: Some risk assessment models are better able to accommodate low data availability

Reality:
Strong data + strong model = most meaningful/useful results
Weak data + strong model = uncertain results
Weak data + weak model = meaningless results

First, let's address the myth that low information suggests the use of a simple risk assessment—one that does not really quantify risk. Using a lesser risk assessment process in an attempt to compensate for low information is an error. Pairing weak data with a weak model generates nothing useful. The proper approach is to begin with a full risk assessment structure, make conservative assumptions where necessary, and then work on 'back-filling' the data that will ultimately drive the risk management.

So, we should use a robust risk assessment, regardless of the current data availability. There are two choices. Let's compare how the statistics-based and the physics-based approaches in solving a typical risk assessment problem: how often will a specific segment of pipeline experience failure from outside excavator force (third party damage)?

Statistics-centric Approach:

In this approach, we focus on historical event frequencies. Let's say that a slice of the national pipeline incident database shows that US transmission averages show 0.0003 reportable third party damage incidents per mile per year. With some investigation, we can get averages for ranges of pipe diameters, product type, or other characteristics should we believe that they are discriminating factors for third party damages. We can assume that some of the historical 'unknown' causes of failure (a significant proportion of the data) were also third party damage related. We can further assume that the entire population of third party failures is higher than the reportable-only count. At the end of this exercise, we have a decent estimate of a historical failure rate for an 'average' pipe segment.

Physics Approach:

In this approach, we focus on the physical phenomena that influence pipeline failure potential. We first make a series of estimates that show the individual contributions from exposure, mitigation, and resistance. For exposure, we ask 'how often is there likely to be an excavator working near this pipeline?' We perhaps examine records in planning and permitting departments; take note of nearby utilities, ditches, waterways, public works, etc that require routine excavation maintenance; and tap into other sources of information. Then we estimate the role of mitigation measures as applied to this particular segment of pipe. We ask: "what fraction of those excavators will have sufficient reach to damage the pipe (suggesting the benefit of cover depth)?" "What fraction of excavators will halt their progress due to one-call system use, recognition from signs, markers, and briefings?" "What fraction will halt their work due to intervention by pipeline patrol?" and others.

Finally, we discriminate among the fraction of excavation scenarios with sufficient force potential to puncture the pipe, based on pipe characteristics and the types of forces likely to be applied. This tells us the resistance—how often is there damage, but not failure? This discrimination between damage likelihood and failure likelihood is essential to our understanding.

All of these estimates can come from simple reasoning, at one extreme, to literature searches, market analyses, database mining, finite element analyses, and scenario analyses, at the other extreme. The level of effort should be proportional to the perceived contribution of the issue to the total risk picture.

Approach Comparisons:

Both of these approaches have merit and yield useful insight. But, only the latter provides the location-specific insights we need to truly manage risk. The statistics-only approach yields an average value, suggesting how a population of pipeline segments may behave over time. There are huge differences among all the pipeline segments that go into a summary statistic. Therefore, we cannot base risk management on such a summary value derived from generic historical data. Risk, and hence 'risk management', ultimately occurs at very specific locations, whose risk may be vastly different from the population average. Stated even more emphatically: "using averages will always result in missing the 'generally rare but critical at this location' evidence". For example, most pipelines are not threatened by landslide, but in the few locations where they are, this apparently rare threat may well dominate the risk.

So, we use the physics-based approach to drive risk management. Using the statistics-based approach is very useful in calibrating risk estimates from populations of pipe segments. More about that in a later section.

3.3.4.2 Myth 2: QRA requires vast amounts of incident histories

Reality:

QRA 'requires' no more data than other techniques
All assessments work better with better information

This is related to Myth1 but merits a bit of independent discussion. Some classical QRA does over-emphasize history, as noted in the discussion of statistician-designed risk assessment. Excessive reliance on history is an error in any methodology. The past is a relevant predictor of the future only in certain cases, as is also detailed elsewhere.

3.4 CHOOSING A RISK ASSESSMENT APPROACH

Understanding the differences between tools and assessment models, as well as the strengths and weaknesses of the different risk assessment techniques is important in choosing approaches. A case can be made for using some techniques in certain situations. For example, a simple bowtie analysis approach helps to organize thinking and is a first step towards formal risk assessment. If the need is to evaluate specific events at any point in time—for example, an incident investigation—a narrowly focused scenario risk analysis (event tree or fault tree) might be the tool of choice.

Scoring or ranking type pipeline risk assessments have served the pipeline industry for many years. However, risk assessments are being routinely used today in ways that were not common even a few years ago. For example, many operators are asking questions today such as:

- How to make full use of inspection data in a risk assessment
- How to generate results that directly suggest timing of integrity assessments

- How to quantify the risk reduction benefit of integrity assessment and other mitigations.
- Beyond the prioritization, how big is the risk? Is it actionable?
- How widespread is a particular risk issue?
- How can subjectivity be reduced?
- How to use past incident results in a risk assessment.

These questions are not consistently nor accurately answered by the relative models. As previously noted, these models were designed and created under a different set of questions.

Similarly, classical ORA techniques have been in use in some industries for decades. But, especially in the pipeline industry with so much variation between and within collections of pipeline segments, these solutions are sub-optimal (see previous discussion).

The new roles of risk assessments have prompted some changes to the way risk algorithms are being designed. The changes lead to more robust risk results that better reflect reality and, fortunately, are readily obtained from data that was also used in previous assessments.

3.4.1 New Generation Risk Assessment Algorithms

The focus of this book is on a comprehensive risk assessment methodology that is both robust and cost-effective to establish and maintain.

While the previous generation of relative algorithms served the industry well, the technical compromises made can be troublesome or unacceptable in today's environment of increasing regulatory and public oversight. Risk assessments commonly become the centerpiece of any legal, regulatory, or public proceedings. This prompts the use of assessment techniques that more accurately model reality and also produce risk estimates that are anchored in absolute terms: "consequences per mile year," for example. Fortunately, a new approach to algorithm design can do this while making use of all previously collected data and not increasing the costs of risk assessment. The advantages of the new algorithms are that they can overcome the previously noted limitations in both competing methodologies:

- More intuitive;
- Better models reality;
- Removes much subjectivity;
- Eliminates masking of significant effects;
- Makes more complete and more appropriate use of all available and relevant data;
- Greatly enhances ability to demonstrate compliance with U.S. IMP regulations;
- Distinguishes between unmitigated exposure to a threat, mitigation effectiveness, and system resistance—this leads directly to better risk management decisions;

- Eliminates need for unrealistic and expensive re-weighting of variables for new technologies, emerging/previously-unknown threats, or other changes; and
- Flexibility to present results in either absolute (probabilistic) terms or relative terms, depending on the user's needs.

The breakthrough evolution in overcoming both the scoring system limitations and the classical QRA limitations was the dissection of PoF phenomena into three separately measurable components. This allowed for a physics-based rather than statistics-based approach. It simultaneously ends the need for a secondary scoring system and the need for (often misleading) reliance on historical event rates—generic data.

The risk modeling approach recommended here falls into several common category labels of models. Generally, this methodology is a type of quantitative model since it numerically quantifies risks in a rigorous way (not a simple numerical scoring approach). It is a deterministic (or 'mechanistic') model, since it is a mathematical representation of physical processes that has been constructed from the modeler's understanding of the science underlying the processes. It has probabilistic components since the real-world processes it mirrors are best represented by probabilities. It expresses results in absolute terms.

In this book, the term 'physics-based' is chosen to classify the new generation of risk assessment algorithms. Since physics as a science includes mechanics, energy, force, and even chemistry to some extent, it captures the fact that this type of risk assessment relies on such underlying science. This hopefully carries a connotation beyond terms like 'mechanistic' or 'deterministic', that the methodology is based on widely-accepted first principles of science and engineering.

3.4.2 Risk Assessment Specific to Pipelines

The recommended practice—a new generation of risk assessment algorithms classified as physics-based algorithmsis the result of years of development efforts[2]. Distinguishing characteristics of this risk assessment methodology from other past and current approaches are worth examining. Especially for the practicing risk assessor, some evidence will be sought that justifies a change to his/her current practice.

The primary distinguishing characteristic of these physics-based algorithms is the breakdown of PoF. This is an essential aspect of risk assessment that is not clearly and completely employed in any alternative methodology. It is a differentiating characteristic compared to all other methodologies. It is a critically important aspect of modern pipeline risk assessmenthere and elsewhere in this book.

Additional discriminating features of this recommended methodology, compared to alternative approaches, include the following:

Differences from classical QRA:

---

2 Some may also use the label 'deterministic'.

- Profiles—ie, the ability to pass the 'map point' test of risk assessment sufficiency, described earlier.
- Reduced reliance on generic historical event frequencies.
- Directly integrates more relevant, location-specific information.

Differences from Indexing/Scoring Approaches:

- Only verifiable measurements are used
- Mathematics to fully represent real-world phenomena
- Data-driven segmentation—full resolution
- Improved use of information
- More transparent—no need for protocols to assign point values.

Differences from HAZOPS/SIL/LOPA/FMEA

- Profiles
- Measurements instead of categories
- Improved use of information
- Aggregations for summarizing risk.

Differences from reliability based design (RBD)

- RBD typically relies excessively on classical QRA, see list above.

Note that comparisons are not offered to other techniques that are more appropriately labeled as tools rather than risk assessments. This includes event/fault tree analyses, matrices, checklists, bowtie, dose-response assessments probit equations, dispersion modeling, hazard zone estimations, human reliability analyses, task-based assessments, what-if analyses, Markov analyses, Bayesian statistical analyses; root cause analyses; Delphi technique.

The criteria discriminating tools from complete risk assessments is detailed earlier in this chapter. Comparisons of the modern approach to selected techniques more often labeled as risk assessments, include differences in recommended methodology compared to PHA, HAZOPS, Matrix, Event Tree / Fault Tree Analyses:

- Ability to broadcast a risk assessment over long, complex systems
- Profiles
- Aggregations for summarizing risk
- Only verifiable measurements are used
- Improved use of information.

Most alternative methodologies suffer from an inability to create a risk profile—changes in risk along a pipeline route. While a profile can also mean changes over time, the risk changes along a route is often the limiting factor of competing risk assessment techniques. Techniques that rely on specific cause-consequence pairings without an ability to aggregate all such pairings, cannot produce a complete profile and therefore cannot present an accurate risk picture. Without a risk profile, understanding and hence, optimum risk management, is compromised.

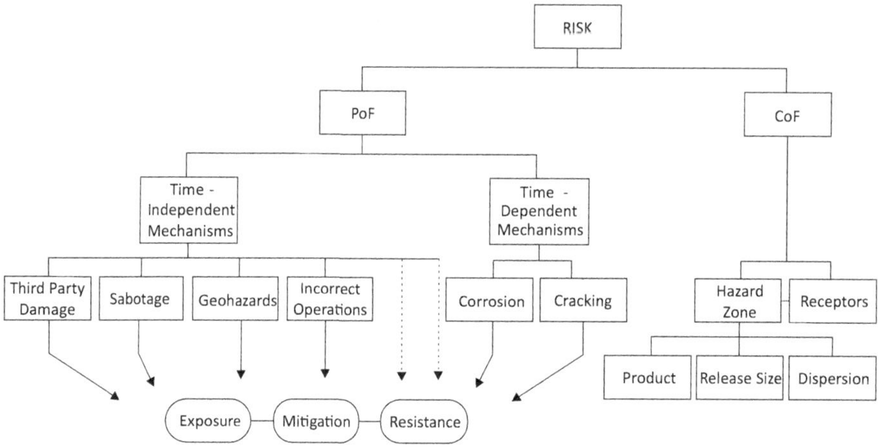

Figure 3.3 Modeling of Pipeline Risk

3.5 QUALITY, RELIABILITY, AND RISK MANAGEMENT

Risk management embodies and overlaps principles of quality assurance, quality control, and reliability. An interesting background discussion on these concepts and their relationship to risk can be found in PRMM.

3.6 RISK ASSESSMENT ISSUES

3.6.1 Quantitative vs. qualitative models

Modeling as a part of the scientific method is discussed in PRMM. As noted, labeling of modeling approaches has caused some confusion. Terms including quantitative, qualitative, semi-quantitative, scoring, indexing, and others have been used to describe types of pipeline risk assessment. There are no standard definitions in common use for these terms. Therefore, they often carry different meanings and different implications to various members of an audience hearing them.

The advice here is to always obtain clarifications when faced with this terminology.

Additional labels of probabilistic, mechanistic, and deterministic are also sometimes seen. These have more standardized definitions but can still cause confusion.

The risk modeling approach recommended here falls into several common categories of models. Generally, this methodology is a type of quantitative model since it numerically quantifies risks in a rigorous way (not a simple numerical scoring approach).

It is a deterministic (or `mechanistic') model, since it is a mathematical representation of physical processes that has been constructed from the modeler's understanding of the science underlying the processes. It has probabilistic components since the real-world processes it mirrors are best represented by probabilities. It expresses results in absolute terms.

3.6.2 Absolute vs. relative risks

> FOCUS POINT
> There is no longer any valid reason to settle for relative risk assessment results. Absolute risk estimates can now be generated more reliably and with less effort.

Closely paralleling the quantitative vs qualitative distinction is the issue of risk presented in absolute vs relative terms. Unlike the previous discussion highlighting potential confusion arising from terminology, the absolute vs relative adds clarity—giving strong indications to any audience of what type of assessment has been performed.

Risks can be expressed in absolute terms—risk estimates expressed in fatalities, injuries, property damages, or some other measure of consequence, in a certain time period and for a specific collection of components such as a pipeline system. For example, "number of fatalities per mile-year for permanent residents within one-half mile of pipeline…". This requires concepts commonly seen in probabilistic risk assessments (PRAs), also called numerical risk assessments (NRAs) or quantitative risk assessments (QRAs), or deterministic or mechanistic models. Absolute risk assessment generates a frequency-based measure that estimates the probability of a specific type of failure consequence at a point in time and space. Also available is the use of a relative risk assessment methodology, whereby risks can make comparisons among components that have undergone the same assessment. Common relative risk measurement systems have been called scoring or indexing systems. Ref [PRMM] presented such a system. The relative risk measurement models are no longer recommended since they have many limitations and no advantages over a properly crafted absolute risk assessment.

The term 'absolute' should not be construed to indicate a level of certainty. It only speaks to the units with which risk estimates are produced.

The "absolute scale" offers the benefit of comparability with other types of risks and a more accurate representation of actual risk, while the "relative scale" was historically used as a compromise solution to avoid what was previously considered a challenging quantification of pipeline risks.

The absolute scale previously suffered from its heavy reliance on historical estimates. This criticism has been mitigated by the methodology presented here.

The absolute and relative risk assessment scales are not mutually exclusive. The absolute scale can be readily converted to relative scales by simple mathematical relationships, should this be deemed worthwhile. For instance, 1.6E-7 failures per year can be normalized to, say 15, on a 100 point PoF scale. A relative risk scale can theoretically be converted to an absolute scale by correlating relative risk scores with appropriate historical failure rates or other risk estimates expressed in absolute terms. In other words, the relative scale can be calibrated to some absolute numbers. This is often a problematic exercise given the mathematical and other limitations commonly associated with the relative risk models. For instance, orders of magnitude differences in real risk are difficult to show on simple point scale. However, when much information has been collected into an older relative risk assessment, that information can be salvaged and efficiently used in a migration to a modern absolute risk assessment. See Chapter I.4.2 Migration from previous methodologists.

A possible consideration underlying the presentation of any numerical modeling result is a common misconception that a precise-looking number, expressed in scientific notation, is more accurate than a simple number. A numerical scale can imply a precision that is simply not available. This effect has been called 'the illusion of knowledge'.

A good risk assessment will require the generation of sufficient scenarios to represent all possible event sequences that lead to possible damage states (consequences). Each event in each sequence is assigned a probability—actually, an expected future frequency. The assigned probabilities are best assigned in absolute terms, leading to final risk estimates that are also in absolute units, for example: leaks per mile-year, dollars per km-year, fatalities per year, etc. The expression in absolute terms widens the uses of risk results and avoids the complications of the relative scales.

A damage state or consequence level of interest is identified and becomes part of the measurement units in an absolute estimate of risk. Most risk acceptability or tolerability criteria are based on fatalities as the consequence of interest.

3.7 VERIFICATION, CALIBRATION, AND VALIDATION

Given enough time, a risk assessment can be proven by comparing predicted pipeline failures against actual. This is the basis of the testing of the risk assessment as a diagnostic tool, as discussed elsewhere. Pipeline failures on any specific system are usually not frequent enough to provide sample sizes sufficient to test the assessment performance. In most cases, initial examination of the assessment is best done by ensuring that risk estimates are consistent with all available information (including actual pipeline failures and near-failures) and consistent with the experiences and judgments of the most knowledgeable experts. The latter can be at least partially tested via structured testing sessions and/or model sensitivity analyses (discussed in Chapter 3.7.4 SME Validation and Chapter 3.7.12 Diagnosing Disconnects Between Results and 'Reality'). Additionally, the output of a risk model can be carefully examined for the behavior of the risk values compared with our knowledge of behavior of numbers in general.

More formal examinations of the risk assessment are also possible. The processes of verification, calibration, and validation are likely not familiar to most readers and, based on a brief literature search, are not even standardized among those who more routinely deal with them. Some background discussion to these processes, especially as they relate to pipeline risk management, are warranted.

In this text, a distinction is made between verification, validation, and calibration. Verification is the process of 'de-bugging' a model—ensuring that functions operate as intended. Calibration is tuning model output so that it mirrors actual event frequencies. This is a practical necessity when knowledge of underlying factors is incomplete (as it almost always is in natural systems). Validation is ensuring consistent and believable output from the model by comparing model prediction with actual observation. Defining these terms in the context of this discussion is important since they seem to have no universally accepted definitions.

An important aspect of proving a risk assessment is agreement with SME beliefs. Users should be vigilant against becoming too confident in using any risk assessment output without initial and periodic 'reality checks'. But users should also recognize that SME beliefs can be wrong. Disconnects between risk assessment results and SME beliefs are opportunities for both to improve, as is discussed in Chapter 3.7.4 SME Validation.

Note also that the conclusions of any risk assessment can be no stronger than the inputs used. Especially when confidence in inputs is low, calibration to a judged performance is warranted.

3.7.1 Verification

Especially where software is used, verification ensures that the model has been programmed correctly and is, to the extent tested, error-free (no bugs). In a pre-acceptance review of a risk assessment, confirmation of calculations should be performed. Therefore, verification—checks to ensure that intended results are produced by the risk algorithms—confirms that the intended routines are functioning properly.

To ensure that all equations and point assignments are working as intended, some tools can be developed to produce test results using random or extreme value inputs.

3.7.2 Calibration

Risk assessment should be performed on individual pipe segments due to the changes along a pipeline route. These individual risk estimates can be combined (into 'populations') and compared to the known behavior of similar populations. For a variety of reasons, discrepancies in predicted population behavior will usually exist. Calibration serves to rectify the inappropriate discrepancies by adjusting the individual estimates en masse so that credible population characteristics emerge.

The process of calibrating risk assessment results begins with establishing plausible future leak rates of populations based on relevant historical experience, adjusted for relevance and other considerations. These rates become 'targets' for risk assessment outputs, with the belief that large populations of pipeline segments, over long periods of time, would have their overall failure estimates approach these targets. The risk assessment model is then adjusted so that its outputs do indeed approximate the target values for behavior of populations.

The choice of representative population is challenging. It is difficult to find a collection of components similar enough to the system being assessed and with a long enough history to make comparisons relevant. A selection of a population that is not sufficiently representative will weaken the calibration process.

Calibration is done using both a representative population and a target level of conservatism. Both are required as illustrated by this thought exercise. Imagine you could run experiments on real pipelines over long periods of time. Say you chose a 70 mile pipeline operating for 50 years. You would run multiple, maybe hundreds or thousands, of trials to see how the 70 mile pipeline performs over many different 50 year lifetimes. Each trial—that is, each 50 year lifetime—is influenced by random influences of exposures, mitigations, resistance, and consequence scenarios over its 50 years in service. In some of those lifetimes, there would be no incidents, so no actual consequences at all. Choosing these trial results as representative of future behavior of the next trial might reflect a P10 level of conservatism. Some of your multiple trials would result in dozens of leaks and ruptures, some producing very consequential results. Using this set of trials to represent future behavior would be choosing a P90 or so level of conservatism. The results of the majority of your trials would form the P50 portion of the distribution of all results, perhaps the center point of a normal or bell-shaped distribution.

With an appropriate comparison population, the chief goal of a calibration will often be the removal of unwanted conservatism. As discussed, conservatism plays an important and useful role in risk assessment for individual components. P90+ inputs are recommended for many initial risk assessments. However, the need for estimates as close as possible to actual risk levels is also important, especially for populations—collections of individuals. A decision-maker gains more insight from a P50 type risk assessment of a pipeline system than a system summary incorporating multiple P90+ inputs. The P50 estimate can become a part of company-wide strategic planning while the P90+ estimates ensure proper attention to risk management for each component.

In a simple calibration exercise, we seek a single factor representing the amount of conservatism included in a risk assessment's P90+ estimates. This factor can be used to reduce the conservative estimates of each component's risk to best-estimates of risk. The resulting collection of 'best estimates' should be close to the representative population's historical risk levels.

One can track differences between P50 and P99 to see, at least partially, reduction in uncertainty. P50 and P90+ have both natural variability (apparent randomness) and uncertainty. Each PXX produces a distribution. The difference between, for instance, the midpoint of the P50 and P90+ distributions can be called the conservatism bias multiple.

Both P50 and P90+ risk assessments will often be needed—the former to represent likely system wide behavior and the latter to use in risk management. Some practitioners choose to run parallel P50 and P90+ assessments. Others perform the P90+ assessment, estimate the conservatism bias, and then use it to 'back calculate' P50 results.

Once calibrated, estimates could represent a wide range of possibilities. For example, a US natural gas transmission pipeline may have components with P50 PoF estimates from perhaps 0.00001 to 0.1 reportable events per mile-year, assuming that segments' actual PoF's could range from about 100 times higher or lower than the US average for reportable incidents on natural gas pipelines.

A similar process can be performed on overall risk values or any intermediate calculations. More calibration—calibrating to lower level algorithms—should produce more confidence in the overall correlation. This essentially provides more intermediate correlating points from which a correlation curve can be better developed.

3.7.3 Validation

Validation of a model is achieved by ensuring that appropriate relationships exist among input data and that produced outputs are representative of real-world experience. Validation seeks to authenticate or verify that the model produces risk estimates that are accurate.

While pipeline industry documents do not generally detail these processes, examples of how the pipeline industry typically uses the term 'validation' are noted in PHMSA and PRCI documents:

US Gas IMP Protocol C.04

> Verify that the validation process includes a check that the risk results are logical and consistent with the operator's and other industry experience. [§192.917(c) and ASME B31.8S-2004, Section 5.12] (**http://primis.phmsa.dot.gov/gasimp/QstHome.gim?qst=145**)

From PRCI, discussing validation of a risk-based model for pipelines:

> The fault tree model and basic event probabilities were validated by analyzing a representative cross-country gas transmission pipeline and confirming that the results are in general agreement with relevant historical information.

Validation of risk assessment is also noted in US IMP documents.

ASME B31.8s

"*...experience-based reviews should validate risk assessment output with other relevant factors not included in the process, the impact of assumptions, or the potential risk variability caused by missing or estimated data.*"

As a part of the validation effort, the general relationship between model output and reality should be examined. When new or altered theories are proposed as part of a model, examination of those must be included in the validation process.

Theories applicable to pipeline risk assessment typically include:

- Metallic corrosion
- Mitigation of metallic corrosion—coatings and cathodic protection
- Stresses in a shell structure (pipe)
- Effect of wall loss on pressure-containing capability
- Component rupture potential
- Probability theory
- Probability distributions as applied to observed phenomena
- Structural theory
- Materials science
- Plastics and coatings performance.

The risk assessment methodology described in this book does not propose new theories of failure mechanisms. It relies upon thoroughly documented models of the above theories including widely accepted beliefs about impacts of certain factors on certain aspects of risk; for example, 'increases in Factor X lead to increased risk'.

3.7.4 SME Validation

Similar to the use of a benchmark for model calibration, a carefully structured interview with SME's can also identify model weaknesses (and also often be a learning experience for SME's). If an SME reaches a risk conclusion that is different from the risk assessment results, a drill down (for example, a deeper examination) into both the model and the SME's basis of belief should be done. Any disconnect between the two represents either a model error or an inappropriate conclusion by the SME. Either can be readily corrected. The key is to identify exactly where the model and the SME first diverge in their assumptions and/or conclusions.

An important step in validation is therefore to identify and correct 'disconnects' between subject matter experts' beliefs and model outputs. This is similar to calibration discussed previously but differs in that validation should occur after calibration has been done. In the absence of calibration of risk results, validation can still be performed on intermediate calculations but the role of conservatism must be factored in. For relative, scoring models, validation can only be done in general terms, where SME's can agree in relative changes to risk accompanying certain changes in inputs.

SME concurrence with assessment outputs should be a part of model validation. Risk assessment-identified higher—and lower—risk segments should comport with SME-identified higher— and lower—risk segments.

SME review should include concurrence with aspects such as:

- Direction and magnitude of risk changes accompanying changes in factors and groups of factors
- identified locations of higher- and lower-threats, considering each threat independently
- identified locations of higher- and lower-consequences.

A good objective of risk assessment should be to have the risk assessment model capture the collective knowledge of the organization—anything that anyone knows about a pipeline's condition or environment, or any new knowledge of how risk variables actually behave and interact, can and should be included in the analysis protocol.

3.7.5 Predictive Capability

Implicit in the notions of validation and verification is the idea of predictive capability. A good risk assessment always produces some estimate of failure probability. Theoretically, this can forecast, to some degree of accuracy, future failures on specific pipeline segments. Except in extreme cases, this is not a realistic expectation. A more realistic expectation is for the assessment to forecast behavior of populations of segments rather than individuals. A good risk assessment will, however, highlight areas where probability and consequence combinations warrant special attention.

Leak/break rate is related to estimated failure probability. In most transmission pipelines, insufficient system-specific information exists to build a meaningful prediction model solely from leak/break rate—events are so rare that any such prediction will have very large uncertainty bounds. Distribution systems, where leaks are precursors to "failures," are often more viable candidates for producing predictions directly from leak/break rates.

A leak/break rate assessment may show both time-dependent failure mechanisms such as corrosion and fatigue and more random failure mechanisms such as third-party damages and seismic events. The random events will normally occur at a relatively constant rate over time for a constant set of conditions.

A leak/break rate is called a "deterioration" rate by some, but that phrase seems to be best applied specifically to time-dependent failure mechanisms only (corrosion and fatigue).

Even though they are commonly expressed as a single value, each failure probability estimate really represents an underlying distribution—all possible failure rates with associated probability of occurrence—with an average, median, and standard deviation. This distribution describes the range of failure rates that would accompany any pipeline section with a particular predicted failure rate.

Nonetheless, to test the predictive power of the risk assessment model, the incident and inspection history in recent years could be examined. Knowing what the risk assessment 'thought' about the risk on the day before the incident (or the day before an inspection) would provide insight into the predictive power of the assessment. Given the role of probability, spot samples from individual segments may appear to show inaccurate predictions, but actual accuracy can only be verified after sufficient data has been accumulated to compare the predicted versus actual long term behavior of a large population.

3.7.6 Evaluating a risk assessment technique

Note: Locating this discussion in this book was challenging. On one hand, a reader is often not terribly interested in this aspect until he is an active practitioner. On the other hand, a reader who has an existing risk assessment approach may need early incentivization to investigate alternative approaches. This latter rationale has obviously determined the issue for purposes of organizing this book. The early discussion has a further advantage of setting the stage—arming the reader with criteria that will later determine the quality of his assessments, even as he works his way through this text to learn about pipeline risk assessment.

In general, proving or confirming a risk assessment methodology addresses the extent to which the underlying model represents and correctly reproduces the actual system being modeled. Another view is that validation involves two main aspects:

1) ensuring that the model correctly uses its inputs and
2) model produces outputs that are useful representations of the underlying real-world processes being modeled.

Ref [1046] focuses on the need for transparency in any risk assessment:

> Transparency provides explicitness in the risk assessment process. It ensures that any reader understands all the steps, logic, key assumptions, limitations, and decisions in the risk assessment, and comprehends the supporting rationale that lead to the outcome. Transparency achieves full disclosure in terms of:
>
> a. the assessment approach employed
> b. the use of assumptions and their impact on the assessment
> c. the use of extrapolations and their impact on the assessment
> d. the use of models vs. measurements and their impact on the assessment
> e. plausible alternatives and the choices made among those alternatives
> f. the impacts of one choice vs. another on the assessment
> g. significant data gaps and their implications for the assessment
> h. the scientific conclusions identified separately from default assumptions and policy calls
> i. the major risk conclusions and the assessor's confidence and uncertainties in them;

j. the relative strength of each risk assessment component and its impact on the overall assessment (e.g., the case for the agent posing a hazard is strong, but the overall assessment of risk is weak because the case for exposure is weak)

Process transparent and the risk characterization products clear, consistent and reasonable" (TCCR) became the underlying principle for a good risk characterization. [1046]

To properly support risk management, the superior risk assessment process will have additional characteristics, including:

- QA/QC and error checking capabilities, perhaps automated
- Ability to rapidly integrate new information and refresh risk estimates
- Be able to rapidly incorporate new information on emerging threats, new mitigation opportunities, or any other changing aspect of risk.
- Seamless integration with other databases and legacy data systems
- Accessible, understandable to all decision-makers.

3.7.7 Diagnostic tool—Operator Characteristic Curve

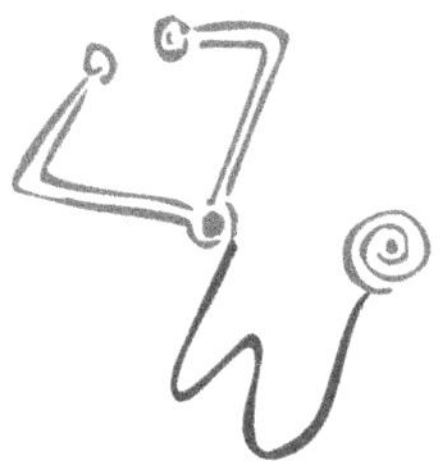

For those seeking a more structured approach to proving a risk assessment, techniques are available. A pipeline risk assessment is really a diagnostic tool. Similar to a diagnostic test used by a doctor, the idea is to determine, with the least amount of cost and patient discomfort, whether the patient has the disease or doesn't. He knows that in any population, a certain fraction of individuals will have the disease and most won't. For a diagnosis to be successful, he must correctly determine into which group to place his patient. In making this determination, the doctor can choose a whole battery of expensive and intrusive tests and procedures in order to have the highest confidence of his diagnosis. On the other hand, he can choose minimal tests and accept a higher error rate in diagnoses. The most accurate test or set of tests will minimize the rate of false positives and false negatives. But there is a cost associated with such testing.

In the case of pipeline risk management, the manager is trying to determine which pipeline segments and components have the 'disease' of higher risk among the hopefully many which do not. His choice of tests to help in the diagnosis goes beyond the risk assessment itself. He can request surveys and inspections to improve the diagnosis, but with an accompanying expense and the potential for inefficient use of resources. The latter occurs when expensive 'tests' do not add much certainty to the assessment.

Both the doctor and the risk manager will be balancing the costs of the diagnostics and the costs of being wrong—false positives and false negatives.

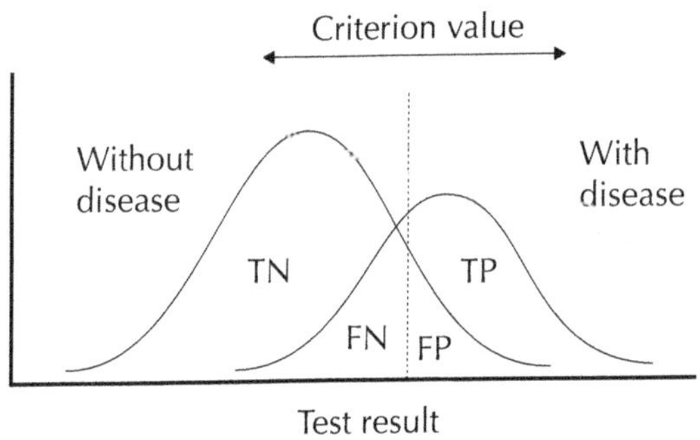

Figure 3.4 Risk assessment as a diagnostic tool; trade-offs among true positives, true negative, false positives, false negatives (TP, TN, FP, FN)

3.7.8 Possible Outcomes from a Diagnosis

The tuner of a leak detection system is well aware of the false alarm phenomena. In order to find smaller leaks, it is necessary to alarm and investigate apparent smaller leaks that later prove to be only transient conditions. After too many false alarms, the investigators grow weary of responding and are less attentive, thereby increasing their error rate when an actual leak does appear. It is standard to sacrifice some leak detectability in order to avoid too many nuisance alarms. A modern leak detection system will state a probability associated with the indication, to assist the investigator in setting his response urgency.

Advanced applications of these ideas is found in signal detection theory, receiver operating characteristic curves, artificial intelligence (machine learning), and others. For our purposes here, it is useful to simply bear in mind the diagnostic intent behind a risk assessment and the corresponding ability to test its diagnostic power over time.

3.7.9 Risk model performance

Some sophisticated routines can be used to evaluate risk assessment outputs. For instance, a Monte Carlo simulation uses random numbers as part of the assessment inputs in order to produce distributions of all possible outputs from a set of risk algorithms. The resulting distribution of risk estimates might help evaluate the "fairness" of the assessment. In many cases a normal, or bell-shaped, distribution would be expected since this is a very common distribution of properties of materials and engineered structures as well as many naturally occurring characteristics. Alternative distributions are also common, such as those often used to represent rare events. All distributions that emerge should be explainable. If some implausible distribution appears, further examination may be warranted. For instance, excessive tails or gaps in the distributions might indicate discontinuities or biases in the results being generated.

3.7.10 Sensitivity analysis

The algorithms that underlie a risk assessment model must react appropriately—neither too much nor too little—to changes in any and all variables. In the absence of reliable data, this appropriate reaction is gauged to a large extent by expert judgment as to how the real-world risk is really impacted by a variable change.

A single variable can play a role as both risk increaser and risk reducer. A casing protects a pipe segment from external force damage but complicates corrosion control; in the offshore environment, water depth is a risk reducer when it makes anchoring damage less likely but it is a risk increaser when it heightens the chance for buckling. So the same variable, water depth, is a "good" thing in one part of the model and a "bad" thing somewhere else.

See discussion of data collection in Chapter 4 Data Management and Analyses for a deeper examination into types and roles of information

Some variables such as pressure and population density impact both the probability (often linked to lower resistance and higher activity levels) and consequence (larger hazard zone and more receptor damage) sides of the risk algorithm. In these cases, the impact on overall risk is not always obvious. When a variable is used in a more complex mathematical relationship, such as those sometimes used in resistance estimates, then influences of changes on final risk estimates will also not be apparent.

Sensitivity quantifications can be utilized for evaluating effects of changing factorsbut require fairly sophisticated analyses procedures. It is important to recognize that many variables will *usually* play lessor roles in overall risk but may occasionally be the single greatest determinant of risk.

3.7.11 Weightings

FOCUS POINT
The use of weightings in a risk assessment will almost certainly result in serious analyses errors.

The use of 'weightings' should be a target of critical review of any risk assessment practice. Weightings have been used in some older risk assessments to give more importance to certain factors. They were usually based on a factor's perceived importance in the majority of historical pipeline failure scenarios. For instance, the potential for AC induced corrosion is usually very low for many kilometers of pipeline, so assigning a low numerical weighting appeared appropriate for that phenomenon. This was intended to show that AC induced corrosion is a rare threat.

Used in this way, weightings steer risk assessment results towards pre-determined outcomes. Implicit in this use is the assumption of a predictable distribution of future incidents and, most often, an accompanying assumption that the future distribution will closely track the past distribution. This practice introduces a bias that will almost always lead to very wrong conclusions for some pipeline segments.

The first problem with the use of weightings is finding a representative basis for the weightings. Weightings were usually based on historical incident statistics—"20% of pipeline failures from external corrosion"; "30% from third party damage"; etc. These statistics were usually derived from experience with many kilometers of pipelines over many years of operation. However, different sets of pipeline kilometer-years shows different experience. Which past experience best represents the pipeline being assessed? What about changes in maintenance, inspection, and operation over time? Shouldn't those influence which data sets are most representative to future expectations?

It is difficult if not impossible to know what set of historical population behavior best represents the future behavior of the segments undergoing the current risk assessment. If weightings are based on, say, average country-wide history, the non-average behavior of many miles of pipeline is discounted. Using national statistics means including many pipelines with vastly different characteristics from the system you are assessing.

If the weightings are based on a specific operator's experience, then (hopefully) only a very limited amount of failure data is available. Statistics using small data sets is always problematic. Furthermore, a specific pipeline's accident experience will probably change with the operator's changing risk management focus. When an operator experiences many corrosion failures, he will presumably take actions to specifically reduce corrosion potential. Over time, a different mechanism should then become the chief failure cause. So, the weightings would need to change periodically and would always lag behind actual experience, therefore having no predictive contribution to risk management.

The bigger issue with the use of weightings is the underlying assumption that the past behavior of a large population will reliably predict the future of an individual. Even if an assumed distribution is valid for the long term population behavior, there will be many locations along a pipeline where the pre-set distribution is not representative of the particular mechanisms at work there. In fact, the weightings can fully obscure the true threat. The weighted modeling of risk may fail to highlight the most important threats when certain numerical values are kept artificially low, making them virtually unnoticeable.

The use of weightings as a significant source of inappropriate bias in risk assessment is readily demonstrated. One can easily envision numerous scenarios where, in some segments, a single failure mode should dominate the risk assessment and result in a very high probability of failure rather than only some percentage of the total.

Consider threats such as landslides, erosion, or subsidence as a class of failure mechanisms called geohazards. An assumed distribution of all failure mechanisms will almost certainly assign a very low weighting to this class since most pipelines are not significantly threatened by the phenomena and, hence, incidents are rare. For example, to match a historical record that shows 30% of pipeline incidents are caused by corrosion and 2% by geohazards, weightings might have been used to make corrosion point totals 15 times higher than geohazard point totals (assuming more points means higher risk) in an older scoring methodology.

But a geohazard phenomenon is a very localized and very significant threat for some pipelines. It will dominate all other threats in some segments. Assigning a 2% weighting masks the reality that, perhaps 90% of the failure probability on this segment is due to geohazards. So, while the assumed distribution may be valid on average, there will be locations along some pipelines where the pre-set distribution is very wrong. It would not at all be representative of the dominant failure mechanism at work there. The weightings will often completely mask the real threat at such locations.

This is a classic difficulty in moving between behaviors of statistical populations and individual behaviors. The former is often a reliable predictor—hence the success of the insurance actuarial analyses—but the latter is not.

In addition to masking location-specific failure potential, use of weightings can force only the higher weighted threats to be perceived 'drivers' of risk, at all points along all pipelines. This is rarely realistic. Risk management can become driven solely by the pre-set weightings rather than actual data and conditions along the pipelines. Forcing risk assessment results to resemble a pre-determined incident history will almost certainly create errors.

Since weightings can obscure the real risks and interfere with risk management, their use should be discontinued. Using actual measurements of risk factors avoids the incentive to apply artificial weightings (see previous column on the need for measurements). Therefore, migration away from older scoring or indexing approaches to a modern risk assessment methodology will automatically avoid the misstep of weightings.

3.7.12 Diagnosing Disconnects Between Results and 'Reality'

FOCUS POINT
A 'gut check' is a reasonable and prudent aspect of validation

PRMM provides a useful discussion of types of disconnects between reality and assessed results that may arise in a risk evaluation. Disconnects discussed there include those that may emerge from:

- New inspection results, including visual inspections
- Incident investigations, including root cause analyses
- Leak history analyses
- Populations vs individuals disconnects.

An important step in validation is to identify and correct 'disconnects' between sources such as subject matter experts' beliefs and risk assessment outputs. Two types of potential disconnects should be explored. The first is comparisons of populations—the behavior of an assessed collection of components (for example, a pipeline system) with a representative population of similar components (other pipeline systems). The representative population will be called a benchmark for purposes here. Common benchmarks include average incident rates for many km of pipelines over several years, often country-wide (for example, US, Canadian, European, etc).

The second comparison disconnect type involves a risk assessment of a component or several components whose risk estimates do not comport with SME beliefs or other evidence. Other evidence includes results of inspections not available prior to the risk assessment.

If assessment results are not consistent with a benchmark believed to closely represent future performance of the system or when a discrepancy arises in a comparison of a component- or location-specific assessment with an SME belief or other evidence, any of several things might be happening:

- Benchmark is not representative of the assessed segments
- Effects of conservatism are not being fully considered
- Both are correct (ie, within the range of expectations), but probability effects make them appear contradictory
- Exposure estimates were too high or too low,
- Mitigation effectiveness was judged too high or too low,
- Resistance to failure was judged too high or too low.
- Consequences estimates were too high or low
- SME belief or contrarian evidence is flawed.

The distinction between PoF and probability of damage (damage without failure) can be useful in diagnosing where the assessment is not reflecting perceived reality. If damages are predicted but not occurring, then the exposure is overestimated and/or the mitigation is underestimated. Alternatively, consider a situation where damage potential is modeled as being very low but an inspection (perhaps ILI) discovers certain damages. It is often difficult to determine which estimate—exposure or mitigation—was most contributory to the damage underestimate, but insight has been gained nonetheless.

Mitigation measures have several aspects that can be tuned. The orders of magnitude range established for measuring mitigation is critical to the result, as is the maximum benefit from each mitigation, and the currently judged effectiveness of each. More research is becoming available and can often be used directly in judging the effectiveness of a mitigation measure.

Note that calibration might also be contributing to such disconnects. Calibrating to a target population of pipeline segments includes ‘outliers’ in the target distribution. So, disconnects involving very few segments may be only due to the outlier effect. More widespread disconnects may indicate that the target population used in calibration is not representative of the pipeline segments being assessed.

A trial and error procedure might be required to balance all these aspects so the assessment produces credible results for all inputs.

3.7.13 Incident Investigation

Incident investigation is both a useful input into a risk assessment and a consumer of risk assessment results. In the former, learnings from the incident are almost always relevant to other portions of other pipelines. In the latter, especially when responsibility (blame) is to be assigned, what should have been known, via risk assessment, prior to the incident is almost always relevant. From this, the risk management decision-making will normally be challenged by parties having suffered damage from the incident.

Retro-fitting a risk assessment for this type of application uses the same steps as any other risk assessment. Care must be exercised to not introduce hindsight, if the assessment is to truly reflect what was/should-have-been known immediately prior to the incident.

When evaluating what should have or 'could have' been known and what should have (or 'could have') been done prior to an accident, the investigation often seeks to determine if decision-makers acted in a reasonable and prudent manner. For more extreme behavior, the legal concept of negligence may also be applicable and some investigations will seek to demonstrate that.

The risk aspect of the investigation can focus on these issues by including the following:

1. List of evidence available prior to incident. This includes information that was readily available to decision-makers prior to the incident. Less available information—determining to what extents research, data collection, investigation, etc, should have been done—is a later consideration.
2. Risk implications of this evidence. This can be demonstrated via a translation, showing how each piece of evidence is translated into a measurement of exposure, mitigation, resistance, or consequence.
3. P50 and P90+ risk assessments prior to incident, using all available information, again, prior to incident. The assessment should model uncertainty as increased risk, reflecting a prudent decision-making practice of erring on the side of over-protection.
4. Decision-making context. Here, the risk report puts the assessment results into context for the reader. This can include at least two types of context:

   Relative: how did the risk of the subject segment—the failed component—compare to other risks under the control of the risk manager, immediately prior to the incident? Should this have been a priority segment for the decision-makers? Did the failure mechanism that actually precipitated the event appear as a dominant threat? Should it have, given the information available at the time?

   Acceptability Criteria: immediately prior to the incident, would the risk from this segment have been deemed 'acceptable' by any common measure of risk acceptability? Even when numerical criteria for 'risk acceptability' or 'tolerable risk' are unavailable for a specific pipeline, inferred and comparative criteria are always available. Examples are numerous and include:

- Risk criteria used in similar applications; for example, siting of pipelines near public schools [1048].
- General industrial risk criteria used in other countries; for example, ALARP
- Land use and setback criteria suggested in some guidelines [1047] and applied in some municipalities
- Risk criteria employed in other industries
- Suggested target reliability levels. [95, 333]

Risk criteria often use fatalities as the consequence of interest. So, even if not directly applicable to the subject pipeline, the fact that a fatality-based risk level is tolerable (or not) in a similar area or for a similar application, may be relevant to the subject incident.

Care should be exercised to emphasize the probabilistic nature of a risk assessment. A risk assessment can easily fail to highlight a threat that later turns out to cause the next failure. But that does not mean that the assessment is incorrect. A 1% probability event can occur before a 90% probability event, but they may still be accurately depicted as 1% and 90% probability events, respectively. Of course, if several events assessed at 1% each happen before the 90% event, the assessment results should become increasingly suspect.

5. Mitigation options prior to the incident. A listing of all risk reduction opportunities available to decision-makers prior to the incident will be useful to the analyses. The reasonableness of each should not be a consideration at this stage—rather the focus should be on a comprehensive list.
6. Cost/benefit analyses of available mitigation prior to the incident. This addresses reasonableness and is also captured in ALARP. See Chapter 13 Risk Management. While spending to prevent consequences that are difficult to monetize (for example, fatality, threatened and endangered species harm, etc) evokes emotionalism in decision-making, there is nonetheless a concept of reasonableness in spending to prevent any type of potential loss. Monetization of all types of consequence is becoming more common. But even expressed in qualitative (non-monetized) ways, the costs of opportunities for consequence avoidance prior to the incident, will still be of use in the investigation.

3.7.14 Use of Inspection and Integrity Assessment Data

The first and primary use of inspection and integrity assessment data, including investigations from failures and damage incidents, is in determining resistance. This is detailed in Chapter 10 Resistance Modeling. A secondary, but also very important use of this information is in revisiting previous assumptions used in the risk assessment. Since this latter use permeates so many inputs into a risk assessment, this topic is explored here in an early chapter.

When inspection does not find damages where they had been predicted by the risk assessment, a common cause is conservatism in the risk estimates. However, one should not discount the possibility of damages present but undetected by the inspection. In the case of ILI, such disconnects may warrant a re-examination of factors such as:

- Assumed detection capabilities to various ILI types regarding various anomaly types and configurations.
- Assumed reductions in detection capabilities to various types of ILI excursions.

When an inspection detects corrosion or cracking damage, it is logical to conclude that damage potential existed at one time and may still exist. When there is actual damage, but risk assessment results do not indicate a significant potential for such damage, then a conflict seemingly exists between the direct and the indirect evidence. Such conflicts are discussed in Chapter 3.7 Verification, Calibration, and Validation, especially Chapter 3.7.12 Diagnosing Disconnects Between Results and 'Reality'.

Identifying the location of the inconsistency is necessary. The conflict could reflect an overly optimistic assessment of effectiveness of mitigation measures (coatings, CP, etc.) or it could reflect an underestimate of the harshness of the environment. Another possibility is that detected damages do not reflect active mechanisms but only old and now-inactive mechanisms. For instance, replacing anode beds, increasing current output from rectifiers, eliminating interferences, and re-coating are all actions that could halt previously active external corrosion. Finally, the apparent disconnect might not be a disconnect at all. It could simply be an actually very rare occurrence whose time had come. Even very low probability events will occur eventually.

The degradation estimates in a risk assessment should always include the best available inspection information. The risk assessment should preferentially use recent direct evidence over previous assumptions, until the conflicts between the two are investigated.

For example, suppose that, using information available prior to an ILI, the assessment concluded a low probability of subsurface corrosion because both coating and CP were estimated to be fully effective. If the ILI recent inspection, indicates that some external metal loss has occurred, then the subsurface corrosion assessment would be suspect, pending an investigation. The previous assessment based on indirect evidence should probably be initially overridden by the results of the ILI pending an investigation to determine the cause of the damage—how the mitigation measures may have failed and how the risk assessment failed to reflect that.

If the risk assessment is modified based upon un-verified ILI results, it can later be improved with results from more detailed examinations, that is, excavation, inspection, and verifications that anomalies are present and represent loss of resistance. If a root cause analysis of the detected damages concludes that active corrosion is not present, the original risk assessment may have been correct. The root cause analysis might demonstrate that corrosion damage is old and corrosion has been mitigated and values may have to again be revised.

A similar approach is used for integrity assessments such as pressure tests. If test results were not predicted by the risk assessment, investigation is warranted.

Techniques to assimilate ILI and other direct inspection information into risk estimates are discussed in Chapter 10 Resistance Modeling.

3.8 TYPES OF PIPELINE SYSTEMS

> FOCUS POINT
> While there are differences among pipeline system types, the similarities are numerous and allow a single method of risk assessment to be employed.

An underlying premise in this book is that only one risk assessment methodology should be used, regardless of variations in system type and components within each system--regardless as well of variations in product transported, geography, pressures, flowrates, materials, etc. This methodology should be consistently applied. This way, even the most diverse collection of system types, components, products transported, geographies, etc. can be compared and managed appropriately. Even very specialized, rare pipeline designs, such as long, encased pipe—pipe-in-pipe configurations or special materials—are efficiently assessed by the same methodology.

The following chapters of this book discuss system-specific difference when such differences require special consideration in the assessment. In the following paragraphs, facility types are discussed and some general differences among pipeline systems are highlighted. Again, this does not suggest that alternate risk assessments are required to deal with these differences. A robust risk assessment framework readily handles all differences.

Differing definitions of 'failure'—a key thing being measured in the risk assessment—for both integrity-focused risk assessment and service interruption risk assessments, may be desirable. Again, the same methodology is still efficiently applied to all asset types, components, and risk/failure definitions.

3.8.1 Background

The following definitions are offered as general discriminators of pipelines based on their differences in service. These definitions are not universally recognized. Regulatory definitions are often more specific, sometimes linking definitions to stress level or other factors. 'Product' generally refers to hydrocarbon products—oil and gas—but also generally apply to water and other substances moved by pipeline.

Conceptually, pipelined product travels from a wellhead to end consumers through a series of pipelines. These pipelines — including flowlines, gathering lines, transmission lines, distribution lines, and service lines — carry product at varying volumes, flowrates, and pressures. Related pipeline type terminology includes the following:

- Feeder lines move products from batteries, processing facilities and storage tanks to the long-distance haulers of the pipeline industry, the transmission pipelines.
- Flowlines connect to a single wellhead in a producing field. Flowlines move product from the wellhead to nearby storage tanks, transmission compressor stations, or processing plant booster stations.
- Gathering lines collect product from multiple flowlines and move it to centralized points, such as processing facilities, tanks, or marine docks.
- Distribution pipelines, also known as "mains," are the middle step between high pressure transmission lines and low pressure service lines.
- Service pipelines connect to a meter that delivers product to individual customers-the end users.

Many examples in this book are directed towards transmission pipelines. As typically the more regulated and higher stressed of the pipeline systems, risk management efforts have been very focused on these systems, especially more recently. There are many similarities between transmission and other pipeline systems, but there are also important differences from a risk standpoint. A transmission pipeline system is normally designed to transport large volumes of product over long distances to large end-users such as electrical power plants, oil refineries, chemical plants, and distribution systems. The distribution system delivers received product to numerous users in towns and cities e.g., natural gas for cooking and heating or water for multiple uses is delivered to homes and other buildings by the distribution system within a municipality. Gathering systems typically have lower pressures and volumes than transmission, are geographically constrained, and often less regulated. The similarities between transmission and other systems arise because a mostly sub-terrain, pressurized pipeline will experience common threats. All pipeline systems have similar risk influences acting on their risk profiles—changes in risk along their routes. All are vulnerable to varying degrees from external loadings, corrosion[3], fatigue, and human error. All have consequences when they fail. When the pipelines are in similar environments (buried versus aboveground, urban versus rural, etc.) and have common materials (steel, polyethylene, etc.), the similarities become even more pronounced. Similar mitigations techniques are commonly chosen to address similar threats.

3 Even plastics, concrete, and specialized metals have some exposure to corrosion, in the general use of the word.

Differences arise due to varying material types, pipe connection designs, interconnectivity of components, pressure ranges, leak tolerance, and other factors. These are considered in various aspects of a risk assessment. In this section, the focus is primarily on the differences among steel pipelines. This focus is warranted since many newer pipeline regulations differentiate among steel pipelines based on relatively minor differences in their use.

3.8.2 Materials of Construction

The wide range of materials used in pipelines is discussed in Chapter 10 Resistance Modeling. As noted, the focus in this section is primarily on the differences among steel pipelines. The history of steel in pipelines is useful background information:

> While iron pipe for other uses in the U.S. dates back to the 1830s, the use of pipe for oil transportation started soon after the drilling of the first commercial oil well in 1859 by "Colonel" Edwin Drake in Titusville, Pennsylvania.
>
> The first pipes were short and basic, to get oil from drill holes to nearby tanks or refineries. The rapid increase in demand for a useful product, in the early case kerosene, led to more wells and a greater need for transportation of the products to markets. Early transport by teamster wagon, wooden pipes, and rail rapidly led to the development of better and longer pipes and pipelines.
>
> In the 1860s as the pipeline business grew, quality control of pipe manufacture became a reality and the quality and type of metal for pipes improved from wrought iron to steel.
>
> Technology continues to make better pipes of better steel, and find better ways to install pipe in the ground, and continually analyze its condition once it is in the ground. At the same time, pipeline safety regulations become more complete, driven by better understanding of materials available and better techniques to operate and maintain pipelines.
>
> They continue to play a major role in the petroleum industry providing safe, reliable and economical transportation. As the need for more energy increases and population growth continues to get further away from supply centers, pipelines are needed to continue to bring energy to you.
>
> From the early days of wooden trenches and wooden barrels, the pipeline industry has grown and employed the latest technology in pipeline operations and maintenance. Today, the industry uses sophisticated controls and computer systems, advanced pipe materials, and corrosion prevention techniques. [1049]

3.8.3 Product Types Transported

The type of product in the pipeline impacts certain failure mechanisms as well as consequence potential. See listing of typical pipeline products and discussion of associated hazards in CoF, Chapter 11 Consequence of Failure.

3.8.4 Gathering System Pipelines

Gathering systems are normally comprised of low-capacity pipelines[4]—typically less than 8 inches in diameter—that move produced fluids from subsurface wells to high-capacity transmission pipelines. Before leaving a hydrocarbon production field, the product is often processed to remove excess water, gases, and sediments as required to meet the quality specifications of transmission pipelines and the refineries they access.

Gathering pipelines are somewhat different from transmission pipelines in design, maintenance, operations, and in the quantity and quality of the liquids they carry. Historically designed, built, and operated under less regulations, gathering systems often have more leaks than transmission pipelines. They are generally lower stress systems, often located in less populated areas so consequences are usually less than in transmission and distribution.

It is not unusual for products such as natural gas being produced and transported through a gathering network to vary in composition from one section of pipeline to another, according to the production from each well.

3.8.5 Transmission Pipelines

Transmission pipelines are typically large-capacity pipe, usually 8 inches or more in diameter and generally transporting fluids over long distances and at relatively high pressures. They typically originate at one or more inlet stations, or terminals, where custody of a product shipment is transferred from the owner (shipper) to the pipeline operator. Accordingly, inlet stations can be access points for truck, rail, and tanker vessels as well as other pipelines, including gathering lines from production areas. Along with pumping stations, storage tanks, sampling and metering facilities can be located at inlets to ensure that the hydrocarbons injected into the pipeline meet the quality control requirements of the pipeline operator and intended recipients.

3.8.6 Distribution Systems

For purposes of this discussion, a distribution pipeline system will be considered to be the piping network that delivers product from the transmission pipeline to (or 'from' in the case of sewer systems) multiple final users (i.e., the consumer) in the same geographical area. This includes the low-pressure segments that operate at pressures close to those of the customers' needs as well as the higher pressure segments that require pressure regulation to control the pressure to the customer. The most common distri-

4 With notable exceptions such as the Alaska North Slope gathering system with large diameter, high pressure systems.

bution systems transport water, wastewater[5], and natural gas, although steam, propane, and other product systems are also in use.

An easy way to picture a distribution system is as a network or grid of mains, service lines, and connections to customers. This grid can then be envisioned as overlaying (or at least having a close relationship with) the other grids of streets, sewers, electricity lines, phone lines, and other utilities.

Some operators of natural gas distribution systems have been more aggressive in applying risk management practices, specifically addressing repair-and-replace strategies for their more problematic components. These strategies incorporate many risk assessment and risk management issues, including the use of models for prioritizing replacements or assessing risk. Many of these concepts will also generally apply to water, wastewater, and any other pipeline systems operating in predominantly urban environments.

Since they are generally comprised of components of smaller volume with less pressure-containment requirements, a wider range of materials and appurtenance designs have been available to distribution systems. Many systems have evolved over many decades, with operators routinely changing from previous materials and practices in favor of better or more economical designs.

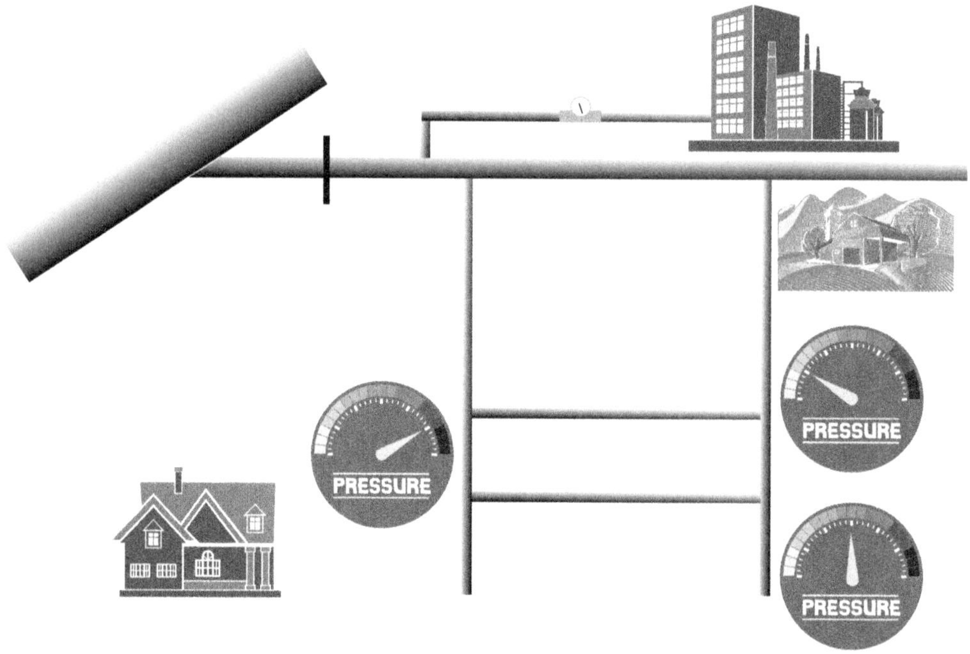

Figure 3.5 Distribution

5 Although technically a collection system, a wastewater systems shares characteristics with distribution as well as gathering systems.

3.8.6.1 Comparisons

Historical accident/incident data offer important insights into what causes pipeline failures. Municipal distribution systems, both water and gas, usually have much more documented leak data available than other pipeline systems. This is due to a higher leak tolerance in distribution systems compared to transmission and an often better (although still historically weak in most) attention to record keeping compared to gathering systems.

System characteristic data—even the basic specifications of pipe material, size, and exact locations—are, however, often less available compared to transmission pipelines. A common complaint among most distribution system operators is the incompleteness of general system data relating to material types, installation conditions, and general performance history. This situation is changing among operators, most likely driven by the increased availability and utility of computer systems to capture and maintain records as well as the growing recognition of the value of such records.

The primary differences, from a risk perspective, of and among distribution pipeline systems include:

- Materials and components
- Pressure/stress levels
- Pipe installation techniques
- Leak tolerance.

Distribution systems also differ fundamentally from transmission systems by having a much larger number of end-users or consumers, requiring specific equipment to facilitate product delivery. This equipment includes branches, meters, pressure reduction facilities, etc., along with associated piping, fittings, and valves. Curb valves are additional valves usually placed at the property line to shut off service to a building. A distribution, gas, or water main refers to a piece of pipe that has numerous branches, typically called service lines, that deliver the product to the final end-user. A main, therefore, usually carries more product at higher pressure than a service line. Where required, a service regulator often controls the pressure to the customer from the service line. In increasingly rare scenarios, customers are directly connected to long lengths of piping that are protected by common pressure control devices, rather than customer-specific control.

Although there are many overlaps, the typical operating environments of distribution systems are often materially different from that of most transmission pipeline segments. Normally located in heavily populated areas, distribution systems are generally operated at lower pressures, built from different materials, and installed under and among other infrastructure components such as roadways, Many distribution systems are older than most transmission lines and employ a myriad of design techniques and materials that were popular during various time periods. They also generally require fewer pieces of large equipment such as pumps and compressors (although water distribution systems usually require some amount of pumping). Operationally, significant differences from transmission lines include monitoring (SCADA, computer-based leak detection, etc.), right-of-way (ROW) control, inspection opportunities, and some aspects of corrosion control.

Because of the smaller pipe size and lower pressures, leak sizes are often smaller in distribution systems compared to leaks in transmission systems; however, because of the environment (e.g., in towns, cities, etc.), the consequences of distribution pipe breaks can be quite severe. Also, the number of leaks seen in distribution systems is often higher. This higher frequency is due to a number of factors that will be discussed later in this chapter.

3.8.6.2 Distribution System integrity

Pipeline system integrity is often defined differently for hydrocarbon transmission versus gathering and distribution systems. In the former, leakage of any size (beyond the microscopic, virtually undetectable amounts) is usually intolerable, so integrity normally means "leak free." The intolerance of even the smallest leak in a transmission pipeline is due to several factors, including economics of product transport and potential consequences from integrity breaches in higher-stress systems. Many distribution systems, on the other hand, tolerate some amount of leakage—system integrity is considered compromised only when leakage becomes excessive.

The higher leak tolerance leads to a greater incidence of leaks in a distribution system. These are often documented, monitored, and placed on "to be repaired" lists. Knowledge of leaks and breaks is often the main source of system integrity knowledge. It, rather than inspection information, is usually the first alert of systemic issues of corrosion of steel, graphitization of cast iron, loss of joint integrity, inferior material performance (eg, lack of brittle failure resistance in certain plastics), and other signs of system deterioration. Consequently, risk modeling in urban distribution systems has historically been more focused on leak/break history. Coupled with the inability to inspect many portions of an urban distribution system, this makes data collection for leaks and breaks even more critical to those risk management programs.

Chapter 3.7 Verification, Calibration, and Validation of this book discusses the application of leak/break data to risk assessment and risk management.

When only certain types of integrity loss are of interest, a change in definition of 'failure' is in order. By simply changing from 'loss of integrity' to something like 'significant loss of integrity', the same methodology can be applied to generate a risk assessment for the desired types of failures.

3.8.6.3 Data

Since distribution systems typically evolve over decades of design, installation, maintenance, and repair practices, they typically harbor much more variety than does any transmission system. Note that portions of many urban distribution systems were designed in the absence of any industry standards governing material selection, quality control, installation techniques, and other practices that are part of a modern pipeline design effort.

The value of record keeping was typically unrecognized in previous decades. This has resulted in large information gaps, even regarding such basic information as exact locations, material types, connector types.

3.8.7 Offshore Pipeline Systems

The often-dynamic environment of pipeline operations offshore can make risk assessment more challenging than for onshore operations. The assessment for the offshore risks follows the same approach as the assessment for onshore facilities. These same risk assessment concepts will also apply to pipeline crossings of all water bodies, including rivers, lakes, and marshes.

Some additional considerations for certain risk aspects will be necessary to account for differences between the onshore and offshore pipelines. Common differences include external forces related to bottom stability —including hydrodynamic forces (inertia, oscillations, lateral forces, debris loadings, etc.) caused by water movements, an often higher potential for pipe spans and/or partial support scenarios, and storm implications—and activities of others (anchors shipwrecks, dropped objects, etc), availability of inspection data, and potential consequences.

Risers, platforms, and all other portions of the offshore systems are readily evaluated by this same risk assessment approach.

3.8.8 Components in Close Proximity

Components in 'close proximity' include those in facilities and shared corridors. Risk assessments for facilities—from large and complex tank farms, pump stations, compressor stations, gas processing plants, etc to simple valve and meter sites—can be conducted in exactly the same way as are risk assessments for simple lengths of pipeline. Likewise, congested pipeline corridors also require no change in methodology. There are, however, some nuances that, while readily accommodated in the suggested risk assessment methodology, warrant some discussion.

Modeling of components within shared corridors and facilities presents an interesting interplay when assessing risks. Each component endangers its neighbors. Neighboring components add to both PoF and CoF. The PoF from component #1 and component #2 add to the PoF for their neighbor, component #3. Conversely, the CoF's from #1 & #2 also add to the CoF for #3. That is, if a failure in #3 damages #1 and/or #2, then the associated losses from those neighbors' damages are additive to the losses arising just from #3.

The PoF aspect is often called a successive or sympathetic reaction and is discussed in Chapter 5 Third-Party Damage. The CoF aspect is discussed in Chapter 11.7 Consequence Mitigation Measures.

3.8.8.1 Facilities/Stations

Note definition of facility or 'station' as used in this book: Facility or station refers to one or more occurrences of, and often a collection of, equipment, piping, instrumentation, and/or appurtenances at a single location, typically where at least some portion is situated above-ground (unburied) and usually situated on property controlled by the owner.

A facility can be as small as a single valve site—perhaps a simple, uninstrumented mainline block valve, in an area covering only a few square feet. A facility can also be as large as a combined tank farm, underground storage field, truck-, rail, and marine-loading facilities, major pump station, electrical substation, and all associated appurtenances, situated on a site covering many acres of land surface. In between are all sizes of meter stations, city gate stations, pump stations, compressor stations, manifolds, and many others.

Comparisons between and among facilities is often desirable in risk management. Operators often want to compare risks associated with portions of pipeline with stations or parts of stations—components within stations. This might be for reasons of general risk management, project prioritization, or to assist in design decisions such as pipeline loops versus more pump stations.

3.8.8.2 Background

As noted, pipeline systems typically have surface (above ground) facilities in addition to buried pipe and include pump and compressor stations, tank farms, truck- rail- and marine loading appurtenances, metering and valve locations. Facilities must be included in most decisions regarding risk management.

Groups of components within a station facility to be evaluated in a risk assessment might include:

- Atmospheric storage tanks (AST)
- Underground storage tanks (UST)
- Sumps
- Racks (loading and unloading, truck, rail, marine)
- Additive systems
- Piping and manifolds
- Valves
- Pumps
- Compressors
- Subsurface storage caverns.

3.8.8.3 Sectioning and Summarizing Risk

For purposes of risk summarization, the contribution from each in-station section of piping, each valve, each tank, each transfer pump, each connector, etc. is aggregated.

This allows any number of summarizations by sub-facility type, geographic location, or other grouping. For example, due to the potential increased hazard associated with the storage of large volumes of flammable liquids, one station risk summarization may consist of all components located in a bermed storage tank area, including tank components (floor, walls, roof), transfer pumping components, manifolds and other piping, safety systems, and secondary containment. This grouping would show a risk estimate reflecting the risks specific to that portion of the station. The risk evaluations for each grouping can be combined for an overall station risk summary or kept independent for comparisons with similar groupings in other stations.

In the design phase of a facility, understanding the risks of each grouping allows more strategic placements within the facility, perhaps relative to populations, roadways, and other risk-influencing features.

Segmenting a component such as pump, loading arm, compressor, etc will also be necessary, at least at a conceptual level. When a component is comprised of multiple parts and materials, failure potential is not consistent among those parts and materials. A tank bottom will be exposed to different failure mechanism severities compared to its sides and roofs. The pump casing has different resistance characteristics than does its suction piping, seals, and mechanical connectors. The most rigorous risk assessment will assess each sub-component for all possible failure mechanisms and consequences. This is not unlike PPM where each subcomponent carries its own maintenance requirements, except that many PPM's focus on the potential for equipment unavailability rather than all consequences.

When a complex component is to be treated as a single component, compromises are required, similar to a manual segmentation strategy on a long pipeline. In either case, averages or worst-case subcomponents will dictate the component's assessed values, potentially masking true risks. The loss of accuracy in a facility component will however normally be much less than the comparable loss for long pipeline segments. The conceptual segmentation will, for each failure mechanism, use the most vulnerable sub-component's characteristics to characterize the entire component. For instance, a pump seal will often govern the leak potential for the entire pump assembly, and a mechanical coupling will often dictate the external force resistance for the entire assembly (discounting instrumentation connections).

3.8.8.4 Unique Risks

While the same risk assessment methodology is appropriate for both stations/facilities and ROW pipe, the differences must be accounted for. Examples of these differences include the following aspects, more commonly found inside fence limits (ie, in facilities, especially where some form of material processing occurs):

- Materials handling and transfer. Adds risk issues associated with loading, unloading, and warehousing of materials.

- Enclosed or indoor process units. Adds risk issues associated with enclosed or partially enclosed processes since the lack of free ventilation can increase damage potential. Consideration of effective mechanical ventilation is appropriate.
- Access. Ease-of-access to the process unit by emergency personnel and equipment impacts consequence potential.
- Drainage and spill control. Adds risk factors for situations where large spills could be contained around process equipment instead of being safely drained away. Increased risk, both PoF and CoF, from sympathetic or successive reactions; for example, one failure precipitates others in nearby components.

3.8.8.5 Corridors, Shared ROW

Pipelines are often co-located in common ROW with other pipeline, electric utility lines, or other utilities. While the risk picture is impacted by these scenarios, the risk assessment methodology requires no revision.

Note the similarities in risk assessment for these compared to facilities. In both cases, a component being assessed has some incremental increase in PoF due to the PoF from nearby components. This is normally a small fraction of the neighboring component's PoF since only a fraction of its Pof events can impact the assessed components, especially when distance, earthen cover, or other barriers are involved. As another similarity, the potential consequences from the assessed component are increased by the potential consequences that could arise from neighboring components that fail due to the failure of the assessed component.

> The ideal engineer is a composite... He is not a scientist, he is not a mathematician, he is not a sociologist or a writer; but he may use the knowledge and techniques of any or all of these disciplines in solving engineering problems.
>
> N. W. Dougherty

4 DATA MANAGEMENT AND ANALYSES

Highlights

"Not everything that matters can be counted, not everything that can be counted matters"

Albert Einstein

Data Management and Analyses

SECTION THUMBNAIL
Data collection, use, and management is a critical element of pipeline risk assessment. Understanding the pipeline-specific aspects of data management is essential to an efficient risk assessment.

We begin by noting the importance of information to a risk assessment. The reliance of the risk assessment on full and complete knowledge cannot be overemphasized. While 'full and complete' information is rarely available, it is nonetheless a target.

A great deal of information is usually available in a pipeline operation. Information that can routinely be used to update the risk assessment includes

- All survey results such as pipe-to-soil voltage readings, leak surveys, patrols, depth of cover, population density, etc.
- Documentation of all repairs
- Documentation of all excavations
- Operational data including pressures and flow rates
- Results of integrity assessments
- Maintenance reports
- Updated consequence information
- Updated receptor information—new housing, high occupancy buildings, changes in population density or environmental sensitivities, etc.
- Results of root cause analyses and incident investigations
- Availability and capabilities of new technologies.

See PRMM for an introduction and background to the management, collection and sources of data typically used in a pipeline risk assessment. Additional and pipeline-specific observations are also offered.

4.1 MULTIPLE USES OF SAME INFORMATION

Information importance is even more amplified when it informs multiple elements of the risk assessment at the same time. It is often the case that individual pieces of data impact several different aspects of risk. For example, pipe wall thickness is a factor in almost all potential failure modes: It determines time to failure for a given corrosion rate, partly determines ability to survive external forces, and so on. Population density is a consequence variable as well as a third-party damage indicator (as a possible measure of potential activity). Inspection results yield evidence regarding current pipe integrity as well as possibly active failure mechanisms. A single detected defect can yield much information. It could change our beliefs about coating condition, CP effectiveness, pipe strength, overall operating safety margin, and maybe even provides new information about soil corrosivity, interference currents, third-party activity, and so on. All of this arises from a single piece of data (evidence).

Many companies now avoid the use of casings. But casings were put in place for a reason. The presence of a casing is a mitigation measure for external force damage potential, but is often seen to increase corrosion potential. The risk model should capture both of the risk implications from the presence of a casing.

Additional examples—a few among many—are shown below:

Table 4.1
Examples of Multiple Usages of Information

Information	Application in Risk Assessment
Product flowrates	corrosion, erosion, surge
AC powerlines	corrosion, impacts from falling object
ILI results	resistance (degradation mechanisms, manufacturing/construction weaknesses, etc), corrosion exposure, corrosion mitigation, crack exposure, outside force damages

4.2 SURVEYS/MAPS/RECORDS

Maps and records of older pipeline system components are not normally as complete as operators would like. Many are faced with very limited information, given the past practices of record-keeping, and are engaged in decades-long efforts to capture critical data. Modern tools and techniques are available to support these efforts. Examples of these along with their applications are discussed in PRMM. The role of this information in risk assessment is multi-faceted, as is noted throughout this book.

4.3 INFORMATION DEGRADATION

Information often has a finite useful life span. Corrosion, for example, is time-dependent, and the timing of corrosion surveys can therefore introduce uncertainty and thus risk. The age of information should therefore be a consideration in any determination based on inspections, surveys, or tests such as pressure testing, where the objective is to identify presence or progression of damages. Since conditions should not be assumed to be static (in a conservative risk assessment), these types of information become increasingly less valuable as they age.

The best way to account for inspection/test age is to account for what might have happened since that inspection/test. This is effectively a measure of information degradation since older inspections/tests with their accompanying higher opportunity for 'things' to happen over the years, will automatically be less useful to the risk assessment. This approach also appropriately shows where inspections/tests might not necessarily need frequent refreshment. That is, where not many 'things' are apt to happen, there is less incentive to repeat the inspection. The value of inspection/test is readily quantified in terms of risk reduction.

Note that this is one of the two ways that age plays a role in risk assessment. The other has to do with era of manufacture and/or construction, as discussed next.

4.4 TERMINOLOGY

As we get into specifics of data collection, let's agree on some terminology that will be useful to following discussions. Several terms might be used in manners unfamiliar to the reader. Terminology is not consistent among all risk modelers so these definitions are more for convenience in describing risk assessment steps here. These definitions mostly relate to the use of a database as a data repository, as will be the case for almost all modern risk assessments.

In common database terminology, each row of data in a table or dataset is called a *record* and each column is called a *field*. So, each record is composed of one or more fields of information and each field contains information related to each record. A collection of records and fields can be called a *database*, a data set, or a data table. Information will usually be collected and maintained in a database (a spreadsheet can be a type of database). Results of risk assessments will also normally be put into a database environment.

Structured Query Language (SQL) is a commonly used programming language for databases. SQL can be used to cull information from the database or to render information in meaningful ways, such as applying algorithmic rules to disparate pieces of data to create estimates of risk. Creating risk assessment processes using SQL can be very efficient since they are readily deployed to numerous software environments.

Geographical Information Systems (GIS) have become an essential tool for managing pipelines. They combine database functionality with geographical, or spatial, information – maps in particular. These systems can be programmed to extract and analyze spatial data according to user-defined algorithms. Typical risk applications would be identification of pipeline intersections with roads, railroads, densely populated areas, etc. More advanced uses include modeling for flowpath or dispersion distances and directions, surface flow resistance, soil penetration, and hazard zone calculations. In simple terms, the GIS draws from data that has a spatial component—connected to points on the planet. While often displayed against a map environment, the data can also be tabulated. Most engineering data related to a pipeline will be tabulated and will have a link to spatial data via a stationing system (see definition). The database housing the tabulated data is not necessarily part of the GIS software—it may be only linked. A modern GIS can interface with a variety of databases, spreadsheets, and other files that house tabulated or spatial data. A linear representation of a pipeline is usually called a *centerline.* All data about the pipeline and its surroundings are tied to the centerline via a linear referencing system.

Using SQL or its own calculating language (sometimes called scripting language), a GIS can be the engine for calculating risk estimates. Programming risk assessment calculations with SQL is an option that allows the risk assessment to draw from multiple data sources and be portable—moved to different database environments.

Each *record* in a database must have an identifier that ties it to some particular element of the system, including facilities that are a part of that system. That is to say, a unique system identifier is needed. This identifier, along with a beginning *station* and ending *station* (or beginning/ending '*measures*'), uniquely identifies a specific component or group of components on a specific pipeline system. It is important that the identifier-stationing combination does indeed locate one and only one point on the system. An alphanumeric identification system, perhaps related to the pipeline's name, geographic position, line size, or other common identifying characteristics, is sometimes used to increase the utility of the ID field.

For purposes here, *stationing* refer to a linear referencing system commonly used in land surveying and in pipeline alignment drawings. It is designed to show fixed distances from beginning points. A stationing system is designed to be unchangeable except through the use of equations that adjust for additions or deletions of lengths. Benefits of stationing as linear reference points are that the station values are persistent over time. They can reference old records based on the same stationing system. The main disadvantage is that true distances are unknown when using station values, until all station-equation adjustments are taken into account.

The term *measures* is commonly used in GIS and is also a linear referencing system. It is similar to stationing except that it represents a continuous system, free from intermediate adjustment equations or other aspects preventing a simple calculation of distance between two points on the pipeline. The continuous centerline distances required in risk assessment are usually based on measures. Unlike stationing, measures are dynamic. When a pipeline is modified—ie, pieces added or removed—measures downstream of the event will change. A GIS can readily maintain both stationing and measures in order to retain references to legacy data sets as well as enjoy the benefits of a centerline free from intermediate station-equation adjustments. An *event* is the common term for a risk variable in GIS jargon. As variables in the risk assessment, events can be named using standardized labels. Several industry database design standards are available. Standardization is necessary for the coherent and consistent exchange of information with external parties such as service companies, other pipeline companies, and regulators. Attributes is the GIS term for an event's unique characteristics. Each event must have and attribute assigned, even if that attribute is assigned as 'unknown'. Some attributes can be assigned as general defaults or as a system-wide characteristic. Each event–attribute combination defines a risk characteristic for a portion of the system.

For example, for the event 'population density', an attribute, perhaps in units of 'persons/m^2', is assigned. In some cases, there will only be specific values that would be appropriately assigned. For the event 'pipe diameter', the possible attributes are the available pipe sizes. For the event 'pipe coating type', a restricted vocabulary list of possible coating types would be the bases of the attributes assigned to the event.

The better GIS applications use a *restricted vocabulary* in which terms are pre-defined, and only those terms may be used. This avoids variation or inconsistent labeling of the same thing. SW for seam-weld, and not "seam weld" or "S weld," for example.

All risk variables and their underlying sources are itemized in the *data dictionary*. The data dictionary should characterize and quantify the attributes of each event. This is the master reference document for the risk assessment, and it should identify the person who oversees the data (the "owner) as well as all other relevant records-management details, or *metadata* such as current last revision date, frequency of updates, accuracy, etc.

Additional terms relating to data preparation are discussed next. These include *events tables*, *LUT*'s, and *point data vs continuous.*

Sidebar

Data Availability

"I don't have enough data to quantify risk"

I hear this often and have concluded that it is actually a short hand phrase reflecting two possible beliefs:

- I don't understand how to use the data I do have
- I think that quantifying risk assessment means that I need large datasets of historical event frequencies.

The truth is, you can perform a credible risk assessment even with only a very limited amount of information. If you only know a product being transported, pressure, diameter, and general location, you could make plausible estimates—very coarse, but at least reasonable.

This reminds me of a lesson learned during a court room proceeding:

Attorney to expert witness, asking a slightly off-topic question: *"Mr Expert, how often might there be a vehicle collision at this intersection each year?"*

Expert: *"I have no idea. I don't have any data for that."*

Attorney, while winking to jury: *"Ok, since you have no idea, then we can speculate that it can happen 1,000 times per year."*

Expert, surprised: *"Oh no, it wouldn't happen that often."*

Attorney: *"Ah, so you DO have 'some idea'. Ok then, let's say it happens 500 times per year."*

Expert, beginning to see the hole he has fallen into: *"Oh no, that also is way too high."*

Attorney: "How about 100 times a year?"

Expert, now somewhat apologetic: *"Well, even that is too high because... ."*

This went on until the attorney had obtained, for the court record, the expert's high and low estimates, even when the expert claimed insufficient data and knowledge to speculate. The attorney knew that it is a simple reasoning exercise to 'know' that, say 2-3 vehicle incidents every day at the same place would not be long tolerated. Even 1-2 per week would probably prompt action. This illustrates that, even in the absence of hard data, reasoning can at least bound an estimate.

Direct reasoning is often overlooked as a source of data. When it comes to probability and risk, we sometimes forget that we have a strong, physics-based understanding of real-world phenomena. Instead of using that understanding in our risk estimates, we tend to simply delegate the risk problem to the statisticians. The statisticians use event frequencies in their work so they base their estimates on historical events. They tell us 'low data'—meaning low historical event frequencies—equates to low predictive power. True enough, especially from a statistics perspective.

But we forget that we still have the underlying physics. Physics tells us how much metal loss can be tolerated before leak or rupture, how much voltage is needed to halt corrosion, how much backhoe bucket force until the pipe breaks, how much landslide a length of pipe can withstand before yielding. We can estimate the numbers needed to calculate these things—often with great accuracy. We don't have to rely on historical events to tell us how often a thing can happen. We are certainly remiss if we ignore history—it must definitely be used in our analyses whenever it is available. But we are also remiss if we ascribe too much relevance to the past or claim we are helpless without that history.

Let's discuss low data availability when we're performing a physics-based risk assessment. It is sometimes not apparent just how much info is readily available. Let's say you know something simple about the soil type—where it's rocky and where it's mostly clay. Some of the risk factors that can be strongly influenced by just this simple piece of information include:

- Potential soil moisture content, impacting corrosivity estimate
- Likelihood of past coating damages during installation
- Propensity of future coating damages to occur
- Dispersion of liquid spills—infiltration vs surface flow

- Amount of potential harm to certain receptors (for example, aquifers vs surface flow)
- Exposure to third party excavation damages
- Exposure to certain geotechnical phenomena (for example, subsidence, shrink/swell, landslide, etc)

Perhaps you can think of more. The point is that you may have more information than you first thought. In this example, a single piece of information—a simple soil characteristic; rock vs clay—has influenced seven different risk variables.

There are many other examples of how simple knowledge of surroundings leads to relevant and important risk information. This also emphasizes why dynamic segmentation—the creation of a risk profile—is essential. We would not understand changes in risk along a pipeline route if we failed to take note of changing soil conditions and integrated the implications of those changes.

The second part of the "I don't have enough data" statement emerges from beliefs about how risk can be quantified. When the underlying belief is something like "we can't quantify risk because we don't have the data", what is often implied is that databases full of incident frequencies—how often each pipeline component has failed by each failure mechanism—are needed before risk can be quantified. That's simply not correct. To quantify how often a pipeline segment will fail from a certain threat, we don't necessarily have to have numbers telling us how often similar pipelines have failed in the past from that threat. This myth is often a carryover from the old—let's say 'classical'—practice of QRA. That practice can be an almost purely statistical exercise. It relies heavily on data of past events as predictors of future events, as is standard practice in statistical analyses. While such data is helpful, it is by no means essential to risk assessment. And when it is used, it must be used carefully. The historical numbers are often not very relevant to the future—how often do conditions and reactions to previous incidents remain so static that this history can accurately predict the future?

With or without comparable data from history, the best way to predict future events is to understand and properly model the mechanisms that lead to the events. A robust risk assessment methodology forces SME's to make careful and informed estimates based on their experience and judgment. With only minimal effort, a group of SME's, in a properly facilitated meeting, can generate credible, defensible estimates of all manner of damage and failure potential along pipelines they know. From these estimates reasonable risk estimates emerge, to be confirmed or updated as actual events are tracked.

Another Aspect of Data Availability

However, let's not dismiss the bona fide 'absence of key information' scenario. It is not uncommon for an operator to have inherited a system with a genuine lack of basic data. Perhaps a gathering or distribution system, assembled over decades, with very poor records has been acquired. Even basic location and materials of construction data might be missing. This is frustrating for a prudent operator wanting to understand risk. He might also encounter resistance in moving resources towards improving the information status.

Information acquisition can be considered risk reduction, when uncertainty is modeled as increased risk. Therefore, a cost-benefit for the information collection efforts can be shown. This is of use in demonstrating the value of information collection.

Here is one approach to, over time, remedy the absence-of-information situation using risk management techniques:

- First, formalize and centralize ALL available information—collect and digitize every scrap of paper in every file cabinet and every piece of information in the minds of all the experienced personnel and all information that becomes available in the course of O&M. This means building a robust database and establishing processes to make it's upkeep a part of day-to-day O&M processes.
- Next, perform a risk assessment using all of this information plus conservative defaults to fill in the knowledge gaps. This will produce risk estimates based on both actual risk and risk driven by the conservative defaults.
- Finally, use these risk estimates to drive an information collection process. This might require that resources be initially spent specifically on filling knowledge gaps—conducting surveys, inspections, tests, etc solely to gain the information that can replace the conservative defaults and thereby reduce the 'possible' risks.

In this approach, the risk assessment itself identifies the most critical information to collect. This is an efficient and defensible strategy to tackle the 'lack of data' issue.

4.4.1 Data preparation

Useful data can come in come in variety of different forms and formats. Some data may be in paper only and will need to be digitized. Location data may be derived from varying sources, such as mileposts, fixed-point measurements, or GPS. Location identifiers from alignment sheet stationings may be inconsistent with linear measurements due to the equations used to record route changes on the alignment sheets. In these cases translation routines or some other standardization technique will need to be employed to correlate the data for accuracy. As a rule, if alignment sheets or other legacy systems are in place and in common use, establishing a translation that preserves the old stationing system is worthwhile. When the older systems are not in common use, they can be replaced by newer, GPS-GIS based formats.

4.4.2 Events Table(s)

Much of the topic of data management will be subject to personal preferences. There will usually be several ways to accomplish the same result. However, experience has shown that one particular data collection format, often overlooked by even more advanced practitioners, has proven to be unexpectedly useful in data preparation, diagnostics, and risk management. This tool is referred to as an 'events table' in our discussion here. It is simply a complete listing of all data along the pipeline, with only 4 essential columns—the pipeline ID, the beginning station or measure, the ending point, the event (diameter, soil type, depth cover, population density, etc) and then the value assigned to the event between the begin and end points. See PRMM for additional explanation of how an event table is constructed.

An events table is a very useful tool for diagnostics and for QA/QC, and perhaps even for directly maintaining the data. It is often the most easily-researched source of information regarding changes along a route. In answering the inevitable questions of 'what makes the risk change at location x?', the events table is easily filtered to show all data inputs associated with location x. While other drill downs are possible, this is often the quickest method to determine fundamental reasons for changing risk estimates.

The events table also proves useful in summarizing the ranges of all data inputs as well as changes to input data over time. The table can readily show, for instance, that in the prior period's assessment, 21 casings had been identified and now there are 23; or previous soil corrosion estimates ranged to a high of 19.5 mpy and now the maximum is 21.1 mpy. As part of QA/QC of input data, such changes should be understood and defensible, so identifying them efficiently is important.

Data events should determine segmentation based on a dynamic segmentation procedure as described later. Therefore, the events table is the input into the dynamic segmentation process.

4.4.3 Look Up Tables (LUT)

A modern risk assessment requires the assignment of a numerical value to each input that is to be included in an algorithm. For example, the event 'coating type' with attributes such as FBE, coal tar, asphalt, tape, etc, is not usable in a calculation until some value is assigned to each attribute type. Qualitative descriptors are often 'translated' into the numbers needed. It is also useful to preserve the descriptive value of the attribute (FBE, tape, etc). When conversions from a descriptor to a numerical value will be routinely needed, a cross reference matrix, called a look up table (LUT) here, is a convenient tool.

Some examples of LUT's include:

- Assigning detection capabilities to various ILI types.
- Assigning reduced detection capabilities to various types of ILI excursions (from ideal inspection conditions).

- Assigning probability of manufacturing defects to various combinations of manufacture date and pipe mill.
- Converting a USGS landslide or flood ranking category into a event frequency.

Spreadsheet and database software programs provide tools to efficiently use LUT's. The LUT is accessed during the calculation routines to obtain the numerical equivalents to the qualitative terms.

As one of their benefits, LUT's provide a simple means to document, preserve, and maintain the relationships between qualitative and quantitative interpretations. If changes are needed, they can be done in the LUT and will then be used in all subsequent calculations. A revision log should be used to track changes to any LUT since alternations will often have far-ranging implications.

4.4.4 Point events and continuous data

All data used in the risk assessment needs to have a dimension of length—a 'from' and 'to' along the pipeline. Some data will not always have this dimension, at least initially. Examples include overline surveys for soil resistivity, depth of cover, and many others, as well as calculated values at points along the pipeline such as for pressure profile, drain down volumes, and others. In these instances, the length dimensions will need to be added. 'Rules' such as 'half the distance between points' or fixed lengths either side of a data point are common ways to assign length. See detailed discussion in PRMM. See also a related discussion on 'eliminating unnecessary segments', in the following section.

4.4.5 Data quality/uncertainty

Additional data preparation issues are discussed in PRMM, including:

- Creating categories of measurements
- Assigning zones of influence
- Countable events
- Spatial analyses
- Data quality/uncertainty.

For a discussion of QA/QC as it applies to data collection and preparation for pipeline risk assessment see PRMM.

See also the general discussions of uncertainty and conservatism in assigning defaults in Chapter 2 Definitions and Concepts and throughout this book.

4.5 SEGMENTATION

Since data collection and segmentation go hand-in-hand, it is appropriate to detail the concepts of segmentation here, in the midst of the discussion on data management.

The conditions along a pipeline route are variable – the hazard potential is not constant – and for this reason a pipeline's risk must be evaluated by examinations of individual components' risks.

A mechanism is required to document the changes along a pipeline and assess their combined impact on failure probability and consequence. Lengths of pipeline (or other components) with similar characteristics are identified and assessed. A new segment is created when any risk condition changes, so each pipeline segment has a set of conditions unique from its immediate neighbors. A segment is not necessarily unique within the population of segments—only different from each of its adjacent neighbors.

Each segment will receive its own risk estimate, based on its conditions and characteristics. Therefore, segmentation plays a critical role in risk assessment. Segmentation supports the creation of profiles—a critical element of risk management, as described in Chapter 12.2 Segmentation and Chapter 2.17 Risk Profiles.

The risk evaluator must decide on a strategy for creating these sections in order to obtain an accurate risk picture. Breaking the line into many short sections increases the accuracy of the assessment. Longer sections, created by ignoring changes in risk, reduce accuracy because average or worst case characteristics must be used to approximate the changing conditions within the section, rather than assessing the actual changes within the section.

Historically, the creation of shorter segments to gain accuracy sometimes resulted in higher costs of data collection, handling, and maintenance. This is no longer the case. Especially with modern computing environments, a dynamic segmentation approach, as described later, is both more accurate and usually more efficient.

4.5.1 Segmentation Strategies

Segmentation is a key part of pipeline risk assessment. Three segmentation strategies have historically been used in pipeline risk assessment: fixed-length, manual, and dynamic segmentation. Only the last, dynamic segmentation, is appropriate for a modern risk assessment. The others are noted here, for perspective, but produce inappropriate section breaks leading to often serious weaknesses in a risk assessment.

Inappropriate section break points limit the model's usefulness and hide risk hot spots if conditions are averaged in the section, or risks will be exaggerated if worst case conditions are used for the entire length. It will also interfere with an otherwise efficient ability of the risk model to identify risk mitigation projects.

If long segments are artificially created, then each pipeline segment would usually have non-uniform characteristics. For example, the pipe wall thickness, soil type, depth of cover, and population density might all change within a segment. If the segment was evaluated as a single entity, the non-uniformity had to be eliminated. This was typically done by using the average or worst case condition within the segment. This obscured actual risks. This significantly weakens the assessment. As an example, consider a 1,000 ft segment to be assessed with one 100 ft cased crossing within. Under a older segmentation strategy, the assessment must assume either all 1,000 ft is cased or all is uncased. Either is incorrect. The reality is that 90% of this segment is uncased and 10% is cased and the only way to fully assess the situation is to treat the uncased differently from the cased.

4.5.1.1 Fixed-length approach

In the first of the three historical segmentation approaches, an artifact of old risk assessment practice, some predetermined length such as 1 mile or 1000ft or even 1 ft is chosen as the length of pipeline that will be evaluated as a single entity. A new pipeline segment will be created at these lengths regardless of the pipeline characteristics. A fixed-length method of sectioning also included lengths based on rules such as "between pump stations" or "between block valves". This was a popular method in the past and is sometimes proposed even today. While such an approach may be initially appealing (perhaps for reasons of consistency with existing accounting systems or corporate naming conventions), it will reduce accuracy and increase costs in risk assessment.

Attempts to avoid errors inherent to this approach by using short, but still fixed, lengths also resulted in inefficiencies, albeit less serious than inaccuracies produced when using longer lengths. If a shorter segment length was used, then processing inefficiencies resulted, with commercial software packages requiring days of continuous processing time to perform risk estimates even for relatively few miles of pipeline. The analyses had to deal with many unnecessary segments based on an arbitrarily chosen short segment length selected, for example, 1 ft, while still requiring averaging or worst-case compromises when even shorter features, such as ILI-detected anomalies, were present.

4.5.1.2 Manually establishing sections

Another previous approach, now also outdated, involved using a pre-determined list of criteria by which to create segments. Modern computational power has eliminated the need to segment the pipeline manually, but a look at the process is useful in understanding of the need for the superior technique that has replaced it.

In a manual segmentation, the risk evaluator would choose factors that he thinks are most impactful on risk in the pipeline system being studied and rank those items with regard to magnitude of change and frequency of change. This ranking would be subjective and incomplete, but it could serve as a basis for sectioning the pipeline(s).

Sections were then divided based on their priority rank of risk factors beginning from the top of the list. The resulting number of sections may have become too large; however, in which case the number of factors on the list was reduced by eliminating some of the low-ranking factors until a cost-effective sectioning—accommodating the computing power of the time—had been achieved.

See PRMM for an example manual segmentation.

4.5.1.3 Dynamic segmentation approach

The third strategy is the most robust approach while also being the most efficient. The modern segmentation strategy, and the only really correct approach, is dynamic segmentation. The idea is for each pipeline section to be unique, from a risk perspective, from its neighbors. When any characteristic changes, a new segment is created. This ensures that every risk variable, and only the risk variables themselves, determine segment breaks.

Since the risk variables measure unique conditions along the pipeline they can be visualized as bands of overlapping information. Under dynamic segmentation, a new segment is created every time any condition or characteristic changes, so each pipeline segment has a set of conditions unique from its neighbors. The data determines the number and location of segment breaks. The length of a segment depends on frequency of condition change: segments where variables change frequently may be an inch or less; segments with relatively constant conditions may be hundreds of feet in length.

Segments created with a dynamic segmentation process are iso-risk, ie, as far as all collected data and knowledge can determine, there are no changes in risk along a segment's length. So, within a pipeline section, we recognize no differences in risk, from beginning to end. Each foot of pipe is the same as any other foot, as far as we know from our data. Should changes be later identified, then the segment should be further subdivided.

We also know that the neighboring sections do differ in at least one risk variable. It might be a change in pipe specification (wall thickness, diameter, etc.), soil conditions (pH, moisture, etc.), population, or any of dozens of other risk variables, but at least one aspect is different from section to section.

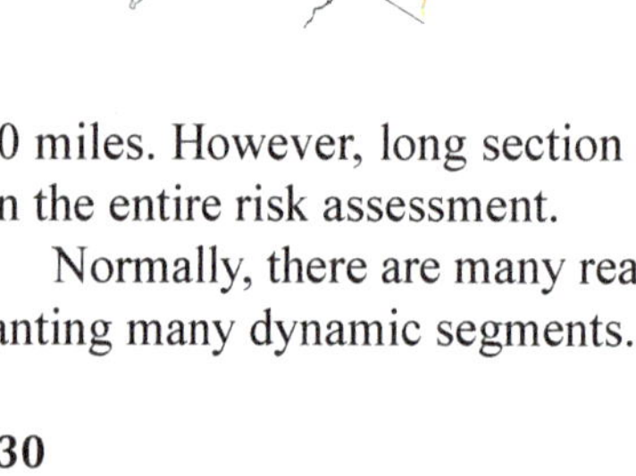

For some aspects of a risk assessment, conditions will remain constant for long stretches, prompting no new section breaks. Aspects such as training or procedures are generally applied uniformly across the entire pipeline system or at least within a single operations area. Section length is not important as long as characteristics remain constant. There is no reason to subdivide a 10-mile section of pipe if no real risk changes occur within those 10 miles. However, long section lengths suggests incomplete data and casts suspicion on the entire risk assessment.

Normally, there are many real and significant changes along a pipeline route, warranting many dynamic segments.

For purposes of risk assessment, dividing the pipeline into segments based on any criteria set other than ***all risk variables*** will lead to inefficiencies in risk assessment. Use of any segmentation strategy other than full dynamic segmentation compromises the assessment.

A computer routine can replace a rather tedious manual method of creating segments under a dynamic segmentation strategy. Related issues such as persistence of segments and cumulative risks are also more efficiently handled with software routines. A software program to be used in risk assessment should be evaluated for its handling of these aspects. Modern GIS software typically has this type of functionality built in. Alternatively, simple programming code performs this task in a variety of software environments.

4.5.2 Eliminating unnecessary segments

PRMM notes instances where data, collected at regular intervals (for example, pipe-to-soil voltages in a close interval survey, pressure changes every 100 ft, soil resistivity readings, depth of cover, etc), have changes that are insignificant from a risk standpoint. Capturing every minor change as a new dynamic segment is not necessary and leads to inefficiency. A useful 'rule of thumb' for when a minor change can be ignored is:

> If an SME would not be interested in the minor difference between two measurements, then the risk assessment probably also should not react to the difference. Therefore, the data should be grouped or categorized to minimize unnecessary segment breaks.

For instance, typical pipe-to-soil voltage readings (a measure of CP performance) measurements such as 0.879, 0.882, and 0.875, could fall into a category of "0.850 to 0.900" and only values falling into categories outside of this range, warrant special attention. This does not eliminate all unnecessary segments, since values very close to boundaries of categories are arguably also not requiring discrimination. Nonetheless, such 'bucketizing' of values can improve data processing efficiency.

4.5.3 Auditing Support

> RULE OF THUMB:
> - Without dynamic segmentation, accuracy is compromised.
> - Dozens to thousands of segments per kilometer should be expected in a modern risk assessment.

Statistics on segment length are also useful auditing tools. As previously noted, long average lengths or maximum lengths of segments are suspicious. A pipeline in a natural environment would logically have conditions changing regularly along its length solely from changes in its surroundings—soil types, creek crossings, elevation changes, road crossings, population density changes, etc. Additional changes due to design specifications, hydraulic profile, installation specifics, and others, suggest that at least dozens of segments per kilometer would be expected for most pipelines. It is not unusual for a modern assessment to generate thousands of segments per kilometer when detailed inspection data such as from ILI is available. A high segment count should not be worrisome. It results in increased accuracy, normally without increased data or modeling costs. It should also not be viewed as excessive. After all, it is actually only a few millimeters of pipeline component that actually fails in most incidents, sometimes a few meters, when the failure forces are exceptional. When inspection data identifies a few millimeters of possible weakness, such as a metal loss feature, that information should be integrated into the risk assessment.

4.5.4 Segmentation of Facilities

Facilities also require segmentation in order to fully assess risk. Geographical or functional groupings (for example, tank batteries, pump houses, manifold area, truck loading area, injection facility, etc.) are commonly used for aggregation of risk results. However, individual components and even sub-components will still require risk assessments. For example, a pump can fail in a variety of ways, involving its casing, impeller, flanges, shaft, and any other component. Which subcomponent failed and the manner in which it failed may have a significant impact on the subsequent consequences of the pump failure. A full understanding of risk requires knowledge of pump failure potential which requires at least cursory attention to the failure potential of each sub-component of the pump.

4.5.5 Segmentation for Service Interruption Risk Assessment

When failure is defined as service interruption risk, some new dynamic segmentation considerations appear. Consistent with all risk assessments, the data collected to assess the risk will also inform the dynamic segmentation. However, since this expanded definition of 'failure' can make the risk assessment considerably larger and more complex, some segmentation shortcuts such as grouping leak/rupture PoF values, might be appropriate. See Chapter 12 Service Interruption Risk.

4.5.6 Sectioning/Segmentation of Distribution Systems

Dynamic segmentation is the preferred approach for assessing all types of pipeline systems including distribution systems and other networked components.

Due to sometimes weak data availability for older pipeline systems, it may not be practical to identify and assess each component, at least not for an initial risk assessment. Since dynamic segmentation is based on location-specific data, temporary alternative segmentation strategies might be needed, pending more data availability. This is especially true for older gathering and distribution systems.

As work-arounds to lack of location-specific information, screening approaches have historically been used to focus resources on portions of the system believed more likely to harbor higher risk. Therefore, areas with a history of leaks, materials more prone to leaks, and areas with higher population densities often already have more resources directed toward them.

Such screening approaches should not be considered to be complete risk assessment foundations. They are based on an initial bias—the pre-determined list of perceived priority risk elements—and will often miss important, but rare and non-obvious failure and consequence potential. A detailed, location-specific risk assessment can identify subtle interactions between many risk variables that will often point to areas that would not have otherwise been noticed as being higher risk. High level screening approaches should be thought of as only intermediate steps, sometimes required pending more data availability, towards the full risk assessment. Some of the possible, interim segmentation strategies such as a non-contiguous, characteristic-based or a geographical segmentation strategy are discussed in PRMM.

4.5.7 Persistence of segments

Under a dynamic segmentation strategy, segments are subject to change with each change of data. This results in the best risk assessments, and does not interfere with tracking changes in risk over time. The risk associated with any stretch of pipeline can always be determined and compared with previous estimates. The user simply picks the 'from' and 'to' boundaries of the section of interest and then obtains the total risk, the total PoF, the maximum CoF, or any other aspect of interest. This involves a summarization or roll up of the dynamic segments that make up the section of interest.

4.6 RESULTS ROLL-UPS

SECTION THUMBNAIL

Significant error potential accompanies improper aggregation of risk estimates (for example, calculating valve-to-valve risk). Decision-making will be flawed if results include masking of extremes and/or insufficient consideration of non-extremes.

Having employed the modern dynamic segmentation approach, the risk assessment is ready to produce estimates of risk at many specific locations along the pipeline. However, any stretch of pipeline can now also be represented by summary risk values. The risk details—sometimes hundreds of segments per mile—will need to be summarized for many risk management activities. Valve-to-valve, trap-to-trap, accounting-based sections, and any other segmentation scheme, can be readily applied to the full risk assessment results in order to produce summary values for many management purposes. See Chapter 2.8 Probability of Failure and Chapter 13.8.2 Profiling.

It is common practice to report risk results in terms of fixed lengths such as "per mile" or "between valve stations," after a dynamic segmentation protocol has been applied. This "rolling up" of risk assessment results is necessary for summarization, reporting, establishing risk management strategies, and perhaps linking to other administrative systems such as accounting or geographic responsibility boundaries.

Summarizations of risks, if not done properly, can be very misleading. Many summarizing strategies will mask important information. Masking occurs when the important details of a collection of numbers is hidden by a summary value that purports to characterize that collection. Several masking scenarios are possible. One simple example is a short section of pipe with an extraordinarily high PoF—perhaps in a landslide zone or a location of CP interference causing corrosion. This problematic segment will often be masked in the summation of the other segments. Viewing a single value purporting to represent the risk of the entire length of pipe (collection of pipe segments) will not reveal to the observer the presence of the extraordinarily high PoF of the short segment unless an aggregation strategy is designed to avoid the masking.

It can be tempting to use an average risk value to summarize. This will clearly mask higher risk portions when most portions are lower risk. Length-weighted averages will also be misleading. A very short, but very risky stretch of pipe is still of concern, but the length-weighting masks this.

For example, the risk per mile of a 10 feet long component might be much higher than the risk per mile for any other segment. Since it is only 10 feet long, it's contribution to overall risk is perhaps tolerable. But it is important to know that a high rate of risk is indeed being tolerated.

It may also be tempting to employ a 'weakest link in the chain' analogy and simply choose the maximum risk segment to represent the risk for the entire collection of segments. As a sole method of aggregation, this is not satisfactory strategy. Examples of difficulties include:

> Seg A max = Seg B max but Seg A has only 1% of its length showing that high risk while Seg B has 80% of its length showing 'high risk'.
>
> Seg A max = Seg B max and each have the same length with the higher risk, but the rest of Seg A is only 1% better while the rest of Seg B is 50% better than its 'high risk' length.

Similar difficulties arise if averages or other summary statistics are used—masking of extremes and/or insufficient consideration of non-extremes are both errors in analyses. Simple summations of risk scores from certain older risk assessment methodologies are especially unsatisfactory since they often do not consider lengths of individual segments.

A system of calculating cumulative risk that will avoid all masking, all under-reporting, and over-reporting of risk, is needed. That system is simply an aggregation of all of the underlying segments comprising the section of interest. The aggregation is done by simple summation when elements are additive, such as EL and frequencies, or the application of OR gate summation when probabilities are combined, as in PoF.

See also the discussion of Cumulative Risk in Chapter 2 Definitions and Concepts.

4.7 LENGTH INFLUENCES ON RISK

For long, linear systems like pipelines, risk is sensitive to length. When all other aspects are equal, a longer pipeline segment will always show higher risk than a shorter.

The total risk generated by a segment uses the actual length. That is important to risk management decisions. However, the rate of risk—risk per unit length—is also important to decision-makers. It is important to understand that Segment A is higher risk than Segment B because Segment A is longer. Subtly different is the critical understanding that Segment B may be less risky ONLY because it is shorter; for example, Segment B actually has a higher risk-per-unit length (for example, risk per km), but its short length makes its total risk low.

The segment with the highest risk value will often not be the same pipe segment when reported on a unitized basis versus a length basis. The riskiest length of pipe in the system is not necessarily the segment with the highest rate of risk, ie, risk per foot. It may actually have very low risk per foot, but simply be longer than other segments.

For example, the risk per mile of a 10 feet long component might be much higher than the risk per mile for any other segment. Since it is only 10 feet long, its contribution to overall risk is masked, unless the rate of risk is examined. As previously noted, a very short, but very risky stretch of pipe is still of concern, even if the length-weighting masks this.

This is why both the segment's risk and its risk-per-unit-length values should be reported by the risk assessment. This is also true for all of the risk sub components since decision-making will also eventually focus on each PoF individually.

CoF is an element of risk that is not pipe length sensitive. CoF in 'per incident' units (for example, $/incident, fatalities/incident, etc) makes CoF a length independent measurement. The maximum CoF in a collection of segments (ie, a stretch of pipeline) will be of interest since it shows the worst consequences that could occur (to a certain PXX) in that collection. It may also be of interest to know when a system has a higher proportion or more overall length of higher CoF values than a system with lower CoF's and/or less length of high CoF. In this case, a length-weighted average CoF, used to supplement the maximum CoF, is meaningful.

4.8 ASSIGNING DEFAULTS

Any gaps in information must be filled prior to calculating risk values. Typical gaps could be lack of information regarding the depth of cover or coating condition on an older pipeline. To fill the knowledge gaps, the risk assessor must select a default input that is consistent with the desired level of conservatism of the assessment. Each event along the pipeline must have an assigned attribute – a value must be provided for the missing data. This is often most efficiently done in two steps. In the first, values are assigned based on SME knowledge of a specific region or system characteristics. For example, hurricane damage potential in Aspen, Colorado, US can confidently be assigned very low probabilities by SME's, as can frost heave phenomena in the islands of the Caribbean. In the second phase, values are assigned in the absence of any available SME information. For instance, until an SME is able to say that landslides will not happen along a stretch of pipeline, then a very conservative default—perhaps 1 to 10 landslides per year for every mile of pipe—should be assigned as an exposure in a conservative risk assessment. After all, if no SME can say such numbers are not possible, then the assessment, especially the P90+ assessments, must assume that they are plausible.

This two-step approach completes a hierarchy of data input into the assessment, as shown by the following list:

1. Location-specific data measurements.
2. Location-specific data estimates.
3. Values assigned to general areas by SME's.
4. Conservative defaults assigned when no other info is available.

These are in order of progressive uncertainty, with defaults carrying the highest level. Defaults are the values that are to be assigned in the absence of any other information. There are implications in the choice of default values and an overall risk assessment default philosophy should be established.

It is not possible to assign a default to all variables: pipe diameter and type of product are examples. Here, the missing data should lead to a non-assessed segment.

All defaults should be contained in one list. This makes the process of retrieving comparing, modifying and maintaining the default assignments simpler. Note that assignment of values might be governed by rules also. These rules can infer the default from some associated information. Conditional statements ("if X is true, then Y") are especially useful. For example, the numerical equivalents of statements such as these may be used to assign values when direct information is unavailable:

If (land-use type) = "residential high" then (population density) = 22 persons/acre

If (pipe date) < 1970 AND if (seam type) = "ERW" OR "unknown" then
(pipe manufacture) = "LF ERW"[1]

Other special equations by which defaults will be assigned may also be desired.

When event frequencies are to be assigned by default for events that have never occurred, a useful exercise may be to quantify the intuitive 'test of time' aspect. See Chapter 2.8.6 The Test of Time Estimation of Exposure That is, if x miles of pipeline have existed for y number of years and the subject event has never occurred, this is useful evidence. Absent any other information, it can be assumed that if the event were to occur now, the historical rate thus created represents a useful predictive rate, at some PXX level of conservatism.

For example, an evaluation team wishes a quick, initial risk assessment and seeks the frequency of ground subsidence events along a pipeline. They believe that the land above their 200 miles of pipeline in this area has never shown any indication of land subsidence in the 20 years the pipeline has existed. Were subsidence to occur somewhere along the pipelines now, the frequency of occurrence could be estimated to be 1 event per 200 miles x 20 years = 0.0025 events/mile-year. Pending the acquisition of better information—perhaps via soils analyses and geotechnical calculations—the team chooses to use this value for their P70 estimate in this initial risk assessment. Given that other threats to system integrity may have estimates that far surpass this value, it may be that additional analyses to produce a better estimate is never warranted. The team could decide that this rough estimate alone is sufficient, unless some future evidence emerges suggesting the need for a better evaluation. This, in itself, is another exercise in risk management—choosing where resources are best applied.

Conservatism in assigning defaults will be appropriate in most risk assessments. A danger in assigning non-conservative values is that they are no longer noticed by risk managers. They are discovered to be non-conservative once an incident happens. At that point, many outside parties will legitimately question the value of an assessment that does not cause gaps in knowledge to be highlighted (ie, via use of conservatism). Credibility will have been lost in addition to the missed opportunity to better manage the risk.

Adhering to a practice of conservatism in defaults requires discipline. It is sometimes difficult to, for instance, use a default of 18" or 24" of cover for all portions of a pipeline that was installed with 36" of cover just 5 years ago. However, with a real chance that some short section has indeed lost cover, the default value reflects real uncertainty, perhaps prompting a depth of cover survey to verify the more likely 36" depth everywhere.

1 Reference to 'low frequency ERW pipe manufacture', historically more problematic than most other pipe types

Sidebar

There are two ways to be wrong when assigning a default in the absence of information:

Call it 'good' when its really 'bad'
Call it 'bad' when its really 'good'

The first is the more expensive of the two possible errors. It masks the fact that something might be wrong and causes the whole risk assessment to lose credibility when its seen to have assumed that everything is 'ok'. The second error prompts investigation which, arguably may misdirect resources occasionally, but reducing uncertainty is more often a valuable exercise.

4.8.1 Quality assurance and quality control

For a discussion of QA/QC as it applies to data collection and preparation for pipeline risk assessment, see PRMM

4.9 DATA ANALYSIS

Much has been written about analyses of numerical data, including the roles of statistics and visualization tools (for example, charts and graphs). For a discussion of data analyses opportunities specific to pipeline risk assessment and risk management, see PRMM and also texts covering more general data analyses options.

5 THIRD-PARTY DAMAGE

Highlights

Third-party interference is the most common cause of pipeline failures on land.

Third-party Damage

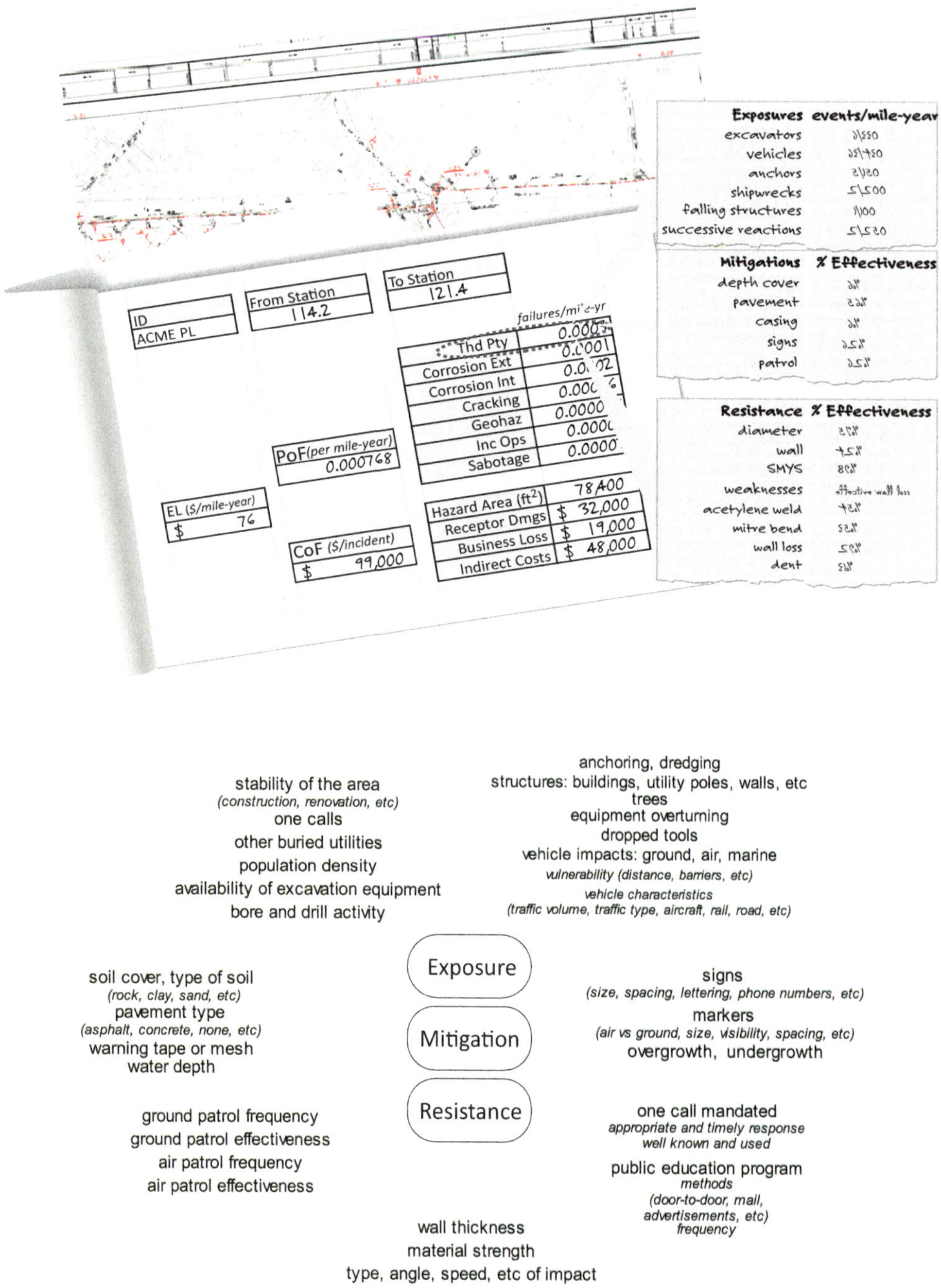

Figure 5.1 Assessing third-party damage potential: sample of data typically used

5.1 BACKGROUND

Much attention has been directed towards preventing third-party damages to pipelines in many industrialized countries. Nonetheless, recent experience shows that this remains a major threat in many places, despite often mandatory protective measures such as one-call systems.

The US pipeline regulator reports that third-party interference is the most common cause of pipeline failures on land, accounting for 20 to 40 percent of failures within most time periods as well as most of the casualties and pollution. [71]

The majority of offshore pipeline accidents are not caused by third-party damages, but this failure mechanism seems to result in more of the deaths, injuries, damages, and pollution [71]. Consequently, this is a critical aspect of the risk picture for offshore facilities also.

See PRMM for a discussion of underlying causes of third-party damage.

Many do not realize the susceptibility of apparently-strong pipeline component to eventual failure from even minor contacts. A simple scratch on the pipeline can, over time, be as serious as an actual puncture, damaging the coating, accelerating corrosion and/or cracking, and leading to eventual failure. A deep-enough scratch can set up a stress concentration area that at some future point could cause failure from fatigue or a combination of fatigue and corrosion-induced cracking.

While a pipeline operator understands the dangers posed by any interference, some contractors and the general public may not. Communication with any and all parties who may need to excavate will increase safety. Hence, the mitigative benefits of public education.

5.2 ASSESSING THIRD-PARTY DAMAGE POTENTIAL

Third-party damage, as the term is used here, refers to any accidental damage done to a component as a result of activities of personnel not directly associated with the pipeline (ie, not as employees or contractors). This failure mechanism is also sometimes called outside force, mechanical damage, or external force, but those descriptions would presumably include damaging earth movements, water impingement, and others. Third-party damage is chosen as the descriptor here to focus the analyses more[1] on damage caused by people not associated with the pipeline. Potential earth movement damage and impacts not directly related to human action (but often indirectly related) are addressed elsewhere in the assessment. Intentional damages are covered in the sabotage assessment.

1 But not exclusively. It may be more efficient to include, for instance falling trees along with falling utility poles in the same part of the risk assessment, as well as dropped tools and toppled equipment caused by first and second parties.

Accidental damages done by pipeline personnel—first- and second-party damages, not third parties—could be covered either here or alternatively in the incorrect operations assessment. Including first- and second-party impact potential here, rather in human error event estimation is usually more intuitive. Since mitigations are often the same for many excavation and vehicle impact exposures, regardless of who is atop the equipment. This is often important for exposures inside facilities/stations where third party activities are improbable but first- and second-party activity levels are generally higher.

Accidental damage includes impacts on unburied components. The argument can be made that aboveground components enjoy the benefit of being visible, thereby avoiding damages (reducing risk[2]) caused by not knowing exactly where the pipeline is (as is often the case for buried sections) and having less threat from corrosion. The opposite would be true for an above ground component in an environment with higher impact damage threats and less corrosive soil environment (ie, this threat trade-off results in an increased risk).

5.2.1 Pairings of Specific Exposures with Mitigations

Although an often-justifiable short cut in risk modeling is to collect many types of exposures and pair them with a single collection of mitigations, it is more correct to pair specific exposures with pertinent mitigations. Where differing exposure-specific mitigations are employed and/or where mitigations have varying effectiveness depending on the type of exposure, pairings of specific exposures with pertinent mitigations will be essential, as previously noted.

For example, the following exposures are often paired with customized mitigations to better reflect real-world threats:

- Excavation—agriculture; treated differently from other types of excavation, ie, shallower but more frequent exposure events from agricultural activities.
- Excavation—construction; often characterized by deeper, infrequent events
- Impacts—vehicles; impacts are different forces than excavation damages in many key aspects; vehicle impact is different from falling objects
- Impacts—falling objects; discrimination among types of objects—trees, buildings, anchors, etc—is often appropriate.

The mitigative benefit of depth of cover, public education, barriers, and others is different for each of these. Other pairings may be equally appropriate. For instance, drilling and boring excavations are materially different from many other types of ex-

2 However, ref [67] reports that, due mainly to the greater chance of impact and increased exposure to the elements, equipment located above ground has a risk of failure approximately 100 times greater than for facilities underground [67]. This will, of course, be situation specific.

cavation and may warrant independent treatment in the risk assessment. See further discussion under mitigation.

5.3 EXPOSURE

In measuring or estimating third party damage exposure, it is important to first list all potential damaging activities and events that could occur at the subject location. Then, numerical frequency-of occurrence values should be assigned to each event. Pre-dismissal of threats should be avoided—the risk assessment will show, via low PoF values, where threats are insignificant. It will also serve as documentation that all threats are considered. A frequency of zero or nearly zero can be assigned to extremely remote exposures. For instance, the exposure from falling trees where no trees are present, is obviously zero. Recording this 'zero' value demonstrates completeness in assessing exposure.

The exposure level will often change over time, but is usually relatively unchangeable by the pipeline operator. Relocation is often the only means for the pipeline operator to change this exposure, and even then, relocation may not result in a permanent reduction in exposure.

Recall that all exposures are evaluated in the absence of mitigation. This is important since it adds clarity and completeness to the assessment. For example, the unmitigated exposure from falling trees might be estimated to be on the order of several times per year—perhaps coinciding with severe storm—wind, ice, flood, etc— frequency. It is only after adding mitigation—notably depth of cover—that the threat appears as small as most intuitively believe it is. Failure to separate the exposure from the mitigation risks an inappropriate dismissal of a threat, especially if conditions change, for example, the pipeline is re-located to an above-ground location under large trees.

It is important to maintain a discipline of assessing exposure separately from mitigation and resistance, avoiding any temptation to short-cut the assessment to a perceived outcome that may not adequately reflect true risk. The 'unprotected beverage can' analogy puts the proper perspective to the exercise of producing the exposure estimates.

Recall also the discussion of mitigation by others. If additional speed control is initiated on a roadway, that action is better modeled as a reduction in exposure or rather than an addition to mitigation. It is generally more efficient in a risk assessment to establish a protocol whereby mitigative actions taken by the pipeline owner are modeled as mitigation while mitigative actions taken by others are modeled as reduced exposures.

Recall the early discussion of nuances of exposure, mitigation, and resistance estimation. Potential damages to the load-carrying capability of the component are exposures while damages to coatings are modeled as reductions in corrosion mitigation. Recall also that an exposure is defined as an event which, in the absence of any mitigation, can reduce the load-carrying capacity. Under this definition, even a minor scratch or gouge is damage since, if a stress concentrator arises, the ability of the component to carry long term fatigue loadings may be reduced. An exposure, therefore, is an activity that when unmitigated would result in a damage that causes a reduction in load-carrying capacity—both immediate and long-term—of the component.

5.3.1 Area of Opportunity

Implicit in a probabilistic risk assessment is the concept of '*area of opportunity*'. Third-party damage potential increases as the area of opportunity for accidental contact increases. The area of opportunity is strongly affected by the level of activity near the pipeline. More activity near a component logically increases the opportunity for a strike. The lowest exposure is associated with scenarios where there is virtually no chance of any digging or other harmful third-party activities near the line.

Population density is therefore typically a consideration in the risk assessment. More people in an area generally means more activity: fence building, gardening, water well construction, ditch digging or clearing, wall building, shed construction, landscaping, pool installations, etc. Many of these activities could disturb a buried pipeline. The disturbance could be so minor as to go unreported by the offending party. As already mentioned, such unreported disturbances as coating damage or a scratch in the pipe wall are often the initiating condition for a pipeline failure sometime in the future.

An area that is being developed or is experiencing a growth phase will often require frequent construction activities. These may include soil investigation borings, foundation construction, installation of buried utilities (telephone, water, sewer, electricity, natural gas), and a host of other potentially damaging activities. Planned or observed development is therefore a good indicator of increased activity levels. Local community land development or planning agencies might provide useful information to forecast such activity.

Excavation damage potential includes drilling/boring and impact driving operations. These are often sensitive to pipeline contacts--it is possible for the equipment operator to hit a facility without being aware of the hit. The drill bits or driver points, designed to go through rock, may experience little change in resistance when going through plastic pipe or cable and can cause much damage to steel pipelines. These are unique forms of excavation with different damage potentials compared to surface excavation. Some mitigation measures may also have differing effectiveness on this type of excavation—for example, marking/locating accuracy requirements, benefits of signs/markers, etc. there may also be other unique exposure aspects. For example, with no visibility from the surface, there will typically be fewer opportunities for a last minute intervention.

The presence of other buried utilities logically leads to more frequent digging activity as these systems are, maintained, inspected, and repaired. This increased exposure is perhaps partially offset by a presumption that utility workers are better versed in potential excavation damages than are some other industry excavators. If considered credible evidence of increased risk, the density of nearby buried utilities can be used as another variable in judging the activity level.

A high activity level nearby normally accompanies a distribution system. Often though, a more experienced group of excavators works near these systems, sometimes to the exclusion of 'amateurs'. Consider excavators working in densely populated or commercialized urban areas. These excavators are owners of or contractors to other utilities, have more experience working around buried utilities, expect to encounter more buried utilities, are often working under strict procedures and permitting systems, and are more likely to ensure that owners are notified of the activity (usually through a one-call system). Consistent use of a one-call system by local contractors can be an indication of informed excavators. Nonetheless, errors are possible. It is still often advisable to conservatively assume that more activity near a pipeline offers more opportunity for unintentional damage to a pipeline.

Other considerations include nearby rail systems and high volumes of nearby traffic, especially where heavy vehicles such as trucks or trains are prevalent or speeds are high. Aircraft traffic should also be included. Aboveground facilities and even buried components are at risk because a vehicle impact can have tremendous destructive-energy potential.

Offshore facilities, including those under streams, rivers, lakes, oceans, etc, are often exposed to damage potential from anchoring, fishing, and dredging activities, along with dropped objects. New water-crossing pipeline installations by open-cut or directional-drill methods may also pose a threat to existing facilities. Offshore dredging, shoreline fortifications, dock and harbor constructions and perhaps even offshore exploration/production drilling activities may also be a consideration. Debris movement along sea bottoms involving man-made objects, from normal current flows and especially during offshore storms have also damaged components.

Also important to some assessments is the potential for sympathetic reactions—failures in a nearby component creating forces sufficient to damage the subject component. Shared pipeline ROW's and many above-ground facilities harbor such threats.

5.3.2 Estimating Exposure

Quantifications of exposure can be done very simply or, alternatively, with a high level of associated research and calculation. When direct measurement of exposure rates are unavailable or when plausible exposure levels are not deemed significant enough to warrant the research, simple reasoning exercises can be used to assign values.

Exposure estimates involve predicting future events. Indicators of past activity can inform estimates of future exposures but with varying degrees of relevance, sometimes to the point of being a contrarian indicator. Consider a high historical frequency of vehicle 'leaving the roadway' type incidents. An abnormally high frequency will often prompt actions such as reduced speed limits or installation of barriers to reduce the exposure. During construction of a new residential area, activities will be high. But once established, activity levels in the new neighborhood may fall to below-average levels. In both examples, the high past rate portends a ***decreased*** future rate.

Nonetheless, historic rates are normally good starting points from which to quantify location-specific exposures. Even with questionable direct relevance, at least understanding the range of values that have occurred elsewhere is useful.

Various commonly-available records show past exposure levels. These records may come from public data sources, direct observation by pipeline personnel, patrols by air or ground, incident records, and telephone reports by the public or by other construction companies. The one-call systems (these are discussed in a later section), where they are being used, provide an excellent database for assessing the past level of excavation activity. Roadway and railway owners, as well as government agencies, typically keep records of vehicle incidents. Aircraft and marine vessel incident rates are also commonly available.

Note however, that all such measures of activities are only a lagging indicator; that is, they may show where past activity has occurred but not necessarily be indicative of future activity. Current and past activity are very relevant to estimates of where damages may have already occurred. Perhaps one of the best indicators of the defect introduction rate—for coatings and pipe wall—is the frequency of excavation activity reports. This is considered in the PoF assessment aspects such as coating effectiveness and resistance. For third party damage exposure, used to predict future failure potential, past and current activity levels may be less relevant as indicators of future activity.

Advance notice of pending excavation activity is especially useful for predicting exposures. Regulatory permitting for land development indicating the impending use of the area—development in progress or planned—is a potential source of information on longer term activity levels. Evidence of more immediate activities arise from pre-excavation indications such as survey markings.

5.3.3 Excavation

The quantification of the risk exposure from excavation damage requires an estimate of the number of potential excavations that present a chance for damage. Excavation occurs frequently in the United States. The excavation notification system in some states record hundreds of thousands of calls per month and millions of excavation markings per year, averaging of thousands per day in some areas [64].

As noted in PRMM, it is estimated that gas pipelines in the US are accidentally struck at the rate of 5 hits per every 1,000 one-call notifications.

An examination of historical excavation damage accidents supports the hypothesis that a higher population density means more accident potential.

Figure 5.2 Potential Excavator Damages

In 1995 the Gas Research Institute (GRI), now known as the Gas Technology Institute (GTI), conducted a useful study on excavation risk-exposure for the gas industry. Results showed rates of 58 third party strikes per 1,000 miles of all types of pipelines, with transmission pipelines receiving only 5.5 hits per 1,000 miles and distribution lines suffering 71 hits per 1,000 miles [64]. Such data from studies such as this can be used as inputs to exposure estimates. See PRMM and ref [64] for a summary on this.

Ref [9988] cites excavation rate values of 0.076 per km/year for agricultural areas and 0.52 per km-year in commercial and industrial areas. With 'typical prevention measures', these rates were thought to lead to hit rates of between 0.004 per km year for undeveloped areas and 0.05 per km-yr for developed areas. This reference also notes that 75% of the excavation impacts are by backhoes which are too small to cause 'serious damage' to the larger diameters typical of transmission pipelines. If the risk assessor concludes relevance to the components being evaluated, then this information can be useful in estimating exposure rates, mitigation (for example, see discussions on depth of cover benefit and patrol effectiveness based on excavation evidence), and resistance.

Even in fairly short distances, exposure rates can vary widely. Indicators such as new construction, repeated work on nearby facilities, anchoring and dredging areas offshore, etc can be very location-specific. Higher exposure rates (perhaps on the order of 0.1 to over 100 events/year at certain locations) and lower exposure rates (perhaps less than 0.01 events per mile year) may be associated with common indicators of exposure level. A more complete listing of such indicators is found in PRMM.

5.3.4 Impacts

General categories of impacts include those from vehicles and falling objects, as discussed below.

5.3.4.1 Vehicles

Type and speed of vehicles are determinants of damage potential. Various traffic impact scenarios are possible for many components. Considerations include moving object congestion, frequency, duration, direction, mass, speed, and distance to facilities. The impact potential is often informed by historical accident frequency, severity and damage caused by cars, trucks, rail cars, offshore vessels, and/or plane incidents.

Vehicle impact potential can be assessed by considering categories of 'momentum', where momentum is defined in the classic physics sense of vehicle speed multiplied by vehicle mass (weight). High speed, lightweight vehicles can cause damages comparable to low speed, heavy vehicles. Momentum exposures can be assessed in a quantitative way by estimating the frequency of occurrence around the component being assessed. For example, a high frequency of light aircraft at a small airport might be two or three planes per hour, whereas a high frequency for heavy trucks on a busy highway might be hundreds per hour. For each type of vehicle, the frequency can be combined with the momentum to yield an exposure estimate. Where the potential for more than one type of vehicle impact exists (and mitigations for each are equivalent), the frequencies are additive.

Most roadways in most Western countries will have occasional vehicle excursions but high incident rates at specific locations will not long be tolerated. A section of road that experiences numerous vehicle excursions every week will normally prompt action. Safeguards such as speed control and barriers will be employed by roadway owners and/or owners of exposed facilities to control the rate. This informs estimates of exposure since rates of, for instance, 100 incidents per week at a single road location, would not seem plausible, at least for most industrialized countries.

The type of vehicular traffic, the frequency, and the speed of those vehicles determine the level of exposure. Vehicle movements inside and near aboveground facilities should be especially considered, including

- Aircraft
- Trucks
- Rail traffic
- Marine traffic
- Passenger vehicles
- Maintenance vehicles (lawn mowers, etc.)

The potential damages caused by various vehicle impact scenarios can be challenging to estimate without detailed calculations of many combinations of component characteristics and impact specifics. This is further discussed in resistance, Chapter 5.5 Resistance.

5.3.4.2 Falling Objects

Objects dropped or falling from heights above the component being assessed are a potential source of damage. The potential for toppled structures nearby should also be included in the assessment. Falling trees, buildings, walls, utility poles, aircraft, meteors, cranes, tools, pipe racks, etc are often overlooked in a risk assessment. This is an understandable result of discounting such threats via an assumption that a buried component is virtually immune from such damage. While this is normally an appropriate assumption, the risk assessment errs when such threat dismissal occurs without due process. The independent evaluation of exposure and mitigation ensures that, should depth of cover condition change, ie, the component is relocated above grade; or a particular falling object indeed can penetrate to the buried pipeline; are not lost to the assessment.

Many of these exposures can be tied to weather phenomena such as windstorm and ice loadings. Therefore, exposure estimates can be tied to location-specific data on recurrence intervals of such phenomena. For instance, most locations along the Gulf of Mexico have hurricane recurrence intervals of around once every 25 years. This suggests a hurricane-induced wind storm event frequency of 1/25 per year, with perhaps only a fraction of those events actually generating wind-borne debris loadings of sufficient magnitude and direction to potentially cause damage to the component being assessed. These loadings could alternatively be considered in the Geohazard portion of the PoF assessment.

Similarly, aircraft crash rates are well documented and even meteorite strike rates have been approximated.

Objects can be dropped from some surface activity (construction, fishing, platform operations, mooring close to platforms, cargo shipping, pleasure boating, etc.) can endanger submerged facilities.

The risk of subsea equipment being damaged by dropped objects can be assessed and used to ensure that proper levels of physical protection are provided in the design phase. Drops per lift, based on UK offshore historical data, suggest rates ranging from 10-5 to 10-7. Coupled with lift frequencies ranging from 10^4 to 10^8 per year, results in ranges of historical exposure rates, possibly appropriate for an offshore segment being assessed. These general rates can be made more scenario specific with knowledge of equipment types, loads being lifted, and many other factors.

In offshore scenarios, dropped objects may travel large horizontal distances before reaching sea bottom. This is dependent upon currents and depth and can be included in the probability that a dropped object from a certain location will strike the component being assessed. A buffer distance around fixed sources such as platforms can provide a zone within which components are threatened from that source. Similarly, for moving sources such as vessels and aircraft, an additional probability of proximity can be added to the assessment.

Damage potential is related to energy imparted which in turn is related to object weight, height, and the acceleration of gravity. For subsea installations, the object terminal velocity as it travels through the water will determine the energy imparted. This is a function of the object's weight, shape, water displacement, and resistance to flow, or drag.

Table 5.1
Example of Exposure Estimate Compilation

	Frequency (events per mile-year) at Location XYZ								
	Cause of Fail								
	Human Error	Geohazard	Geohazard	Vehicle Impact	Weather	Weather	Other	Total	
Object Falling		Seismic	Flood		freeze	wind			
Tree	0.01	0.01	0.001	0.01	0.01	0.01	0.001	0.052	
Building, 1-story	0.005	0.001	0.05	0.0005	0.0002	0.001	0.001	0.0587	
Building, multi- story	0.0001	0.0001	0.0001	0.000001	0.000001	0.00001	0.0001	0.000412	
Wall	0.001	0.01	0.01	0.001	0.0001	0.001	0.001	0.0241	
Tools	0.02	0.02	0.02	0.02	0.02	0.0002	0.001	0.1012	
Heavy Equipment	0.05	0.0005	0.0005	0.001	0.0005	0.0005	0.001	0.054	
								0.29	events/ mile-year
							an event every	3.4	years

5.3.5 Station Activities

Surface facilities will often have different types and frequencies of activities compared to ROW segments. Attention to the frequency and duration of normal vehicle movements, in-station excavations, facility modifications, and visitor traffic is usually warranted. Controlled access, third-party facilities present, and continuous work inspection are considerations. While damage caused by employees of the pipeline owner/ operator is not technically third-party damage, all such station activities may be more efficiently addressed as part of excavations and impacts assessment.

For surface facilities, there is often need for additional emphasis on internal traffic, including loading operations involving trucks, rail, marine vehicles as well as the potential for successive reactions.

5.3.6 Successive reactions

The potential for successive reactions warrants more discussion. A successive, or sympathetic, reaction is the damage caused to one part of a facility by a nearby event (for example, rupture, fire, explosion, etc) on another part of the facility. Accidental rupture and explosion of a vessel containing combustible material can cause heat and/or projectile damage to a other parts of the facility or to neighboring facilities. Debris, projectiles, and impulse loadings from nearby explosions are readily apparent potential causes of damage. More subtle damage scenarios include neighboring pipelines, even when both the assessed and the neighbor components are buried.

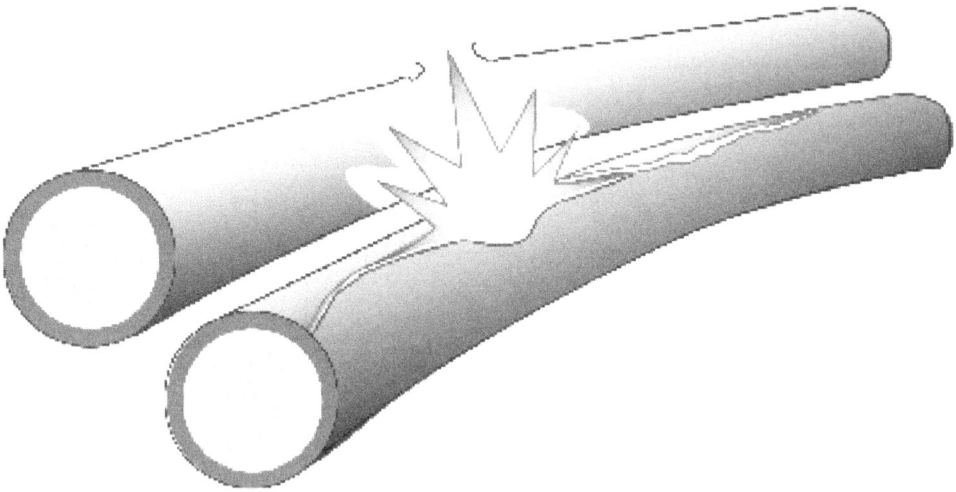

Figure 5.3 Successive or Sympathetic Failures

Segments that are susceptible to such secondary effects will show a higher risk, even if only a very minor increase. Because this event depends on the occurrence of another, the level of exposure for this kind of external force is low. The probability of the initial event is normally low and the successive reaction event, as a fraction of the initial failure probability, should usually be very low.

The damage potential is a function of what is being transported or stored, and the volume and pressure. The potential can be quantified by calculating or estimating the thermal and/or overpressure effects from failure of a neighboring component.

Factors such as barriers, shielding and distance reduce the threat, and estimated[1] exposure or mitigation values should reflect this.

Ideally, the likelihood of failure of the causal event is based on its own complete PoF assessment. This additional assessment might not be possible if the causal event can occur from a neighboring facility that is not under company control. If so, an estimate, perhaps based on generic component information (for example, average natural gas transmission pipeline failure rates), consistent with the PXX level specified, is appropriate.

Table 5.2
Example of Successive Reaction Exposure Estimates

Initiating Event Location	Frequency of Failure per year	Fraction Potentially Damaging to Neighboring Equipment	Exposure (events/ year)	
Pipeline AB, Sta 110 to 145	0.00001	0.1	0.000001	
Valve 1 on Pipeline XY	0.00005	0.05	0.0000025	
Vessel 1	0.001	0.01	0.00001	
Vessel 2	0.05	0.2	0.01	
Truck Loading	0.006	0.5	0.003	
			0.013	exposure from successive reactions (events/year)
		an event every	76.8	years

5.3.7 Offshore Exposure

Anchoring, fishing equipment impacts, shipwrecks, platform sinkings, debris transport by moving waters, shoreline constructions, dredging are some external forces unique to the offshore environment. Vortex shedding and loadings due to moving waters are aspects of risk captured elsewhere in the assessment (ie, as contributors to cracking failure). As with an onshore assessment, in an offshore third-party damage exposure estimate, the evaluator assesses the probability of potentially damaging activities occurring near the pipeline. A complete list of plausible activities will be necessary for a full assessment. Higher activity logically increases the exposure. Each exposure should be assigned an exposure rate: events per mile-year, for example. Where exposure levels are higher or multiple exposures co-exist, PoF increases.

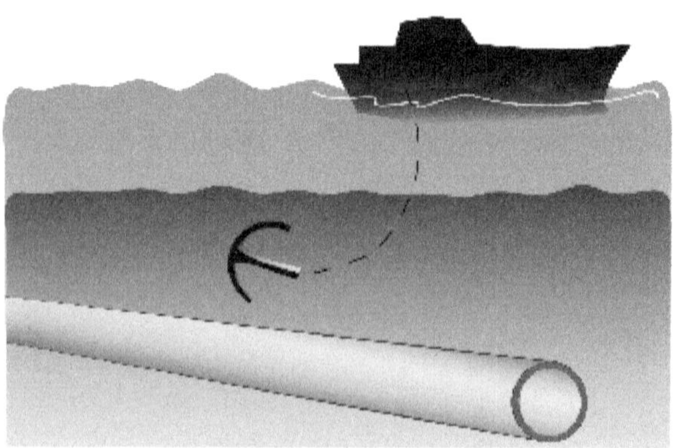

Figure 5.4 Anchor Damage Potential

Table 5.3
Sample of Exposure Assignments to Offshore Component

Exposure	P95 Exposure Rate (events per mile-year)	Comments
Storm debris movement[1]	0.02	1/10 x severe storm freq
Anchor drag	0.1	Based on ship tr
Anchor drop	0.02	Based on ship
Foreign pipeline construction	0.05	
Trawling	0.1	
Dropped object from vessel	0.05	
Ship wreck (sinking)	0.001	
Dropped object from platform	0.1	

5.3.8 Other Impacts

Detonations, including subsurface detonations from seismograph, mining, or construction, can damage pipeline components and should be included in exposure estimates. See PRMM for more discussion.

Damage from wildlife is not uncommon in some areas. Large animals can damage coatings and instrumentation and sometimes even directly threaten the integrity of pressurized components. Even birds and insects can cause damage that eventually contributes to a failure.

External impacts related to geohazard events with little to no man-made materials involved—landslides, rock falls, sea bottom movements, etc.—are normally considered in the Geohazard assessment.

5.4 MITIGATION

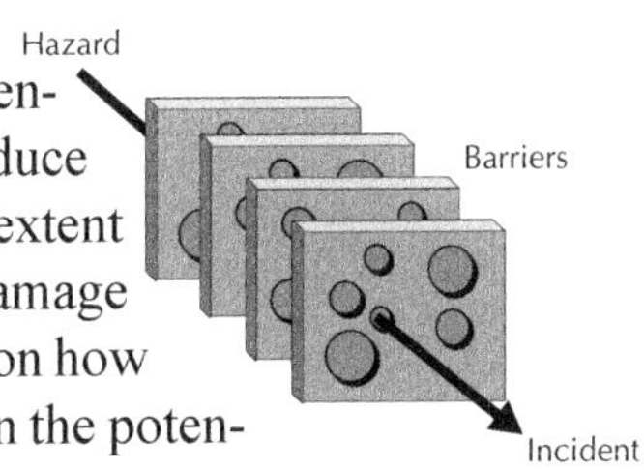

Given the prevalence of accidental third party damage potential, pipeline operators usually take significant steps to reduce the possibility of damage to their facilities by others. The extent to which mitigation is effective is related to how many damage incidents are avoided. Avoidance of damage in turn depends on how readily the system can be damaged by an event and how often the potentially damaging event occurs.

Continuing the earlier discussion, specific pairings of exposure with mitigation effectiveness is part of a more robust assessment. This recognizes that the same mitigation will often have different effectiveness on different exposure types. Examples include:

- 1 foot depth of cover is generally more protective against agricultural equipment damage than against excavation equipment damage
- Depth of cover may have little mitigative benefit against subsurface boring operations
- Patrol is more effective against exposure scenarios that are slower to manifest, such as cross country pipeline construction and residential developments.

Some assumptions commonly used in assessing mitigation effectiveness for this threat include the following:

- One-call effectiveness is generally an AND gate between sub-variables such as system type, notification requirement, and response. The AND gate is applicable since all sub-variables together represent the effectiveness of the mitigation. If any single aspect is deficient, then the overall effectiveness is suspect.
- The mitigation of patrol is normally an AND gate between patrol type and frequency. Patrol type implies an effectiveness and includes combinations of different types—ground-air, for example. But regardless of the effectiveness of the each patrol, if not done at sufficient time intervals, overall mitigation effectiveness is suspect.
- External damage protection is typically an OR gate between cover, warning mesh/tape, exterior protection since each measure can act independently to reduce the chance of damage.
- Casing is a mitigation (against external forces) as it is something added to a pipeline system. For risk assessment purposes, slabs, casings, and even concrete coatings are considered to be distinct from the component and therefore best treated as mitigation measures. Under this view, the component is not damaged when only the protection against another threat is damaged. Some loss of mitigation may have occurred, but not direct damage to the component.

Component wall thickness or strength, even when ‘excessive’, is not a mitigation. It does not prevent damage. If the wall thickness/strength is greater than what is required for anticipated pressures and external loadings, the ‘extra’ is available to provide additional protection against failure from external damage or corrosion. Mechanical protection that may be available from extra pipe wall material or strength is accounted for in Chapter 10.4.3 Effective Wall Thickness Concept.

5.4.1 Depth of Cover

The depth of cover is the amount of earth, or equivalent protection, over the pipeline that serves to prevent damage to a buried component from third-party activities and impacts. In general, deeper and stronger cover —more resistant to penetration, ie rock or pavement versus sand—provides greater protection. Interestingly, protection does not really begin with the first amount of cover. A small amount of cover, enough to conceal the component but not enough to protect the line from even shallow earth moving equipment can increase risk beyond what a 'no cover' scenario would present.

A relationship between cover depth and mitigation effectiveness will be needed. This relationship is most robustly established by obtaining an accurate distribution of types of equipment potentially active in the area and then considering potential reaches and forces from such equipment.

It is often appropriate to employ different relationships for different classes of equipment and practice. For instance, as previously noted, agricultural equipment will often not penetrate the ground to the same extent that many construction excavations will. In this case, more mitigation effectiveness is achieved sooner—at shallower depths—when potential damages from agricultural excavations are assessed.

It may also be appropriate to apply different factors or different relationships between depth (and equivalences) and mitigation benefit for impacts not related to excavations. Two feet of cover prevents damage from falling telephone poles more reliably than from train derailments. The robust solution is to estimate distributions of possible forces from potential impact events and calculate the fraction of those events that are nullified by various mitigation measures. Some studies are available to assist in the determination of an appropriate relationship between depth and mitigation effectiveness. Research from similar pipeline environments can also be useful.

In the absence of more definitive research, an appropriate relationship can be theorized by rationalizing changes in effectiveness when depth of cover changes, considering the types of excavation practice involved. From such rationalizations, equations can be posited and employed in the risk assessment.

A schedule or simple formula can then be posited to assign mitigation effectiveness based on cover. For instance:

12in. of cover = 10% mitigation effectiveness

36in. of cover = 65% mitigation effectiveness

A sample relationship, with exponentially increasing protection as depth increases, is as follows:

Effectiveness = 1 – exp(-[amount of cover in inches] x [factor]

Where the factor chosen reflects the assessment designer's view of the rate of change in the effectiveness variable as the depth variable changes.

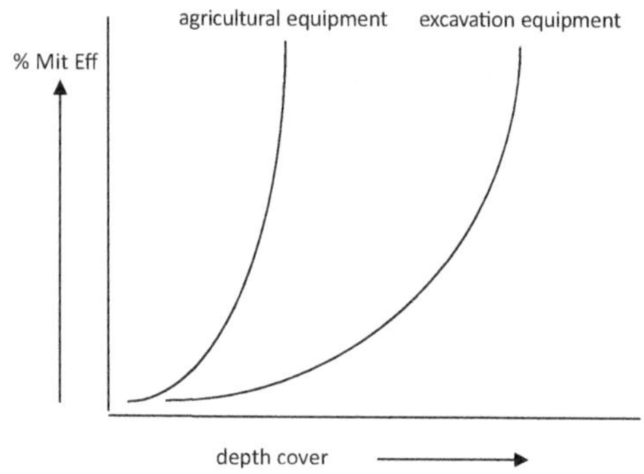

Figure 5.5 Conceptual Relationships Between Depth of Cover and Mitigation Effectiveness

Mitigation credit should also be given for comparable means of protecting the line from mechanical damage including slabs, casings, roadway pavements, etc. Cover for a distribution system often includes pavement materials such as concrete and asphalt as well as sub-base materials such as crushed stone and compacted earth. These are more difficult materials to penetrate and offer more protection for a buried pipeline. Additionally, most municipalities own rights of way and control excavations on public property, especially when penetrating pavements. This control suggests reduced third party damage exposure to a pipeline buried beneath a roadway, sidewalk, etc.

Casing pipe was historically installed to carry anticipated external loads and to protect road and railroad structures from damage if releases occur. A casing pipe is merely a pipe larger in diameter than the carrier pipe whose purpose is to protect the carrier pipe from external loads. Casing pipe can cause difficulties in corrosion control as is discussed later. When the casing carries the external load and protects the section being evaluated from outside forces, it acts as a mitigation.

A robust assessment will determine the benefits of these barriers by quantifying the reduction in PoD that is achieved by the additional protection—how many otherwise damaging events will be interrupted by this protection? This requires estimates of types of equipment and associated forces potentially making contact with the barrier as well as the equipment operator's response. When such rigor is unwarranted, a simple schedule can be developed for these barriers by equating the mechanical protection to an amount of mitigation effectiveness. For example, depending on the types of excavation equipment, values such as the following are plausible:

2in. of concrete coating	= 50% mitigation
4in. of concrete coating	= 80%
Pipe casing	= 99.99%
Concrete slab (reinforced)	= 90%
4in. asphalt roadway	= 85%

It is not only the physical strength of the barrier, but also the implication to the excavation equipment operator of the presence of the barrier. An excavator will normally react differently to a casing pipe than to additional depth of cover. Ideally, he will react to any unexpected encumbrance as an indication that the area is not free of buried structures and should be treated more carefully. This is the idea behind buried warning markers such as highly visible strips of warning tape or mesh with imprinted warnings. Either will logically reduce excavation damage potential and can be valued in terms of its ability to independently protect the component in the location being assessed.

Consider the following sample assignments of mitigation effectiveness:

- Warning tape assessed as 90% effective suggests that nine out of ten excavation scenarios will be halted by this mitigation measure alone (remnant exposure = one hit out of ten excavations).
- Warning mesh assessed as 95% effective suggests that nineteen out of twenty excavation scenarios will be halted by this mitigation measure alone (remnant exposure = one hit out of twenty excavations).

Sea bottom (and lake-, river-, creek-, etc bottom) cover and equivalents (for example, concrete mattress, rip rap rock deposits, concrete coatings, etc) provide mitigation from offshore exposures. Just as with other natural barriers, the water depth can also be treated as a mitigation in the risk assessment.

After assigning a mitigation effectiveness to each protection type independently an OR gate is used to obtain the combined effectiveness. For example, if a component is 60% protected by 30" of earth cover and also is encased by a steel casing pipe providing an additional 98% protection, then the combined mitigation from these two methods is 1 – (1-60%)*(1-98%) = 99.2%.

5.4.2 Impact Barriers

Since the presence of aboveground components is something that is often difficult to change—their location is usually based on strong economic and/or design considerations—preventive measures must be taken to reduce their vulnerability to any exposures that may accompany the site. Additional types of protection from mechanical damages include barriers against impacts on these unburied components. Exposures from vehicular collision, falling objects, vandalism, and sabotage may be offset by the mitigative benefit of barriers other than burial and those previously discussed.

Barriers and protections, both man made and natural, around aboveground facilities should be identified and assigned mitigation effectiveness in general or for specific exposure types (for example, varying by type and speed of vehicle impact, etc.):

- Exposure
- P95 Rate (events per mile-year)
- Comments

Sample Listing of Protective Measures to be Combined for Mitigation Effectiveness:

Area surrounded by 6-ft chain-link fence
Protective railing (4-in. steel pipe or equivalent)
Trees, wall, earthen berm, or other substantial structure(s) between vehicles and facility
Ditch (minimum 4-ft depth/width) between roadway and facility
Waterbodies, min 10 ft +
Waterbodies, < 10 ft
Concrete traffic control barriers
Water filled traffic control barriers

Figure 5.6 Offshore protection by articulated concrete mattress

Distance from vehicular traffic on roads, railroads, flight paths, and ship activity can be treated as a type of 'barrier mitigation' or as part of the exposure assessment. As distance increases, the frequency of exposure events logically decreases.

Assignment of mitigation effectiveness can have a basis ranging from simple, SME-based judgments to robust calculations specific to each exposure-mitigation-type scenario. Note the potential for some barrier types to exacerbate an exposure, for example, an earthen berm serving to launch a fast moving vehicle so it becomes an airborne impact threat.

5.4.3 Protection for aboveground facilities.

Security measures that protect against vandalism or other intentional damage may also provide mitigative benefits to accidental damages. Surveillance systems, barriers, lighting, etc. may also offer some protection from accidental impacts under some scenarios.

5.4.4 Line locating

A key to avoiding accidental excavation damage is the line-locating process, undertaken by an operator when notified of pending excavations by others. It typically involves a notification system, line locating equipment and procedures, marking practices, levels of supervision during activates, and others. As a multi-faceted process, it is challenging to assess, from a risk assessment mitigation effectiveness perspective. A robust risk assessment examines the various aspects of typical programs and potential criteria for use in assessing overall effectiveness.

Often called 'One Call' systems, excavation notification systems have become commonplace and their use often mandated by law. They have varying amounts of mitigative effectiveness, as evidenced by many operators' experiences.

The pipeline company's response to a report of third-party excavation activity is the next critical step in this mitigation measure. Notifications without proper response in a timely manner negate the effects of reporting. Response includes the efficiency and accuracy of the locating equipment and procedures employed as well as the clarity of the markings. Finally, the communications between all parties and the amount of oversight during nearby excavations are important contributors to the effectiveness of these programs.

See PRMM for further details on all aspects of line locating programs.

The assigning of error prevention rates—or mitigation effectiveness, 'success rate of mitigation'—to the process of line locating is important for risk assessment. Since the accuracy of maps/records is only one facet of the entire locate process, error rates associated with the other aspects must also be included in the assessment. Using a scenario-based analysis tool such as event trees or LOPA is often useful in assessing mitigation effectiveness.

Integrating all of the above considerations into an effectiveness estimate is challenging. The risk assessment requires this estimate of the overall program at each location assessed, recognizing that this effectiveness may vary along a pipeline, from season to season, and over time in any area (for example, change of management results in change in focus). Human error potential is high in these multi-faceted programs.

Company SME's have typically assigned ***maximum*** effectiveness values in the range of 20% to 80%, for one-call/locate programs, based on their experiences with specific pipeline segments. For perspective, the higher end of this range assumes that 8 out of 10 otherwise damaging events are avoided solely (assuming no depth cover, no signs, etc) through the one-call and line locating program while the lower end assumes only 2 out of 10 events are avoided, even with a very good program. Actual effectiveness values are then assigned based on differences from the idealized, perfect program.

5.4.5 Signs, Markers, and Right-of-way condition

Establishing a clear, well-marked ROW is an interesting mitigation practice. On one hand, the more recognizable and inspectable a ROW is, the less the likelihood of accidental interference. This is also helpful in leak detection. However, a manicured path through otherwise difficult-to-traverse terrain, may also invite unwanted activity. Therefore, the risk assessment should fairly evaluate the benefits, if any, to mitigation offered by a clear ROW alone versus a potential increase in exposure.

Most will agree that signs and markers do provide some mitigation benefit. However, pipeline accident photos showing burning excavation equipment immediately adjacent to a warning sign, show that the protective benefit is clearly not complete, at least in some locations. Various types of signs and markers, including curb markers in paved areas and painting of fence posts, are used to mark ROW's.

Subtleties of marker position, frequency, size, colors, lettering fonts, languages, etc are logically related to effectiveness. However, against the backdrop of initial human reaction, desensitization, and other behavioral issues, such considerations are difficult to quantify.

It is usually impractical to mark all locations of a distribution system. Many components are under pavement or on congested private property. Nonetheless, in some areas, markers are used and believed to reduce third-party intrusions.

Where mitigation benefit is believed to increase with increased identifiability as a ROW, the evaluator can establish a schedule of mitigation effectiveness associated with various levels of marking/clearing; for example, "...two or more markers visible from all points on the ROW' provides 10% mitigation".

In an offshore environment, this mitigation may only be effective at shore approaches or shallow water where marking is more practical and third-party activity levels are higher. At such locations, marking of offshore pipeline routes provides a measure of protection against unintentional damage by third parties. Buoys, floating markers, and shoreline signs are typical means of indicating a pipeline presence. On fixed-surface facilities such as platforms, signs are often used. When a jetty is used to protect a shore approach, markers can be placed. The use of lights, colors, and lettering enhances marker effectiveness.

Company SME's have typically assigned ***maximum*** effectiveness values in the range of 2% to 20%, based on their experiences with specific pipeline segments. For perspective, the higher end of this range—a rating of 'excellent'—assumes that 2 out of 10 otherwise damaging events are avoided solely through the markers (assuming no depth cover, no public awareness, etc) while the lower end of the range, again with a rating of 'excellent', assumes only 2 out of 100 events are avoided. Actual effectiveness values are then assigned based on differences from the idealized, perfect program.

5.4.6 Patrol

Patrol is an important part of pipeline protection and consequence minimization (leak detection, primarily). There is a myriad of patrol types, effectiveness, and frequencies, making the assessment more complex. For instance, air patrol includes the obvious variable of frequency, but also the less obvious considcrations of speed, altitude, use of spotter to assist the pilot, use of unpiloted aircraft (for example, drones) and others. See PRMM for a background discussion.

The assessment may also wish to give credits for patrols during activities such as close interval surveys (see Chapter 6.4 Corrosion—General Discussion) or even daily commutes by employees. For instance, formal patrols might not be part of a distribution system owner's normal operations. However, informal observations in the course of day-to-day activities are common and could be included in this evaluation, especially when such observations are made more formal. Much of an effective system patrol for a distribution system will have to occur at ground level. Company personnel regularly driving or walking the pipeline route can be effective in detecting and halting potentially damaging third-party activities. Routine drive-bys, however, would need to be carefully evaluated for their effectiveness before credit is awarded. Training or other emphasis on the drive-by inspections could be done to heighten sensitivity among employees and contractors.

It is not unusual for operators to conduct formal patrols at frequencies much greater than regulatory requirements. In some instances, daily patrols are perhaps justified and provide a measurably greater safety margin. Frequencies greater than once per day (once per 8-hour shift, for instance) could even be justified by a risk-based cost-benefit analysis.

5.4.6.1 Patrol Effectiveness

The effectiveness of any patrol frequency can be determined from an analysis activities to be detected or at least a reasoning exercise simulating such an analysis. Historical data of findings on previous patrols will often follow a typical rare-event frequency distribution. Once the distribution of findings per patrol is approximated, the curve will have some predictive capabilities, to the extent that the types of activities remain constant.

An effectiveness corresponding to the actual patrol frequency should consider the types of activities likely to occur and the ability to intervene. An analysis of the "opportunity to intervene" in various common excavation activities is a necessary aspect of the effectiveness.

The most thorough intervention opportunity analyses begins with a list of expected third-party activities compared to a continuum of opportunity to detect. Estimating detection probability requires an understanding of how long prior to and after the activity occurs, evidence of its presence can still be seen. Since third-party activities can cause damages that do not immediately lead to failure, the ability to inspect when there is evidence of recent activity is important. Effectiveness changes depend on the type of third party activity. It seems reasonable, for instance, to assume that activity involving heavy equipment requires more staging, is of a longer duration, and leaves more lasting evidence of the activity. All of these promote the opportunity for detection by patrol. The frequency of the various types of activities will be very location- and time-specific.

Sample probabilities of non-detection for typical patrol frequencies from ref [50] are as follows:

Twice a day	13%
Daily	30%
Every other day	52%
Weekly	80%
Biweekly	90%
Monthly	95%
Semi-annually	99%
Annually	99.6%

Detection by 'other than patrol personnel' is 1/3 as likely as by patrol

Intervention opportunity analyses can be the basis of optimizing patrol frequency in addition to assessing the probability of detection for any given frequency. For example, management may decide that the appropriate patrol frequency should detect, with a 90% confidence level, at least 60% of all threatening events. This might be based on a cost/benefit analysis. Patrol frequencies required to achieve this goal can be estimated from the analysis.

5.4.7 Damage Prevention / Public Education Programs

A damage prevention program encompasses many of the mitigation activities listed here, but it is often more associated with public education programs—ensuring that all potential excavators understand pipelines and how to avoid damage to them. See PRMM for a full discussion of such programs.

Transmission pipeline company SME's have typically assigned maximum effectiveness values in the range of 5% to 30%, based on their experiences with public education along specific pipeline segments. For perspective, the higher end of this range assumes that 3 out of 10 otherwise damaging events are avoided solely through the line locating program (assuming no depth cover, no signs, etc) while the lower end assumes only 5 out of 100 events are avoided. Actual effectiveness values are then assigned based on differences from the idealized, perfect program.

5.4.8 Other Mitigation Measures

Other emerging technologies that will likely play increasing role in third party damage mitigation include satellite observation, ground vibration monitoring, acoustical sensors on the pipe, buried sensor cables, motion detectors, infrared activated cameras, and others. Some of these, such as sensors, can be included in a risk assessment as a form of patrol, perhaps a continuous patrol. Others may warrant an independent place as a mitigation measure in the risk assessment model. This is no problem for the risk assessment model proposed here since any type of additional mitigation opportunity is readily combined with all previous measures. Mitigation estimates are additive (OR gate), reflecting the real-world cumulative benefits, as well as diminishing returns, associated with multiple layers of protection.

5.5 RESISTANCE

Adding estimates of resistance to exposure and mitigation moves the assessment from PoD to PoF. Recall that PoD measures the potential for any type of damage that threatens near term or long term load-carrying capacity of the component. PoF is the fraction of damaging events that result in immediate failure.

Factors that make a component less susceptible to failure if damaged—more resistive—include material type, wall thickness, component geometry, toughness, and stress level. Possible weaknesses from past damages, including corrosion, as well as manufacturing and construction issues can also play a role.

The pipe wall thickness and material strength/toughness are among the most important considerations in assessing puncture resistance. The geometry, diameter and wall thickness, influence resistance to buckling and bending. Since internal pressure induces longitudinal stress in the pipe, a higher internal pressure can indicate reduced resistance to certain external forces. Other longitudinal stresses such as caused by lack of uniform support can similarly impact load-carrying capability.

Potential damage to the component depends on characteristics of the striking object and the impact scenario. Force, contact area, angle of attack, velocity, momentum, and rate of loading are among these characteristics. Potential effects include damages to coating, weights, anodes, and component walls, possibly leading to rupture immediately or after some other contributing event.

To better estimate resistances to possible loadings that could be placed on the pipeline, exposures such as excavation, vehicle impact, fishing, and anchoring can be grouped based on the types of equipment, vehicles, engine power, type of anchors or fishing equipment, and others. Fishing equipment and anchors that dig deep into the sea bottom or which can concentrate stress loadings (high force and sharp protrusions) present greater threats—they can be more challenging to resist. Analyzing the nature of the exposures will allow resistance distinctions to be made involving types of excavators, vehicle impacts, anchored vessels, fishing techniques, and others. Such distinctions, however, may not be warranted for simpler risk assessments that use conservative assumptions.

See Chapter 10 Resistance Modeling for modeling options for including resistance in the risk estimates.

6 TIME-DEPENDENT FAILURE MECHANISMS

Highlights

The challenge will often be the prediction of very small areas of degradation among large areas that are damage free.

Time-Dependent Failure

FOCUS POINT
Assessing damage potential and failure potential from time-dependent mechanisms of corrosion and cracking requires estimations of TTF.

Time-dependent failure mechanisms involve some form of degradation—loss of material or other form of weakening over time. These threats are efficiently assessed via the PoF triad (see Chapter 2.8 Probability of Failure). Under this protocol, exposure is measured as unmitigated material loss or crack progression rates (normally units of mpy or mm/yr), mitigation is a reduction in exposure (a reduction in exposure rate), and resistance is the effective[1] wall thickness of the component. Under this assessment protocol, PoD will be >0 unless a material impervious to any time-dependent failure mechanism is assessed or mitigation is 100%—both being unusual possibilities. There are however, pipeline materials with very low susceptibility to certain types of time-dependent failure mechanisms. In those cases, the assessment will show very low PoD and PoF values.

While the discussion here often focuses on steel transmission pipelines, the same time-dependent failure mechanisms are possible in a gathering, distribution, offshore system and on facility components. Even when very different materials are involved—for example, plastic vs steel—the mechanisms are modeled exactly the same way. The same risk assessment techniques apply to all types of pipeline components and, indeed, to any object.

The production of an intermediate calculation, TTF, in an assessment for time-dependent failure mechanisms, reflects the time aspect of degradation type threats and distinguishes them from time-independent failure mechanisms. The TTF is used to produce PoF but is also useful as a stand-alone value for decision-making. It is an essential determinant of inspection and integrity assessment frequencies.

Most pipeline materials are chosen for their ability to have unlimited life spans, so long as deterioration mechanisms are avoided. For most materials, the deterioration mechanisms are corrosion and cracking, with sub-classifications such as UV degradation and creep included for certain materials. In some pipeline systems, such as gathering pipelines intended for finite service lives, some amount of degradation (corrosion) is accepted. See discussion in Chapter 13 Risk Management.

Degradation does not usually progress uniformly on all components of a pipeline or even on a single component. In the assessment of time-dependent threats, the measurement of interest is: probability and potential severity of one or more degrading

1 See Chapter 10

locations per unit surface area. The challenge will often be the prediction of very small areas of degradation among large areas that are damage free.

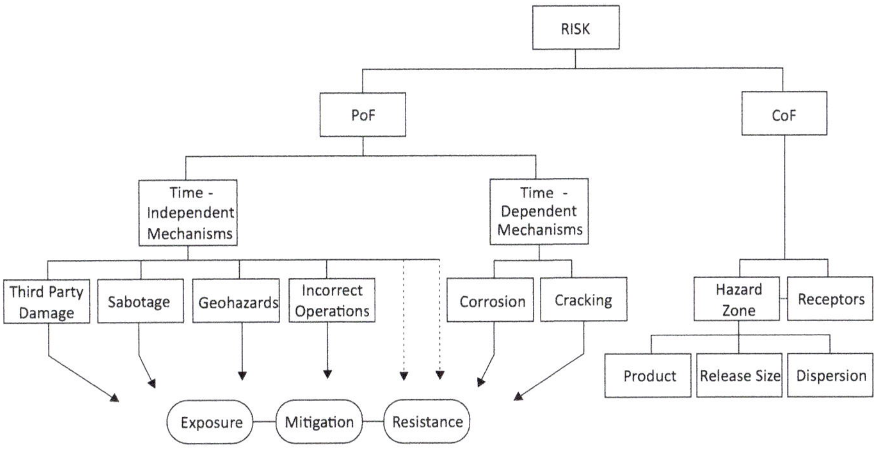

Figure 6.1 Basic risk assessment model.

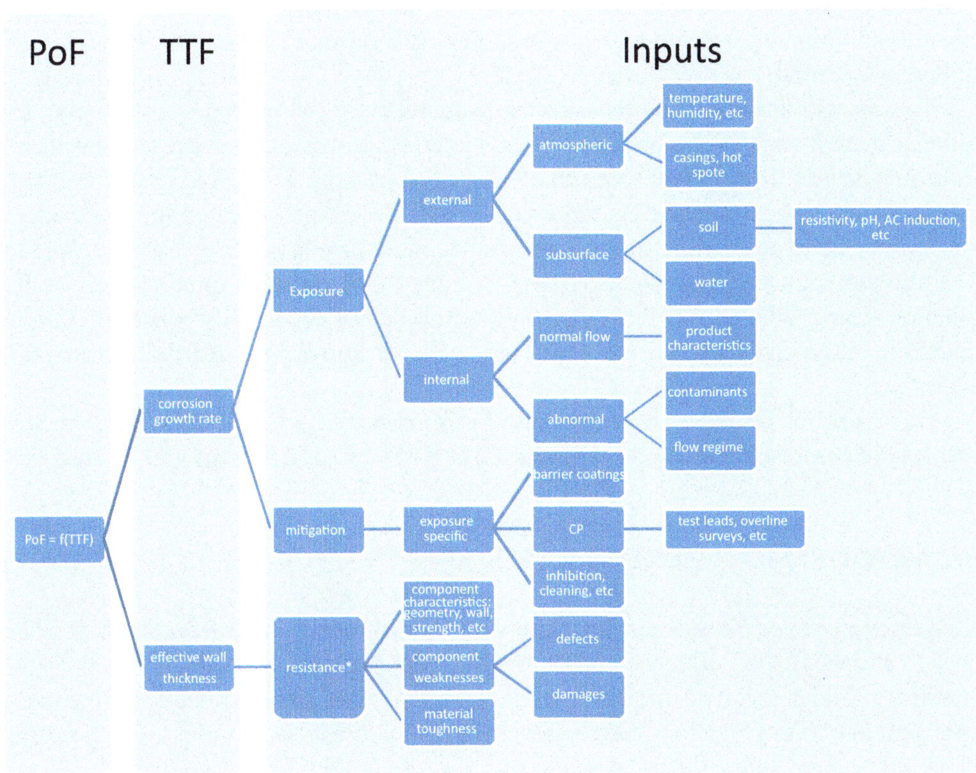

Figure 6.2 Assessing corrosion potential: sample of data used

6.1 POF AND SYSTEM DETERIORATION RATE

The risk assessment described in this chapter is measuring/estimating the probability and aggressiveness of phenomena such as corrosion and cracking as time-dependent mechanisms. Unmitigated exposure is the first element of the measure of PoF, as it is for all threats.

For time-dependent mechanisms, damage potential, as measured by exposure, may be very low in many instances, making PoF low, even with minimal mitigation and resistance. Examples include corrosion potential for steel pipe in dry, sandy, benign soils; components well protected by coatings and cathodic protection; and plastic or concrete lines in dry, neutral pH soils. When exposure is low, long TTF is expected, even when mitigation is weak. The appearance of long TTF periods for some low degradation estimates may at first seem excessive. However, they are not inconsistent with research including one study that uses 220+ years as a median life expectancy for the normally corrosion-vulnerable material of cast iron [2].

On the other hand, aggressive degradation conditions may exist—high exposures. Examples include corrosion mechanisms involving acidic, contaminated soils; steel pipe with a high potential to become anodic to other buried structures; concrete pipe in high chloride soils; MIC activity; AC induced current; and high-stress, high fatigue cycle conditions. In extreme cases, a high degradation rate can lead to through-wall failure of a component in a matter of days.

To translate damage potential into the probability of failure for a component, additional factors such as the wall thickness, material properties, and stress levels need to be considered. In simplest terms and with some assumptions, given an initial wall thickness and a degradation rate, the time to corrode or crack through the component wall can be estimated. More minor wall loss depth over a larger area can also lead to a failure in a higher-stress scenario (metal volume loss leading to rupture), and, at the other extreme, pinhole leaks through the pipe wall do not necessarily constitute failure under the "excessive leakage" definition often used or implied for distribution pipeline systems.

Age-based or historical leak-rate based risk estimates generally play a more limited role in risk management, as is discussed in Chapter 2 Definitions and Concepts.

6.2 MEASUREMENTS VS ESTIMATES

Risk assessments of time-dependent failure mechanisms rely on both materials science based estimates of possible degradation—for example, soil corrosivity based on its chemistry--and actual measurements of degradation, often extrapolated from the measurement site to the location being assessed. This manifests as parallel analyses paths in the assessment where the most recent and most accurate information plays the larger role in the assessment. See the earlier discussion in Chapter 2.14 Measurements and Estimates.

6.3 USE OF EVIDENCE

When degradation mechanisms are not directly observable, the assessment must use mostly indirect evidence to infer damage potential. This is consistent with the historical practice of corrosion control on buried pipelines. Any detection of degradation damages or direct measurements of actual degradation rate can then be used to calibrate the previous assessment results and/or tune the risk model.

Where a degradation rate is actually measured, the risk assessment can be calibrated with this information. A finding of 'no damage' however, must be used carefully. Caution must be exercised in assigning favorable rates based solely on the non-detection of damages at certain times and at limited locations. It is important to note that the potential for some corrosion or cracking damages can be high even when no active damage is detected during a sampling process, especially a random sampling process.

See detailed discussion of use of inspection and integrity assessment information, Chapter 10.3 Inspections and Integrity verifications.

6.4 CORROSION—GENERAL DISCUSSION

6.4.1 Background

As a common cause of failure in most metallic structures, including metallic pipelines, corrosion often plays a large role in risk assessment. Even for non-metallic pipeline components, the expanded definition of 'corrosion' as any degradation mechanism, brings corrosion into the risk assessment. Background discussions on types of corrosion can be found in PRMM.

6.4.2 Assessing Corrosion Potential

As with other failure modes, evaluating the potential for corrosion follows logical steps, replicating the thought process that a corrosion control specialist would employ. This involves (1) identifying, at all locations, the types of corrosion possible, both on internal, external surfaces; (2) identifying the vulnerability of the pipe material—how probable and how aggressive is the potential corrosion; and (3) evaluating the corrosion prevention measures used.

Quantifying this understanding is done using the same PoF triad that is used to evaluate each failure mechanism: exposure, mitigation, and resistance, each measured independently. This will result in the following measurements, ready to be combined into a TTF estimate from which a PoF estimate can emerge:

- Aggressiveness of unmitigated corrosion at contact point between internal contents and component (units of mpy or mm/yr)

- Aggressiveness of unmitigated corrosion at contact point between external environment and component. (units of mpy or mm/yr)
- Effectiveness of mitigation measures (units = %)
- Amount of resistance (units = equivalent wall thickness, inches or mm)

The independent measurements of exposure and mitigation are critical to the understanding of corrosion damage potential. For example, a subsurface environment of Louisiana swampland may present a very corrosive environment, while a dry Arizona desert environment typically produces a very low corrosion rate. The mitigation—the coating system and the cathodic protection system—are obviously more critical to damage prevention in Louisiana. Perhaps, the damage potential in the Louisiana system with very robust corrosion prevention could be made roughly equivalent to the Arizona desert situation where minimal corrosion preventions are needed since the environment is very benign. But it is important to understand when the damage potential is low because of exposure versus due to mitigation.

The two factors that must be assessed to define the corrosion exposure are the material type and the environment. The environment includes the conditions that impact the pipe wall, internally as well as externally. Because most pipelines pass through several different environments, the assessment must allow for this by sectioning appropriately.

Corrosion mechanisms are among the most complex of the potential failure mechanisms. There are a wide variety of available mitigation methods and supporting inspection techniques. As such, many more pieces of information are efficiently utilized in assessing this threat. Because corrosion is usually a highly localized phenomenon, and because inspection opportunities often provide only general information, uncertainty is often high.

6.4.3 Corrosion rate

The time to failure is related to the resistance of the material and the aggressiveness of the corrosion mechanism—the mitigated corrosion rate. The material resistance is a function of material strength and dimensions, most notably wall thickness and the stress level. This chapter examines the process of estimating first the unmitigated- and then the mitigated corrosion rate. A separate estimate is produced for internal and external corrosion potential.

The unmitigated corrosion rate is the 'exposure' in the exposure-mitigation-resistance modeling triad. Exposure estimates should consider the formation of a protective layer or film of corrosion by-products that often occurs and precludes or reduces continuation of the damage. Similarly, temperature effects, rare weather conditions, releases of chemicals, or any other factors causing changes in the corrosion rate should be considered.

Corrosion is a volumetric loss of material but common convention states corrosion rate in terms of depth penetration (pitting). Mils per year (mpy, one mil = 1/1,000 inch) and mm/year are common units of pitting corrosion rates in metals.

While plastics are often viewed as corrosion proof, sunlight and airborne contaminants (perhaps from nearby industry) are two degradation initiators that can affect certain plastic materials and can be efficiently modeled as corrosion in a risk assessment.

6.4.4 Unmitigated Corrosion Rates

As the phrase implies, an unmitigated corrosion rate is a measure of the corrosion progression that may occur in the absence of any corrosion control actions. Normally, a pitting rate is used as the most conservative measure, since pitting rates are usually the most aggressive. When a general (non-pitting) corrosion rate is also active, the resistance measurement (see "Normalizing Exposures with Resistance and Consequences") should take into account loss of component integrity by loss of metal, in addition to loss of integrity by a pitting-induced leak.

There is much research available showing corrosion rates under various laboratory scenarios. Even though laboratory results are often not directly transferable to field conditions, they nonetheless provide valuable insight into plausible corrosion rates, especially when extreme conditions, unlikely to be seen in actual field characteristics, are simulated in the laboratory and suggest maximum rates.

Note that corrosion rates are very situation specific. Any type of corrosion might lead to a failure under the right circumstances, even when history suggests it to be relatively rare failure mechanism.

Recall the previous discussion of measurements versus estimates arising from inferential information. Proper risk assessment uses all available information. In the case of corrosion rates, information often appears in both general forms—measurements and inferences. The final estimate emerges from an examination of both, after adjustments for information age and accuracy have been made. The assessment chooses the best estimate based on the strength of evidence—newer and more accurate information is chosen over older, less accurate information. Note the nuances that have to be considered. For example, highly accurate measurements, but taken some distance from the point of interest where conditions may not be consistent (ie, internal corrosion coupons); or measurements taken at a point in time no longer reflective of recent conditions.

See also PRMM.

6.4.5 Types of corrosion

Many types of corrosion are possible. All can be efficiently modeled in the same way. The discussion here will focus on corrosion of carbon steel. Regardless of material or specific corrosion mechanism, the corrosion assessment should recognize the two locations where corrosion can occur—the external surface of the component or the internal surface. Since these two are significantly different both in terms of exposure and mitigation, they are usually best assessed independently.

For evaluation purposes, the two corrosion (types) locations are further broken down as follows:

External Corrosion:

- Exposure to Atmosphere
- Burial in Soil
- Submersion in Water
- AC-induced
- Interferences.

Internal Corrosion:

- Stream-based
- Under-deposit.

MIC is a potential exacerbating factor in most of these and is therefore appropriately assessed within each, instead of as an independent aspect.

From a chemistry perspective, these corrosion processes are often very similar.

6.4.6 External Corrosion

A pipeline component can be susceptible to external corrosion damage via atmospheric corrosion, subsurface corrosion (including submerged conditions), or both.

Atmospheric corrosion deals with pipeline components that are in contact with the atmosphere. It is a normally a less aggressive corrosion mechanism but there are dramatic exceptions. Alternately wet and dry areas, such as splash zones near water bodies or an annular space inside buried casings, have caused aggressive corrosion and pipeline failures. Failure potential due to atmospheric corrosion are lower in most segments because 1) most pipelines are predominantly buried and, hence, have few portions exposed to the atmosphere, 2) atmospheric corrosion rates are usually low, and 3) there are increased inspection opportunities for above-ground components.

Subsurface corrosion includes both onshore and offshore installations and is the result of potentially very aggressive mechanisms, including various types of galvanic corrosion cells and interference potential from electrical sources and other buried structures. There are also challenges in gaining knowledge of actual corrosion on subsurface components. Subsurface pipe corrosion is often the most information-rich area of risk assessment, reflecting the numerous data-collection practices and the complicated mechanisms underlying this type of corrosion.

Modern metallic distribution pipeline systems (steel and ductile iron, mostly) are installed with coatings and/or cathodic protection when soil conditions warrant. This is equivalent to practices in modern transmission pipelines. However, in many older metal systems, especially older urban distribution systems, little or no corrosion barriers were put into design considerations.

As a special form of subsurface external corrosion, AC-induced corrosion is best examined independently. Also warranting special attention in subsurface systems are nearby sources of DC electricity that can interfere with protective systems or generate new corrosion potential.

Erosion can be thought of as an external corrosion mechanism (in the broad definition of 'corrosion'). Often due to moving water, it is most often included in geohazards. The potential for undermining (loss of support), impingement forces, and others is normally more likely than material loss due to erosion. However, one can envision scenarios involving susceptible component materials in an aggressive flowing (or even stagnant) fluid environment that warrants assessment as a bona fide external degradation mechanism. Erosion or abrasion by wind borne particles is an example. UV degradation of plastics and other material-property changing mechanisms can be included here and/or in the resistance estimations. See Chapter 2.8.12 Nuances of Exposure, Mitigation, Resistance and Chapter 10 Resistance Modeling.

6.4.7 Internal Corrosion

Internal corrosion deals with the potential for corrosion originating within the pipeline. Some significant pipeline failures have been attributed to internal corrosion. Internal corrosion results in wall loss and is caused by a reaction between the inside pipe wall and the interior environment, ie, the product being transported and its flow regime. Internal corrosion may not be the result of the product ***intended*** to be transported, but rather a result of impurities in the product stream. Erosion is a possible internal corrosion mechanism (again, in the broad definition of 'corrosion') as is discussed in a later section.

6.4.8 MIC

The term microbiologically-influenced corrosion (MIC) is used to designate the localized corrosion affected by the presence and actions of microorganisms. MIC was described in a previous section.

External corrosion manifestations of MIC are typically characterized by pitting and crevice corrosion, according to some experts. Soils with sulfates or soluble salts are favorable environments for anaerobic sulfate-reducing bacteria [69]. Also additional discussion in PRMM.

6.4.9 Erosion

As noted previously, erosion is also considered here as a potential time dependent mechanism for both internal and external surfaces. For instance, an exposed concrete pipe in a flowing stream can be subject to erosion as well as mechanical forces. Erosion on an interior component wall is caused by high velocity flow streams containing abrasive particles and can be particularly damaging at impingement points such as elbows.

6.4.10 Corrosion Mitigation

Corrosion mitigation is specific to the type of corrosion and, often, to the location. Details are discussed in subsequent sections. Here, the philosophy of modeling corrosion mitigation is discussed.

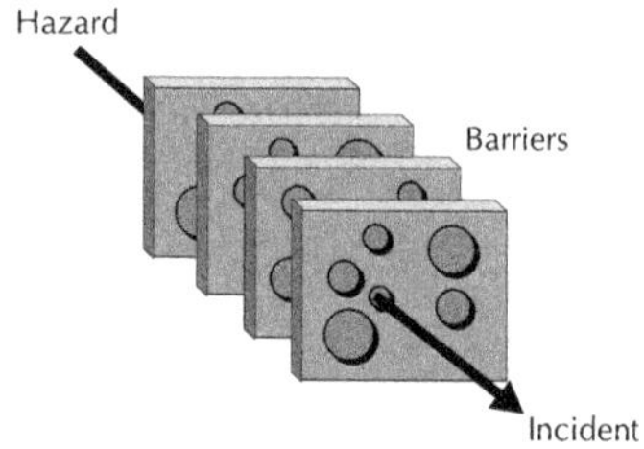

Similar to other mitigation where an OR gate can combine mitigation measures acting independently, a multi-layer defense against corrosion uses the same modeling approach. The common mitigation against external corrosion for a buried metal pipeline is a two-part defense of coating and cathodic protection (CP). These two are usually employed in parallel and provide redundant protection. Some practitioners rate these measures as equally effective, in theory at least. Since each can independently prevent or reduce corrosion, an OR gate can be used in assessing the combined effect. The notion of independence here refers to a modeling protocol, not to an idea that the two are not related in considerations of real world design, economics, maintenance, etc.

An effective modeling approach quantifies external corrosion potential by coupling exposure (corrosion aggressiveness) with the probability of one or more active corrosion points on the pipeline segment. This probability is based on an estimate of the frequency of active corrosion locations, derived from estimates of coating holiday rates plus the efficiency with which CP prevents those holidays from experiencing corrosion.

Underpinning this procedure is the belief that the simultaneous occurrence of multiple defects is appropriately modeled as the product of the independent defect rates. That is, the probability of both 1 and 2 occurring simultaneously is Probability 1 x Probability 2.

6.4.11 Corrosion Failure Resistance

The resistance to failure by corrosion is efficiently measured as an effective wall thickness. The wall thickness is a critical part of all stress-carrying capacity calculations that underpin resistance estimates in thin-shelled, pressure containing structures. This wall thickness, taken with the mitigated corrosion rate, yields a time to failure, or remaining life estimate. For example, a 0.250" effective wall thickness, experiencing 10 mpy pitting corrosion, would be expected to leak in 25 years. A rupture could occur sooner, depending on the lateral corrosion damage and the stress level. This is efficiently modeled in a parallel analysis—'growing' the corrosion damage laterally as well as in depth. The shorter of the leak-driven or the rupture-driven TTF estimate provides the final TTF value for use in generating the PoF estimate.

The 'effective' adjective in front of 'wall thickness' allows inclusion of any weaknesses (previous damages, manufacturing or construction defects, stress concentrators, etc) or vulnerabilities (selective seam corrosion, heat affected zones of welds, etc) which, when modeled as equivalent reductions in pipe wall thickness, show reduced remaining life estimates and corresponding increases in PoF. This is fully discussed in Chapter 10 Resistance Modeling.

6.4.12 Sequence of eval

Regardless of the type of corrosion or its location on the pipeline system, the risk assessment protocol is the same. That protocol is as follows:

1. Estimate Exposure (assuming no mitigation).
2. Estimate Mitigation Effectiveness.
3. Combine the above into an estimate of degradation (PoD, typically expressed as mpy or mm/year).

6.4.12.1 Estimate exposure levels

The unmitigated corrosion rate for the PXX level of conservatism desired is first estimated.

This first step involves evaluating the pipe's internal and external environments. For each corrosion type, external and internal, and their associated sub-types (AC induced, MIC, etc), an assessment is made of the corrosivity at the material's interface with its immediate environment, if no mitigation is employed. Once a database of location-specific characteristics of the pipeline and its surroundings is built, this process can be at least partially automated. The following discussion illustrates a typical approach to characterizing each component's environmental exposures (the threats to the pipe from its immediate environment).

To differentiate two general types of external corrosion, typically with quite different pitting rates, contacts with the atmosphere are first identified. These include locations with depth of cover = 0, casings, tunnels, spans, valve vaults, manifolds, and meters. Under an assumption of a mostly-buried pipeline, these occurrences are rarer and represent potential for atmospheric corrosion.

Next, location-specific characteristics that typically harbor more aggressive atmospheric corrosion rates are identified. These include supports, hangars, splash zones, tree sap depositions, and many others. These are treated as external corrosion 'hot spots'.

If the pipe is ***not*** exposed to the atmosphere, then the typical assumption is that it is immersed in soil or water and should be treated as being in a subsurface corrosive environment. As with atmospheric corrosion, location-specific info is needed. Soil corrosion rates are measured or estimated at all points along the pipeline. Provisions can also be added to capture scenarios where a component is exposed to both atmospheric and soil corrosivities, such as a pipeline laid atop the ground, at ground/air interfaces, in splash zones, and others.

For internal corrosion, the normal assumption is that all portions of the system are exposed to the product being transported and, hence, to any internal corrosion potential promulgated by that product. Therefore, all portions have general exposure to internal corrosion. Especially where corrosion rates can change over both time and space—for example, where contaminant and velocity excursions impact internal corrosion —a probability-weighted corrosion rate can be used.

Next, location-specific characteristics that exacerbate internal corrosion such as areas of accumulations of solids and liquids, are identified, perhaps by elevation profiles, velocity profiles, and product stream analyses. These are 'hot spot' locations for increased internal corrosion rates, analogous to the external corrosion hot spots.

6.4.12.2 Estimate mitigation effectiveness

For barrier-type mitigation, such as coatings, and certain chemical inhibitors, the probability of a gap in protection per unit of surface area to be protected is estimated. For subsurface corrosion, both soil burial and submersion in water, the probability of unprotected surface area is similarly estimated. Then, the role of secondary mitigation measures such as CP, inhibitor injection, cleaning, etc are overlaid with the barrier effectiveness.

These mitigation effectiveness estimates can be very challenging to produce. Much information is often available, but inferential and/or location-specific in nature, requiring interpretation and extrapolation to assessed areas with less information. Overline surveys provide very useful but only indirect evidence.

6.4.12.3 Estimate Degradation Rate

Exposure and mitigation estimates are then combined to yield probabilistic damage rates, after mitigation, typically in units of mpy or mm per year. All combinations of unmitigated exposure and mitigation effectiveness are considered along the assessed pipeline. Hot spots—ie, aggressive unmitigated corrosion—at locations with weak mitigation will show highest damage rates.

The mitigated damage rate estimates are now combined with estimates of effective pipe wall thickness to estimate TTF. This is value, again usually changing along the pipeline, is often of more interest than the final PoF. TTF can more effectively drive risk management decision-making, including integrity re-assessment intervals.

Finally, choosing and applying a representative relationship between TTF and PoF yields the estimate for corrosion PoF for the future year of interest.

6.5 EXTERNAL CORROSION

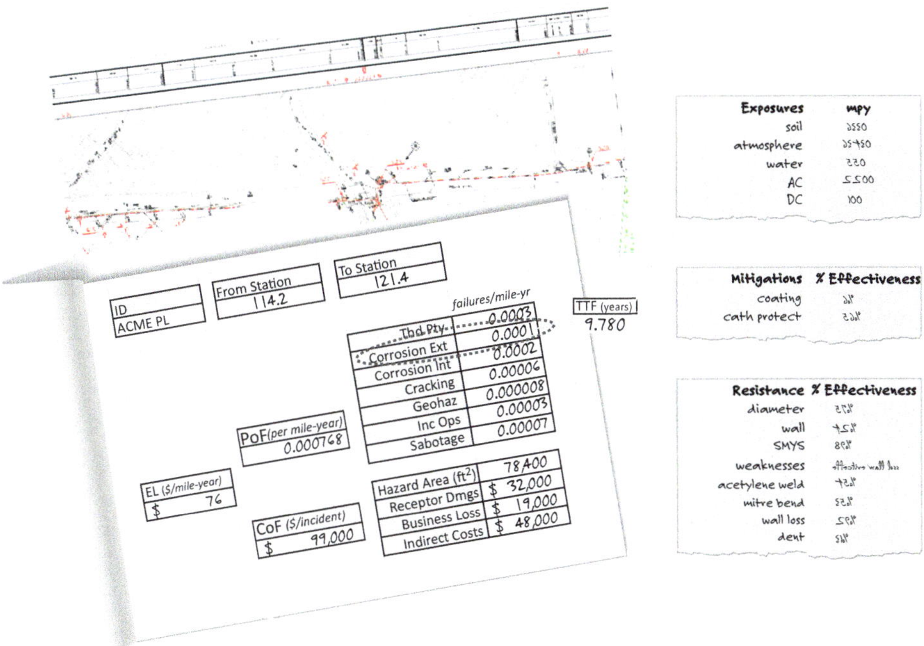

As with all PoF analyses, we begin with an assessment of the exposure level and then consider mitigation measures and finally, the ability to absorb damage (resistance). Measuring these independently is an essential aspect of understanding the corrosion threat to component integrity.

A very benign environment, from a corrosion threat perspective, can be seen as roughly equivalent to a more corrosive environment with effective mitigation, but calls for a significantly different risk management approach. For example, loss of corrosion mitigation on a component buried in a desert may not significantly increase failure potential—due to already-low exposure levels. The same loss—effectiveness reduction—for a component buried in a swamp would be much more serious. This important distinction is apparent in a risk assessment that measures exposure separately from mitigation.

6.5.1 External Corrosion Exposure

A major aspect of assessing external corrosion potential is an evaluation of the environment surrounding the component. See PRMM for a brief introduction to galvanic corrosion.

Different pipe materials have differing susceptibilities to damage by various conditions. Potential deterioration of cement-containing materials such as concrete or asbestos-cement pipe, plastics, metals, and others may need to be included here. Any and all knowledge of pipe material susceptibility to degradation should be incorporated into the exposure estimates.

Exposure estimation entails imagining a completely unprotected surface. Unmitigated corrosivity is primarily a measure of how well the external environment can act as an electrolyte to promote galvanic corrosion on the pipe. Additionally, aspects of the external environment that may otherwise directly or indirectly promote corrosion mechanisms should be considered. These include bacterial activity, the presence of corrosive-enhancing chemicals, and stray electrical effects.

Coating systems, most commonly paint, are often used to protect corrodible metallic surfaces but are not to be considered when assessing exposure. Because a coating system is always considered to be an imperfect barrier, the external electrolyte—usually soil, water, or atmosphere—is assumed to be in contact with the pipe wall at some points and hence requires an estimate of its aggressiveness (exposure).

The evaluator should be alert to instances where the external conditions change rapidly along the pipeline route. Changes in soil type, water table (for example, low elevation creek crossings), and the presence of casings are obvious examples for buried components. Less obvious are certain road bed materials, past waste disposal sites, imported foreign materials, etc. that can cause highly localized corrosive conditions. In an urban environment, the high number of construction projects leaves open the opportunity for many different materials to be used as fill, foundation, road base, etc. Some of these materials may promote corrosion by acting as a strong electrolyte, attacking the pipe coating, or harboring bacteria that add corrosion mechanisms. A lower resistivity soil will promote graphitization of low ductility cast iron pipe as well as corrosion of carbon steel.

The assessment should also consider situations where piping of different ages and/or coating conditions is joined. Dissimilar metals, or even minor differences in chemistry along the same piece of steel pipe, can cause galvanic cells to form and promote corrosion.

If it can be demonstrated that corrosion is not possible in a certain area, exposure (corrosion rates) are essentially zero. The evaluator should ensure that adequate tests of all possible corrosion-enhancing conditions at all times of the year have been made.

6.5.1.1 Atmospheric type

Atmospheric corrosion is the chemically driven degradation in a material resulting from interaction with the atmosphere. The oxidation of metal in the air is the most common manifestation. The estimated annual loss due to atmospheric corrosion is estimated to be billions of dollars [31].

Even predominantly-below-ground cross-country pipelines are not immune to this type of damage. Components are exposed to the atmosphere when they are installed above ground level or are in subsurface enclosures such as vaults or casings. In the risk assessment, it is appropriate to capture an atmospheric corrosivity value for all areas of the pipeline, even when contact with the atmosphere is not occurring. This is the same for soil corrosivity. As part of the dynamic segmentation process, portions that are actually unburied will use the atmospheric corrosion value rather than the soil corrosivity values.

Certain characteristics of the atmosphere can enhance or accelerate corrosion. For steel, this is the promotion of the oxidation process. Oxidation of metal is the primary mechanism examined here although the process is identical for any other corrosion scenario of a pipeline material in an atmosphere.

The most common atmospheric characteristics influencing metallic corrosion include:

- Moisture. Higher air humidity or other moisture contact is usually more corrosive.
- Temperature. Higher temperatures tend to promote corrosion.
- Airborne chemicals: naturally occurring airborne chemicals such as salt or CO_2 or man-made chemicals, often considered pollutants, such as chlorine and compounds containing SO_2 typically accelerate oxidation (corrosion) processes.

Marine atmospheres are usually highly corrosive, and the corrosivity tends to be significantly dependent on wind direction, wind speed, and distance from the coast. An equivalently corrosive environment is created by the use of deicing salts on the roads of many cold regions.

Dew and condensation can exacerbate corrosion. A film of dew, saturated with sea salt or acid sulfates, and acid chlorides of an industrial atmosphere provides an aggressive electrolyte for the promotion of corrosion. Also, in humid regions where nightly condensation appears on many surfaces, the stagnant moisture film can promote corrosion. Frequent rain washing which dilutes or eliminates contamination can help reduce otherwise aggressive corrosion rates.

Temperature plays an important role in atmospheric corrosion in two ways. First, there is the normal increase in corrosion activity which can theoretically double for each ten-degree increase in temperature. Secondly, the temperature differences of metallic objects from the ambient temperature promotes condensation. This temperature difference may be due to lags in temperature equalization due to the metal's heat capacity. As the ambient temperature drops during the evening, metallic surfaces tend to remain warmer than the humid air surrounding them and do not begin to collect condensation until some time after the dew point has been reached. As the temperature begins to rise in the surrounding air, the lagging temperature of the metal structures will tend to make them act as condensers, maintaining a film of moisture on their surfaces. The period of wetness is often much longer than the time the ambient air is at or below the dew point and varies with the section thickness of the metal structure, air currents, relative humidity, and direct radiation from the sun. Differences in temperature between pipe wall due to flowing product and ambient conditions can cause similar effects.

Cycling temperature has produced severe corrosion on metal objects in tropical climates, in unheated warehouses, and on metal tools or other objects stored in plastic bags. Since the dew point of an atmosphere indicates the equilibrium condition of condensation and evaporation from a surface, a temperature below the dew point enables corrosion by condensation on a surface that could be colder than the ambient environment.

Airborne pollutants are another source of corrosion. Sulfur dioxide (SO_2), which is the gaseous product of the combustion of fuels that contain sulfur such as coal, diesel fuel, gasoline and natural gas, has been identified as one of the most important air pollutants which contribute to the corrosion of metals. Less recognized as corrosion promoters, are the nitrogen oxides (NO_x), which are also products of combustion. A major source of NO_x in urban areas is the exhaust fumes from vehicles. Sulfur dioxide, NO_x and airborne aerosol particles can react with moisture and UV light to form new chemicals that can be transported as aerosols. [**1026**]

In the absence of direct corrosion rate measurements, a schedule can be devised to show not only the effect of a corrosion promoter, but also the interaction of one or more promoters. For instance, a cool, dry climate is thought to minimize atmospheric corrosion. If a local industry produces certain airborne chemicals in this cool, dry climate, however, the atmosphere might now be as severe as a tropical seaside location.

See PRMM for an example list of relative corrosivities for different types of atmospheres. To utilize such lists in a modern risk assessment, corrosion rate estimates should be assigned to each. For instance, an atmosphere characterized by industrial pollutants and/or a marine environment, especially when surfaces are alternately wet and dry, may support corrosion rates of 10 to over 50 mpy. A cool, dry, desert environment may support virtually negligible rates—0.1 mpy or less.

It should be apparent by now, that proper segmentation is required in a modern risk assessment. Components with atmospheric exposures must be distinct from those that have no such exposures. A cased piece of pipe will be an independent section for assessment purposes since it has a distinct risk situation compared with neighboring sections with no casing. The neighboring sections will often have no atmospheric exposures and hence no atmospheric corrosion threat at all. Similarly, within a facility, components located near to emissions of pollutants and/or high heat, may suffer radically different corrosion rates than other components in the same facility.

6.5.1.2 Subsurface Corrosion

Although subsurface components of many materials can be susceptible, this part of the corrosion exposure assessment will most commonly apply to metallic pipe material that is buried or submerged. If the component being evaluated is not vulnerable to subsurface corrosion, as may be the case for a plastic pipeline, this exposure goes to zero. If the component is totally aboveground (and flood potential is ignored), the segmentation process allows this component to also have zero exposure to subsurface corrosion.

More than one corrosion mechanism may be active on a buried metal structure. Complicating this is the fact that corrosion processes are mostly detected indirectly, not by direct observation.

6.5.1.3 Soil corrosivity

Because a coating system is always considered to be an imperfect barrier, the soil is always assumed to be in contact with the pipe wall at some points. Soil corrosivity is often initially a qualitative measure of how well the soil can act as an electrolyte to promote galvanic corrosion on the component. Aspects of the soil that may otherwise directly or indirectly promote corrosion mechanisms should also be considered. These include bacterial activity and the presence of other corrosion-enhancing substances.

The possibly damaging interaction between the soil and the pipe coating is not a part of this variable. Soil effects on the coating (mechanical damage, moisture damage, etc.) should be considered when judging the coating effectiveness as a mitigation variable.

The importance of soil as a factor in the galvanic cell activity is not widely agreed on. Historically, the soil's resistance to electrical flow has been the measure used to judge the contribution of soil effects to galvanic corrosion. As with any component of the galvanic cell, the electrical resistances play a role in the operation of the circuit. Soil resistivity or conductivity therefore seems to be one of the best and most commonly used general measures of soil corrosivity. Soil resistivity is a function of interdependent variables such as moisture content, porosity, temperature, ion concentrations, and soil type. Some of these are seasonal variables, corresponding to rainfall or atmospheric temperatures. Some researchers report that abrupt changes in soil resistivity are even more important to assessing corrosivity than the resistivity value itself. In other words, strong correlations are reported between corrosion rates and amount of change in soil resistivity along a pipeline [41].

As the environment that is in direct contact with the pipe, soil or water characteristics that promote corrosion must be identified. The evaluator should list those characteristics and assess all locations accordingly. Resistivity is widely recognized as a variable that generally correlates with corrosion rate of a buried metal. Additional soil characteristics that are thought to impact metallic and concrete pipes include pH, chlorides, sulfates, and moisture. Some publicly available soils databases (such as USGS STATSGO) have ratings of corrosivity of steel and corrosivity of concrete that can be used in a risk evaluation.

Even within a given pipeline station, soil conditions can change. For instance, tank farm operators once disposed of tank bottom sludges and other chemical wastes on site, which can cause highly localized and variable corrosive conditions. In addition, some older tank bottoms have a history of leaking products over a long period of time into the surrounding soils and into shallow groundwater tables. Some materials may promote corrosion by acting as a strong electrolyte, attacking the pipe coating or harboring bacteria that add corrosion mechanisms. Current soil conditions should ideally be tested to identify placement of non-native material and soils known to be corrosion promoting.

A schedule can be developed to assess the average or worst case (either could be appropriate—the choice, however, must be consistently applied) soil resistivity. This is a broad-brush measure of the electrolytic characteristic of the soil.

Table 6.1
Sample Pitting Corrosion Rates, mpy

Type		
Atmospheric	**Salt Water**	**Soil**
0.001-5	1-50	1-20

6.5.1.4 Subsurface corrosion of nonmetallic pipes

The same methodology is used to assess the damage potential from buried pipe corrosion for nonmetallic materials. For nonmetallic pipe materials, the corrosion mechanisms may be more generally described as degradation mechanisms

AC-Induced Corrosion
If a pipeline becomes energized by AC current, perhaps by passing through a magnetic field, a sometimes-very-aggressive type of corrosion can occur. This is typically seen on steel pipelines located near to higher power AC transmission systems. See PRMM pages for more information.

No AC power in proximity of a component will usually be the lowest risk scenario followed by various exposure levels created by AC power being nearby, its configuration, soil resistivity, coating condition, and numerous other factors, with various possible levels of preventive measures being used to protect the pipeline

6.5.1.5 EAC

Environmentally assisted cracking (EAC) occurs from the combined action of a corrosive environment and a cyclic or sustained stress loading. Combining a crack growth rate with a corrosion growth rate is one way to model the potentially more aggressive nature of EAC. While corrosion significantly contributes to this failure mechanism, it is discussed and modeled as a cracking phenomenon in Chapter 6.8 Cracking.

6.5.2 External Corrosion Mitigation

SECTION THUMBNAIL
Assess the common two-part defense against corrosion of buried steel pipelines by estimating the chances of a coincident gap (ie, non-performance) in both coating and CP.

The most common form of prevention for external corrosion on metallic surfaces is to isolate the metal from the offending environment. This is usually done with coatings. If this coating is perfect, the corrosion process is effectively stopped—the electric circuit is blocked because the electrolyte is no longer in contact with the metal. It is safe to say, however, that no coating is perfect. If only at the microscopic level, defects will exist in any coating system.

For a buried or submerged metallic pipeline, common industry practice is to employ a two-part defense against galvanic corrosion on components. The first line of defense is a coating over all metallic surfaces, as discussed above.

The second line of protection typically employed in a buried steel pipeline is called cathodic protection (CP). Creating an electrical current on a metallic component that is immersed in an electrolyte (such as soil or water) provides a means to reverse the electrochemical process that would otherwise cause corrosion.

As would be expected, corrosion leaks are seen more often in pipelines where no or little corrosion prevention steps are taken. It is not unusual, to find older metallic components that have no coating, cathodic protection, or other means of corrosion prevention. In certain countries and in certain time periods in most countries, corrosion prevention was not undertaken.

Most transmission pipeline systems in operation today have cathodic protection systems, even if they were not initially provisioned with them. The presence of unprotected iron pipe and non-cathodically protected steel lines, is found in older distribution systems. As would be expected, these locations are statistically correlated with a higher incidence of leaks [51] and are primary candidates in many "repair-and-replace" decision-support models.

In some older buried metal station designs, little or no corrosion prevention provisions were included. If the station facilities were constructed during a time when corrosion prevention was not undertaken, or added after several years, then one would expect a history of corrosion-caused leaks. In the US, lack of initial cathodic protection was fairly common for buried station piping constructed prior to 1975.

Corrosion prevention requires a great deal of continuous attention in most pipeline systems. This should be a part of assessing a program's effectiveness. This requires evaluation of various corrosion control measures including program appropriateness and adequacy for conditions, coverage, and PPM. A good PPM program includes inspection programs on tanks and vessels, for atmospheric corrosion, hot-spot protection, and overline surveys for buried portions.

For buried pipeline components, the general form of the mitigation estimate will be the combined effectiveness of the coating and the CP. This is conceptually an OR gate since each is an independent means of mitigation, at least theoretically. The effectiveness of each is measured in defect rate or gap rate; fraction unprotected per unit surface area, e.g. coating holidays per square foot of coated area, CP gaps per sq meter of protected surface, etc. The probability of both a coating holiday and a CP occurring simultaneously is the probability of an active corrosion location.

Consideration of changing mitigation effectiveness over time is an important aspect of risk assessment. This includes not only coating degradation and damages, and changes in CP, but also changes in inspection and remediation practices throughout the history of the segment. For example, a segment may be assigned three different mitigation effectiveness estimates:

1. Prior to installation of CP (coating effectiveness only, if any)
2. From installation of CP to when overline coating surveys (and subsequent remediations) became common practice
3. Future years for which risk estimates are sought.

Each of these time periods suggests differences in mitigation which result in different modeled mpy degradation rates. Coupling the mitigated mpy rates with their respective time periods produces estimates of remaining wall thickness postulated for today and future times.

Actual measurements of remaining pipe wall thickness will 're-set the clock' by overriding these estimates—replacing estimates of 'what might have happened' with 'what actually did happen'. A pressure test can also be used to re-set the clock by confirming that a certain level of metal loss did not occur. The role of inspection and testing as is detailed in Chapter 10 Resistance Modeling.

6.5.2.1 Coating

Discounting its role in supporting the economics of CP, coating effectiveness is appropriately assessed in terms of its barrier effectiveness or defect rate.

Common pipeline system coatings include paint, tape wraps, waxes, hydrocarbon-based products such as asphalts and tars, epoxies, plastics, rubbers, and other specially designed coatings. For aboveground components, painting is the most common technique with many different surface preparation and paint systems being used. Some different coating materials might be found in distribution systems compared with transmission pipelines (such as loose polyethylene bags surrounding cast iron pipes), but these are still appropriately evaluated in terms of their suitability, application, and the related maintenance practices. See PRMM for more on this.

Some coating defects are more severe, not simply from a 'larger is worse' size perspective but from a variety of secondary effects. A smaller coating defect can sometimes create more consequential damage. Since corrosion results in a volumetric loss of metal, a small area of corrosion can create deeper defects sooner, compared to a larger corroding area. Coating effectiveness is the complement of coating gap rate, ie, coating effectiveness = (1 – coating gap rate).

To assess the present coating condition, several things should be considered, including the original installation process.

Coating evaluations—measurements

A directly measured coating defect rate—in units such as defects per square foot or square meter—is the most useful input to the risk assessment. A rigorous evaluation of coating condition would involve specific measurements of defects found, adjusted by the time that has passed since the inspection and the detection/identification abilities of equipment used during the inspection.

Several overline survey technologies have been developed to provide coating condition information for buried pipelines. Direct examinations, usually requiring excavations, also provide opportunities to directly measure coating defect rates and possibly extrapolate those findings to similar unexcavated segments.

Cathodic protection is designed to compensate for coating defects and deterioration. One way to assess the condition of the coating is to measure how much cathodic protection is needed per unit of surface area. Cathodic protection requirements are related to soil characteristics and the amount of exposed steel on the pipeline. Coatings with defects allow more steel to be exposed and hence require more cathodic protection. Cathodic protection is generally measured in terms of current consumption. A certain amount of electrical potential halts the electrochemical forces that would otherwise cause corrosion, so the amount of current generated while maintaining this required voltage is a gauge of cathodic protection. A corrosion engineer can make some estimates of coating condition from these numbers. This is often expressed as a % bare value, suggesting the coating gap rate.

Finally, metal loss inspections, such as from ILI, also provide evidence of coating defect rates. The coating defect rate can be significantly underestimated by the confounding role of CP. While external metal loss by corrosion certainly confirms that <u>both</u> coating defect and gap in CP exists, a finding of no external metal loss does ***not*** confirm coating integrity/effectiveness (for example, coating could failed but CP is protective).

Coating evaluations—estimates

In the absence of a direct measurement of coating effectiveness, an estimate can be generated. This will usually be much less certain but may be the only information available to the risk assessment. How effectively the coating is able to reduce corrosion potential at any point in time can be assessed in terms of defect rate and shielding potential. The defect rate—at any point in time depends on four factors, each of which should contribute to the estimate:

1. Quality of the coating system itself
2. Quality of the coating application
3. Damage/degradation rate since installation
4. Effectiveness of the inspection and defect correction program.

The first two address the fitness of the coating system—its ability to perform adequately in its intended service for the life of the project, given its material properties and its application. A quality coating is of little value if the application is poor. The second two consider the maintenance of the coating. When the last 2 factors are sufficiently quantified, perhaps by an inspection process, then a measured defect rate is available and the inferred estimate is not needed.

For estimation purposes, each of these components can be quantified based on their contribution to defect rate. The last factor—ability to remedy defects—will indicate a reduction in defect rate, while the others are usually assumed to add to current and future defect rates. There will be dependencies. A high initial quality coating is of reduced value if the application is poor; protection during installation and service is weak, or when the inspection and defect correction program is poor.

In the absence of measured coating defect rates (via overline survey or ILI or inferred by CP current demand), an estimation model for buried components could take the following general form:

([base defect rate] + [in service damage rate]) x [application factor] x [remediation factor]

Where

Base defect rate is defects per surface area per year, expected from this coating in this environment when application is perfect.

Application factor = multiplier, >= 1.0, showing increase in defect rate due to non-perfect application of the coating. This should account for both increased defect rates when initially applied plus increased defects due to application-related accelerated degradation of the coating in service.

Damage rate = additional defects (beyond those expected with an aging but perfectly applied original coating of this type), in units of defects per unit surface are per year; expected at this location since last inspection, since original installation, or in the future, depending upon risk being measured.

Remediation factor = multiplier, <=1.0, showing the decrease in defect rate due to effectiveness of inspection and remediation practices. This is an offset to the damage rate. It captures the general effectiveness of addressing coating defects, either since last inspection,

> since original installation, or in the future, depending upon risk period being measured. This reflects both the rigor of the remediation intention and the error rates in finding and adequately correcting the defects. This factor would not appear in a detailed risk assessment since location-specific remediation, usually re-coatings associated with excavations, would be directly included in the risk assessment. These locations can often be assumed to be initially defect free, and will have different date of coating installation and inspection, often a different type of coating, and updated quality control of application, compared to neighboring segments. All of these should combine to show increased coating effectiveness and reduced PoF at the remediated location.

The inputs should be supported by formalized reasoning analyses and be expressed in measurement units like defects per square meter. Such subjective SME estimates should be verified or modified based on subsequent data collection such as coating surveys or experience gained from excavation inspections.

Different analyses are usually warranted for mill-applied versus field-applied (for example, girth welds) coating types and application qualities since application is more problematic for field-applied coatings.

Considerations that should inform these estimates are further discussed below.

Coating fitness

An evaluation of the coating in terms of its ability to perform, ie, its appropriateness in its present application, is usually appropriate. Where possible, an SME should use data from coating stress tests or actual field experience to rate the quality. When these data are not available, drawing from any similar experience or from judgment will be required. See PRMM for a sample list of qualitative descriptors which are used only to better ground the analyses. These descriptors should include era of manufacture issues—ie, an older coating that, at the time of selection was believed to be fit for the application, but which was later revealed by the passage of time in service, to be inappropriate.

Coating Application

Most designed coating systems will experience relatively long service lives when properly applied. Part of the fitness assessment should include the likelihood of proper application. When installation is potentially more problematic, error rates are expected to increase. A good example is the use of polyethylene 'shrink sleeves' to cover and protect girth welds. This field-applied coating system has a good track record for some operators and very poor for others. Since its success is very sensitive to the environmental factors during installation and the skill of the installer, there is a wide disparity in experience. It is not uncommon for the same pipeline to have a section, installed by one crew, to have no issues while another section, installed by a different crew, experiences widespread and system failures of girthweld coatings. Under certain application errors, these sleeves are additionally susceptible to disbondment and subsequent shielding of CP currents, making their presence even more problematic to those owners with the more poorly applied sleeves. Certain field-applied tape coatings used on girth welds have similar experiences and problems.

The quality of the coating application process can be judged in terms of attention to pre-cleaning, coating thickness as applied, the application environment (control of temperature, humidity, dust, etc.), and the curing or setting process. See PRMM.

Coating condition

Ideally, sufficient inspection information will exist to inform estimates of coating effectiveness along a buried pipeline. Where coating inspections and repairs are performed but data for a subject segment is not available, the practice of the past and future inspection can be evaluated for thoroughness and timeliness. Distinctions may be necessary for various types of coating defects, eg, disbondments may not be detectable by certain inspection methods. Documentation should also be an integral part of the best possible inspection program—absence of complete documentation leading to reduced confidence in inspection effectiveness.

From any level of examination or testing, the current coating properties can be compared against design or intended properties to assess the degradation or other inconsistency with design intent.

Inspection results should lead to assignments of increased or decreased defect rates with consideration of the time periods in between inspections. A PXX defect rate—new holidays emerging per length of pipe per year—should be applied to the time periods between inspections. The inspection, once conducted, serves to re-set the clock—overriding the previously estimated defect rate (with consideration for inspection capabilities, including error rates).

When a direct measurement of defect rates is unavailable, estimates based on the above considerations must be made. Coating defect rates can range from 100%, a value used for uncoated surfaces, to only one in tens of thousands of square feet of surface area. One study [1051] estimated 7.38 coating defect sites per linear km for a 30 yr old pipe based on UK data. This study also estimated the proportion of coating defect sites with active corrosion = 1%, thereby giving insight into estimating a CP gap rate, discussed in the following section.

Example: 6.1 Migrating from Qualitative Descriptors:

In the absence of good coating defect rate information for a particular pipeline, a rate can be inferred by 'recycling' some previously collected coating information. Utilizing previously assigned, qualitative descriptions of coating condition, perhaps from an older risk assessment, can be useful.

A simple thought-exercise can provide plausible coating defect rates from the descriptors. A scale based on these qualitative descriptors can be generated as illustrated below:

Table 6.2
Sample Linkage of Coating Descriptors to Coating Defect Rates

Coating Evaluation	Assumed %Effective	Resulting Defect Rate	Estimated Defect Rate
excellent	0.9999999	1E-07	5.0E-07
good	0.99999	1E-05	2.5E-06
fair	0.99	0.01	3.2E-04
poor	0.9	0.1	4.0E-02

Per square foot of coated surface

This links the qualitative descriptor—perhaps carried over from a previous risk assessment—to a defect rate implied by that descriptor. This is obviously a very coarse assessment and should be replaced by better knowledge of the specific pipeline being evaluated.

To better visualize the implications of this simple relationship, and perhaps to help SME's derive such a relationship, consider the following 'defect rate estimates' for a sample pipe diameter of 12". For various lengths of the 12" pipe, the probability of a coating defect is estimated. This can then be used to help validate and tune the coating assessment protocols since records and/or SME's can often relate actual experiences with a particular coating to such defect rates.

Table 6.3
Visualizing Coating Defect Rates

Coating Evaluation	Defect Rate per sq ft per year	Probability of Defect in Segment, per year				
		L = 1 ft	L = 10 ft	L = 100 ft	L = 1000 ft	L = 5280 ft
excellent	5.0E-07	0.00%	0.00%	0.02%	0.16%	0.83%
good	2.5E-06	0.00%	0.02%	0.18%	1.75%	8.91%
fair	3.2E-04	2.46%	22.1%	91.8%	100%	100%
poor	4.0E-02	11.8%	71.4%	100%	100%	100%
absent	1.0E+07	100%	100%	100%	100%	100%

In the above table, a mile of "excellent" coating has about a 15% chance of having at least one defect. A 'fair' coating under this system is almost certain to have at least one defect every 1000 ft. These results might seem reasonable for a specific pipeline's coating. The probability of a coating defect is assumed to be proportional to both the quality of coating and the length of the segment (length as a surrogate for surface area of the segment). If the results are not consistent with expert judgment—perhaps ratings for "fair" are too severe, for instance, for the intended level of conservatism—then the modeler can simply modify the equation that relates the coating descriptor to defect rate.

Of course, this model is using many assumptions that might not be reasonable for many pipelines. In addition to the highly arguable initial assumptions, many complications of reality are ignored. For instance coatings fail in many different ways, so the meaning of coating "failure" (shielding vs increased conductance vs. holiday, etc) should be clarified.

Nonetheless, these estimates capture a perceived relationship between coating quality and surface area in estimating probability of coating damage or defect. Note that in this application, the probability of a defect diminishes rapidly with diminishing segment length. As segments are combined to show PoF along longer stretches of the pipeline, the small defect counts must be preserved (and not rounded). The modeler should be cautious that, through length-reduction and rounding, the probabilities are not accidentally masked.

Example: 6.2 Estimating coating condition

For a pipeline system being assessed, installation records indicate that a high-quality paint was applied per detailed specifications to all aboveground components. The operator sends a trained inspector to all aboveground sites once each quarter, and corrects all reported deficiencies at least twice per year. Pending field inspection and additional SME input, the evaluator makes a preliminary, experience-based estimate of one coating defect every 10 square feet of pipe at hot spots such as supports and air/ground interfaces; and an estimated defect rate of 0.001 per square foot elsewhere.

In a subsequent examination, a different pipeline system contains multiple locations of aboveground components at metering stations and other surface facilities. Minor coating repair—touch-up painting—is done occasionally at these locations at the request of the local operating personnel. No formal painting or inspection specifications exist. The regional field personnel request paint work whenever it is deemed necessary, based solely on personal, but experienced, opinion.

The evaluator feels that the utilized paint system is appropriate for the conditions. Application is suspect because no specifications exist and the painting contractor's workforce is subject to regular turnovers. Inspection is providing some assurance because the foremen do make specific inspections for evidence of atmospheric corrosion and are trained in spotting this evidence. Defect remediation is suspect because defect reporting and correction is not consistent.

Given the higher uncertainty and the desire to produce conservative estimates (P90), the risk evaluator assigns ***coating defect rates ten times higher*** on this pipeline than those used in the previous example. The evaluator also assigns an even higher coating defect rate to segments inside buried casings. This recognizes a known hot spot where coating damages are common and corrosion exposure is also often higher (due to alternating wet/dry conditions).

These values will next be used to estimate the number of active corrosion points which will be paired with corrosion rate estimates for each location, leading to a preliminary quantification of external corrosion failure potential for the above ground portions.

6.5.2.2 Cathodic protection

FOCUS POINT

The effectiveness of CP is usually based on inferred rather than direct information. Issues requiring more inference should be modeled as increased uncertainty—reduced CP effectiveness—and include:

Pipe-to-soil voltage readings that:

- Are at increased distance from component being assessed
- Are not recent
- Do not include robust criteria for acceptability
- Do not take into account rectifier interruptions.

As previously noted, CP is one of the two commonly used defenses against metallic corrosion; the other is coatings. Cathodic protection employs an electric current to offset the electromotive force of corrosion. A current is applied to the metallic component and electrochemical reactions take place at the anode and cathode.

As a modeling convenience, CP effectiveness can be assessed as an all-or-nothing mitigation measure. Its role is technically to slow down a corrosion process but, in practical application, meeting a CP criteria is believed to effectively halt any corrosion. The science suggests that every 100 mV shift from native potential results in an order of magnitude less corrosion rate. Attempts to only reduce rather than halt corrosion by CP are not found.

The CP demand is related to the characteristics of the electrolyte, anode, and cathode. Older, poorly coated, buried steel facilities will have quite different CP current requirements than will newer, well-coated steel lines. Old and new sections must often be well isolated (electrically) from each other to allow cathodic protection to be effective. Given the isolation of buried piping and vessels, a system of strategically placed anodes may sometimes be more efficient than a rectifier impressed current system at pipeline stations. It is common to experience electrical interferences among buried station facilities where electrical short circuiting, shorting (unwanted electrical connectivity) of protective current occurs with other metals and may lead to accelerated corrosion.

Distribution systems and buried piping at larger facilities are often divided into zones to optimize cathodic protection. Given the isolation of sections, the grid layout, and the often smaller diameters of distribution piping, a system of distributed anodes—strategically placed anodes—is sometimes more efficient than a rectifier impressed current system.

Offshore pipelines and structures also employ CP. Because of the strong electrolytic characteristics of water, especially seawater, adequate cathodic protection is often achieved by the direct attachment of anodes (sometimes called bracelet anodes) at regular spacing along the length of the pipeline. Impressed current, via current rectifiers, is sometimes used to supplement the natural electromotive forces. The design life of the anodes is always important since the anodes deteriorate over time.

See PRMM for more background discussion.

CP system effectiveness

A CP test lead is an accessible connection to a buried pipe component, usually a wire attached to the component and brought above ground. The test lead provides an opportunity to measure the pipe-to-soil voltage to determine the effectiveness of the CP application. Although major cathodic protection problems can be caught during normal readings of widely-spaced test leads, localized problems are harder to detect and can be significant.

The use of test lead readings to gauge cathodic protection effectiveness has some significant limitations since they are, in effect, only spot samples of the CP levels. Nonetheless, monitoring at test leads is the most commonly used method for inspecting adequacy of CP on pipelines. The role of the test leads as an indicator of CP effectiveness should be based on an estimation of how much piping is being monitored by test lead readings. We can assume that each test lead provides a measure of the pipe-to-soil potential for some distance along the pipe on either side of the test lead. As the distance from the test lead increases, uncertainty as to the actual pipe-to-soil potential increases. Uncertainty increases with increasing distance because the test lead reading represents the pipe-to-soil potential in only a localized area. Because galvanic corrosion can be a localized phenomenon, the test leads provide only limited information regarding CP levels distant from the test leads. How quickly the uncertainty increases with distance from the test lead depends on factors such as soil conditions (electrolyte), coating condition (CP demand), and the presence of other buried metals (interference sources). According to one rule of thumb, the test lead reading provides good information for a lateral distance along the pipe that is roughly equal to only the depth of cover,

As a risk assessment modeling approach, a linear scale in length of pipe between test leads for transmission pipelines while a percentage of pipe monitored might be more appropriate for a distribution piping grid. For preliminary and less detailed risk assessments, an effective zone of 'influence' for information obtained at the test lead may be more useful in understanding risk.

Offshore, the effectiveness of the cathodic protection can also be assessed by pipe-to-soil voltage readings although these systems normally provide few opportunities to install and later access useful test leads. When pipe-to-electrolyte readings are taken by divers or other means at locations along the pipeline, this can be treated in the risk assessment as a type of survey—either test lead or CIS, depending on the spacing of readings.

Closely-spaced pipe-to-soil voltage reading surveys (CIS) provide more definitive indications of CP effectiveness, as detailed in PRMM. These surveys are performed in a variety of ways, both onshore and off.

One obstacle to obtaining a complete overline survey is the presence of pavement over the pipeline often limiting the access to the electrolyte. More permeability and other characteristics impact the loss of data, with older asphaltic pavements sometimes having minimal impact and newer concrete pavements making readings impossible. Inaccessible locations caused by encroachments, landowner issues, and others, also create gaps in a survey.

Varying amounts of post-survey analyses are applied following a CIS. Some companies simply react only to instant off readings less negative than -0.85V. Others use NACE criteria to identify numerous types of anomalies based on severity of dips (trending) in continuous readings and combinations of trending behaviors of the ON and OFF readings. Further analyses opportunities include gaining insights into coating performance and possible deterioration rates.

CP effectiveness

CP effectiveness is the complement of CP gap rate, ie, CP gap rate = (1 – CP effectiveness).

Removing age and criteria considerations for a moment, let us focus on the distance-from-reading aspect of estimating CP effectiveness. According to the above beliefs, the evaluator has options for interpolating between readings from the annual test lead survey.

The relationship between confidence and probability of detection can be formalized. Mathematical checks can also be employed to ensure that gap rates are capped to realistic values, even when confidence is extremely low. By dividing the gap rate by the confidence, the final gap rate increases with decreasing confidence—0.01 gaps/ft^2 with 50% confidence yields 0.02 gaps/ft^2; with 10% confidence, 0.1 gaps/ft^2; and so forth. Then, the risk assessor can assign to all locations, a gap emergence rate (x gaps/ft^2 per year) to account for new interference sources, shielding effects, coating deterioration, and other causes of diminished CP. By one strategy, pipeline segments within 10 ft of a test lead, receiving annual confirmations of acceptable CP levels, will show essentially complete CP effectiveness—100% effective mitigation. As distance from test leads increases and/or time between readings increases, CP gaps are modeled to emerge.

6.5.3 Monitoring Frequency

The role of age in CP surveys is the same as for other types of inspections. The time between inspections and the rate of emergence of anomalies during this interval combine to show the benefits of any inspection timing protocol.

Continuous CP monitoring via existing SCADA systems is becoming more common. Report-by-exception systems are used, monitoring pipe-to-soil voltages and even differences from native potentials when buried coupons, not under the CP circuit influence, are included. The cost/benefit analysis of continuous monitoring is dependent upon the damages that could result from CP outage periods. When more frequent and longer duration outages in more corrosive environments are being avoided by improved monitoring, benefits increase and costs are justified. See also a relevant discussion under Sabotage threat assessment.

Interference and Shielding

CP effectiveness can be reduced when interferences change aspects of the intended protective galvanic cell. Changes to anode, cathode, or electrolytic pathways can all cause interference. Common interferences situations arise where electrical shorting occurs with other metals or shielding essentially blocks protective currents from reaching the surface to be protected. Extreme interferences may create anodic regions on the metal intended to be protected. This accelerates corrosion rates and should be considered in the exposure estimates.

Two types of mitigation interference are appropriately evaluated in a pipeline risk assessment: DC-related and shielding effects. AC effects, and especially attempts to mitigate them (unintentionally blocking some protective DC while controlling AC), can impact CP systems but are more recognized as adding to corrosion potential rather than reducing mitigation. Where there is believed to be a stronger influence on mitigation, AC effects should certainly appear in both places in the risk assessment.

In assessing the interference potential, the assessment should consider the isolation techniques used in separating protected surfaces from other CP systems, sources of electrical power, and metals, including nearby pipelines, casings, foundations, junkyards, offshore platforms, shore structures, and many others. When isolation is not provided, joint cathodic protection of the structure and the protected metal should be in place.

Because distribution systems are often co-located in areas congested with other buried utilities, often with their own CP systems, special operator methods by which interference could be detected and prevented may be employed. Examples include strict construction control, strong programs to document locations of all buried utilities, close interval surveys, extensive use of test leads and interference bonds, and increased inspections.

In a more robust risk assessment on a buried pipeline, the presence of a nearby buried feature—metallic or potential source of shielding—causes a new dynamic segment. Each occurrence of a casing, foreign pipeline or utility crossing, electric railroad crossing, buried metal debris, concrete structure, and others would be an independent pipeline segment for purposes of risk assessment. Such segments would carry the risk of interference (including shielding effects) whereas neighboring segments might not. For transmission pipelines in corridors with foreign pipelines, higher threat levels of interference may exist, although it is common for pipeline owners in shared corridors to cooperate, perhaps bonding their systems together, and thereby reduce interference potentials.

The two potential mitigation interference phenomena, shielding and DC-related interference are discussed in PRMM.

6.5.4 Combined Mitigation Effectiveness

When a coating holiday or a CP coverage anomaly is located, that location may be treated as having reduced mitigation or at least reduced reliability of mitigation. When both coating and CP gaps coincide, an active corrosion location is assumed.

In the absence of location-specific coating and CP gap information, general gap rates for each are used to determine the probability of coincident gaps. SME's can normally produce credible gap rate estimates for lengths of pipeline or areas of facilities. Based on older surveys, experience with excavations on the assessed system and similar systems, they can estimate how often they would expect a coating defect or an area unprotected by CP. This is of course not as reliable as an overline survey or ILI-based information, but is sometimes all that is available to an assessment.

With coating holiday rates and CP coverage gap rates, an estimate of the number of active corrosion points can be made. This is illustrated in the following example:

6.5.4.1 Example

The risk assessor has conducted a facilitated SME meeting and has obtained, for the preliminary P90 risk assessment of 20 miles of pipeline, estimates of coating and CP gap rates. He has chosen units of 'per mile' to help SME's produce their estimates (he can later convert to per ft2, a more appropriate unit to account for varying pipe diameters and associated surface areas). Results are SME estimates of 30 coating holidays per mile and 2 CP gaps per mile.

Since he seeks a probability of coincident gap in a very small area, he chooses one linear foot of pipeline as representative of the size of each hypothetical gap, converts the SME estimates to a 'per foot' unit, and multiplies them together to arrive at a frequency of coincident gap locations:

30/5280 gaps/ft x 2/5280 gaps/ft x 5280 ft/mile
= 0.011 coincident gaps/mile

For the 20 miles being assessed, he finds a small chance of active corrosion locations, expressed as a frequency of occurrence of:

20 miles x 0.011gaps/mile = 0.22 gaps in the assessed segment, or
22% probability of an active corrosion location somewhere in this 20 mile length of pipeline.

This approach reflects the reality of the complex corrosion control choices commonly encountered in pipeline operations. It is not uncommon for the corrosion specialist to have results of various types of surveys of varying ages and be faced with the challenge of assimilating all of this data into a format that can support decision making. Mirroring the SME's valuations provides the additional benefits of showing the value of some techniques over others as well as the value in increased survey frequencies. Additional adjustments for survey accuracy (including conditions under which the survey took place), operator errors, and equipment errors are also relevant. Such adjustments should play a role in assessments (even though they are not illustrated here) because they are important considerations in evaluating actual CP effectiveness. The assessment scheme is patterned after the decision process of the corrosion control engineer, but is of course considering only some of the factors that may be important in any specific situation.

6.5.5 External Corrosion Resistance

As noted, the resistance to failure by corrosion is efficiently measured as an effective wall thickness. This thickness, taken with the mitigated corrosion rate, yields a time to failure, or remaining life estimate. In the case of external corrosion and pressure containment, hoop stress carrying capacity is affected since, it is the extreme fibers of the vessel (pipe or component) that are being degraded. This has the effect, at least theoretically, of causing more loss of strength than an equivalent amount of interior wall loss.

See Chapter 10 Resistance Modeling.

6.6 INTERNAL CORROSION

6.6.1 Background

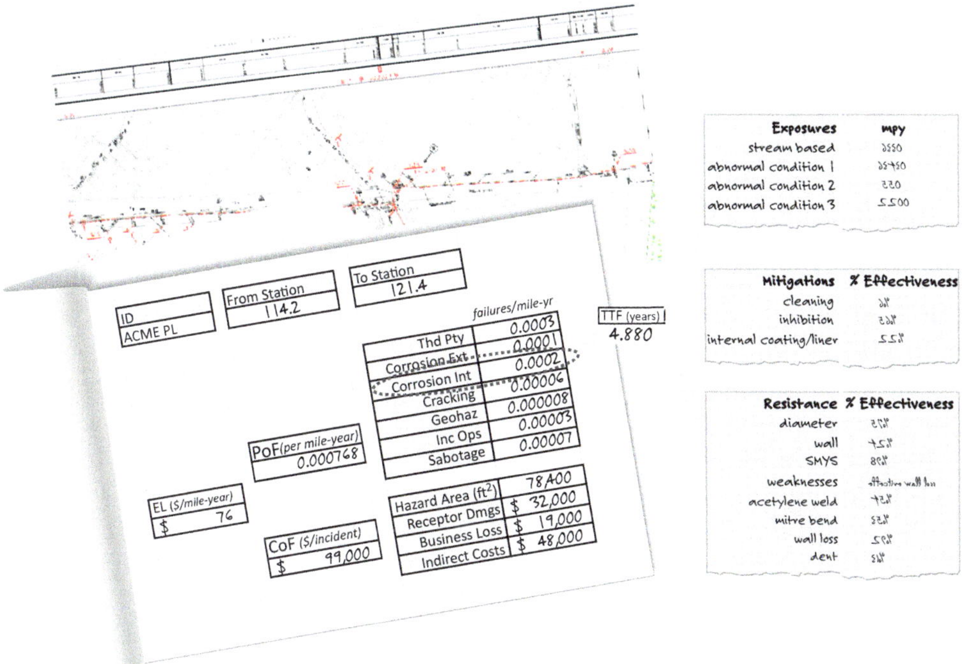

An assessment of the potential damage by internal corrosion is appropriate for most risk assessments. Internal corrosion results in pipe wall loss and is caused by a reaction between the inside pipe wall and the interior environment, ie, usually the product being transported and the influences of its flow regime. As with most analyses presented in this book, the focus is on steel components but the principles apply to all pipe materials. The assessment of the threat from internal corrosion is conducted by an examination of the product stream characteristics and the preventive measures being taken to offset corrosion potential. Presentation of the chemistry underlying internal corrosion mechanisms is beyond the scope of this text.

Corrosive activity may not be the result of the product ***intended*** to be transported, but rather a result of impurities in the product stream. Water and solids intrusion into a natural gas stream, for example, is not uncommon. As with other hydrocarbons, the natural gas (methane) will not harm steel, but water and other impurities can certainly promote corrosion.

The electrochemical process that causes steel to corrode from products transported involves anodic and cathodic reactions just as in external corrosion. Substances that commonly contribute to corrosion in pipelines are dissolved acid gases such as carbon dioxide (CO_2) and hydrogen sulfide (H_2S) as well as organic acids. For the electrochemical reactions to occur, an ionizing solvent must be present, which in the pipeline environment is usually water. Salts, acids, and bases dissolved in the water create the necessary electrolyte. Influencing factors can be very complex. For example, CO2 exacerbated corrosion of carbon steel varies with changes in velocity, pH, temperature, and shows significant changes to various combinations of these factors.

Internal corrosion commonly causes damage to the bottom portions of the pipe. In theory, a pipeline carrying hydrocarbons and a small amount of water will not experience internal corrosion if the water is dispersed and suspended in the product stream rather than flowing as a separate phase in contact with the bottom of the pipe.

Depending on the definition of 'failure' used in the risk assessment, reactions occurring inside pipe components that do not threaten integrity of those components may be excluded. An example of this is the buildup of wax or paraffin in some oil lines. While such buildups cause operational problems, they do not normally contribute to the corrosion threat unless they support or aggravate a mechanism that would otherwise not be present or as severe. See also the discussion under Chapter Risk of service interruption.

Some of the same measures used to prevent internal corrosion, are used not only to protect the pipe, but also to protect the product from impurities that may be produced by corrosion. Jet fuels and high-purity chemicals are examples of pipeline products that are often carefully protected from such contaminants.

Certain facilities can be exposed to corrosive materials in higher concentrations and for longer durations. Sections of station piping, equipment, and vessels can be isolated as "dead legs" for hours, weeks, or even years. The lack of product flow through these isolated sections can allow internal corrosion cells to remain active allowing accumulations of corrosion damages over time. Also, certain product additive and waste collection systems can also concentrate corrosion promoting compounds in station systems designed to transport products within line pipe specifications.

6.6.2 Exposure

To more efficiently model the various types of internal corrosion, it can be categorized into general classes, depending on the exposure scenario. One categorization uses two scenarios, corrosion under normal versus abnormal (or 'special') conditions. In the first, the transportation of a product always corrosive to the pipe (or other component) wall, is examined. In the second are scenarios where the product is corrosive to the pipe wall only under abnormal conditions. The distinction between the two becomes blurred in some scenarios, but is still a useful way to ensure both classes are addressed in an assessment.

The corrosivity of the pipeline contents that are routinely in immediate contact with the pipe wall are first assessed. The greatest threat exists in systems where the product is inherently incompatible with the component material and is also in continuous contact. This can be termed 'general' since it is the corrosivity that is most obvious and potentially occurs over the majority of the pipeline.

Another threat arises when corrosive impurities can get into the product stream (ie, an 'upset' scenario) or become concentrated/combined to create a more corrosive condition. This can be called a 'special' corrosion rate since it is abnormal, occurring infrequently over time and/or in only very few locations along the pipeline. These two scenarios can be assessed separately and then combined for an assessment of product corrosivity:

Corrosion Rate = [general product stream corrosivity] +
[corrosivity under special conditions]

These are additive since the worst case scenario would be a scenario where both are active in the same pipeline at the same location—both a corrosive product and potential for additional corrosion through special conditions. The balance between the two is situation specific, but because hydrocarbons are inherently non-corrosive to most pipe materials and most transportation of hydrocarbons strives for very low product contaminant levels, special corrosion rates might dominate for many hydrocarbon transport scenarios. In water transport, by contrast, general corrosion would be expected to dominate.

To begin, we assess the general corrosion potential from normal contact between flowing product and the component wall, based on product specifications and/or product analyses. Next, the potential for abnormal contacts between component wall and contents is assessed. Higher concentrations and contact durations of dropout contaminants such as water and solids accumulations in low spots can occur during no-flow, low-flow, or steep inclination conditions. Scenarios of offspec product receipts are included as special corrosivity. In either case, the term contaminant is used here to mean some transported substance that is beyond the agreed upon product purity specification limits and is corrosive to the pipe wall, even though the specification may allow some amounts of the substance.

Each of the two general scenarios of internal corrosion are assigned an unmitigated corrosion rate—the exposure—normally in units of mpy or mm/year, and a probability that such a corrosion rate manifests at the location being evaluated. This parallels the approach used to evaluate external corrosion. The locations of coincident loss of protective coating and CP, thereby allowing external corrosion, is analogous to the locations of sufficient contact time between corrosive substances and internal pipe wall that allow internal corrosion.

In many cases, assigning a mpy (or mm per year) exposure value will be a very generalized approximation. Rarely will an actual or even potential corrosion rate at a particular location be fully understood. Sometimes actual corrosion rates on similar components in similar conditions will be known. Sometimes, laboratory corrosivity rates in laboratory conditions will be known and may be extrapolated to field conditions. Use of either in estimating potential corrosivity at other locations will be problematic, but may be the only basis for an estimate. Since actual rates will be very site-specific, a plausible range of rates, rather than a single value, may be more useful.

From such a range, especially if an underlying probability distribution is also known or can be reasonably theorized, P50 and P90+ values for location-specific corrosion rates can be assigned.

Recall an earlier discussion of the 'test of time' as providing some evidence regarding the level of exposure. In the case of internal corrosion potential, however, 'years in service without findings of corrosion' may not be very compelling evidence, given the typical ranges of transported fluids and the often-changing operation and maintenance practices that many pipelines experience. See Chapter 2.8.6 The Test of Time Estimation of Exposure.

6.6.2.1 General Corrosion (Flow stream characteristics)

It is often economically advantageous to transport substances through a pipe that has some susceptibility to corrosion by the substance. This implicitly accepts the damage potential as manageable and/or the threats to integrity as tolerable.

There can be varying degrees of corrosivity tolerated by a transporter. These are examined in PRMM. Rates of 200 mpy or more have been seen in actual operating hydrocarbon pipelines and are clearly intolerable for most operations. Unmitigated corrosion potential approaching 10 mpy or more would be considered aggressive by most pipeline operators. Rates of 0.1 to 2 mpy are often treated as mildly corrosive and sometimes even as inconsequential. Note however, that over many years, even mild corrosion can threaten integrity.

Transportation of products by pipeline is normally governed by contracts that state delivery specifications. Most specifications will state the acceptable limits of product composition as well as the acceptable delivery parameters. When formal contracts do not exist, there is usually an implied contract that the delivery will fit the customer's need and be compatible with the transportation process. See additional discussion of specifications under Chapter 12 Service Interruption Risk.

The product specification can be violated when the composition of the product changes. This will be termed off-spec and will cover all episodes where the product deviates sufficiently from the intended specification to cause corrosion.

Most transmission pipelines require, via transportation specifications, that transported products are non-corrosive to the pipeline materials. Distribution systems, as receivers of transmission-quality product, similarly expect only non-corrosive products. Gathering systems generally do not carry such specifications and some amounts of corrosive products are expected. Regardless of the existence of a specification, episodes that create corrosion potential are possible in all types of pipeline systems and commonplace in some.

While very specific corrosion chemical processes can be modeled, it is often within the realm of desired accuracy to simplify corrosivity estimates. In one such simplification, the flow stream characteristics can be efficiently divided into two main categories—water related and solids related—for purposes of evaluating corrosivity [94].

Internal corrosion is also a common threat in hydrocarbon gathering pipelines where mixtures, including water and solids, and multiphase fluids are transported. Microorganism activities that can promote internal corrosion should also be considered. Sulfate-reducing bacteria and anaerobic acid-producing bacteria are sometimes found in oil and gas pipelines. They produce H_2S and acetic acid, respectively, both of which can promote corrosion [79].

Water is a pipelined product that presents special challenges in regard to internal corrosion prevention. Metallic water pipes often have internal linings (cement mortar lining is common) to protect them from the corrosive nature of the transported water. Raw or partially treated water systems for delivery to agricultural and/or landscaping applications are becoming more common. Water corrosivity might change depending on the treatment process and the quality of the transported water.

6.6.2.2 Special Corrosion (Upset potential and/or Abnormal Situations)

This is a measure of the potential for an increase in corrosion activity due to abnormal situations like contamination of the transported product or flow pattern changes. For instance, low flow rates can increase the chance of solid or liquid deposition and accumulation, while higher flow rates can cause erosion. Anything that leads to increased corrosive contaminant contact with pipe walls will logically increase corrosion potential and rate. Relevant phenomena can be rare and hard to assess. For instance, drag-reducing agents that are sometimes added by pipeline operators to enhance throughput can lower the ability of flowing hydrocarbon to entrain water by dampening turbulence. On the other hand, a return to higher flow rates can remove accumulations and could therefore be seen as preventative as described later.

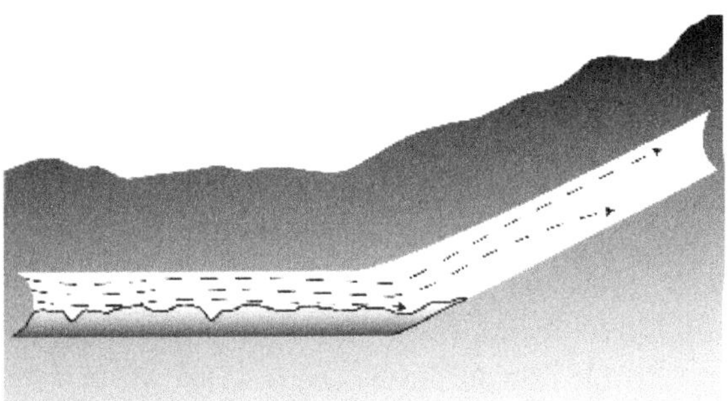

Figure 6.3 Liquid and solids holdup when critical angle exceeded

Under an assumption that a pipeline is designed to receive products and sustain a flow regime that minimize corrosion, accumulations of corrosive fluids and solids are considered to be abnormal conditions and hence modeled as part of the 'special corrosion' rates. Changes in flow patterns including stagnant flow conditions that lead to increased corrosion potential can usually be considered to be special conditions for most portions of most pipelines under the assumption that such scenarios are rare. If this assumption is not valid—ie, accumulations, higher contact time with pipe wall, etc are normal, then corrosion-accelerating flow patterns can form the basis of the 'general' corrosion rate. This involves treating accumulations of allowed-by-specification constituents as part of the 'general corrosion' rate, perhaps to better distinguish from accidental introductions of contaminants (constituents not allowed by specification).

A critical inclination angle calculation can be used to supplement and support exposure estimates at specific locations. This can be extended into applications of flow modeling that predict corrosive locations along a pipeline based on fluid stream, flow regime, elevation profile, pressures, temperatures, and other factors, continues to improve. Sophisticated models with high quality input data can more reliably predict depositions, accumulations, and subsequent special corrosion rates, in addition to corrosion potential from 'normal' contact of product stream with pipe wall.

When such modeling includes the effects of inhibitors, biocides, and perhaps other mitigation measures, the predictions generated should be treated as mitigated corrosion rates in the risk assessment. Therefore, provisions for failure of a mitigation measure should be considered.

The overall assessment of upset potential or abnormal conditions, as contributing factors to special internal corrosion potential, can be accomplished through an evaluation of the product stream and the items listed in PRMM.

Recall that estimates of exposure assume no mitigation. When mitigation is defined as only measures taken by the pipeline owner, then mitigation taken by others becomes part of the exposure evaluation. So, the likelihood of an error in a supplier's delivery system, leading to a contamination episode, is a part of the exposure estimate. Alternatively, a full exposure-mitigation analyses could be conducted on the product delivery system into the pipeline with the results feeding into the pipeline's product corrosivity exposure estimates.

When foreign material enters the pipe from external sources (not product stream sources), product contamination and internal corrosion are possible. With the lower pressures normally seen in distribution systems, infiltration can be a problem. Infiltration occurs when an outside material migrates into the pipeline. Most commonly, water is the substance that enters the pipe. While more common in gravity-flow water and sewer lines, a high water table can cause enough pressure to force water into even pressurized pipelines including portions of gas distribution systems. Conduit pipe for fiber optic cable or other electronic transmission cables is also susceptible to infiltration and subsequent threats to system integrity.

Special corrosion rates can be extremely aggressive. An operator installed a new hydrocarbon gathering system (oil and condensates) which, after only a few years in service, experienced internal corrosion leaks. Upon investigation, corrosion rates in excess of 200 mpy were discovered—far exceeding what was thought plausible in such systems. MIC was identified as a prime contributor. MIC rates up to about 10 mm/year (about 400 mpy) have been shown to be possible under laboratory conditions. Special pitting corrosion rates that do not involve MIC can also be very high.

6.6.2.3 *Probability of Corrosion Rates*

The most robust internal corrosion assessments will use surface areas in the probability estimates. Analogous to the assessment of external corrosion potential, this approach allows the assessment to highlight specific locations where internal corrosion is more likely, for example, at bottom of pipe at low spots with lower velocities and greater likelihood of contaminants having been introduced.

Recall the discussion of measurements and inferences as a means to model the disparity of information commonly seen along a typical pipeline. In some locations, direct measurements of corrosion rates will be available. In other locations, only relatively weak inferential evidence will be available and must be used to create an estimate of corrosion rate.

Monitoring is a key aspect of estimating the exposure, recognizing that many monitoring activities will be measuring a ***mitigated*** corrosion rate. Monitoring can be either direct or indirect. In either case, extrapolation from the monitored locations to all unmonitored locations will be required.

Monitoring and measurements that can be useful in the assessment include the use of probes and coupons, scale analysis (product sampling), inhibitor residual measurements, dewpoint control results, monitoring of critical points by ultrasonic wall thickness measurements, and effluent examinations from pigging programs.

It is not uncommon for pipelines to experience changes in service conditions over their lifetimes. In the oil and gas industry, product streams and excursion potentials change as new wells are tied in to existing pipelines and the stream experiences changes in composition, pressure, or temperature. While an internal corrosive environment might have been stabilized under one set of flowing conditions, changes in those conditions may promote or aggravate corrosion. Liquids settle as transport velocity decreases. Cooling effects might cause condensation of entrained liquids, further adding to the amount of free, corrosive liquids. Liquids may now gravity flow to the low points of the line, causing corrosion cells in low-lying collection points. Reduced velocities and increased depositions may prevent sweeping of accumulated solids and liquids.

Inspection for Corrosion Damages

Repeated wall thickness measurements at the same location, usually by ILI or NDE, offer a means of direct corrosion monitoring. The high inaccuracies associated with locating and sizing the often tiny, pin-hole type corrosion features means that uncertainty should be a part of the findings.

An alternate method is to use a spool (test) piece of pipe that can be removed and directly inspected for corrosion damage.

Any inspection program must consider inaccuracies and limitations of extrapolations of results and be repeated at appropriate intervals.

Caution must be exercised when assigning benign corrosion rates based solely on the non-detection of internal corrosion at certain times and at limited locations. It is important to capture where the potential for corrosion might be high, even when no active corrosion has yet been detected.

Indirect Corrosion Monitoring

Spot monitoring of internal corrosion is often done by either an instrumented probe or by insertion and subsequent inspection of a coupon designed to corrode when exposed to the transported product. Both methods require an attachment to the pipeline to allow the probe or coupon to be inserted into and extracted from the flowing product. More advanced configurations of probes or coupons, such as provisions for accumulations and simulations of stagnant pitting potential, add more credibility to any extrapolations from location-specific monitoring.

Monitoring of product streams also presents opportunities to infer corrosion potential. Product stream composition measurements range from simple moisture analyzers to full chromatograph analyses and from monthly composite sample 'bombs' or occasional 'grab samples' to nearly continuous analyses.

Monitoring of the materials displaced from a steel pipeline during maintenance pigging may include a search for corrosion products such as iron oxide—mentioned as a 'direct' monitoring method—or fluids and solids that are corrosive. Since contact time with the pipe wall is an important aspect of corrosion rate, the presence of corrosive materials alone is not the full picture. Nonetheless, examination of pigging effluent will help to assess both the corrosion potential and the extent of damage in the line. Examinations of filters and traps for corrosion by-products like iron oxide yields similar useful information, both direct and indirect.

Extrapolations

A probability estimate will normally be required to incorporate a potential corrosion rate into the risk assessment. The probability of a certain corrosion rate at a specific location on the pipeline arises from an understanding of all the elements previously discussed—corrosion mechanisms, product stream characteristics, and results from inspection and monitoring. Since much of this information will not be precisely known at all points along the pipeline, extrapolations from where it is known will be necessary. Furthermore, since conditions often change over time, both time- and location-uncertainties arise. Therefore, uncertainty over time (ranges of possible corrosion rates at the same location over time) and space (distance from locations of known or better-estimated corrosion rates) are both included in the probability values.

A probability or confidence level assigned to corrosion rate estimates captures the amount of uncertainty of the extrapolations as well as the uncertainty in the measurements of corrosion rates at the known locations.

Example: 6.3

Pipeline XYZ relies upon ACME Production Company to deliver a hydrocarbon stream substantially free of any corrosive component. Historical performance data from product stream analyzers and an examination of ACME's potential error rates associated with processes related to product delivery lead to estimates of general product stream corrosivity and possible contaminate drop out potentials at the delivery point and locations farther downstream. Then, flow patterns are studied to estimate contaminant accumulation potentials at the location being assessed. Combining both leads to estimates of 0.1 mpy 90% of the time and 10 mpy 10% of the time, at the location of interest. Pipeline XYZ estimates product corrosivity to be 0.1 x 0.9 + 10 x 0.1 = 1.1 mpy as a probability-weighted (also potentially viewed as a time-weighted corrosion rate) summation of corrosion rates at this location. An alternative approach would be to use 0.1mpy for P50 and 10 mpy for P90 estimates of internal corrosion exposure at this location. The internal corrosion mitigation practices would then be used with these estimates to arrive at the potential damage rate estimates.

6.6.3 Mitigation

Having assessed the potential for a corrosive product stream, the evaluator can now examine and evaluate mitigation measures being employed against potential internal corrosion. The probable effectiveness of mitigation measures is used with the exposure estimates to assess damage potential, modeled as a reduction in unmitigated damage rate. Estimating mitigation effectiveness will be challenging in many cases. The goal is to understand, for each unit of surface area (square inch of internal surface area), the ability of the mitigation measure to at least partially block corrosion that would otherwise occur.

With both exposure and mitigation varying along the pipeline, the probability of worst-case corrosion is directly related to the probability of mitigation gaps coinciding with the higher corrosion rates. Gaps in mitigation effectiveness at contamination accumulation points are more threatening than gaps occurring elsewhere.

Typical internal corrosion mitigation measures include:

- Internal coatings,
- Inhibitor injection,
- Regular cleaning,
- Operational measures such as flowrate modifications to sweep of liquid/solid accumulations, and
- Product treatments.

Monitoring via coupon or other probe is a common supporting activity although it is not a direct mitigation itself.

Although there are real-world dependencies among these measures (for example, inhibitors may not be effective without mechanical removals of buildups by pigging), they can generally be modeled as independent measures and can be related using OR gate math.

6.6.3.1 Pigging

It is common practice in many pipelines to use pigs to prevent long-term accumulations of liquids and solids and clean internal surfaces of a pipeline. Types of pigs, frequency of cleaning, and characteristics of the cleaning 'run' are all important to the program effectiveness (see background discussions in PRMM). Components such as sharp bends may reduce the cleaning effectiveness and, when coupled with a relative low spot—ie, critical angle exceeded—then this location may become a hot spot for internal corrosion.

Monitoring of the materials displaced from the pipeline following a cleaning pig should include a search for corrosion by-products such as iron oxide in steel lines. This will help to assess the extent of corrosion in the line and therefore the effectiveness of the pigging. A reduction in contaminant residence time—contact time with pipe wall—may be the appropriate measure of effectiveness. This will 'reward' more frequent and more thorough cleaning operations.

6.6.3.2 Inhibitor injection

Corrosion-inhibiting chemicals can be injected into the pipeline to prevent or reduce corrosion damage. Inhibitors are applied at intervals or continuously. Inhibition programs can be very expensive. Inhibitor effectiveness is often partially verified by an internal monitoring program as described above.

Formulations may have "oxygen-scavenging" properties that allow them to bond with the oxygen in the fluid and prevent its reacting with the pipe (oxygen being the primary corrosion agent with steel). Other chemical formulations create a film or barrier between the metal and the fluid. Biocides can be added to address micrbiologically-induced corrosion.

In some applications, another benefit of these additives is that they usually contain surface-active compounds that decrease oil and water interfacial tension so as to make it more difficult for water to separate from the oil flow. Conversely, chemical demulsifiers that are added to oil to remove water during processing before delivery to the pipeline can have the undesired effect of increasing the interfacial tension and thus causing easier separation of oil and water in the pipeline flow.

The risk assessment should consider whether the inhibitor injection equipment is well maintained and injects the proper amount of inhibitor at the proper rate.

Generally, it is difficult to completely eliminate corrosion through inhibitor use alone. A pigging program is usually necessary to supplement inhibitor injection. The pigging is designed to mechanically remove free liquids, solids, or bacteria colony protective coverings, which might otherwise interfere with inhibitor or biocide performance. Experience in some company's internal corrosion programs is that chemical inhibition is virtually ineffective without supplemental mechanical cleaning via pigs.

Even with both inhibition and mechanical cleaning, effectiveness is uncertain. When pitting corrosion is prevalent, mechanical cleaning and inhibitor effectiveness in narrow, deep corrosion features is problematic. Challenges are even more pronounced in multi-phase or multi-velocity flow regimes. Any change in operating conditions must entail careful evaluation of the impact on inhibitor effectiveness.

Recall that accumulation points are typically hot spots for internal corrosion. Therefore, gaps in inhibition effectiveness at contamination accumulation points are more threatening than gaps occurring elsewhere.

6.6.3.3 Internal coating/liners

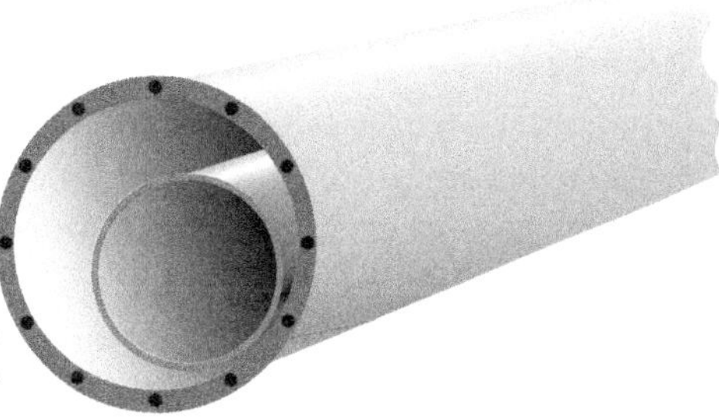

Internal coating has not been common practice for many pipelines but is growing in popularity due to advancements in liner materials, the deterioration of valuable pipelines, and their high replacement/repair costs. Internal coating includes the use of liners inserted into existing pipes, spray-on concrete or mortar, plastic, or other material, and the manufacture of multi-material composite pipes. A common concern in such systems is the detection and repair of a leak that may occur in the liner. Such leaks may accelerate corrosion at locations far from the leak location.

If an internal coating system is employed as defense against internal corrosion, its role in mitigation can be assessed in the same way as an external coating system. Its effectiveness can be judged by the same criteria as coatings for protection from atmospheric corrosion and buried metal corrosion described in this chapter. A holiday or defect rate per unit area shows the effectiveness of the coating. The probability of a defect coinciding with a corrosivity event—which is 100% of the surface area for general corrosivity and often <100% of the surface area for special corrosivity—yields the probability of that corrosion rate manifesting. Coating defects at internal corrosion hot spots points are more threatening than defects occurring elsewhere.

Note that an internal coating/liner that is applied for purposes of reduction in flow resistance might be of limited usefulness in corrosion control.

Operational measures

Dehydration, filtering, and other methods are commonly used to address corrosion potential prior to the product contacting the internal pipe wall, especially when the pipe material and product are not incompatible but where concentrations of impurities could lead to corrosion. Temperature, pressure, and flow rate control are other operational measures typically used to reduce corrosion potential, especially where duration of contact between product and material surfaces is a critical determinant of damages. The effectiveness of such measures is dependent upon many factors, including equipment design and maintenance, monitoring, and operator skills and procedures.

Example: 6.4 Assessing internal corrosion:

A section of a pipeline carrying natural gas from offshore production wells is being examined. Drying and sulfur-removal treatment takes place offshore. The line has been designed for flow rates that limit contaminant deposition or, if deposition does occur, residence time. Variance to design flow rates is common but unquantified.

Inhibitor is injected to manage corrosion from associated liquids that get past the treatment process. It has been determined that the inhibitor injector had failed for several weeks prior to correcting the malfunction. Pigs are run bi-monthly to clean out any accumulations. Both liquids and solids are removed.

Corrosion rates are monitored continuously via probes. Because the probes are located at the onshore receiving station, it is not possible to use the data to simulate corrosion resulting from deposition.

The highest corrosion rates observed at these coupons is 2.1 mpy but most readings are less than 0.1 mpy.

The evaluator requires a quick initial assessment and quantifies the damage potential as follows:

Exposure: Product corrosivity

The line is exposed to corrosive components only under upset conditions, but upset conditions appear to be rather frequent. The unmitigated general corrosion rate is estimated from experience with similar pipelines, to be 5 mpy, at the P90 level of conservatism. Corrosion probes normally show virtually no corrosion but are not deemed to provide representative corrosion rates at the more critical locations.

A critical angle calculation is performed and locations with inclines exceeding the critical angle are identified. These locations are assigned a P90 special corrosion rate of 10 mpy—additive to the general corrosion rate—due to deposition/accumulation potential. Therefore, some locations along this pipeline are modeled to have 5 + 10 = 15 mpy of corrosion potential prior to mitigation.

Mitigation

Inhibitor injection the inhibitor injection program is designed to limit corrosion to 1 mpy anywhere in the treated segment. Since effectiveness is difficult to achieve at all locations and the risk assessment is to be conservative, 50% effectiveness is the initial SME estimate, based on changes from pre-inhibition observations, ie 2.1 mpy observed in coupon analysis. This also captures the idea that inhibition alone, without prevention of accumulations, is more problematic.

Operational measures SME's assign a P90 value of 20% effectiveness for operational procedures alone, in acknowledgment that design flow rates should minimize depositions and sweep accumulations, but there do not appear to be devices or procedures to strictly control flowrates. SME's estimate that relatively low flow conditions manifest about 10% of the year. This value is doubled to arrive at a P90 estimate of 20%.

Pigging 50% effectiveness is assumed as an initial SME estimate based on trial-and-error applications of pig types and pigging frequencies used over several years.

Effectiveness of each of the preventive measures (inhibitor injection, operational measures, and maintenance pigging) is limited because of difficulties in continuously achieving corrosion control with the actions in a real-world production environment.

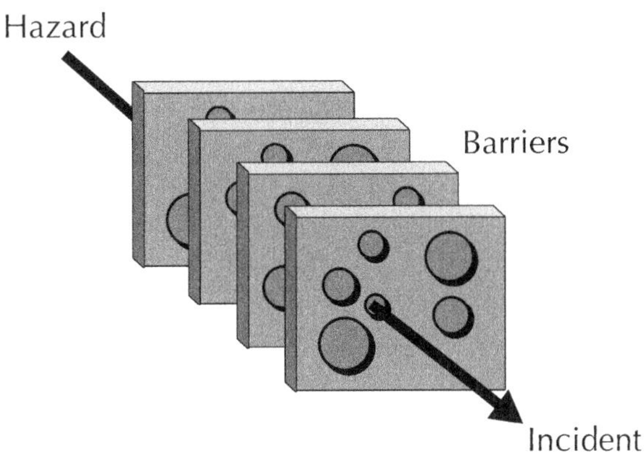

Figure 6.4 Swiss Cheese Analogy: More Slices and/or Fewer Holes Reduces Event Probability

Total, using OR gate math: 1-(1-0.5)(1-0.2)(1-0.5) = 80% initial estimate of combined mitigation effectiveness.

6.6.3.4 Damage Rates

Based on this initial P90 evaluation, mitigated corrosion rates are estimated to range from 1 to 3 mpy along the pipeline: 5 mpy x (1-80%) = 1 mpy to 15 mpy x (1-80%) = 3 mpy at low spots. These values are next used with best estimates of current wall thicknesses at all locations to obtain estimates of TTF. The extreme damage rate—15 mpy is plausible at low spots if mitigation fails—is also used to help establish the relationship between TTF and PoF by calculating a worst-case damage rate.

6.7 EROSION

Erosion, usually as a form of internal corrosion, can also considered a time dependent failure mechanism. Erosion can be thought of as 'mechanical corrosion' (recall the roots of the word 'corrosion'). Erosion is the removal of a component's wall material caused by the abrasive or scouring effects of substances moving against the component. It is a form of corrosion in the most general definition of the word. Abrasive particles moving at high velocities and impinging on an internal surface are the normal causes of erosion. Since internal erosion is generally avoided through design and operational measures, the potential for erosion can be treated as a special corrosion rate under internal corrosion. It often warrants an independent evaluation in the overall risk assessment, however.

While commonly associated with internal wall loss due to product stream characteristics, it can also occur on external surfaces. Wind born sand particles can cause significant damages to certain component materials, for example.

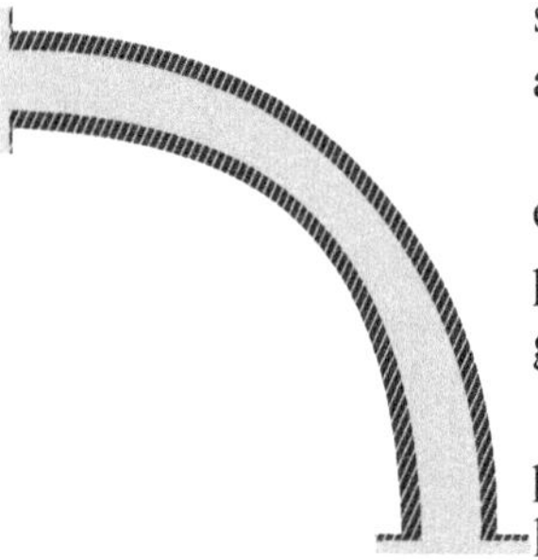

Erosion of pipe or component wall thickness is considered in this part of the risk assessment, while erosion of support, such as soil erosion during a flood, is captured under geohazards and resistance.

Interior wall erosion is a real problem in some oil and gas production regimes. Production phenomena such as high velocities, two-phase flows, and the presence of sand and solids create the conditions necessary for damaging erosion.

If occurring in the product stream, impingement points such as elbows and valves are the most susceptible erosion points. Gas at high velocities may be carrying entrained particles of sand or other solid residues and, consequently, can be especially damaging to the pipe components.

Historical evidence of erosion damage is of course a strong indicator of susceptibility. Other evidence includes high product stream velocities (perhaps indicated by large pressure changes in short distances) or abrasive fluids. Combinations of these factors are the strongest evidence. If, for instance, an evaluator is told that sand is sometimes found in filters or damaged valve seats, and that some valves had to be replaced recently with more abrasion-resistant seat materials, he may have sufficient reason to suspect significant exposure to this threat in certain components, especially those with impingement points. Calculations are available to help determine susceptibility when parameters such as velocity, particle size, and liquid contents are known or can be estimated.

A PoF for erosion is generated in the same way as for corrosion and cracking. First, an unmitigated erosion rate is estimated and normally expressed in mpy or mm/year. If mitigation such as liners or injected fluids are used to protect pipe surfaces, their effectiveness is estimated. The mitigated erosion rate is then used with an effective wall thickness (see Chapter 10.4.3 Effective Wall Thickness Concept) in TTF estimates. The TTF estimates lead to PoF estimates. As with corrosion, a probability aspect is usually needed, especially when a gap in mitigation—such as a hole in a liner—must coincide with an impingement point before damage occurs.

6.8 CRACKING

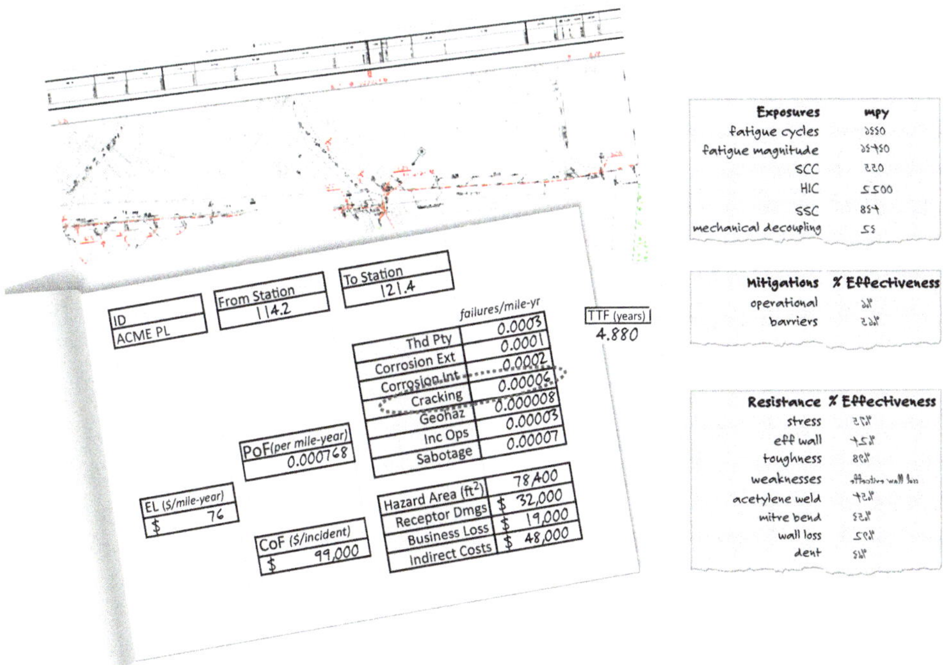

Cracking as a failure mechanism has not been a dominant source of accidents for most pipeline systems. However, for susceptible systems, failure modes can be dramatic and have resulted in serious incidents. Examples include fatigue failures in metallic components and rapid crack growth phenomena in plastics.

For all pipeline materials in common use, cracking can be evaluated in the same fashion as for steel. This is a major benefit for risk assessment.

As with other failure modes, evaluating the potential for cracking follows logical steps, replicating the thought process that a specialist would employ. This involves (1) identifying, at all locations, the types of cracking possible, both on internal, external surfaces; (2) identifying the vulnerability of the pipe material—how probable and how aggressive is the potential cracking; and (3) evaluating the prevention measures used.

As with corrosion potential, quantifying this understanding is done using the same PoF triad that is used to evaluate each failure mechanism: exposure, mitigation, and resistance, each measured independently. This will result in the following measurements, ready to be combined into a TTF estimate from which a PoF estimate can emerge:

- Aggressiveness of unmitigated cracking at any point on the component (units of mpy or mm/yr).
- Effectiveness of mitigation measures; a reduction in crack growth rate that would otherwise occur (units = %).
- Amount of resistance (units = equivalent wall thickness, inches or mm).

For purposes of risk assessment, the potential for cracking can be evaluated in two general categories: fatigue and environmentally assisted cracking (EAC). This categorization is useful since the two, while similar and sometimes overlapping, require slightly different analyses.

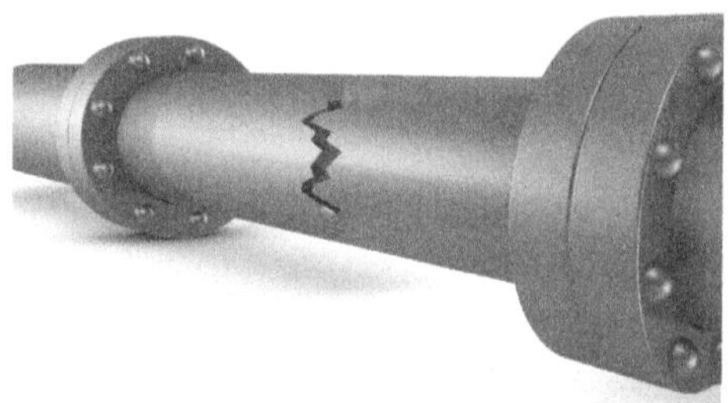

6.8.1 Background

Defects and flaws are found in all materials. They may be invisible to the naked eye but, when subjected to sufficient stress, may enlarge to critical dimensions, ie, dimensions that precipitate failure. Predicting the initiation and subsequent rate of growth accurately is usually not possible; cracks may emerge and grow over decades or virtually instantly depending on the circumstances.

Stress concentrators are another common contributing factor in crack related failures. Any discontinuity in a material, such as a sharp edge, slot, gouge, scratch, or dent, can increase the stress level. Fatigue lives of components can be significantly altered by corrosion damages. In corrosion fatigue, the acting stresses sufficient to cause failure can be less severe because pipe strength is diminished as a result of corrosion. For example, corrosion pits can become stress concentrators that allow routine pressure fluctuations to cause the formation and growth of cracks in the pit. When cracking is accelerated by environmental factors such as corrosion, the term *Environmentally Assisted Cracking (EAC)* is used to describe the phenomena.

Other phenomena influence crack potential by changing material properties. The metallurgy of steels or properties of non-metallic components can change from, for instance, exposure to excessive heat sources such as open flames as well as excessive cold. Changes in non-metallic materials can parallel the discussion of steel components. For instance, UV degradation, when causing brittleness in some plastics, can impact failure potential in ways similar to the HAZ in steel.

Fatigue loads will further the susceptibility to crack-type failures. Crack progression advancing solely through repeated cycles of mechanical effects is called fatigue cracking in this discussion.

In some larger, high-pressure gas pipelines, catastrophic fractures have been observed where the cracks propagate for miles along the pipeline. In these cases crack growth is rapid, exceeding the depressurization wave and potentially causing a violent release over considerable distance.

These kinds of failures increase the size of the product-release point but not necessarily the volume of the release. There is certainly an increased threat from mechanical damage—projectile debris for example. Steel sleeves can be used to arrest the crack growth until the depressurization wave passes, and crack-resistant materials, heavier-walled or duplex pipe are also preventive measures.

Catastrophic or "avalanche" failures are further discussed under 'exposure.'

6.8.2 Crack initiation, activation, propagation

Some modelers of cracking identify three distinct phases of crack progression through a material: initiation, activation, and propagation (fracture). All three are required before material failure by cracking occurs. This is a useful model for pipeline risk assessment since each of the three can be influenced by different factors whose identification and assessment leads to better understanding of failure potential and failure avoidance.

In this simple model, the crack potential and cracking avoidance can be understood as follows. If initiators—defects, stress concentrators, etc—can be avoided, then concerns for subsequent activation, propagation, and cracking failure are reduced. If initiators are present, then activation may be avoided by control of fatigue and/or stress. Propagation potential is impacted by flaw characteristics and stress, with the latter influenced by component thickness, allowing for the use of crack arrestors to prevent propagation.

6.8.3 Assessment Nuances

More so than most other failure mechanisms, cracking analyses bring shades of gray to the assignment of exposure, mitigation, and resistance. Factors such as material characteristics that influence the rate of cracking through a component wall can logically be classified as either an exposure variable or resistance variable. So, if material degradation or change (for example, creation of a HAZ) causes the material properties to change, is that better modeled as increased crack propagation rate (ie, more exposure)? or rather as reduced effective wall thickness (ie, less resistance)? Either will work—mathematically, there will be no difference in PoF estimates—but the latter may be more intuitive from a modeling perspective. See the discussion in Chapter 2.8 Probability of Failure.

Additional nuances appear in determining whether risk reduction actions are more appropriately modeled as changes to mitigation (blocking an exposure) versus resistance (absorbing forces), as is detailed in the discussion of mitigation and resistance later in this chapter.

Recall also the example of other modeling choices (reduced exposure or increased resistance?) for the role of an expansion loop or a span in a pipeline discussed in Chapter 2.8 Probability of Failure and Chapter 10.4 Resistance Modeling.

6.8.4 Exposure

6.8.4.1 Fatigue

Although historical pipeline accident data does not indicate that cracking is a dominant failure mechanism in most pipelines fatigue failure has been identified as the largest single cause of metallic material failure [47] and is certainly a real threat to some pipeline components. Fatigue is the weakening of a material due to repeated cycles of stress and is dependent on the number and the magnitude of the cycles. (See PRMM)

Fatigue cracking occurs as a result of repetitive, or cyclic, stress loadings on a pipe. Cyclic stresses can be axial (parallel to the axis of pipeline), circumferential (hoop stress in the tangential direction), or radial (perpendicular to the axis). Hoop stress is usually the most important source of cyclic loadings in pipelines because stress created by internal pressure is normally the largest stress the pipe experiences.

Fatigue is characterized by the formation and growth of microscopic cracks on one or both sides of the pipe wall. The first stage in the fatigue process is crack initiation, or nucleation. While nucleated cracks do not cause a fracture, some may coalesce into a dominant crack as the variable amplitude loading continues. In the second stage, the dominant crack grows in a more stable manner, and may eventually reach the thickness of the wall to produce a leak. Alternatively, the dominant crack may exceed a critical length or depth that the pipe steel can no longer endure. In this potential third stage, the crack becomes unstable and rapidly grows to a size that can produce a fracture and rupture.

Because the most highly stressed points are normally on the outer surface of a pressurized component, fatigue cracks usually originate on the exterior of the pipe and progress inwardly.[2] Pipe segments most vulnerable to fatigue cracking are those with pre-existing flaws or dents and other surface deformities caused by mechanical forces during installation or while in service. Stresses can concentrate at these damage sites, enabling cracks to form and grow after a relatively small number of load cycles, a phenomenon sometimes called low-cycle fatigue.[3] Other locations on a pipe susceptible

2 According to the Canadian National Energy Board (NEB), there have been no reported cases of internal SCC in North American transmission pipelines (NEB 2008).

3 Conversely, high-cycle fatigue occurs under a low-amplitude loading in which a large number of load cycles is required to produce failure.

to stress concentrations include discontinuities at grain boundaries and voids formed during pipe manufacturing.

Ref [1027] summarizes factors affecting fatigue life of metals as follows:

- Magnitude of stress including stress concentrations caused by part geometry.
- Quality of the surface; surface roughness, scratches, etc. cause stress concentrations or provide crack nucleation sites which can lower fatigue life depending on how the stress is applied.
- Surface defect geometry and location. The size, shape, and location of surface defects such as scratches, gouges, and dents can have a significant impact on fatigue life.
- Significantly uneven cooling, leading to a heterogeneous distribution of material properties such as hardness and ductility and, in the case of alloys, structural composition.
- Size, frequency, and location of internal defects. Casting defects such as gas porosity and shrinkage voids, for example, can significantly impact fatigue life.
- In metals where strain-rate sensitivity is observed (ferrous metals, copper, titanium, etc.) strain rate also affects fatigue life in low-cycle fatigue situations.
- For non-isotropic materials, the direction of the applied stress can affect fatigue life.
- Grain size; for most metals, fine-grained parts exhibit a longer fatigue life than coarse-grained parts.
- Environmental conditions and exposure time can cause erosion, corrosion, or gas-phase embrittlement, which all affect fatigue life. [1027]

These influences should be taken into account, as much as is practical, in the evaluation of material resistance.

6.8.4.2 Crack Growth Rate

From a risk assessment modeling point of view, a representative crack growth rate is sought and will be used with an estimate of effective resistance. The counts and magnitudes of stress cycles linked to crack growth rates is a rational way to model exposure, ie crack growth rate. This is admittedly an oversimplification of this complex issue. Fatigue depends on many variables as noted previously. At certain stress levels, even the frequency of cycles—how fast they are occurring—is found to affect the failure point.

It is conservative to assume that any amount of cycling is potentially damaging. Stress magnitudes can be based on a percentage of the tolerable operating stress levels or proportional loadings can be used. (Also see PRMM for discussion on categorizing pairings of cycle magnitude and frequency.)

Less common causes of fatigue on buried components and aboveground connections to equipment include loading cycles from traffic, wind loadings, water impingements, harmonics in piping, rotating equipment, and ground freezing/thawing cycling. Surges, slack line and vapor pocket collapse, and other transients are examples of abnormal initiators of cycles. Modern SCADA systems provide an excellent means of collecting stress cycles (from internal pressure changes) for examination.

A load spectrum is the family of stress-producing cycle counts and magnitudes. An equivalent cycle representing the full spectrum of actual cycles can be determined by a method such as Rainflow counting. S-N curves relate cyclic stress levels (uniaxial stresses, normally) with counts that result in failure, assuming that stresses are well below the material's elastic limit. This is sometimes referred to as the high-cycle fatigue regime, where counts greater than 10^4 can be absorbed before failure occurs. The S-N curve is also used to determine the damage contribution from each cycle. Cumulative damage theories have been developed to relate the spectrums of cycling to failure time via the S-N curves. The Miner's Rule (or Palmgren-Minor Rule) collects the damage contributed by each cycle. The Paris Law equations are also used to relate stress intensity factors to fatigue crack growth.

With tens of thousands of cycles typically required for failure in this regime, representative crack growth rates will usually be very small. For instance, even if a relatively small cycle count of 10,000 cycles is required to fail a component of wall thickness of 0.250", the implied crack rate is 250/10,000 = 0.025 mils per cycle.

Low cycle fatigue occurs when loadings produce stresses beyond the elastic limit and plastic deformation occurs. Pressure testing of components can produce this type of fatigue. Relationships using strain limits have been formulated to predict failure under these high stress loading scenarios. [1028]

A possible source of fatigue stresses is highway and railroad crossings. Research in these scenarios have produced design guides that include considerations for circumferential and longitudinal stresses due to earth loads, traffic loads, and the pipe's internal pressure. For instance, the following variables used in such calculations provide insight into the most critical determinants of imparted stresses:

S_{Lh}	S_{Hh}	Stress	Cyclic stress due to the highway
K_{Lh}	K_{Hh}	Stiffness factor	Dependent on wall thickness to diameter ratio & soil type
G_{Lh}	G_{Hh}	Geometry factor	Dependent on depth & diameter
R		Pavement type factor	
L		Axle configuration factor	
F_i		Impact factor	Dependent on depth according to formula 1.75–0.03·(H-5) for depths of 5′ – 30′and equal to 1.75 for depths < 5′
w		Applied surface pressure	83.3 psi for single axle (12 kips / 144 in^2)
			69.4 for tandem axle (10 kips / 144 in^2)
H	Depth		Depth from top of pavement to crown of pipe
D	Diameter		Nominal outside diameter of the pipe
t_w	Wall Thickness		Nominal wall thickness of the pipe

Example: 6.5 Estimating fatigue cracking rates:

PRMM presents an example point assignment scheme for evaluating fatigue risk in an older risk assessment approach. Modifying that example to reflect the updated risk assessment approach, the following example of assessing fatigue potential is offered.

The risk assessment has identified two types of cyclic loadings in a specific pipeline section: (1) a pressure cycle of about 200psig caused by the start of a compressor about twice a week and (2) vehicle traffic causing an external loading resulting in a 5-psi longitudinal stress at a frequency of about 100 vehicles per day. The section is approximately 4 years old and has an MOP of 1000psig. The traffic loadings and the compressor cycles have both been occurring since the line was installed.

For the first case, the evaluator uses a frequency of (2 starts/week ∞ 52 weeks/year ∞ 4 years) = 416 cycles and a cycle magnitude of (200psig/1000psig) = 20% of MAOP per cycle. Using these values and published crack growth information yields a crack growth rate of 0.1 mpy, using additional conservative assumptions regarding defects present, material toughness, crack properties, and other factors.

For the second case, the to-date cycles are equal to (100 vehicles/day ∞ 365 days/year ∞ 4 years) = 146,000. The cycle magnitude is equal to (5psig/1000psig) = 5% of MAOP. Using these two values even in a conservative analysis results in very small per-cycle crack growth rates, and summarizes into annual estimate of crack growth at 0.02 mpy.

The cracking rates are conservatively assumed to coincide at a single theoretical defect, resulting in a combined crack rate of 0.12 mpy for use in TTF calculations.

6.8.4.3 Vibrations/Oscillations

As an indicator of potential fatigue loadings and a common cause of failure of mechanical couplers, sources of vibration can be included in the risk assessment. Rotating equipment—pumps and compressors—are common sources of vibration. Components on supports, especially when shared with traffic as on a road or railroad bridge, can be subjected to continuous or intermittent vibrations. Vehicle traffic over buried components can impart vibrations in addition to direct fatigue stresses. When vibration is believed to be a separate failure mechanism from fatigue, it can be added to the risk assessment, perhaps most logically as increased PoF from cracking. Failures involving separation of mechanical couplings like threaded or flanged connections, more influenced by vibration effects than classical fatigue, can be considered types of cracking failures.

There are often more opportunities for fatigue type failure mechanisms within more complex facilities including severe pump starts/stops, pressure cycles, fill cycles, traffic loadings, etc. Rotating equipment vibrations, as a prime contributor to vibration effects, can be directly measured or inferred from evidence such as action type (piston versus centrifugal, for example), speed, operating efficiency point, and cavitation potential. Vibration monitoring is a common part of rotating equipment instrumentation, mostly to ensure reliability but also supporting integrity management.

Vibration and oscillations are also possible due to fluid movements around a pipeline, including wind and water: Vortex induced vibration (VIV); wind induced vibration (WIV). Vortex shedding, whether by wind or water, can generate sufficient forces under certain circumstances, to move a pipeline segment. This movement can become rapid and relatively large, causing fatigue loadings in the pipe material. Fluid density, speed, cross sectional area in flow stream, frictional drag across the object and other factors influence the onset and magnitude of movements.

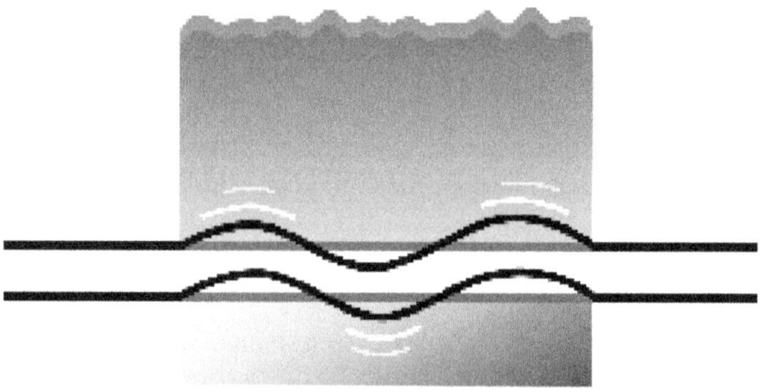

Figure 6.5 Vibrations or Oscillations can Cause Fatigue

Vibration monitoring provides insights into fatigue potential. This helps to identify when a material is subjected to higher vibration frequency (number of events/time), and/or, higher magnitude (change amount), considering duration (time) and proximity to component being assessed (when not the component itself), A robust program would include monitoring of in-service equipment/material's frequency, duration, and level and location of vibration stresses from various sources, including pumps, rotating equipment, wind, throttling valves, surges, temperature changes, ground movements, traffic, etc.

Common practices to minimize vibration effects include compensations designed into equipment supports, PPM practices especially for rotating equipment, the use of pulsation dampers, and the use of high ductility materials operating at relatively low stress levels. The assessment should also consider varying risk reduction effectiveness of programs such as continuous monitoring with automatic shutdown (which shuts down equipment upon exceedance of pre-set vibration limit) versus monitoring with alarm versus manual monitoring (ie, spot sampling).

6.8.4.4 Mechanical Couplers

Separation of mechanical couplers—screwed connections, flanges, etc—can also be modeled as a cracking phenomenon. There is a time-dependency implied in these types of failures since, at one time, no leak was present. The time until sufficient 'loosening' occurs can be treated as analogous to a crack progression rate through a material.

6.8.4.5 EAC

Environmentally assisted cracking (EAC) occurs from the combined action of a corrosive environment (or other material-property-influencing environment), coupled with a cyclic or sustained stress loading. The more common EAC forms include stress corrosion cracking (SCC), hydrogen stress corrosion cracking (HSCC), sulfide stress corrosion cracking (SSCC), hydrogen-induced cracking (HIC), hydrogen embrittlement, and corrosion fatigue. Corrosion fatigue cracking arises from the same pressure-related cyclic stresses that produce fatigue and mechanical cracking but are exacerbated by active corrosion mechanisms. These are all recognized flaw-creating or flaw-propagating phenomena.

Some forms of EAC can be caused or exacerbated by hydrogen-assisted cracking. For instance, when sources of hydrogen are present—such as from agents in a product stream (such as H_2S) or from external sources such as excessive cathodic protection voltage—cracking potential may increase. Hydrogen-assisted cracking can occur as a result of the diffusion and concentration of atomic hydrogen in a crack space or other micro-structural void in a metal. These concentrations may increase the existing stress load on the metal to form a stress concentrator where cracks can develop. Hydrogen can also adsorb to the metal surface to reduce surface energy and migrate to the microstructure reducing interatomic bond strength and providing a nucleation site for cracks. See also the discussion of failures of repair sleeves due to hydrogen permeation through steel (Chapter 10 Resistance Modeling, and ref [1001]).

As perhaps the most common of the EAC forms in pipelines, SCC has been more deeply researched than others, allowing further discussion. While specific to SCC, some of the following discussion is also relevant to the other types of EAC, for example, residual stresses, sensitizing agents on material surface, etc.

6.8.4.6 SCC

Stress corrosion cracking occurs under certain combinations of physical stresses coupled with active corrosion. Accounting for several hundred documented pipeline failures in the United States [52] some investigators think that the actual number of SCC related failures is higher since SCC is often very difficult to recognize.

See PRMM for a background discussion of this most common form of EAC.

Low stress in a benign environment is the condition least likely to support SCC, whereas high stress in a corrosive environment is the most favorable. Maximum SCC rates of over 40 mpy have been reported in both laboratory and field environments.

It is generally accepted that three conditions must be present to support SCC: tensile stress, a susceptible material, and a corrosive environment at the surface.

In addition to the necessary three conditions to support SCC, an additional factor must be present for an SCC ***failure*** to occur. This is the formation of a crack of critical size. Since SCC is characterized by colonies of tiny cracks, the formation of a critical-size crack involves the coalescence of multiple, otherwise-benign tiny cracks. There are many instances of SCC colonies that will not coalesce nor grow and therefore pose no threat to a pipeline. However, there is not currently a reliable way to differentiate these from the fewer scenarios where component integrity is actually threatened by the colonies.

ASME/ANSI B31.8 identifies high risk factors, as discussed in PRMM. An automatic screening incorporating these criteria can be set up in a computer environment. Note, however, that operators report discovery of SCC in locations that do not have all of these characteristics. Therefore, the threat (unmitigated SCC crack growth rate) cannot often be assigned zero.

Stress Tensile stress on the surface of a component is a prerequisite for SCC. A static surface stress may be generated from in-service conditions, such as sustained internal pressures. The acting stress may also be residual in nature, introduced during bending and welding in manufacturing, or it may arise from external soil pressure and differential settlement. At sites of surface damage, such as dents and corrosion pits, stress levels in the circumferential and axial directions are higher than on undamaged portions of the pipe surface. The same locations on the pipe that concentrate cyclic stresses, such as gouges, surface discontinuities, and appurtenances, can concentrate static stresses. In many cases, the stress will be virtually undetectable. Furthermore, breaks in the surface film may occur at these discontinuities to make the area more prone to electrochemical corrosion.[4]

As with most cracking regimes, the higher the stress, the more potential for SCC crack formation and growth. Limiting the introduction of residual stresses during pipe manufacturing, transportation, and installation are important to reduce SCC susceptibility. Internal pressure is the major in-service source of static hoop stress. Lowering the operating pressure of a pipeline would be expected to reduce the potential for SCC. Some sources suggest that a stress level corresponding to design factor of class 2 (per regulations in the US: CFR 49 Part 192), 0.60, could be considered to be a threshold, below which there is

4 At sites of surface damage, such as dents and corrosion pits, stress levels in the circumferential and axial directions are higher than on undamaged portions of the pipe surface.

no evidence of cracking. By this criteria, SCC would not be expected in class 3 or 4 areas (population density categories in US regulations, see *Class Location*) which correspond with design factors of 0.5 and 0.4. However, the specific relationship between SCC and hoop stress is not well established. Evidence from SCC failures show that hoop stresses have varied between 46 and 77 percent of the SMYS of a pipeline.[4]

Environment High pH levels are believed to be a contributing factor in classic SCC on steel surfaces.

Material type In steel, a higher carbon content (>0.28%) is thought to increase the likelihood of stress corrosion cracking.

These necessary conditions for SCC of steel are further discussed in PRMM.

6.8.4.7 Nonmetal EAC

As noted, nonmetal materials are also susceptible to mechanical-corrosion mechanisms such as stress corrosion cracking (SCC). While the environmental parameters that promote EAC in nonmetals are different than in metals, there are some similarities. When a sensitizing agent is present on a sufficiently stressed pipe surface, the propagation of minute surface cracks accelerates. This mirrors the mechanism seen in metal pipe materials. Organic chemicals can also aggravate environmental stress corrosion cracking [2]. For plastics, sensitizing agents can include detergents and alcohols. The evaluator should determine (perhaps from the material manufacturer) which agents may promote EAC. A high stress level coupled with a high presence of contributing soil characteristics would warrant assignment of a relatively high crack exposure in the risk assessment.

6.8.4.8 Avalanche Failure

Avalanche failure potential was previously noted. A crack will move at the speed of sound through a material. If the crack speed is higher than the depressurization wave—where pressure is the driving force creating the failure stress—then cracking continues. Material properties and thickness can each reduce crack speed. Crack arrestors take advantage of this. Less compressible products depressure quickly and therefore do not provide the sustained driving force for continued crack growth. So, changes in either the material or the product can change the potential for crack propagation.

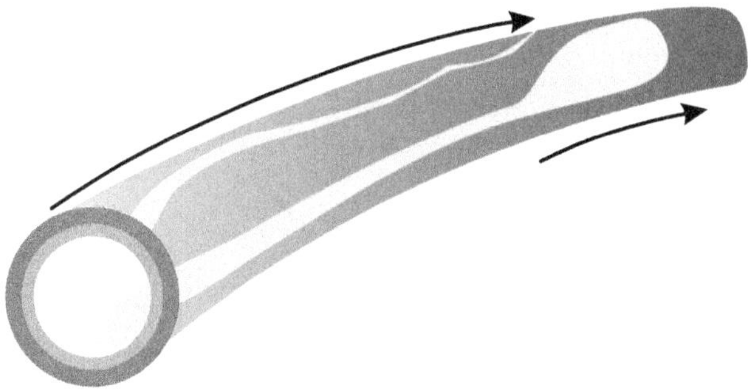

Figure 6.6 Crack propagation vs product depressurization

6.8.5 Mitigation & Resistance

Cracking mechanisms are somewhat unique in that they often involve more complex interactions of cause and effect variables. Many crack scenarios involve a force that causes a movement or strain which generates a stress which grows a crack. Failure potential can be reduced by reducing the initial force, by protecting against the force, by reducing the movement, or by absorbing the strain or stress without damage.

Consider wave induced vibration (WIV). Wind is the initiating force and cannot be changed, but it can be blocked. A windscreen or re-direction would be a measure that changes the movement potential. The pipe movement from WIV could also be prevented by changing span length, pipe weight or profile, or adding dampers. Alternatively, stress levels could be changed by altering the amount of restraint at the supports.

Crack growth rate is the measure of exposure and is usually a function of a component's movements and stresses. The operator often has more control over cracking exposure than exposures from other threats. In cracking, mitigation is therefore sometimes indistinguishable from changes to exposure and resistance. Many measures to reduce failure potential from fatigue are actually changes to exposure rather than defenses against a pre-established exposure. For instance, risk reduction can be achieved through reduction of internal pressure cycles—directly reducing the exposure level.

A challenge is determining which operational changes are 1) actually altering the exposure versus 2) blocking the exposure (mitigation) versus 3) resisting the exposure (resistance). Classifying each change as either exposure, mitigation, or resistance is sometimes not as obvious as for other failure mechanisms. Fortunately, the risk assessment format and mathematics ensures the same final PoF estimate regardless of how the elements are classified. Nonetheless, a brief discussion of some nuances in the cracking PoF assessment is warranted to deepen the understanding of modeling crack failure potential.

Depending on how integrated they are with exposure estimates, some actions and devices can be clearly modeled as independent mitigation measures. Use of pipe casings or other load transfer techniques would reduce the transmission of loads to the pipe and could be considered independent mitigation measures. Most would agree that the wind screen option in the previous WIV example is best modeled as a mitigation.

Vibration dampers, anti-WIV devices, special supports are examples that can be modeled as either mitigation—blocking the movement that causes damage—or resistance—allowing the component to absorb the forces without damage. The forces generating the exposure have not changed, but the component is either more protected from or more able to tolerate their otherwise damaging effects.

Other potential mitigation measures may already be included in the exposure estimates. These include minimization of component vibrations and stress through careful attention to equipment supports, PPM practices, continuous monitoring with automatic shutdown (ie, excessive exposure is being prevented). A pulsation damper is potentially modeled as either a mitigation device or is factored into exposure estimates, perhaps contingent upon the owner of the equipment[5] or the primary intent of the damper.

In EAC, mitigation of corrosion will also reduce the EAC crack growth rates. When the risk assessment combines the corrosion growth rate with the cracking rate into the EAC growth rate, then the corrosion mitigation is accounted for.

In many cases, cracking PoF reduction occurs more directly through resistance influences. Choice of material (even specific steel metallurgy), wall thickness, and stress level reduce crack growth potential from a resistance standpoint. It is common practice to put extra strength components with very high ductility into applications where higher fatigue loadings are anticipated. Use of high ductility materials operating far from their maximum stress levels is a proven method of designing crack resistance into a structure.

5 See Chapter 2.8.12 Nuances of Exposure, Mitigation, Resistance for a discussion of foreign owned/operated risk mitigation systems.

7 GEOHAZARDS

Highlights

The chess-board is the world, the pieces are the phenomena of the universe, the rules of the game are what we call the laws of Nature. The player on the other side is hidden from us.

Thomas Henry Huxley

Geohazards

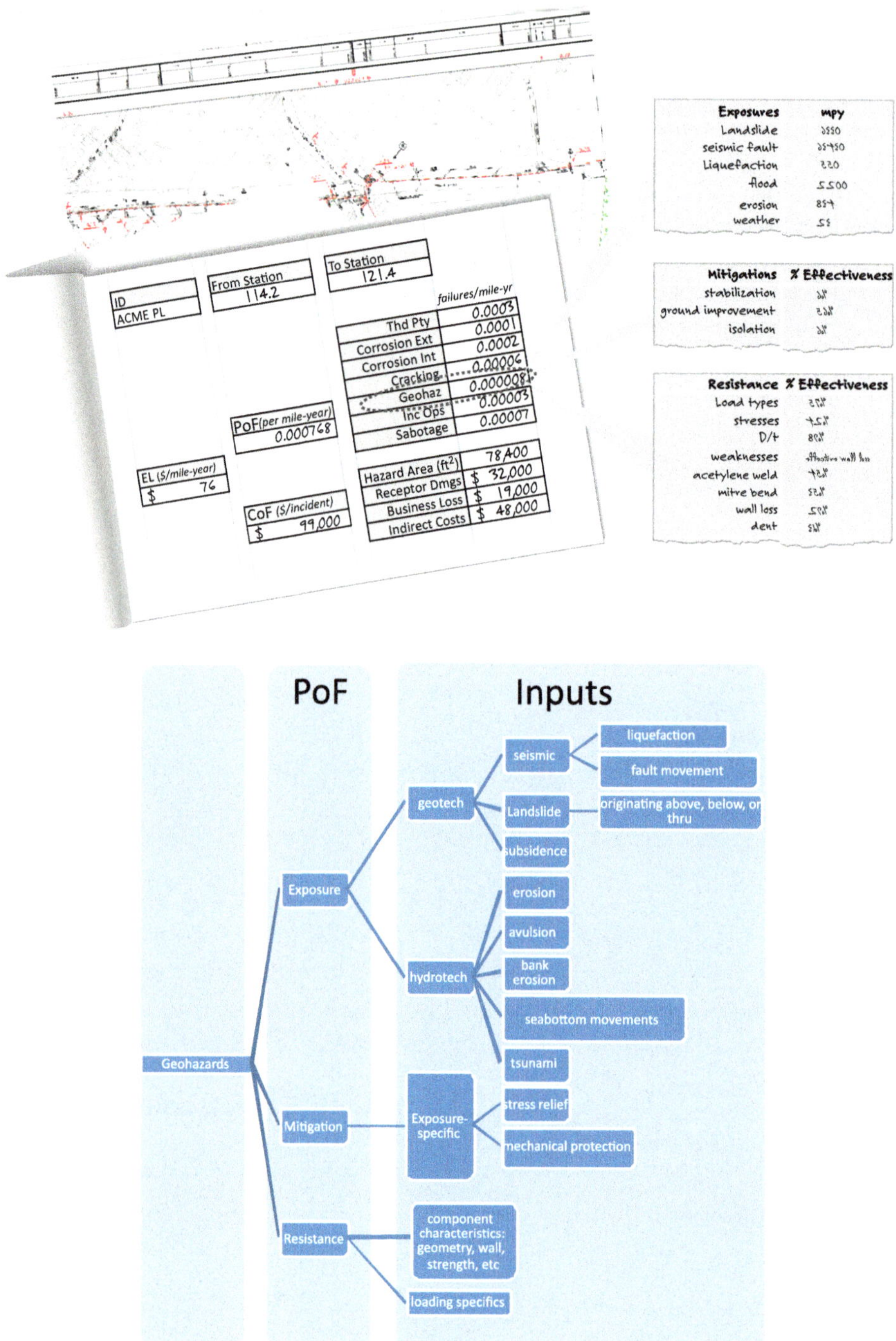

Figure 7.1 Sample of data typically used to assess the geohazard failure potential.

All of our creations are subject to the laws of nature—Mother Nature hates things she did not create.

SECTION THUMBNAIL
How to assess the damage potential and failure potential from geohazard-related forces such as from landslides, floods, and seismic events.

Events that subject a pipeline to injurious loads/stresses due to land movements and/or geotechnical events of various kinds, are termed 'geohazards' in this text. Geohazards may cause sudden and catastrophic movements of large masses of earth or they may be slow-acting forces that induce stresses on the pipeline over a long period of time. They can cause immediate failures or add considerable stresses to the pipeline, limiting its ability to resist other failure mechanisms.

Potentially damaging geohazard events are caused by onshore and offshore phenomena of seismic fault movements and soil liquefaction, aseismic faulting, soil shrink-swell, expansive soil movement, subsidence, erosion, landslide, scour, washout, frost heave, ice berg scour, ice/snow loadings, hail, water/debris impingements, sand dune movements, meteorites, lightning, and others. These terms sometimes describe overlapping phenomena or are different terms for the same phenomenon (for example, erosion and washout) but a full listing ensures that none are overlooked. Many weather-related phenomena can trigger a damaging geohazard event. Freezes and flooding are examples. Events such as falling trees (due to windstorm, ice, etc) can be included either as geohazards or as impacts, covered in third party damage potential (as a modeling convenience as discussed in Chapter 5 Third-Party Damage).

Water/land movements examined in a risk assessment should include all potential for pipeline damage or failure, onshore or offshore, due to triggering events such as tsunami, hurricane, flood, windstorm, rainfall, moisture and temperature changes and others. Again, terminology that includes overlapping events helps ensure complete coverage of initiating mechanisms.

The geohazard threat is usually very location specific. Many miles of pipeline are located in regions where the potential for damaging land/water movements is nonexistent. On the other hand, land movements are the primary cause of failures, outweighing all other failure modes, for sections of other pipelines.

Geohazards logically fall into a failure cause category often called 'external forces'. However, that categorization would have to capture exposures ranging from vehicle impact to excavator contact to landslide and many others, resulting is a non-transparent risk model. Geohazards normally warrants consideration as an independent threat. However, several overlapping elements can be involved and can make categorization of the cause of failure as third-party damage vs. geohazards difficult. For instance, a failure scenario involving man-made structures moving along the seabottom during a storm, has elements of both third party and geohazard. Scenarios of structures overturning during wind and ice storms similarly have both aspects. The assessor should choose a modeling structure that is preferable to his users.

Geohazards may be further categorized to make modeling more efficient. Sub classes may include hydraulic or hydrotech for exposures related to water, especially moving waters, and geotech for phenomena not involving water to any significant extent. (Also see PRMM.)

7.1 FAILURE PROBABILITY: EXPOSURE, MITIGATION, RESISTANCE

Recall that all exposures are evaluated in the absence of mitigation. For example, the unmitigated exposure from falling trees might be estimated to be on the order of several times per year—perhaps coinciding with severe storm frequency. It is only after adding mitigation—notably depth of cover—that the threat appears as small as most intuitively believe it is.

As with all threats, it is important to maintain a discipline of assessing exposure separately from mitigation and resistance, avoiding any temptation to short-cut the assessment to a perceived outcome that may not adequately reflect true risk.

Note also that risk reduction 'credit' for things like extra strong pipe to withstand instability events is recognized in the resistance assessment and should not be a consideration in exposure estimation.

7.1.1 Pairings of Specific Exposures with Mitigations

Although an often-justifiable short cut in risk modeling is to collect many types of exposures and pair them with a single collection of mitigations, it is sometimes more correct to pair specific exposures with pertinent mitigations. This is essential where differing exposure-specific mitigations are employed and/or where mitigations have varying effectiveness depending on the type of exposure. For example, depth of cover plays varying roles in geohazard phenomena of landslide, flood, buoyancy control, and seismic liquefaction and cannot be assigned the same level of mitigation benefit to each.

7.1.2 Spans and Loss of Support

Geohazard phenomena may generate stresses directly or, alternatively, they may change the support conditions, thereby indirectly changing stresses. Therefore, depending on the effect, their role as exposures or resistance reducers or both should be considered in the PoF assessment.

For instance, many of the geohazard events may not directly threaten integrity but rather will indirectly endanger a pipeline via creation of a span. Examples include subsidence, erosion, scour, and even some landslide scenarios. Span modeling is discussed as a nuance of the PoF triad in Chapter 2 Definitions and Concepts.

7.1.3 Component Types

As with many threats, differences in component material properties will complicate, for some systems, the modeling of susceptibility to damage from some land movements. For instance, the many types of materials often found in a distribution system pipeline means that assessment must often accommodate flexible and inflexible pipe, a variety of mechanical couplers, plastics and metals, and other considerations. Each component, in consideration of its performance under various loading scenarios, may have varying damage potentials and resistance capabilities.

For instance, for the same reasons that they are less threatened by spans, larger diameter pipelines made from more flexible materials and joining processes that create a more continuous structure, such as welded steel and fused PE pipelines, have historically performed better in seismic events and when exposed to soil movements from frost action and subsurface temperature changes.

Facilities with taller structures such as tanks, will need to include potential for toppling during some geohazard events such as earthquakes, floods, and landslides.

Offshore components typically have additional considerations for scour, lateral forces when spanning, and buoyant. See discussions of cracking and vortex shedding.

7.2 EXPOSURES

In measuring or estimating exposure to geohazards, it is important to first list all potentially damaging mechanisms that could occur at the subject location. Then, numerical exposure values should be assigned to each. Pre-dismissal of exposures should be avoided—the risk assessment will show, via low PoF values, where threats are insignificant. It will also serve as documentation that all threats are considered.

Geohazards are normally first considered in the design phase where mitigation or resistance is increased to reduce failure potential, as needed. In the risk assessment, each exposure should have a future frequency of occurrence assigned, by imagining that no mitigation nor resistance is available to prevent failure—the imagined 'unprotected tin can' scenario.

Some common geohazard threats to pipeline integrity are discussed here and in PRMM. As previously noted, some nuances of continuous exposure and changes in resistance due to loss-of-support will often need to be considered in defining exposure events and resulting damage and failure potentials.

7.2.1 Landslide

Slope is often an aspect of a damaging land movement. Landslides, rockslides, mudslides, mudflows, creep, and other related events can occur from heavy rain, especially on slopes or hillsides with removed or heavily cut vegetation or where construction or other activities have altered the land. Debris flows—usually involving steep mountain channels and soil liquefaction ('mountain tsunami' in Japanese [1016])—are also included here.

A sometimes used categorization of landslides based on soil movements, geometry of the slide, and the types of material involved results in the following five categories: falls, topples, slides, spreads, and flows [777]. See PRMM Figure 5.5 and Table 5.7.

Landslide events can have frequencies ranging from 'never' to 'multiple times per year' for longer stretches of pipeline. They are logically related to the frequencies of the underlying causal events such as precipitation and seismic events.

Some available public databases provide rankings for landslide potential. As with soils data, these are very coarse—usually missing smaller, but potentially severe scenarios such as embankments and steep creek banks. These datasets are best supplemented with field surveys or local knowledge. Nonetheless, as a preliminary method of assigning initial threat values to long lengths of pipeline quickly, such ranks, converted into event frequencies, can be useful. The conversion from ranks into frequencies should incorporate the protocols underlying the assignment of ranks in the original data. For example, see the discussion of factors used to establish ranks in the US Natural Disaster Study later in this chapter.

7.2.2 Soils (shrink, swell, subsidence, settling)

Earth movements involving localized changes in soil volume can cause shrinkage, swelling, or subsidence. These can be caused by changing temperatures or moisture contents as well as subterranean water movements and other phenomena. These can cause loss of support as well as additional shear forces and bending stresses on a pipeline component.

Changes in soil moisture content and temperature effects, often occurring in seasonal patterns, have been correlated with both water and gas distribution system break rates. These are often related to soil movements which cause changes in stresses on buried components. Where such correlations between break rates and physical phenomena are established, they can be used in risk assessment and break forecasting as well as in comparative risk assessments between regions with differing climates.

Exposure rates to some of these phenomena can therefore be linked to frequencies of triggering events such as temperature changes and rainfall events. In other cases, the potential may be largely unknown. Refer to Chapter 2.8.6 The Test of Time Estimation of Exposure discussion of exposure estimates using 'test of time' evidence.

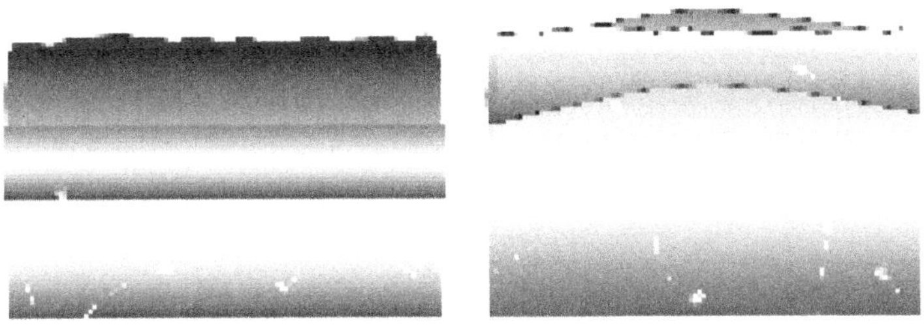

Figure 7.2 Frost heave/uplift exposure

7.2.3 Aseismic faulting

Aseismic faulting is a phenomenon where soil masses move along fault lines but without seismic actions. Depending on specific circumstances, this may be assessed as an exposure or alternatively as a resistance issue (ie, additional loadings causing weakness).

7.2.4 Seismic

Seismic events can pose a threat to pipelines in several ways. High stress/strain can be associated with seismic events in either aboveground or buried facilities. Many different phenomena are generated by seismic activities, including fault movements, soil liquefaction, ground shaking, generation of landslides and tsunamis, soil settlement, and others. See PRMM for more discussion.

Understanding seismic events helps to determine how they should be characterized in a risk assessment. For buried pipelines, seismic hazards can be classified as being either wave propagation hazards or permanent ground deformation hazards. Strong ground motions can damage aboveground structures. Fault movements sometimes cause severe stresses in buried pipe.

Permanent ground deformation (PGD) damage typically occurs in isolated areas of ground failure with high damage rates while wave propagation damage occurs over much larger areas, but with lower damage rates. Wave propagation hazards are characterized by the transient strain and curvature in the ground due to traveling wave effects. PGD (such as landslide, liquefaction induced lateral spread and seismic settlement) hazards are characterized by the amount, geometry, and spatial extent of the PGD zone. The fault-crossing PGD hazard is characterized by the permanent horizontal and vertical offset at the fault and the pipe-fault intersectional angle.

The principal forms of permanent ground deformation are surface faulting, landsliding, seismic settlement and lateral spreading due to soil liquefaction. One type of PGD is localized abrupt relative displacement such as at the surface expression of a fault, or at the margins of a landslide. The second type of PGD is spatially distributed permanent displacement which could result, for example, from liquefaction-induced lateral spreads, or ground settlement due to soil consolidation. For localized abrupt PGD, pipeline damage mainly occurs around the ground rupture trace. On the other hand, breaks for spatially distributed PGD may occur everywhere within the PGD zone.

The types of faults and the expected amount of fault offset can be empirically correlated with earthquake magnitude. Relationships for predicting the occurrence and types of landslides, and the amount of earth flow movement based on seismic event characteristics are also available. Wave propagation hazards are also empirically related to maximum moments. [777]

Liquefaction caused by seismic movements can fluidize soils to a point at which their ability to support the component is compromised. An unsupported condition can lead to additional and sometimes excessive stresses. A pipeline is also potentially subject to horizontal force due to liquefied soil flow over and around the pipeline as well as uplift or buoyancy forces. Pipeline responses to such loadings may need to be considered as failure potential or at least impairment of resistance (ie, reduction in stress carrying capacity).

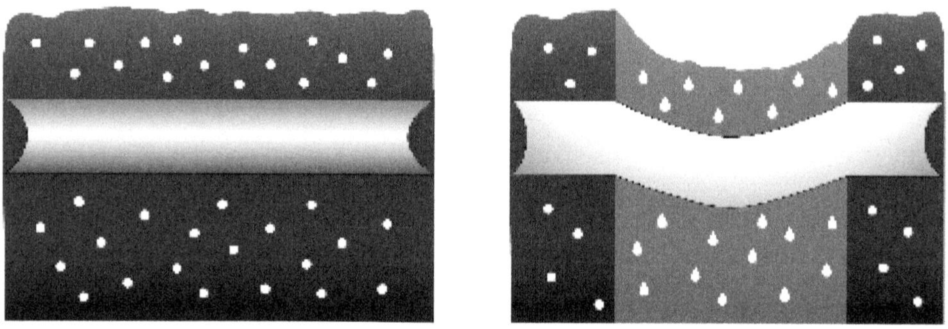

Figure 7.3 Liquefaction of Soils

Modern pipeline design considers seismic potential and will often provide useful input for the risk assessment in terms of event recurrence intervals.

7.2.5 Tsunamis

As a special type of flood or external force event, a tsunamis is a high-velocity water wave, often triggered offshore by major abrupt displacement of the seafloor from initiators such as seismic events or landslides. A seiche is a similar event that occurs in a deep lake [70b]. in deep water, these events are of less concern but have the potential to cause rapid scour, erosion, and flowing water impingements when they occur in shallow areas. Aboveground components can be especially vulnerable to lateral forces and debris loadings. This threat can be quantified by considering the potential for off-shore seismic events, the shore approach geometry and other location-specific factors. A history of such events may be used to inform the exposure estimate although the potential may exist along almost every large, deep water body. It can be included with other flooding events or assessed as an independent threat to the pipeline. Refer also to previous discussion of quantifying exposures and span-creating events.

7.2.6 Flooding

Flood waters can impart abnormal forces onto components, including buoyancy effects and debris loadings, loss of support (ie, scour, erosion), and fatigue from moving waters.

This potential threat has been a specific focus with regard to pipeline integrity. In the US, the pipeline regulator, PHMSA has released several Advisory Bulletins on this subject, each of which followed an event that involved severe flooding that affected pipelines in the areas of rising waters. Three of the more notable events (as of this writing) are briefly described below:

- On August 13, 2011, Enterprise Products Operating, LLC discovered a release of 28,350 gallons (675 barrels) of natural gasoline into the Missouri River in Iowa. The rupture, according to the metallurgical report, was the result of fatigue crack growth driven by vibrations in the pipe from vortex shedding.
- On July 1, 2011, ExxonMobil Pipeline Company experienced a pipeline failure near Laurel, Montana, resulting in the release of 63,000 gallons of crude oil into the Yellowstone River. The rupture was caused by debris washing downstream in the river damaging the exposed pipeline.
- On July 15, 2011, NuStar Pipeline Operating Partnership, L.P. reported a 100-barrel anhydrous ammonia spill in the Missouri River in Nebraska. The 6-inch-diameter pipeline was exposed by scouring during extreme flooding.

This advisory bulletin [1017] continues as follows:

> As shown in these events, damage to a pipeline may occur as a result of additional stresses imposed on piping components by undermining of the support structure and by impact and/or waterborne forces. Washouts and erosion may result in loss of support for both buried and aerial pipelines. The flow of water against an exposed pipeline may also result in forces sufficient to cause a failure. These forces are increased by the accumulation of debris against the pipeline. Reduction of cover over pipelines in farmland may also result in the pipeline being struck by equipment used in farming or clean-up operations.
>
> Additionally, the integrity or function of valves, regulators, relief sets, and other facilities normally above ground or above water is jeopardized when covered by water. This threat is posed not only by operational factors, but also by the possibility of damage by outside forces, floating debris, current, and

craft operating on the water. Boaters involved in rescue operations, emergency support functions, sightseeing, and other activities are generally not aware of the seriousness of an incident that could result from their craft damaging a pipeline facility that is unseen beneath the surface of the water. Depending on the size of the craft and the pipeline facility struck, significant pipeline damage may result.

Though these accidents account for less than one percent of the total number of pipeline accidents, the consequences of a release in water can be much more severe because of the threats to drinking water supplies and potential environmental damage.

7.2.6.1

A further examination of the advisory [1017], issued by the regulator, provides insight into not only regulatory expectations, but also commonly employed risk mitigation measures at waterway crossings.

To: Owners and Operators of Gas and Hazardous Liquid Pipeline Systems.

Subject: Potential for Damage to Pipeline Facilities Caused by Severe Flooding.

Advisory: Severe flooding can adversely affect the safe operation of a pipeline. Operators need to direct their resources in a manner that will enable them to determine the potential effects of flooding on their pipeline systems. Operators are urged to take the following actions to prevent and mitigate damage to pipeline facilities and ensure public and environmental safety in areas affected by flooding:

1. Evaluate the accessibility of pipeline facilities that may be in jeopardy, such as valve settings, which are needed to isolate water crossings or other sections of a pipeline.
2. Extend regulator vents and relief stacks above the level of anticipated flooding, as appropriate.
3. Coordinate with emergency and spill responders on pipeline location and condition. Provide maps and other relevant information to such responders.
4. Coordinate with other pipeline operators in the flood area and establish emergency response centers to act as a liaison for pipeline problems and solutions.
5. Deploy personnel so that they will be in position to take emergency actions, such as shut down, isolation, or containment.
6. Determine if facilities that are normally above ground (e.g., valves, regulators, relief sets, etc.) have become submerged and are in danger of being

struck by vessels or debris and, if possible, mark such facilities with an appropriate buoy and Coast Guard approval.
7. Perform frequent patrols, including appropriate overflights, to evaluate right-of-way conditions at water crossings during flooding and after waters subside. Determine if flooding has exposed or undermined pipelines as a result of new river channels cut by the flooding or by erosion or scouring.
8. Perform surveys to determine the depth of cover over pipelines and the condition of any exposed pipelines, such as those crossing scour holes. Where appropriate, surveys of underwater pipe should include the use of visual inspection by divers or instrumented detection. Information gathered by these surveys should be shared with affected landowners. Agricultural agencies may help to inform farmers of the potential hazard from reduced cover over pipelines.
9. Ensure that line markers are still in place or replaced in a timely manner. Notify contractors, highway departments, and others involved in post-flood restoration activities of the presence of pipelines and the risks posed by reduced cover.

If a pipeline has suffered damage, is shut-in, or is being operated at a reduced pressure as a precautionary measure due to flooding, the operator should advise the appropriate PHMSA regional office or state pipeline safety authority before returning the line to service, increasing its operating pressure, or otherwise changing its operating status.

Flood exposure estimates can arise from several sources, including published flood severity/frequency information, often linked to meteorological events. The subset of flood events that can cause failure to a pipeline component will be a function of the definition of the resistance baseline, as discussed in Chapter 2.8.12 Nuances of Exposure, Mitigation, Resistance.

7.2.7 Scour and erosion

Erosion is a readily recognized threat for shallow or above-grade pipelines close to river banks or other areas subject to higher-velocity flows. Many pipelines are exposed to threats from scour in less apparent situations, such as bridge foundations. A potential integrity threat occurs when cover erodes during flood flows, exposing the pipeline to moving waters and transported debris. The pipeline could become overstressed from lateral forces, buoyancy, or lack of support. Scour potential estimates are available for many waterways, often expressed in terms of maximum scour depths related to storms of certain recurrence intervals.

These scour estimates can directly inform exposure frequency estimates in the risk assessment. For instance, knowing that, say, a 3 foot deep scour potential is associated with a 100 year flood event, provides input for frequencies of loadings such as water/

debris impingement and vortex induced vibration (ie, these exposures are expected every 100 years for components having 3 ft of cover or less). Presumably, more frequent storms can also produce scour, but to lesser depths.

Relationships between frequencies of events that cause various scour depths will also inform mitigation effectiveness estimates for depth of cover and other protections in place or contemplated; for example, rock cover, concrete mattresses, etc.

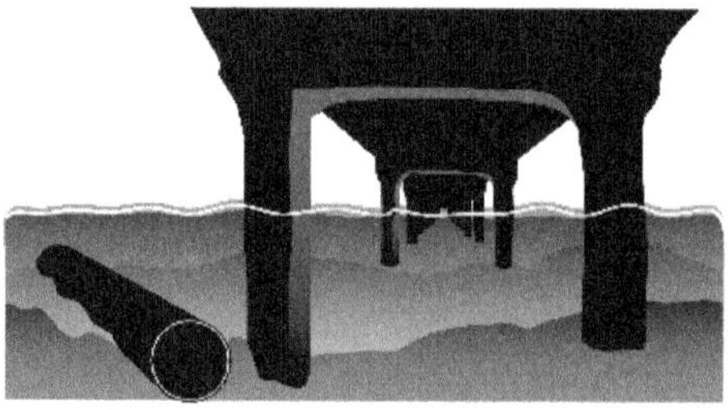

Figure 7.4 Scour at Bridge Piers[1]

7.2.8 Sand movements

The potential for wind erosion, and dune formation and movement, is another possible source of damage or at least span-creation. Seabottom sand ripples, dunes, and other instabilities are the equivalent phenomena in the offshore environment. Any of these may produce changes to loads, depth of cover, and support conditions and should be included in the risk assessment.

Exposure estimates may be based on design-phase studies, when available. In some cases, instability may be almost continuous, for instance in a high wave energy zone offshore, but only rarely severe enough to endanger the pipe component. See discussion of spans and support conditions under Chapter 2.8.12.6 Spans.

7.2.9 Weather

The threats associated with meteorological events should be included, either as damaging phenomena or as triggering events for subsequent damaging phenomena. Events such as a wind storm, tornado, hurricane, lightning, freezing, solar flares or storms, hail, wave action, snow, and ice loadings against unprotected components may be independent damage producers, along with any previously discussed phenomena they may precipitate. Even when the exposure is minimal and/or mitigation will normally eliminate the threat, inclusion into the risk assessment is important.

1 Structures in the flow stream can cause or exacerbate scour.

Electromagnetic pulses (EMP) from lightning or solar storms can damage electronic components. Such damage can lead to 'failures' such as service interruption and, in rare cases, perhaps even loss of integrity—leak/rupture. A sometimes complex chain of events needs to be identified and scrutinized to fully understand certain potential scenarios involving failures of electronic components.

Lightning strikes are a common cause of damages to electronic components as well as initiators of wildfire. US government maps are available showing lightning strike density, expressed in the mean annual number of flashes per square kilometer. Maps have been created with rankings from zero 100 for the country, where 100 represents the highest lightning strike density and zero represents the lowest lightning strike density. With assumptions of some fraction of lightning strikes being potentially threatening to a component, such rankings can inform estimates of exposure rates.

A frequency of occurrence for each possible weather event, in the absence of mitigation, is a logical starting point for exposure estimation. National weather agencies typically have databases that can be consulted. For example, points along the US Gulf of Mexico have a hurricane recurrence interval of about 25 years. This suggest a windstorm and flood exposure of 1/25 per year from hurricanes alone. This value can be refined based on hurricane magnitudes and considerations of surge heights, sustained wind speeds, and other location-specific characteristics that lead to varying damage potentials. Then protective measures, such as depth of cover, are assessed as universal or exposure-specific mitigations.

7.2.10 Fires

While often not a direct threat to integrity of a buried pipeline, fires can lead to increased erosion and landslide potential. Above ground components may be threatened by more intense or longer duration fires or when less heat resistance components (for example, gaskets, tubing, seals, plastics, instrumentation, etc) become exposed. Minor leaks may ignite and blocked-in, liquid-full components may be subject to BLEVE ruptures.

Wildfire prediction models based on factors such as topography, fuel, live shrub moisture content, weather, wind, lightning ignition efficiency are used in the US, with mapped results available from government sources. Exposure estimates can emerge from such sources and others, eg, meteorological data.

7.2.11 Other

Additional threatening phenomena are at least peripherally related to geohazards, as noted here.

Excessive external pressure is a potential threat to some offshore components' integrity, perhaps best included in the assessment as a type of geohazard. Pipelines in deep water are subjected to external forces from the hydrostatic pressure of the water column. Especially when there is reliance on internal pressure to protect the pipe from buckling, this is a source of exposure and/or an element of the resistance estimate.

Onshore scenarios of external pressure are also plausible. In one operator's experience, hydrogen permeation through steel repair sleeves caused numerous buckles to the pipe beneath. The source of hydrogen was high CP levels and the annular space pressure of around 300 psig was reportedly sufficient to cause the buckling. [1001]

Stability issues are inherent in many geohazards. See discussion of spans and support conditions under Chapter 2.8.12.6 Spans.

7.2.12 US Natural Disaster Study

In the US, maps are available showing relative threats to pipelines from some common geohazards. While expressed with a only relative scale, the derivation of the rankings provides a way to generate frequency estimates for many of these phenomena, at least on a coarse—large geographical areas, possibly missing smaller but important features—level. It is useful to examine the methodology of establishing these hazard indices.

Excerpts from this ref [1018] are shown below to assist the risk assessor in determining the usefulness of such relative-scale information into a contemplated assessment. Note that, in the absence of more definitive information, a relative scale itself can be 'grounded' with frequency values and thereby used in preliminary exposure estimates (for example, score of 70 = 0.1 events/year, etc).

As of this writing, US databases area available [1018] showing hazard indexes for:

- earthquake (HER = Earthquake Hazard Rank)
- hurricane (HHR = Hurricane Hazard Rank)
- tornado/storm (TSRR = Tornado/Storm Hazard Rank)
- flood/scour (FHR = Flood Hazard Ranking)
- landslide (LHR = Landslide Hazard Ranking)
- other (lightning and snow depth; OHR = Other Hazard Risk)

This index system also includes a summary layer, produced using the composite rank formula:

$$NPHI = .3(FHR) + .2(EHR) + .2(LSHR) + .l(TSHR) + .l(HHR) + .l(OTHER)$$

Where:
FHR = flood hazard rank
EHR = earthquake hazard rank
LSHR = landslide hazard rank
TSHR = tornado/storm hazard rank
HHR = hurricane hazard rank
OHR = other natural hazards hazard rank

Table 7.1
National Pipeline Risk

Hazard	Variables Included	Methodology	Notes
Hurricane	Historical count	94 year history of hurricanes per coastal county	2
TSRR	Historical count	Number of occurances over 30 years per one degree box	3
Landslide	swelling clays, landslide incidence, susceptibility, subsidence	LSHR = 0.3 (clay) + 0.4 (incidence) + 0.2 (susceptibility) + 0.1 (subsidence)	6
Earthquake	Spectral response acceleration coefficient	Based on single, complex variable ranked 0-100	4
Other	30 year Mean annual lightning strike density; Snow depth with 95% chance of not being exceeded	OHR = .5 (lightning strike) + 5 (snow depth)	5
Flood	Annual flooding frequency, potential scour depth	FHR = 0.5(flooding) + 0.5(scour depth)	1

Table Notes

1. For the Annual flooding frequency layer one-kilometer grid cells were assigned the following values based on the annual chance of flooding:
   Frequent (5O-100%): Flooding = 100
   Rare (O-5%): Flooding = 33
   Occasional (5-50%): Flooding = 67
   No Flooding: Flooding = 0
   These values were then multiplied by the percentage of area they covered for each soil map unit. The percentage values were summed to give the value for each soil map unit. A grid of these values was created and then ranked from 0 to 100. For the Potential scour depth layer one-kilometer grid cells were ranked based on their value (potential scour depth in feet).
   Highest value Scour depth = 100 Lowest Value Scour depth = 0
2. The total number of direct and indirect landfalling hurricanes per coastal county was used from 1990 (assume typo—should probably be "1990") until 1994. From the county baaed polygon coverage, a point coverage was derived. From this point coverage first a Triangulated Irregular Network (TIN) and then a continuous surface grid was created, in order to more appropriately represent the hazard without the use of political boundaries. These numbers were ranked from zero to 100, where 100 represents the highest number of land-falling hurricanes and zero represents the lowest number of land-falling hurricanes.
3. The centroids of the one-degree cell areas were used to generate a Triangulated Irregular Network (TIN). This resulted in a continuous surface that more naturally depicts the distribution of tornado events. A grid was created at a resolution of one kilometer from the TIN. The values were ranked from zero to 100, where 100 represents the highest number of tornadoes and zero represents the lowest number of tornadoes.

4. The spectral response acceleration coefficient is an indicator of the probability of receiving specific intensities of ground shaking from earthquakes. For the EHR the spectral response acceleration coefficient at a period of 0.3 seconds expressed as a fraction of gravity with a 90% chance of not being exceeded in 50 years is used. The data are prepared by the U.S. Geological Survey (USGS) for the NEHRP Recommended Provisions for the Development of Seismic Regulations for new Buildings.
5. Lightning strike density is expressed in the mean annual number of flashes per square kilometer. Contour lines were digitized from a very small scale map. The areas in between the contour lines were given the mid-value of the class. These values are ranked from zero 100 for the country, where 100 represents the highest lightning strike density and zero represents the lowest lightning strike density. The 95% annual nonexceedence probability was calculated for 239 weather station in the United States. From this point coverage first a Triangulated Irregular Network (TIN) and then a continuous surface grid was created in order to more appropriately represent reality. These numbers were ranked from zero to 100, where 100 represents the highest snow depth and zero represents the lowest snow depth.
6. The LSHR values were ranked from zero to 100, where 100 represents the highest ground failure hazard and zero represents the lowest ground failure hazard.

While relative indices like these are not ready for direct inclusion into a modern risk assessment, the underlying methodology provides insights into the phenomenon and current abilities to forecast them.

7.2.13 Offshore

Offshore pipelines, including those crossing inland waterways such as rivers and lakes, are exposed to many of the same forces as those onshore—landslides, rockfalls, seismic events, etc—plus others unique to the offshore environment. The interaction between the pipeline and the seabed or riverbed will frequently set the stage for external loadings offshore. The following discussion focuses on ocean environments, but will often apply, albeit to a lesser extent, to inland creeks, rivers, large lakes, and sometimes even ponds. See also the discussion of stream scour and flooding.

One of the largest differences between the risk assessments for offshore and onshore environments appears in this issue of stability. This reflects the very dynamic nature of most offshore environments under normal conditions and more so with storm events.

Figure 7.5 Ice gouging, ice keel exposure

7.2.13.1 Stability Issues

Offshore bottom conditions are constantly changing by normal forces of moving water. This changes the stability conditions for structures resting directly on the seabottom or with shallow cover. Additional instability events associated with storm-related forces, changes in bottom topography, temporary currents, tidal effects, and ice movements are also often relevant to a risk assessment.

Offshore "high-energy" areas, evidenced by conditions such as strong currents, or tides, are common areas of instability. Seabed and riverbed morphology is constantly changing due to naturally occurring conditions (waves, currents, soil types, etc.). Vortex shedding, lateral loadings, scour, and other forces caused by frequent changes in bottom conditions are commonly associated with wave zones and high steady current environments.

At times, the pipeline itself, as an obstruction that has been introduced into the system, contributes to bottom changes. Sand wave migration—size, direction, and rates—can be predicted with an understanding of bottom conditions. Rare occurrence events, often carrying higher energy, may create greater damage potential. This includes hurricanes, severe storms, and rare ice movements.

Bottom instability generates integrity concerns primarily from issues related to support and/or fatigue-loading. A common conservative assumption in risk assessment is that increased instability of bottom conditions leads to increased potential for pipeline over-stressing and failure.

The pipeline can become an unsupported span despite initial installation and efforts to maintain cover. Once uncovered or spanning, it is subjected to additional stresses due to gravity, buoyancy, and wave/current action. Consider scenarios such as a buried line becoming uncovered by scour or erosion of the seabed/streambed perhaps with uplift forces (for example, an emptied liquid pipeline), and subsequently becoming exposed to flowstream forces and impact loadings from floating debris and material being moved along the seabed or riverbed. Such external forces can damage coatings, both concrete and anticorrosion types, and even damage the pipe steel with dents, gouges, buckling, or punctures.

Pipelines exposed to a flowstream may move due to intermittent lateral forces, buoyancy issues, or vortex shedding. Movements of a free-spanning pipeline, resulting in cycling and fatigue loadings may eventually weaken a component to the point of failure. Fatigue and overstressing threats are amplified by larger span lengths, higher water velocities, and larger profiles (diameter).

Rigid pipelines, because of their diminished capacity to withstand certain external stresses, will be threatened under less severe conditions. Mechanical coupling of pipe joints usually adds rigidity and, hence, reduces resistance.

A full evaluation of any potentially damaging offshore phenomena requires an evaluation of many subvariables such as soil type, seismic event types, storm conditions, cover condition, water depth, current speeds and directions, etc, as discussed in PRMM.

As also noted in PRMM, some of the common instability issues and their dominant factors—ie, a 'function of'—include the following:

- Fault movement damage potential = f{fault type; slip angle; pipeline angle; seismic event}
- Liquefaction damage potential = f{seismic event; soil type; cover condition}
- Slope stability = f{slope angle; soil type; rock falls; initiating event; angle of attack; landslide potential}
- Erosion/scour potential = f{current speed; bottom stability; concrete coating}
- Additional Loadings = f{hydrodynamic forces; debris transport; current speed; water depth}

In new offshore pipeline systems, more threatening areas along the proposed route are normally identified in the design phase studies. The design process is in fact a risk management practice. The risk assessment of a new facility will therefore generally reflect the mitigated threat. The potentially damaging events—the 'exposures' in the PoF analyses—should nonetheless be captured in the assessment, regardless of mitigation measures subsequently employed to offset their presence. Even after design-phase mitigation, some risk remains.

A level of reliability is typically chosen in the design phase and can be used to infer the future damage rate—the remnant risk. For instance, a structure could be designed to withstand a 100 year storm or alternatively, a 500 storm flood; a 50 year recurrence interval seismic event or 100 year. There remains the potential, albeit remote, that a more severe event occurs in the structure's life and produces forces beyond its resistance abilities. That should be reflected in the risk assessment.

For existing systems, seabed and riverbed profile surveys are a useful method to gauge the stability of an area. The effectiveness of the survey technique should be considered as discussed in PRMM.

In summary, offshore pipelines are more threatened in areas where damaging soil movements and/or water movements are more common or more severe. More specifically, this involves scenarios where a high-energy water zone—wave-induced currents, steady currents, scouring—is routinely causing seabed morphology changes; where unsupported pipeline spans are present; where water current action is sufficient to cause oscillations on free-spanning pipelines—fatigue loading potential is high—or impacts from floating or rolling materials; where fault movements, landslides, subsidence, creep, or other earth movements are more probable; and where ice movements are common and potentially damaging.

Risk reduction efforts typically focus on avoidance, correction, or protection techniques. These include reburial as well as various armoring approaches—ie, reinforcing a location using concrete mattresses, grout bags, mechanical supports/anchors, antiscour mats, or rock dumping. Such methods also provide protection against impacts (for example, anchors, shipwrecks, dropped objects, etc) and therefore influence risk from third party activities.

7.2.13.2 River and Stream Scour

Pipeline crossings of inland waterways are threatened by many of the same phenomena previously discussed. The potential threat from scour has been studied with specific regard to pipeline integrity. In the US, a Dec 2012 PHMSA report [1032] to congress on hazardous liquid pipeline crossings of inland rivers, streams, and other waterways offers some insights into frequencies of cover-depletion events at waterways. This report determined that there are ~2,572 hazardous liquid pipeline crossings of water ways >100ft in width (high water mark to high water mark) out of ~2,841 crossings of inland bodies of water in the US. The authors identified 20 accidents at water crossings between 1991 and Oct 2012, 16 of which involved depletion of cover, either from scour or new river channel creations. These 16 incidents were 0.3% of all reported hazardous liquid pipeline accidents and 0.5% of those accidents exceeding the PHMSA threshold of 'significant incidents'.

7.2.14 Induced Vibration

Vortex shedding, whether by wind or water, can generate sufficient forces under certain circumstances, to move a pipeline segment. This movement can become rapid and relatively large, causing fatigue loadings in the pipe material. Fluid density, speed, cross sectional area in flow stream, frictional drag across the object and other factors influence the onset and magnitude of movements. The exposure generated by this phenomenon is most often captured as cracking. See Chapter 6.8 Cracking.

Example: 7.1 Wave-induced pipe movements

To illustrate both a portion of an offshore geohazard assessment as well as the migration into a modern risk assessment approach, an offshore risk assessment example originally presented in PRMM is re-visited here.

An offshore pipeline makes landfall in a sandy bay. The line was originally installed by trenching. While wave action is normally slight, tidal action has gradually uncovered portions of the line and left other portions with minimal cover. With no weight covering, calculations show that flotation due to negative buoyancy is possible if more than about 20ft of pipe is uncovered. This shore approach is visually inspected at low-tide conditions at least weekly. Measurements are taken and observations are formally recorded. The line was reburied using water jetting 8 years ago.

Using rudimentary wave-induced-vibration and fatigue calculations, along with average storm frequencies, the evaluator estimates unmitigated crack growth on this shore approach to be 4 mpy. With the strong inspection program and a history of corrective actions being taken, the effectiveness of cover—eliminates pipe movement when cover is not depleted—is judged to be fully effective except for short periods during storms and between remediation and is assigned a value of 95% effectiveness. This results in a P90 damage rate of 0.2 mpy to be used in subsequent TTF estimates.

7.2.15 Quantifying geohazard exposures

Where geohazards have been rated based on recurrence interval—for example, 100 year flood, seismic event with 10% probability of exceedance in 50 years, etc—those ratings can directly inform an exposure estimate. The land movement potentials from various phenomena can be added so that multiple threats in one location are captured.

Event frequencies of 0.1 to 10 per year or higher may be appropriate for areas where damaging geohazard events are common. Regular fault movements, landslides, subsidence, creep, active earthquake faults, or frost heave are commonly recurring phenomena in some areas.

Event frequencies of 0.001 to 0.1 per year may be appropriate when damaging geohazard events are possible, but when no damage in the subject area has been recorded. A P90+ exposure estimate based on length-time (mile-years, km-years) as described in section Chapter 2.8.6 The Test of Time Estimation of Exposure may be appropriate to capture the notion of pipelines that have withstood the test of time.

When evidence of geohazard events is rarely if ever seen and movement potential is conceptually approaching nonexistent, then rates of less than 0.001 per year may be appropriate.

In keeping with an "uncertainty = increased risk" bias (see discussion of PXX, Chapter 2.16 Conservatism (PXX)), having no knowledge of earth movement potential should register as high risk, pending the acquisition of information that suggests otherwise.

7.3 MITIGATION

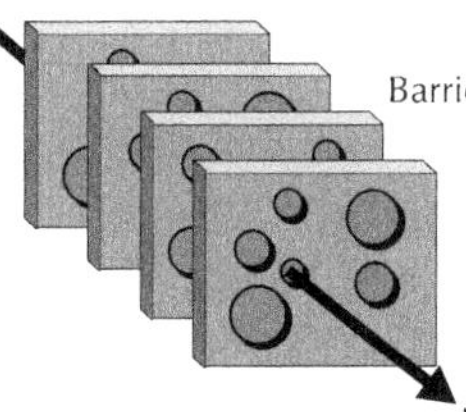

The potential damages resulting from any events involve considerations for mitigation and resistance. Pipeline components are typically designed with protection from a wide variety of geohazards. Depth of cover will be a typical, and usually very effective, mitigation measure for geohazards along most portions of a pipeline.

However, using characteristics such as depth of cover to screen for vulnerabilities will usually result in dismissing threats from certain phenomena such as fire, as well as certain weather events previously noted. Such screening weakens the risk assessment and should be avoided. Each threat is best measured as an exposure to a theoretical, unprotected component.

Once unmitigated exposures are identified and quantified, mitigations are similarly identified and assessed. In areas where multiple damaging events are possible, the assessment should reflect the combined threats, considering the mitigation benefit from each measure as applied to each exposure. Mitigations, as reactions to a perceived threat, typically may include any of the following: (some taken from ref [1033])

- Inspection / survey
- Stabilization (cover condition, anchors, piles, articulated mattresses, various support types, mix of mitigation, changing exposure, etc.)
- Ground Improvements
    - o Drainage to control water access by interception ditches, French drain, ditch plugs, etc
    - o Erosion control vegetation
    - o Soil densification (for example, by surface loadings, dewatering, or vibrations)
    - o Slope re-grading, to reduce soil movement potential
    - o Toe berms, to increase resistance to soil movement
    - o Retaining walls, to halt movements
    - o Surface diversion berms, to prevent erosion
    - o Channel reinforcement by armouring with rock, sandbags, vegetation, etc.
    - o Channel movement control
    - o Re-establish depth of cover
- Pipe isolation
    - o Deep burial to avoid shallow slope movements and frost heave, for example a directional drill
    - o Synthetic geotextile pipe wrap, manufactured backfill, or straw backfill to reduce friction loadings from ground movements
- Avoidance
    - o Pipeline re-route

- o Above ground pipe components
- Ditch modifications
  - o Wider ditch to reduce friction and allow movements
  - o Bedding and padding to prevent contact with rocks/boulders
  - o Excavation to relieve strain loadings.

7.3.15.1 Regular monitoring

When the pipeline and/or the potentially threatening phenomenon is visible or otherwise detectable in advance, monitoring can provide intervention opportunities. Regular, appropriately scheduled surveys that yield verifiable information on pipeline location, depth of cover, land movement, rainfall, moisture content, strain levels, water depth/current velocities for offshore pipelines, and other early-warning characteristics should be included in the risk assessment.

Earthquake monitoring systems alert of seismic activity and magnitude often only moments prior to the time of occurrence. This is nonetheless very useful information because areas that are likely to be damaged can be immediately investigated.

Where movements of icebergs, ice keels, and ice islands are a threat, well-defined programs of monitoring and recording ice movement events can be effective in reducing pipeline risk.

Timeliness of detection will be important. Frequency of surveying should be based on historical issues such as flooding, seabed and bank stability, wave and current action, ice storms, and risk factors specific to the pipeline section. The assessment can consider the basis for survey frequency—ideally, a written report with backup documentation justifying the frequency—to determine if adequate attention has been given to the issue of timeliness.

7.3.15.2 Continuous monitoring

Devices or techniques used in monitoring programs that will alert an operator of a significant change in stability conditions or other threats provide some risk reduction. Indicator devices might include strain gauges on the pipe wall itself, or survey markers to detect soil movements near to any component, and seabed or current monitors near to offshore components. Follow-up inspection and action is an essential aspect of the mitigation benefit. Mitigation that provides intervention opportunities is most beneficial when the monitoring is extensive enough to reliably detect all damaging or potentially damaging conditions before failure occurs.

See PRMM for an example evaluation of potential for earth movements.

Example: 7.2 Potential for earth movements

As another illustration of an update to a scoring-type risk assessment, consider the following modified example originally appearing in PRMM.

In the section being evaluated, a brine pipeline traverses a relatively unstable slope. There is substantial evidence of slow downslope movements along this route although sudden, severe movements have not been observed. The line is thoroughly surveyed annually, with special attention paid to potential movements. Survey results have reportedly prompted remedial actions several times in the previous 10 years, although record-keeping is incomplete. The evaluator makes a preliminary assessment of the exposure to be 0.5—an event once every other year—evidenced by the need for multiple remedial actions in a 10 year period. The surveying and subsequent remediation appears to be protective of the segment but are not formally documented. Mitigation effectiveness for the combined survey-remediation protocol is estimated to be 50% in its current state. This equates to an estimate of damage once every 4 years, from this apparently effective mitigation but with unknown error rates and continuance assurance. The evaluator advises the operator that this estimate can be increased if steps such as the following are taken:

- Formalize the survey procedures
- Establish the survey frequency on the basis of failure/damage probability
- Formalize the remediation procedures, especially regarding action thresholds and timing.

7.4 RESISTANCE

One common reaction to geohazard threats is increased component strength, specifically the ability to resist external loads considering both stress and strain issues. Other measures to add resistance to geohazards will often be phenomena-specific.

Understanding of failure modes is essential to the modeling of resistance. The following discussion, taken from ref [1034], on seismic induced failure modes illustrates this as well as gives insight into many other geohazard phenomena.

7.4.1 Failure modes for buried pipelines subject to seismic loading.

The principal failure modes for corrosion-free continuous pipelines (e.g. steel pipe with welded joints) are rupture due to axial tension, local buckling due to axial compression and flexural failure. If the burial depth is shallow, continuous pipelines in compression can also exhibit beam-buckling behavior. Failure modes for corrosion-free segmented pipelines with bell and spigot type joints are axial pull-out at joints, crushing at the joints and round flexural cracks in pipe segments away from the joints. The principal failure modes for corrosion-free continuous pipeline with burial depth of about three feet or more are tensile rupture and local buckling. Buried pipelines with burial depths less than about 3 feet (i.e., shallow trench installation) may experience beam buckling behavior. Beam buckling has also occurred during post earthquake excavation undertaken to relieve compressive pipe strain.

Intuitively, beam buckling is more likely to occur in pipelines buried in shallow trenches and/or backfilled with loose materials. That is, beam buckling load is an increasing function of the cover depth. Hence, if a pipe is buried at a sufficient depth, it will develop local buckling before beam buckling.

When strained in tension, corrosion free steel pipe with arc welded butt joints is very ductile and capable of mobilizing large strains associated with significant tensile yielding before rupture. On the other hand, older steel pipe with gas-welded joints often cannot accommodate large tensile strain before rupture. In addition, welded slip joints in steel pipe do not perform as well as butt welded joints

Buckling refers to a state of structural instability in which an element loaded in compression experiences a sudden change from a stable to an unstable condition. Local buckling (wrinkling) involves local instability of the pipe wall. After the initiation of local shell wrinkling, all further geometric distortion caused by ground deformation or wave propagation tends to concentrate at the wrinkle. The resulting large curvatures in the pipe wall often then lead to circumferential cracking of the pipe wall and leakage. This is a common failure mode for steel pipe.

For segmented pipelines, particularly those with large diameters and relatively thick walls, observed seismic failure is most often due to distress at the pipe joints. In areas of compressive ground strain, crushing of bell and spigot joints is a fairly common failure mechanism in, for example, concrete pipes.

For small diameter segmented pipes, circumferential flexural failure have been observed in areas of ground curvature.

Axial pull out of segmented pipe such as cast iron or concrete with rubber gasketed joints and bell-spigot is also a common failure mode for seismic events [1034].

A structure's design documentation will often state the geohazard events that the structure is rated to withstand—for example, "maximum scour from a 100 year flood"; "seabottom instability from 100 year storm"; "landslide from 50 year rainfall event". These values are useful in the risk assessment since they suggest a point in the load probability distribution, below which the structure's survival rate should be high. In the absence of unanticipated weaknesses, the structure should be highly resistive to events of lesser magnitude (normally more frequent events are of lesser magnitude) than the stated design intent. Resistance to more severe events (generally more infrequent) will be questionable.

Knowledge of safety factors will be useful in estimating resistance. Technically rigorous structural analyses can be performed where the most robust resistance estimates are required. These will require more combinations of specific loadings onto specific components and comparing resulting calculated stresses against stress carrying capacities.

Full discussion of resistance is found in Chapter 10 Resistance Modeling.

One touch of nature makes the whole world kin.

William Shakespeare

8 INCORRECT OPERATIONS

Highlights

Human error potential: an important but difficult to quantify aspect of risk assessment.

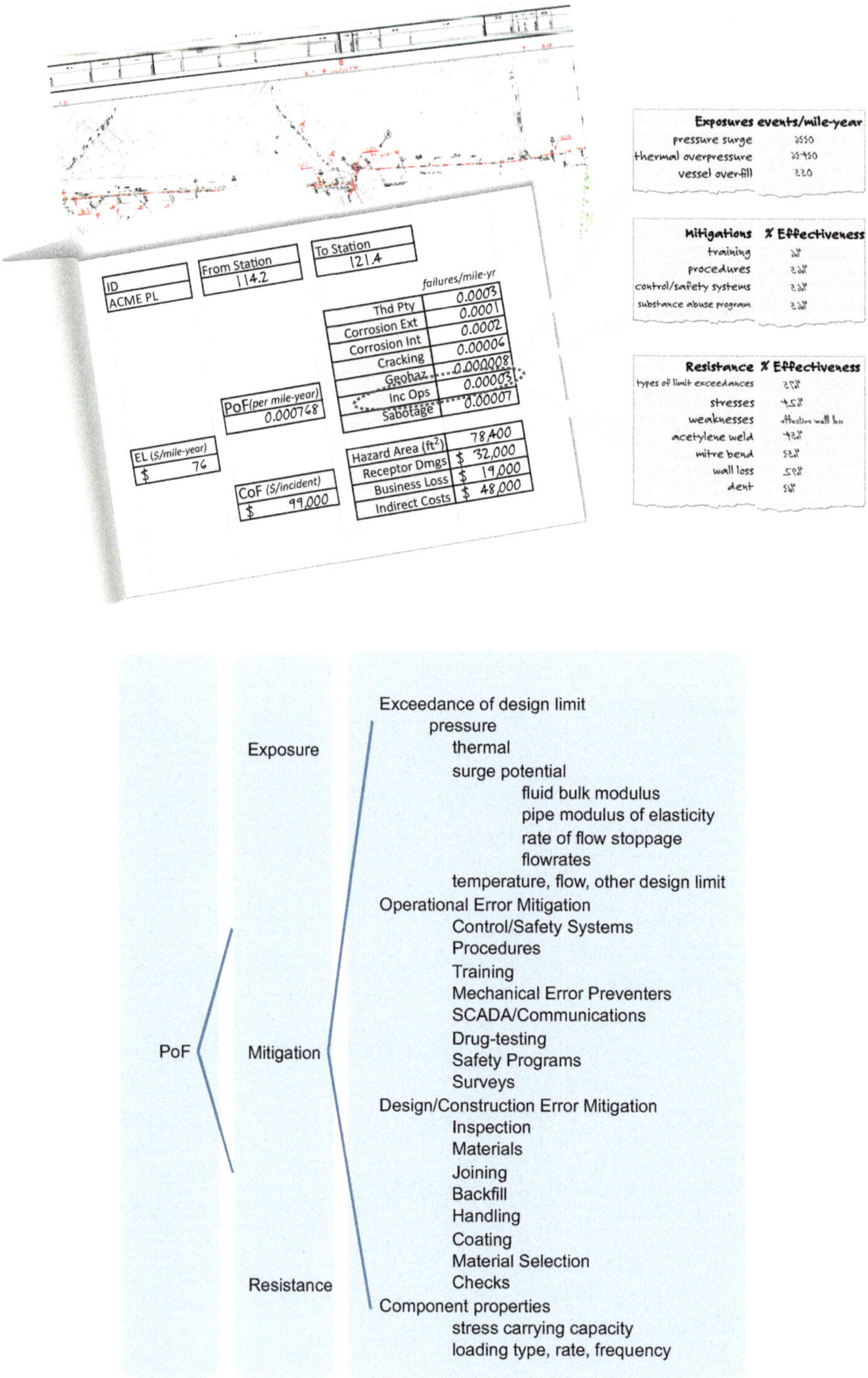

Figure 8.1 Assessing Human Error Potential—Sample of Data Used

8.1 HUMAN ERROR POTENTIAL

The potential for human error is an important aspect of risk but challenging to quantify. It would be remiss to discount this potential threat and thereby diminish the importance of the many types of mitigation employed against it. Large budgets are spent on training, procedures, safety systems, and other mitigations, and that spending continues because it is widely believed that the benefits outweigh the costs. Even though only generalizations and subjective determinations may be available to quantify these benefits and many other aspects of error potential, risk knowledge improves greatly from efforts to measure this.

The error potential focus is often directed more towards stations and facilities. A more complex environment such as a station normally provides many more opportunities for human error—first party and second party—compared to ROW miles. Offshore platforms and their onshore counterparts—pump/compressor stations, tank farms, meter facilities, etc—normally have a high density of components, a more complex design, and more frequent human activities compared to most portions of most pipelines,.

Since human error potential permeates every aspect of risk, it logically influences multiple portions of the risk assessment. Consequence minimization and mitigation effectiveness are often quite sensitive to operator error, with less sensitivity usually associated with exposure rates and resistance factors.

Despite the need to consider error potential in many specific processes, assessing error potential as an independent failure cause has the advantage of avoiding duplicate assessments for many of the pertinent risk variables. This recognizes that the same variables would apply in most other failure mechanisms and it makes sense to evaluate such variables in a single place in the assessment.

Nonetheless, the role of human error should also be considered in all estimates that are used in the risk assessment, especially for exposures and mitigations. For instance, the effectiveness of many mitigation measures are sensitive to error rates—eg, line locates, safety device maintenance, evaluation of CP surveys, and many more have large human interface aspects.

See PRMM for background discussions on many of the risk assessment considerations surrounding human error potential.

8.1.1 Human Error Potential Considered Elsewhere in Risk Assessment

The role of human error in risk requires an understanding of potential for pipeline failure caused by errors committed in designing, building, operating, or maintaining a pipeline.

Human error impacts all of the other probability-of-failure analyses. Active corrosion, for example, suggests an error in corrosion control activities, under an assumption that knowledge and resources to prevent corrosion exist.

As noted above, the human error potential should be captured in the estimation of each mitigation measures' effectiveness. If there are potential differences in human error potential for each failure mechanism or among exposures, one can pair error-reduction mitigation measures with specific exposures. For example, perhaps training and procedures for surge prevention are more robust than those for thermal overpressure events.

The focus in this chapter is on real time operator errors that directly precipitate failures. When failure is defined as leak/rupture, there are usually fewer relevant exposure scenarios, due to the common design principle of 'fail safe' operations. That is, it is normally difficult to accidentally and immediately threaten any pipeline component's integrity solely by miss-operation of the pipeline's devices and equipment. With failure including the often higher potential for service interruption, human error scenarios become more common. In other words, it is easier to interrupt or otherwise compromise a pipeline's operation (by improperly operating devices and equipment) than to cause a leak/rupture.

It is believed that error potential in the operations phase will often be relevant to error potential in other phases, if only in terms of the similar underlying causes of exposure and opportunities for mitigation. Therefore, this centralized approach for examining human error in a risk assessment provides a more efficient means of understanding error potential elsewhere.

Errors by outside parties are more efficiently modeled as part of the exposure rates of other failure mechanisms. This includes vehicle and equipment impacts and explosions from nearby facilities.

Non-operational errors are discussed here but usually better modeled in other portions of the risk assessment. Errors during design and construction tend to introduce weaknesses into the system. These are best considered in the evaluation of resistance. Maintenance errors tend to reduce reliability of equipment, decreasing mitigation when the equipment is protective of integrity; for examle, safety systems, monitoring instrumentation, etc. Design, construction, and maintenance errors are therefore contributors to failure frequency and consequence but not often initiators. If the assessed component has functioned correctly for some period of time under similar stresses prior to a failure, then the original error is a contributing factor but not the final failure mechanism. Operational errors, on the other hand, can and do precipitate failure directly.

Finally, human errors can fail to minimize consequences or even exacerbate them, as is discussed in the CoF assessment.

8.1.2 Origination Locations

Operational human-error scenarios potentially causing damage to a pipeline may originate at a facility far from the damage location. Overpressure at a pump station may cause a rupture only at a weak point in a pipe segment miles away. Station operations typically have more opportunities for errors such as overpressure due to inadvertent valve closures and incorrect product transfer resulting in product to the wrong tank or to overfilled tanks.

Therefore, facilities, especially those with more frequent or complex human interfacing, will play a large role in risk assessment for portions of the system far beyond the facility boundaries. They are often initiating points for a failure manifesting elsewhere along a system.

Recall HAZOPS as a scenario-based analysis tool to identify events and sequences of events that can lead to failures, including operability issues. A HAZOPS will organize the facility into 'nodes'—discreet portions of the facility being evaluated. HAZOPS are often overlooked in a pipeline risk assessment due to a perception that they only apply to a station facility and not to ROW miles. In reality, they usually identify most, if not all, of the potential human error scenarios that could cause failures anywhere, including locations long distances from the station being assessed. When a HAZOP node includes ROW pipe—perhaps shown as a delivery or receipt point on the P&ID schematic of the facility—then the applicability is most apparent. When specified as a node, the HAZOP facilitator should ensure that this node includes more than just the immediate receipt or delivery pipe components. It should include all features along the pipeline—low spots, weaknesses, etc.—even at long distances from the facility.

Figure 8.1 Human error potential during operations

8.1.3 Continuous Exposure

Recall the discussion in Chapter 2.8.12.4 What Constitutes 'Exposure'? Normalizing Exposure and Resistance. Incorrect operations has several examples of this. For instance, suppose that a high pressure source is connected to a pipeline via a pressure regulating (control) valve. The pressure source creates the threat exposure and the regulator is the mitigation in this elementary example. Failure is avoided through the use of control and safety systems. The source represents continuous exposure—the pipe downstream of the regulator is subject to immediate overpressure (and potential failure) if the regulator fails. For modeling purposes, it is an on-going, unrelenting, cause of immediate failure if unmitigated.

Measuring this type of exposure appropriately in a risk assessment model requires the correct coupling of the continuous exposure with a corresponding mitigation effectiveness. A high-demand or continuous exposure requires mitigation with very high reliability. The modeling issue with continuous exposure is the choice of time units in which to express the rate of exposure. The continuous exposure can be counted as one event per day, once per hour, once per minute, once per second, or even less. Any of these is appropriate as long as the corresponding mitigation—the regulator effectiveness—is measured in the same per day, per hour, per minute, etc units of reliability.

Another nuance of exposure measurement involves the baseline for resistance. There could be dramatically increased exposure when zero resistance is assumed. That is, the number of potentially damaging events increases when the threshold for damage is lowered. This is detailed in Chapter 2.8.12 Nuances of Exposure, Mitigation, Resistance.

8.1.4 Errors of omission and commission

Some errors are actually reductions in mitigation effectiveness while others are direct exposure events. As a minimum, this distinction should be made in a risk assessment. A more general PoF assessment may use only these two categories, applying a numerical effectiveness value (or penalty) to all mitigations involving human actions and also estimating a frequency of future error-generating failures (in the absence of any mitigation).

It is important to understand the types of errors possible, perhaps appropriately categorized by their root causes. Error rates associated with each would be estimated in the more robust risk assessments. Then, since certain mitigations have varying effectiveness for each type of exposure, specific pairings would be needed. In preliminary or less robust assessments, satisfactory accuracy may be achieved by treating all exposures the same with all mitigations applied to the collective exposure frequencies.

One possible categorization scheme would group by underlying cause of the error. For example, errors due to:

- Impairment
- Lack of knowledge
- Inattention
- Apathy
- Stress.

Another categorization scheme could group by the type of error, including skill-based errors, (memory lapse, slip of action), mistakes (rule-based, ie incorrect application of a good rule, application of a bad rule, or failure to apply a good rule; or knowledge-based) [1019].

PRMM provides a useful background discussion of stress influences on human error and how to incorporate research concepts into a risk assessment.

8.2 COST/BENEFIT ANALYSES

As with many other elements of a strong risk assessment, an objective and defensible cost/benefit analysis can be conducted for error-prevention practices whose benefits were previously difficult to quantify. Instrument maintenance and calibration, training, procedures, personnel qualification programs, and many others provide measurable benefits in risk reduction. Their value was always recognized, hence their universal use over many decades of industrial application. However, determining the appropriate level of robustness and justifying additional efforts had to be 'sold' rather than demonstrated via objective analyses.

A good risk assessment provides a more objective, consistent, and defensible way to show benefits—avoided losses—obtainable from risk reduction actions.

8.3 ASSESSING HUMAN ERROR POTENTIAL

As with other failure mechanisms, the most detailed assessment will always pair specific exposures with corresponding mitigations. For example, substance abuse programs will logically reduce only exposure events involving impairment factors; training may only reduce errors having 'lack of knowledge' as an underlying cause. However, sufficient accuracy in assessment is often achieved by taking a more general approach, perhaps applying all mitigations equally to all types of exposures.

Although human error is involved in almost every failure, human errors that can directly threaten integrity are relatively rare in most pipeline systems. When service interruption events are included in 'failure' the number of possible human error events increases.

Assessing this failure mechanism begins with examinations of error potential in each phase of pipelining.

8.4 DESIGN PHASE ERRORS

Design phase errors include incorrect equipment sizing, inappropriate assumptions and/or incorrect calculations regarding loads and/or resistance, improper materials selection (considering stresses, fatigue, environmental factors of temperature, corrosivity, and others, etc), and others.

The risk assessment could begin with a baseline representing the completely unmitigated exposure—the error rate associated with designs originating from an uneducated, inexperienced, layman, attempting component designs while working in a harsh environment with no tools (ie, computer, calculator, graph paper, etc). A very high error frequency would be expected—perhaps 50% to 90+% (error rates ranging from one in 'every other designed component' to 'every component')—depending on design complexities and nuances. This error rate would be reduced by the commonplace error reduction measures such as education, training, procedures, certifications, quality checks, etc.

This robust approach has the advantages of valuing each aspect of error-reduction. However, a completely unmitigated error rate may be hard to visualize, given educational, credential, continuing education requirements normally associated with most design practices, not to mention common workplace conditions and tools that further help to improve the processes. A modeling convenience that will often not result to excessive loss of accuracy, might be to begin with an error rate reasonably attributable to a 'standard' design process common to the region and era. This standard process may, for example, be a design team of 2-year technical college designer/drafters overseen by an experienced licensed profession engineer. Perhaps this team produces component designs with serious integrity-threatening errors once every 100 designs. Using this as a baseline case, error-reduction measures such as those discussed in this chapter, would reduce the damage potential from such errors.

In identifying where in the design processes errors were more likely, techniques such as HAZOPS can be effective in re-constructing or at least acting as a surrogate for the original design and operations intents.

8.5 CONSTRUCTION PHASE ERRORS

Potential errors during construction are similarly assessed. Error rates could be assessed for either a completely unmitigated construction scenario or for some 'standard' practice. The latter is more intuitive. Then, additional mitigation measures that are in place can be valued. Risk assessments will require an estimate of weakness probabilities along lengths of the system. The rate of occurrence of weaknesses is logically influenced by error rates during design, manufacture, and installation. The 'test of time' rationale (see Chapter 2.8.6 The Test of Time Estimation of Exposure) may be appropriate for both design and construction errors. Recognize however, that some slow-acting phenomena may be attributable to design errors but have simply not yet manifested. Perhaps installations performed in challenging conditions or with questionable quality control can be associated with rates of weaknesses (increased susceptibilities to certain damages) of one or two every 10 km, despite successful pressure testing and years of operations.

8.6 ERROR POTENTIAL IN MAINTENANCE

Errors in maintaining equipment, control, and safety systems leads to reduced effectiveness of those systems. An otherwise high reliability of a pressure control regulator or relief valve is lessened when proper maintenance is not performed. The device's mitigation effectiveness estimate in the risk assessment should reflect this.

Similarly, errors in corrosion control, marking/locating, patrol, and even public education, all reduce the ability of the mitigation to protect the system.

8.7 OPERATIONAL ERRORS

Error potential during operations is potentially a direct initiator of failure. An immediate damage or failure event is possible since personnel are actively operating equipment such as valves, pumps, compressors, and many others where incorrect actions or sequences produce unintended results and may cause damages. Emphasis therefore is on error prevention rather than error detection.

For estimating error rates during operations, the unmitigated exposure rate may again be difficult to imagine—an operation with no procedures, no training, no control or safety devices, etc. However, there are some very 'stout' systems that, even with no standard mitigations, would still not be damaged or fail by any conceivable operator action, much less an error. For example, if there are no pressure sources that could exceed design limits, including surge potential and blocked-in, liquid-full, heating scenarios, then it may be physically impossible to overpressure any component. In this case, the inherently low risk operation should show very low exposure and perhaps suggest that mitigation is largely unnecessary. Therefore, the estimation of the unmitigated exposure rate to operational errors is important. Distinguishing between systems' exposure rates may be more important to the determination of PoF than all possible mitigation measures.

Most hazardous substance pipelines are designed with sufficient redundancy in control and safety systems that it takes a highly unlikely chain of events to cause a leak/rupture type failure solely by the improper use of system components. A system can be made to be even more insensitive to human error through physical barriers and intervention opportunities. Nonetheless, history has demonstrated that the seemingly unlikely event sequences occur more often than would be intuitively predicted.

As noted, human error potential involves difficult to assess aspects of a working environment. As a starting point, the evaluator can look for a sense of professionalism in the way operations are conducted. Corporate culture typically guides this. Seemingly unrelated aspects such as a strong safety program, housekeeping, or facility attractiveness can all be evidence of attention and standard of care, which usually also translate to improved error prevention.

The mitigation measures commonly employed are intertwined. For example, better procedures enhance training and vice versa; safety systems supplement procedures; mechanical devices complement training.

Activities requiring high levels of supervision are logically more susceptible to error. Better training and professionalism usually mean less supervision is required.

Special product issues are often affected by human actions, especially when assessing service interruption potential, and can be considered here. For example, hydrate formation (production of ice as water vapor precipitates from a hydrocarbon flow stream, under special conditions) has been identified as a service interruption threat and also, under special conditions, an integrity threat. The latter occurs if formed ice travels down the pipeline with high velocity, possibly causing damages. Similarly, pressure surge events are often generated by human actions. Because such special occurrences are often controlled through operational procedures, they warrant attention here.

A manned facility with no site-specific operating procedures and/or less training emphasis may have a greater incorrect operations-related likelihood of human error than one with appropriate level of procedures and personnel training.

8.7.1 Exceeding Design Limits

The possibility of exceeding any threshold for which the system was designed is an important element of a leak/rupture risk assessment. A measure of the susceptibility of the facility to overstressing is modeled here as a part of the incorrect operations failure assessment. While design limits related to temperature, product velocity, and others are used, pressure exceedances are by far the most common integrity threats to a pipeline. Internal pressure is the most important design threshold for most pipelines and is often the primary design limit of interest. Overpressure will be the focus of this discussion while also illustrating the approach to assess any other relevant design exceedance potential. Other limit states such as temperature, level, flowrate, etc, can follow a parallel assessment path as the one outlined here for overpressure potential. For instance, vessel overfill/overflow can be included in leak/rupture scenarios and modeled in a fashion very similar to overpressure.

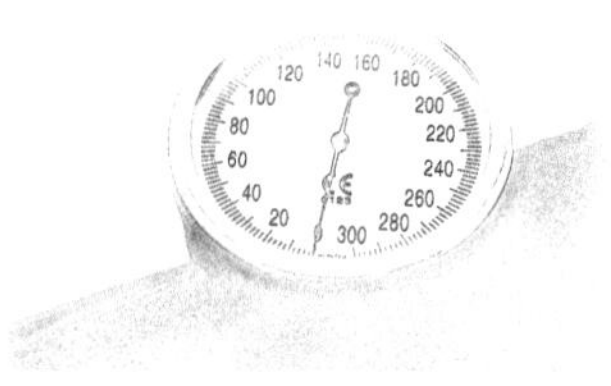

The safest scenario occurs when no pressure source exists that can generate sufficient pressure to exceed allowable limits. A system in which it is not physically possible to exceed the design pressure is inherently safer than one where the possibility exists. A pump that, when operated in a deadheaded condition, can produce a maximum of 900-psig pressure cannot, theoretically, overpressure components designed for 1800 psig. In the absence of any other pressure source (including heat) or scenario, this situation suggests that no overpressure exposure exists.

A pipeline system operated at levels well below its original design intent can also be inherently safe from overpressure. This is a relatively common occurrence as pipeline systems, originally designed for more severe conditions, change service or ownership or as throughputs decline. It is also common for pipeline systems to have pressure sources that can exceed allowable stresses, should control/safety systems fail. Note that the adequacy of safety systems and the potential for specialized stresses such as surges and fatigue are examined elsewhere in this model.

Where pressure sources can overstress systems and control and safety systems are needed to protect the facility, then risk increases. This includes consideration of the maximum pumping head and thermally induced pressure increases. Pumps and compressors are often the primary sources of pressure. Inherent overpressure safety occurs when that prime mover is incapable of creating excessive pressure in the assessed component. Certain pumps and compressors are unable to generate excessive pressures, even under 'deadhead' (pumping against a blockage) conditions.

Allowable stresses may change with changes in environmental factors such as temperature. For instance, extreme heat or cold can change the stress-carrying capacity of a material, making failure under normal operating pressure possible.

8.7.2 Potential for Threshold Exceedance

Required for a complete risk assessment are knowledge of the source pressure (pump, compressor, connecting pipelines, tank, well, the often-overlooked thermal sources, etc.) and knowledge of the system strength. The first includes pump and compressor deadhead limits; foreign pipeline connections; well connections; and even position along the hydraulic profile (where sufficient pressures to exceed limits cannot be generated). A pump running in a "deadheaded" condition by the accidental closing of a valve or a surge created by the rapid introduction of relatively high volumes of compressible fluids are classic examples of overpressure scenarios. It is important to exclude all considerations of pressure control and overpressure safety systems at this point.

Sources of overpressure should include scenarios of 'blocked-in, fluid-full with subsequent heating' (where the fluid has no room to expand) that aren't already captured in elsewhere. For instance, daytime heating of liquid trapped in a pipe segment, valve body, etc, is efficiently captured here, while an external fire scenario is probably better captured in geohazard or sympathetic reaction scenarios.

It is sometimes difficult to obtain the maximum pressure potential as it must be defined for the 'exposure' assignment, ie assuming absence of all safety and pressure-limiting devices. This is especially true when a foreign entities owns and operates a pressure source. Foreign ownership is common when the source is a connecting pipeline, a storage facility, or other non-owned delivery into a system being assessed. When the pressure source is not under operator control, the evaluation can be either more complex or involve more simplifying assumptions. In examining the overpressure potential, the evaluator may have to obtain information from operators of owned-by-others connecting equipment to understand the maximum source pressure potential. When another division, group, company etc controls both exposure and mitigation, their applied mitigation is usually more efficiently embedded in the exposure estimate. See discussion in Chapter 2.8.12 Nuances of Exposure, Mitigation, Resistance.

Ultimately, a simple yes/no answer should be available to answer this first question of 'can a threshold be exceeded?' Rare scenarios should still generate a 'yes' answer. Their improbability will be captured in the exposure value assigned. For instance, in a high volume system transporting highly compressible fluid (gas), overpressure might be conceivable, but only after many hours of 'packing'. This scenario still warrants a 'yes' answer to the 'is overpressure possible?' question, but the high improbability should be considered when assigning exposure rates.

When the answer is 'yes', an exposure is estimated for each plausible scenario. All sources of overpressure (or other threshold exceedances) should first be identified. Then all credible scenarios generating overpressures should be identified. Risk analyses tools such as HAZOPS and PHA are often very efficient in providing an exhaustive list of scenarios. Sometimes, frequencies are also assigned to each as part of those analyses. The frequencies of each scenario is then estimated, under the assumption that there is no mitigation—above that available to resist the MOP—exists. Some scenarios may only manifest under a relatively complex chain of events. In assigning a rate of exposure, the evaluator must sometimes determine the implied time period for an overpressure event to manifest. Would it take only the inadvertent closure of one valve to instantly build a pressure that is too high? Or would it take many hours (and many missed opportunities to intervene) before pressure levels were raised to a dangerous level?

To define the ease of reaching MOP (whichever definition of MOP is used) some qualitative descriptors can be created to envision the possibilities. A range of possibilities is illustrated by the following:

A. Continuous exposure, for example, one exposure per minute occurs[1]

Where routine, normal operations would, absent preventive measures, continuously expose the component to design pressure or higher. Overpressure is prevented by pressure control equipment, procedure, or safety device.

B. Rare exposure, for example, once every few years of operation

Where overpressure can occur only through a combination of multiple procedural errors or omissions or would require long periods of 'packing'. In these cases, exposure estimates may be challenging to produce, perhaps generated from a PHA/HAZOPS type process that quantifies the likelihood of each step in such unlikely scenarios

1 See discussion under Chapter 8.1.3 Continuous Exposure

C. Impossible, for example, essentially zero incident potential per year
Where direct or indirect pressure sources cannot, under any conceivable chain of events, overpressure the pipeline.

Overpressure can occur rather easily in some systems. Overpressure could occur fairly rapidly due, perhaps, to 'packing' a segment of incompressible fluid For example, The only protective measures may be procedural, where the operator is relied on to operate 100% error free, or a simple safety device that is designed to close a valve, shut down a pressure source, or relieve pressure from the pipeline.

If exceedance of some design limit is avoided only through perfect operator performance and one safety device, a higher probability of exceedance—often leading to failure—is being accepted. Error-free work activities are not realistic and industry experience shows that reliance on a single safety device, either mechanical or electronic, inevitably leads to gaps in protection.

In other systems, overpressure is possible and protection is achieved via redundant levels of control or safety devices. These may be any combination of controllers (for example, pressure, flowrate, etc), relief valves; rupture disks; mechanical, electrical, or pneumatic shutdown switches; or computer safeties (programmable logic controllers, supervisory control and data acquisition systems, or any kind of logic devices that may trigger an overpressure prevention action). When at least two independently operated devices are available to prevent overpressure of the pipeline, the accidental failure of at least one safety device, is offset by the backup protection provided by another.

Operator procedures are normally also in place to ensure the pipeline is always operated at levels below design limits. Any safety device can be thought of as a backup to proper operating procedures and, hence, as an independent mitigation measure. Industry experience shows a procedural error coincident with the failure of two or more levels of safety is not as unlikely an occurrence as it may first appear.

In other systems, situation where sufficient pressure could be introduced and the pipeline segment could theoretically be overpressured, but the scenario is extremely unlikely. An example would be a compressible fluid in a larger volume pipeline segment, requiring longer times to reach critical pressures. For example, a large diameter gas line would experience overpressure if a mainline valve were closed but only if the situation went undetected for hours.

In order to assess the exposure rate for a particular design limit exceedance, say, 'overpressure', a measure of tolerable pressures is needed. The most readily available measure of this will normally be the documented maximum operating pressure or MOP. Design pressure and/or maximum allowable pressures values may also be available. These values must be dissected to understand the true strength of the component, free from safety factors and influences of other intermittent loadings and nearby weaknesses. The risk assessor must decide, in the context of desired PXX and trade-offs between complexity and robustness, the extent of simultaneous consideration of changing resistance (for example, from extreme temperature effects reducing material capabilities, unanticipated external loadings such as debris impingement in flowing water, etc) with loadings potentially contributing to overpressure. This is also discussed in Chapter 2 Definitions and Concepts.

8.7.3 Surge potential

The potential for pressure surges, or water hammer effects, is assessed as a form of human error. A background discussion is provided in PRMM.

When surges are possible, operating procedures to prevent surge scenarios are normally in place. Additional mitigation may include mechanical devices such as surge tanks, relief valves, and slow valve closures.

In a robust risk assessment, the surge required to cause damage to the component being assessed (or a hypothetical component without resistance), would be calculated. This would also consider weakness potential since a component, weakened by corrosion, cracking, gouges, additional stresses, or others, may be able to withstand only a fraction of the surge load otherwise tolerable.

Example: 8.1 Assessment of surge potential:

Consider the surge example from PRMM: A crude oil pipeline has flow rates and product characteristics that are supportive of pressure surges in excess of MOP. The only identified initiation scenario is the rapid closure of a mainline gate valve. All of these valves are equipped with automatic electric actuators that are geared to operate at a rate less than the critical closure time. If a valve must be closed manually, it is still not possible to close the valve too quickly—many turns of the valve handwheel are required for each 10% valve closure.

In a preliminary P90 assessment, the evaluator assigns an exposure of about one valve closure event per month with a 98% reliability for each valve actuation; PoD from surge = 5 events/year x (1 – 98%) = 0.1 damages/year (a damage scenario about once every 10 years, involving the failure of an actuator to properly close the valve). Sources of conservatism (P90) in this estimate are documented by the evaluator and include intentional overestimation of aspects such as the expected annual frequency of valve operations, the fraction of the year where flowing conditions are sufficient to generate a significant surge, the number of surges that could cause damage, etc.

8.8 MITIGATION

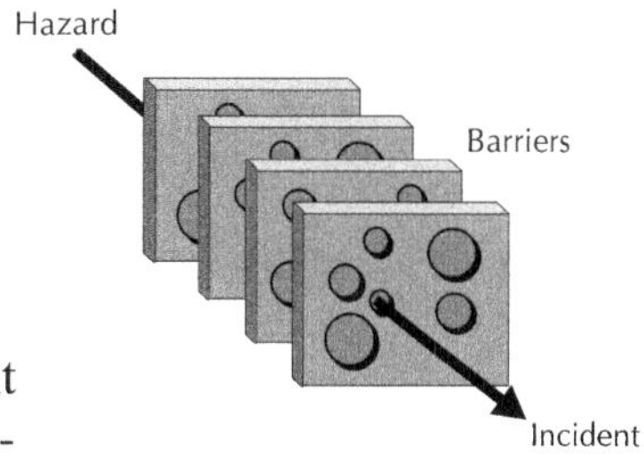

8.8.1 Control and Safety systems

Control systems and safety devices, as an important aspect of the risk picture, are included here in the incorrect operations assessment. This is done under the premise that control systems are a surrogate for human actions—operating the system within design parameters; and safety systems exist as a backup for situations in which human error causes or allows design thresholds to be reached. Both systems therefore impact the possibility of a pipeline failure due to human error.

This discussion will focus on the role of control and safety systems in preventing leaks/ruptures. Their expanded role into preventing or mitigating service interruption scenarios is covered in Chapter 12 Service Interruption Risk. The role of control and safety systems in consequence potential is discussed in Chapter 11 Consequence of Failure.

A control or safety device continuously mitigates against exceedance of a threshold. Control and safety systems can be as simple as a single device—perhaps a regulator, pressure switch, or a relief valve. They can be also extremely sophisticated and complicated: completely orchestrating product movements through multiple prime movers—pumps or compressors—associated with multiple pipeline systems, while monitoring and reacting to all events that may lead to a design parameter excursion, and recording and archiving all events and status conditions. A wide array of sensors, switches, and computers accompany most modern pipeline control/safety systems. Flowrate or pressure regulation valves are examples of devices that often mitigate against overpressure while also ensuring operational efficiencies.

For purposes of this part of the assessment, control and safety systems can both be treated as mitigation. When terminology 'safety system' or 'safety device' is used, the intention is to also include control system and control device.

Control/safety systems that employ computer-based logic are common. These allow more complex actions and sequences to be orchestrated, controlled, and protected but also create additional failure points. A modern risk assessment will need to include an evaluation of all computer permissives programs for all facilities, including PLC, SCADA, and other logic-based processes.

As in other aspects of this risk assessments, it is important to separate mitigation and resistance from exposure for systems under the operator's control, but this separation is often problematic when estimating exposure rates from systems controlled by others. A distinction between safety systems controlled by the pipeline operator and those outside his direct control is usually warranted. Risk assessment expanded into an assessment of non-owned systems is certainly possible, but requires cooperation from the other owner.

8.8.1.1 Safety systems evaluation

Failure potential is reduced as safety systems are able to reliably interrupt a sequence of events that would otherwise result in damage or failure. Understanding of this intervention opportunity began with the identification of exposure scenarios and now requires identification and evaluation of the various actions that initiate, or are initiated by, devices involving, for example, changing level, flow, temperature, and pressure conditions. When devices are established to initiate independent action—without human intervention—to protect systems, they offer direct mitigation benefit. If false alarms can be minimized, then safety systems that automatically close valves, stop pumps, and/or isolate equipment in extreme conditions are very valuable. When complete autonomous action is not appropriate, human action in combination with safety systems provide mitigation. Early warning alarms and status alerts when actions are taken should ideally be sent to a monitored control center. Also valuable is the ability of a manned control center to remotely activate equipment, including isolation and shutdown devices, to avoid or minimize damage scenarios. Less effective, especially for unmanned, infrequently visited sites, but still useful are safety systems that merely produce a local indication of abnormal conditions.

Safety systems that provide increasing station facility overpressure protection beyond specific equipment shutdown and isolation, include equipment lock-out, station isolation, station lock-out, and relief systems. Lock-out typically requires a person to inspect the station conditions prior to resetting trips and restarting systems.

A sometimes complex chain of events needs to be identified and scrutinized to fully understand certain failure scenarios involving failures of control systems, especially when interacting electronic components are involved. Electronic systems can often fail in multiple ways by a variety of effects (for example, EM pulses) that do not threaten most other components.

To ensure the on-going adequacy of safety systems, periodic reviews are valuable. Such reviews should also be triggered by formal management of change policies or anytime a change in made in a facility. HAZOPS or other hazard evaluation techniques as well as instrument-specific techniques such as LOPA, are commonly used to first assess the need and/or adequacy of safety systems. This is often followed by a review of the design calculations and supporting assumptions used in specifying the type and actions of the device. The most successful program will have responsibilities, frequencies, and personnel qualifications clearly spelled out. Many regulations for pipelines require or imply an annual review frequency for overpressure safety devices.

As an early step in the risk assessment, each portion of the pipeline system being assessed must be associated with its potential exposure scenarios and relevant control/safety systems. Each safety device located at a pump/compressor stations, metering facility, storage facility, or control center will often influence, if not protect, many miles of the system. For instance, a pressure regulator impacts all system components downstream of its location and possibly upstream as well. A pump motor shut off switch often impacts miles of system both upstream and downstream of its location.

The next step is to assess the reliability of each safety device, considering all potential device failure modes including loss of power or communications. Some valves and switches are designed to "fail closed" on such interruptions. Others are designed to "fail open," or remain in its last position: "fail last." The important thing is that the equipment fails in a mode that leaves the system in the least vulnerable condition, ie 'fail safe'.

This can be a very complex process, as is detailed in industry standards for SIL and LOPA. Alternatively, reasonable estimates can also be generated with only a few inputs and in a short time. Of course, the latter approach will be less robust and, consequently less defensible, but perhaps sufficient, especially for preliminary risk estimates.

For all control/safety devices, the evaluator should examine the status of the devices under loss of power or communications scenarios.

In a more robust analysis, guidance is available from sources such as ref [1002], as excerpted below:

> Multiple Protection Layers (PLs) are normally provided in the process industry. Each protection layer consists of a grouping of equipment and/or administrative controls that function in concert with the other layers. Protection layers that perform their function with a high degree of reliability may qualify as Independent Protection Layers (IPL). The criteria to qualify a Protection Layer (PL) as an IPL are:
>
> - The protection provided reduces the identified risk by a large amount, that is, a minimum of a 10-fold reduction. The protective function is provided with a high degree of availability (90% or greater).
> - It has the following important characteristics:
>   a. Specificity: An IPL is designed solely to prevent or to mitigate the consequences of one potentially hazardous event (e.g., a runaway reaction, release of toxic material, a loss of containment, or a fire). Multiple causes may lead to the same hazardous event; and, therefore, multiple event scenarios may initiate action of one IPL.
>   b. Independence: An IPL is independent of the other protection layers associated with the identified danger.
>   c. Dependability: It can be counted on to do what it was designed to do. Both random and systematic failures modes are addressed in the design.
>   d. Auditability: It is designed to facilitate regular validation of the protective functions. Proof testing and maintenance of the safety system is necessary.
>   e. Only those protection layers that meet the tests of availability, specificity, independence, dependability, and auditability are classified as Independent Protection Layers.

This reference cites some typical probability of failure on demand (PFD) values for certain independent protection layers.

independent protection layers	PFD
relief valve	10-2
human performance (no stress)	10-2
human performance (under stress)	0.5 to 1.0
operator response to alarms	10-1
overpressure of well maintained vessel	10-4

Some annual failure rate examples are also offered [1002]

Low	a failure or series of failures with a very low probability of occurrence within the expected lifetime of the plant, eg 3 or more simultaeous instrument, valve, or human failures; spontaneous failure of single tank or process vessel	<10-4
Medium	a failure or series of failures with a low probability of occurrence within the expected lifetime of the plant; eg dual instrument or valve failure; combination of instrument failure and operator error; single failure of small process lines or fittings	<10-4 to 10-2
High	a failure can reasonable be expected to occur within the lifetime of the plant; eg process leaks single instrument or valve failure; human errors that result in material releases	>10-2

Alarms and other systems that rely on human intervention are logically more susceptible to failure on demand. Error potential is reduced when the condition-sensing device or permissive limit exceedances automatically initiate a full, or partial, shutdown of affected station equipment, with an alarm to remote/local personnel. In the absence of automatic actions, condition-sensing device or permissive limit exceedances may issue an alarm at a continuously manned location that requires operators to evaluate the conditions and remotely initiate a full, or partial, shutdown of affected station equipment.

The potential for human error to incorrectly/inadvertently isolate the safety device from the component(s) being protected is also an important part of this analysis. Note that some systems provide no plausible scenario where such human error could cause such isolation, for example a three way valve with redundant devices.

The maintenance and calibration protocols used on the safety device should also be included in the analyses. Most published reliability rates would assume adherence to the device manufacturer's recommended maintenance and calibration practice. In practice, however, it is not uncommon for a company to choose a more- or a less-robust protocol. Note that a superior risk assessment can show the value of changes in maintenance/calibration practice by estimating the corresponding changes in device reliability.

Different reliability values are acceptable depending on the criticality of the process being protected. At the highest levels of protection, reliabilities such as the following would be expected:

Low Demand Mode of Operation	High Demand or Continuous Mode of Operation
PFD	Probability of dangerous failure per hour
10-5 to 10-4	10-9 to 10-8

at the lowest protection level, values such as the following may be appropriate:

Low Demand Mode of Operation	High Demand or Continuous Mode of Operation
PFD	Probability of dangerous failure per hour
10-2 to 10-1	10-6 to 10-5

[1003]

Finally, the reliability of each sub-system is combined for an estimate of the overall reliability. Manufacturer's stated reliability values will usually be based on ideal conditions and maintenance practices. Variations from ideal should be considered in the risk assessment. For maintenance, this will require at least some understanding of various control/safety system's "predictive and preventative maintenance" (PPM) programs, including equipment/component inspections, monitoring, cleaning, testing, calibration, measurements, repair, modifications, and replacements. See further discussion of maintenance later in this chapter.

The reliability and timeliness of SCADA dispatch processes would also need to be assessed as part of the overall mitigation effectiveness of safety systems providing alerts only.

Example: 8.2 Assessing a set of safety systems:

Consider a pipeline connected to a pump capable of overpressuring a component. A pressure regulator and multiple safety devices are installed to avoid overpressure. A pressure-sensitive switch halts flow upon high pressure indications; and a relief valve will open and vent the entire pumped product stream to a flare upon an extremely high pressure indication. This facility is remotely monitored by a SCADA system, transmitting appropriate data (including pressures) that is continuously monitored in a control center. Remote shutdown of the pump from the control center is possible. Communications for data received in the control room as well as control instructions generated by the control center are deemed to be 98% reliable.

Exposure is assessed as 'continuous' and quantified as 'every minute':
60x24x365 = 525,600 events/yr.

Note that four levels of mitigation are present (regulator, pressure switch, relief valve, control room monitoring), any of which is capable of providing full protection. With preliminary, conservative reliability values of 99% assigned to each of the first three and 50% to the last (with consideration of human error and communications outage rates), combined mitigation effectiveness is 99% OR 99% OR 99% OR 50% = 99.99995%.

This results in a PoD estimate of 0.26 events/yr [525,600 events/yr x (1 – 99.99995%) = 0.26], a damaging overpressure event, perhaps causing at least a minor permanent deformation, about once every 4 years.

8.8.2 Procedures

The use of procedures to ensure correct operations and avoid errors is well known. As a means of mitigating scenarios that precipitate failure, procedures and their use should be a part of the mitigation effectiveness estimates.

A range of quality, rigor, and utility exists among operators' procedures and often within different functional or geographical areas of the same operator. A list of ingredients that distinguishes the most effective use of procedures can first be created as the program that warrants the highest effectiveness estimate. Perhaps first among the ingredients is a corporate culture that requires the adherence to procedures—ie, their correctness and everyday use. Without this, the desire to follow a procedure correctly may be missing.

Since each mitigation measure is evaluated independently from others, we assume there has been no training on the procedures. Some might think this is an unreasonable position—training and procedures are so intertwined that independent evaluations of the two seems nonsensical to many. But this is not necessarily the case. Procedures alone can be clear and complete enough to produce error-free operations in some cases. Here's an example to illustrate. Good procedures allow the purchaser of a shipped, disassembled table to assemble that table properly and without incident, even though there has been no training on table assembly. The procedure stands on its own merit. However, the desire by the purchaser to correctly complete the assembly is critical to the success rate.

Most would agree that the highest rated, ie, most effective, procedure system would have all of the following ingredients:

- Strong corporate culture mandating their prominent role in day-to-day activities
- Clearly written
- Complete coverage of all tasks in all procedures
- User-friendly format and beyond—perhaps even enticing and entertaining to the user
- Use of video, photographs, illustrations, etc as appropriate for optimum understandability and utility
- Regularly reviewed and refreshed
- Field-tested and verified regularly
- Validated by independent audit
- Readily retrieved and protected (version control) by robust document management system.

Many technical writing 'best practices' could be consulted to provide further guidelines for "what makes an excellent procedure".

In a superior program, there should be evidence that procedures are actively used, reviewed, and revised. Such evidence might include filled-in checklists and procedures in active use in field locations and with field personnel.

Activities near a pipeline, but not actually on it, are also appropriately included when such activities may have risk implications. For instance, nearby excavations can impact a pipeline's support conditions, perhaps increasing exposure from landslide, erosion, or subsidence.

Locating processes—finding and marking buried utilities prior to excavation activities—are important for any subsurface system, but perhaps especially so for distribution systems that often coexist with many other subsurface structures. Such procedures may warrant additional attention in this evaluation.

A protocol should exist that covers procedures maintenance: who develops them, who approves them, how training is done, how compliance is verified, how often they are reviewed, what is the update process, etc. A document management system should be in place to ensure version control and proper access to most current documents. This is commonly done in a computer environment, but can also be done with paper filing systems.

While procedures are normally a mitigation measure, they may alternatively generate exposures, especially in abnormal operations. Procedure execution during operations that can put the system integrity at risk, are a part of the exposure rate in the risk assessment.

Any recent history of station procedure-related problems should be investigated for evidence of procedure effectiveness.

8.8.2.1 Mitigation Effectiveness

Transmission pipeline company SME's have typically assigned maximum effectiveness values in the range of 30% to over 90%, based on their experiences and ideas of how effective the highest quality procedures program could be, as a stand-alone error prevention item. For perspective, the higher end of this range assumes that fewer than 1 out of 10 otherwise damaging events would occur solely through the hypothetical best procedures program (assuming no training or other mitigations)—9 out of 10 are avoided—while the lower end assumes only 3 out of 10 events are avoided by the best program. Actual effectiveness values are then assigned based on differences from the idealized, perfect program.

8.8.3 SCADA/communications

8.8.3.1 Background

A SCADA system allows remote monitoring (of parameters such as pressures, flows, temperatures, and product compositions) and some remote control functions, normally from a central location, such as a control center. Standard industry practice for hydrocarbon transmission pipelines in most western countries is 24-hours-per-day monitoring of "real-time" critical data with audible and visible indicators (alarms) set for abnormal conditions. At a minimum, control center operators normally have the ability to safely shut down critical equipment remotely when abnormal conditions are seen.

Interfaces between the pipeline data-gathering instruments and conventional communication paths such as telephone lines, satellite transmission links, fiber optic cables, radio waves, or microwaves facilitate the delivery of information to and from the control center. Modern communication pathways and scan rates can refresh data at least every few seconds with 99.9% + reliability and often include redundant (sometimes even manually implemented dial-up telephone lines) pathways in case of extreme pathway interruptions.

A SCADA system often serves also as safety devices, when computer logic is used to control critical operational parameters.

In providing an overall view of the entire pipeline from one location, a SCADA system facilitates system diagnosis, leak detection, transient analysis, and work coordination, thereby impacting risk in several ways including:

- human error avoidance,
- surge avoidance,
- leak detection,
- emergency response.

The focus in this part of the risk assessment is on the role of SCADA in human error avoidance; for example, mitigation of incorrect operations.

8.8.3.2 SCADA Capabilities

See PRMM for a discussion of SCADA system concepts.

When the SCADA provides control or safety functions, its role in damage/failure prevention is captured as another level of safety system (see previous discussions). The more technical aspects of kind and quality of data and control (incident detection) and the use of that capability in consequence minimization (ie, leak detection and emergency response), can be assessed in the measure of consequence potential (see Chapter 11 Consequence of Failure).

Figure 8.2 Control Center as part of SCADA

8.8.3.3 Error Prevention

Setting aside for now its role as a safety system and consequence minimizer, the emphasis here is on the SCADA role in reducing human error-type incidents. From the human error perspective only, the major considerations are that a second "set of eyes" is monitoring, is hopefully consulted prior to field operations, is involved with all critical activities, and that more reliable coordination of the system operations is provided. Although human error potential exists in the SCADA loop itself—more humans involved may imply more error potential, both from the field and from the control center—the cross-checking opportunities offered by SCADA can reduce the probability of human error in operations. One emphasis should therefore be placed on how well the two locations are cooperating and cross-checking each other.

Protocols that require field personnel to coordinate all station activities with a control room offer an opportunity for a second set of eyes to interrupt an error sequence. In the best practices, critical stations are identified and must be physically occupied if SCADA communications are interrupted for specified periods of time. Proven reliable voice communications between the control center and field should be present. When a host computer provides calculations and control functions in addition to local station logic, all control and alarm functions should be routinely tested from the data source all the way through final actions.

While transmission pipeline systems are common users of SCADA, these mitigation concepts apply to offshore, distribution, gathering pipelines, as well as tank farms, pumps stations, platforms, etc., even where a standard SCADA is not being used. As a means of reducing human errors, the use any system or protocol of regular coordination of actions between multiple observers, such as field operations and a central control is an intervention point for human error reduction. Some systems and facilities have protocols for communications/coordination producing benefits of multiple eyes and minds confirming actions, although a SCADA type system is not present. Some facilities will have distributed control and monitoring (DCM) systems that act like SCADA albeit in a more limited geographical area.

8.8.3.4 Mitigation Effectiveness

Transmission pipeline company SME's have typically assigned maximum effectiveness values in the range of 5% to 30%, based on their experiences with SCADA systems in human error avoidance. For perspective, the higher end of this range assumes that 3 out of 10 otherwise damaging events are avoided solely through the use of a superior SCADA system while the lower end assumes only 5 out of 100 events are avoided. Actual effectiveness values are then assigned based on differences from the idealized, perfect program.

8.8.4 Substance Abuse

Errors with an underlying cause of 'impairment' can be partially mitigated by programs to manage substance abuse. In some countries, government regulations or common industry practice require drug and alcohol testing programs for certain classes of employees in the transportation industry.

Since these mitigation measures are focused on specific types of human errors—those involving impairments—they are most correctly applied only to those exposures.

8.8.4.1 Mitigation Effectiveness

In transmission pipeline companies that operate free of significant substance abuse issues, SME's have typically assigned maximum effectiveness values in the range of 1% to 5% for exceptional substance abuse programs. For perspective, even the higher end of this range assumes that only 5 out of 100 otherwise damaging events are avoided solely through this program while the lower end assumes only 1 out of 100 events are avoided. Actual effectiveness values are then assigned based on differences from the idealized, perfect program.

8.8.5 Safety/Focus programs

With inattention being an underlying factor in many human error events, company programs that provide focus may act as mitigation, even when not directed specifically at the failure being measured by the risk assessment. A focus on employee safety is an example. An employee safety program[2] is a nearly intangible but still important factor in a risk assessment (although very central to employee safety risk management).

It is intangible in the sense that the impact on human error potential derived from a strong safety program is difficult to quantify. However, most would agree that the extra care and attention to routine tasks that is fostered by a high level of safety awareness

2 A safety program is different than a safety system, with the latter referring to physical devices that prevent exceedances of pressure, flowrates, etc.

and a corporate culture of safety should translate to some benefits in all types of human error avoidance.

Similarly, other peripheral company focuses such as on good "housekeeping" practices can be revealing. Housekeeping can include treatment of critical equipment and materials so they are easily identifiable (using, for instance, a high-contrast or multiple-color scheme), easily accessible (next to work area or central storage building), clearly identified (signs, markings, ID tags), and clean (washed, painted, repaired). Housekeeping also includes general grounds maintenance so that tools, equipment, or debris are not left unattended or equipment left disassembled. All safety-related materials and equipment should be maintained in good working order and replaced as recommended by the manufacturer. Station logs, reference materials, and drawings should be current and easily accessible, in the more effective programs.

8.8.5.1 Mitigation Effectiveness

Transmission pipeline company SME's have typically assigned maximum effectiveness values in the range of 1% to 5%, based on their experience. For perspective, even the higher end of this range assumes that only 5 out of 100 otherwise damaging events are avoided solely through this type program, even the best conceivable, while the lower end assumes only 1 out of 100 events are avoided. Actual effectiveness values are then assigned based on differences from the idealized, perfect program.

8.8.6 Training

Training is a key mitigation measure protecting against human error. PRMM discusses a list of key ingredients in a training program:

Documented minimum requirements

- Testing
- Topics covered:
  - Observed and assessed performance of actions
  - Job procedures (as appropriate)
  - Scheduled retraining
  - Proficiency testing and periodic re-testing
  - Detailed record-keeping
  - Progress/performance tracking.

Training on tasks whose execution can put the system integrity at risk, are especially critical to the risk assessment. A high level of worker turnover makes training even more critical. Both of these aspects should be included in the risk assessment.

For maximum effectiveness as a risk mitigation, written procedures dealing with all operational actions, abnormal and emergency actions, repairs, and routine maintenance should be readily available. Not only should these exist, it should also be clear that they are in active use by the personnel. The recommendation here is to look for checklists, revision dates, and other evidence of their use. Procedures supplement training by helping to ensure consistency. Specialized procedures are required to ensure that original design factors are still considered long after the designers are gone. A prime example is welding, where material changes such as hardness, fracture toughness, and corrosion resistance can be seriously affected by the subsequent maintenance activities involving welding.

The assessment should consider the effectiveness of the retraining schedule and the periodic retesting in terms of their ability to adequately verify employee skills. Higher workforce turnover rates have been correlated to increased error rates, due to loss of experience and training benefits that otherwise accrue to a more stable workforce. This could be an influencing factor when assigning mitigation effectiveness.

8.8.6.1 Mitigation Effectiveness

Transmission pipeline company SME's have typically assigned maximum effectiveness values in the range of 30% to over 90%, based on their experiences and ideas of how effective the highest quality training program could be, as a stand-alone error prevention item. For perspective, the higher end of this range assumes that fewer than 1 out of 10 otherwise damaging events would occur, prevented solely through the hypothetical best training program (assuming no procedures or other mitigations) while the lower end assumes only 3 out of 10 events are avoided by the best program. Actual effectiveness values are then assigned based on differences from the idealized, perfect program.

8.8.7 Mechanical error preventers

The role of mechanical error preventers as mitigation measures should reflect the combined effectiveness of the devices/measures being rated. Examples of common devices/measures are noted in PRMM as:

- Three-way valves with dual instrumentation
- Lock-out devices
- Key-lock sequence programs
- Computer permissives—logic controls that will prevent certain actions from being performed out of sequence
- Highlighting of critical instruments.

Effectiveness should reflect the combined (OR gate addition) of each application. An application is valid only if the mechanical preventer is used in all instances of the scenario it is designed to prevent.

Transmission pipeline company SME's have typically assigned maximum effectiveness values in the range of 30% to over 90%, based on their experiences and ideas of how effective the highest quality mechanical error prevention program could be, as a stand-alone error prevention item. This varies widely based on the type of facility being assessed since, for some, a wide range of mechanical devices are possible and practical, but for others, few devices are available. For perspective, the higher end of this range assumes that fewer than 1 out of 10 otherwise damaging events would occur solely through the hypothetical best program (assuming no training, procedures, or other mitigations) —9 out of 10 are avoided—while the lower end assumes only 3 out of 10 events are avoided by the best program. Actual effectiveness values are then assigned based on differences from the idealized, perfect program.

8.9 RESISTANCE

As discussed here, many of the damage scenarios for leak/rupture that are directly caused by human error involve overpressure. Therefore, internal pressure related stress-carrying capacity is a main consideration for resistance from human errors. The defect-free stress carrying capacity is readily calculated for most pipeline components. Inclusion of possible defects is then added to the analyses as detailed in Chapter 10 Resistance Modeling.

When scenarios such as vessel overflow/overfill are included, an assessment approach directly analogous to overpressure can be efficiently employed. Resistance may be minimal for such scenarios, unless features such as secondary containments are included as resistance rather than as consequence minimizers.

Under an expanded definition of 'failure', a system's resistance to human error is more complex. A system's ability to absorb excursions of contaminants, flowrate deviations, etc can be multi-faceted. Aspects such as time to overpressure (or exceed some other threshold) should be included in either exposure estimates or resistance estimates. For instance, a high volume system transporting a highly compressible fluid (gas), will often have a degree of inherent resistance (or reduced exposure) since overpressure is possible, but only after many hours of 'packing'. See also the discussion in Chapter 12 Service Interruption Risk.

8.9.1 Introduction of Weaknesses

In addition to real time failures, human error contributes to delayed failures via the introduction of unintended weaknesses into a pipeline system. These may occur in any of the four phases introduced previously: design, construction, operations, and maintenance. For each phase, estimates of types and frequencies of weaknesses created should be estimated. For each potential type of weakness, a reduction in load-carrying capacity will be required in order to fully understand the impact on risk. This is fully discussed in Chapter 10 Resistance Modeling. In this chapter, discussion of sources of human-error types of weaknesses is offered, to assist in the estimation of possible weaknesses in any component being risk-assessed.

Potential errors committed during the design/construction can be difficult to assess for an existing pipeline. Historic design processes are often not well defined or documented and are often highly variable. Nonetheless, an assessment resulting in at least a rough estimate of weakness potential is prudent.

Errors or, by today's standards, inferior practices in past design/construction practice, tend to introduce weaknesses and will most often appear as resistance issues in a modern risk assessment. Even though a system with a weakness may operate with sound integrity for many decades, the presence of that weakness, coupled with certain loadings, can eventually precipitate a failure. The risk assessment should identify and quantify the types of weaknesses that may be present.

The suggested approach is for the evaluator to seek evidence that error-preventing actions were taken during the design/construction phases. If design/construction documents are available, a check or certification can be done to verify that no obvious errors have been made. Otherwise, evidence such as from inspections and testimony of SME's may drive the assessment.

Chapter 10 Resistance Modeling details the types of weaknesses commonly encountered in pipelines. Here, the potential for such weaknesses is discussed.

8.9.2 Design

A formal hazard identification process during design helps to ensure that all threats are understood and appropriately mitigated. HAZOP studies and other appropriate hazard identification techniques are discussed in Chapter 3 Assessing Risk. These techniques provide value inputs into estimates of exposure, mitigation, resistance, and consequence. Thoroughness and timeliness are important: if this type of analysis is not available from original design, it can be performed at any time and results used to strengthen the risk assessment.

Potential design errors include flaws revealed during operations and maintenance practices. While often more 'real-time', apparent O&M errors can also conceivably manifest long after the actual error-introducing activity has occurred. For example, a mis-designed flow/pressure control system that operates satisfactorily for years until a rare combination of factors causes the controls to overpressure a component.

8.9.3 Material selection

The assessment should consider that the rigor with which proper materials were identified and specified with regard to all plausible stresses.

Notably in distribution systems, a wide range of pipe and appurtenance materials have been used with a variety of different joining techniques. Some of these choices have later proven to be problematic, from an integrity standpoint. Certain installations of cast iron, plastics, tees, and couplings have generated a disproportionate amount of failures for some operators.

Given that a certain amount of care and prudence is associated with the 'standard' practice, risk reduction for this item can be based on the existence and use of ***additional*** control documents and procedures—beyond standard practice— that govern all aspects of pipeline material selection and installation. Superior practices can influence the risk assessment via reduced incidences of weaknesses.

8.9.4 QA/QC Checks

The risk assessment should consider the extent to which design calculations and decisions were checked for errors at key points during the design, material procurement, and installation processes.

Given a certain amount of error-checking in the 'standard' practice, assignment of additional mitigation effectiveness would be warranted for systems whose design process was more carefully monitored and checked. This would be reflected in a reduced rate of weaknesses to be associated with the components/segments benefiting from the more aggressive quality assurance programs.

8.9.5 Construction/installation

Typical construction-error risk elements are discussed in PRMM and here in Chapter 8.5 Construction Phase Errors. When a mitigation or an exposure exceeds the norm assumed in the error rate produced by 'standard' practice, the influences of these factors should be included in the assessment.

For assessing the potential for construction phase weaknesses in a system, the evaluator should seek evidence regarding the steps that were taken to ensure that the pipeline section was constructed correctly. This includes the construction specifications as well as checks on the quality of workmanship during installation.

Challenging installation conditions are logically linked to potentially higher error rates. Offshore, arctic and tropical environments, congested urban areas are a few examples of more difficult conditions. When it can be determined that an installation period involved difficulties due to weather, labor disputes, resistance from outside parties, excessively aggressive time urgencies, and other influences, error rates would similarly be expected to increase. Delayed effects from sabotage activities can also be included here. For instance, an intentionally drilled hole partially through a pipe wall can be treated as a resistance reduction just as a defective girth weld.

Construction errors on distribution systems may be more common due to the increased level of continuous construction activity coupled with the variability of construction crews, and materials used, all often spanning several decades of installation.

Weaker inspection practices during construction suggest higher incidence rates of errors; for example, an assumption that more weaknesses were introduced.

Questionable materials purchase, receipt, or installation practices should result in higher estimates of weaknesses in a system.

Less than 100% inspection of all joints, failure to meet minimum industry-accepted practices, questionable practices, or other uncertainties should lead to higher estimated incidences of weaknesses when conducting a conservative risk assessment.

Uncertain practice of backfill/support techniques during construction warrants consideration of higher rates of coating defects as well as strength reductions such as dents and gouges.

High levels of residual stresses due to improper handling have played a role in historical failures. Transportation fatigue—the growth of cracks in larger diameter pipes, transported by rail prior to improved handling protocols—is another example of a handling-related failure contributor.

The evaluator may assume reduced incidences of weaknesses when he sees evidence of superior materials handling practices and storage techniques during and prior to construction. Calculations can be performed to assess the susceptibility of certain pipe specifications to damage by improper handling. When susceptible, weaker handling practices warrants higher incidences of weaknesses.

Field-applied coatings (normally required for joining) are problematic because quality control, including the effects of ambient conditions, are difficult to manage. Careful control of temperature and moisture is normally required and all coating systems will be sensitive to some extent to surface preparation.

A major integrity threat to some pipelines is the presence of CP-shielding, poorly applied coatings over girth welds. When just one of these issues is present, the pipeline may still experience a long life with few extraordinary considerations needed. However, the presence of both shielding coating and disbondment creates a systemic threat to integrity that is challenging to manage.

Because overpressure protection is identified as a critical aspect in a many pipeline systems, maintenance of regulators and other pressure control devices is critical. The evaluator should seek evidence that regulator activity is monitored and periodic overhauls are conducted to ensure proper performance. Other pressure control devices should similarly be closely maintained.

The care of an odorization system in a gas distribution system should also be considered, with questionable maintenance practices leading to reduced leak detection capabilities.

Severe weather preparatory programs are common for many facilities and are logically included in a risk assessment. These might include hurricane, windstorm, flood, ice/hail, wildfire, and extreme temperature events such as freeze protection programs.

Other preparatory events can be examined in a similar fashion, with results informing risk assessment inputs.

9 SABOTAGE

Highlights

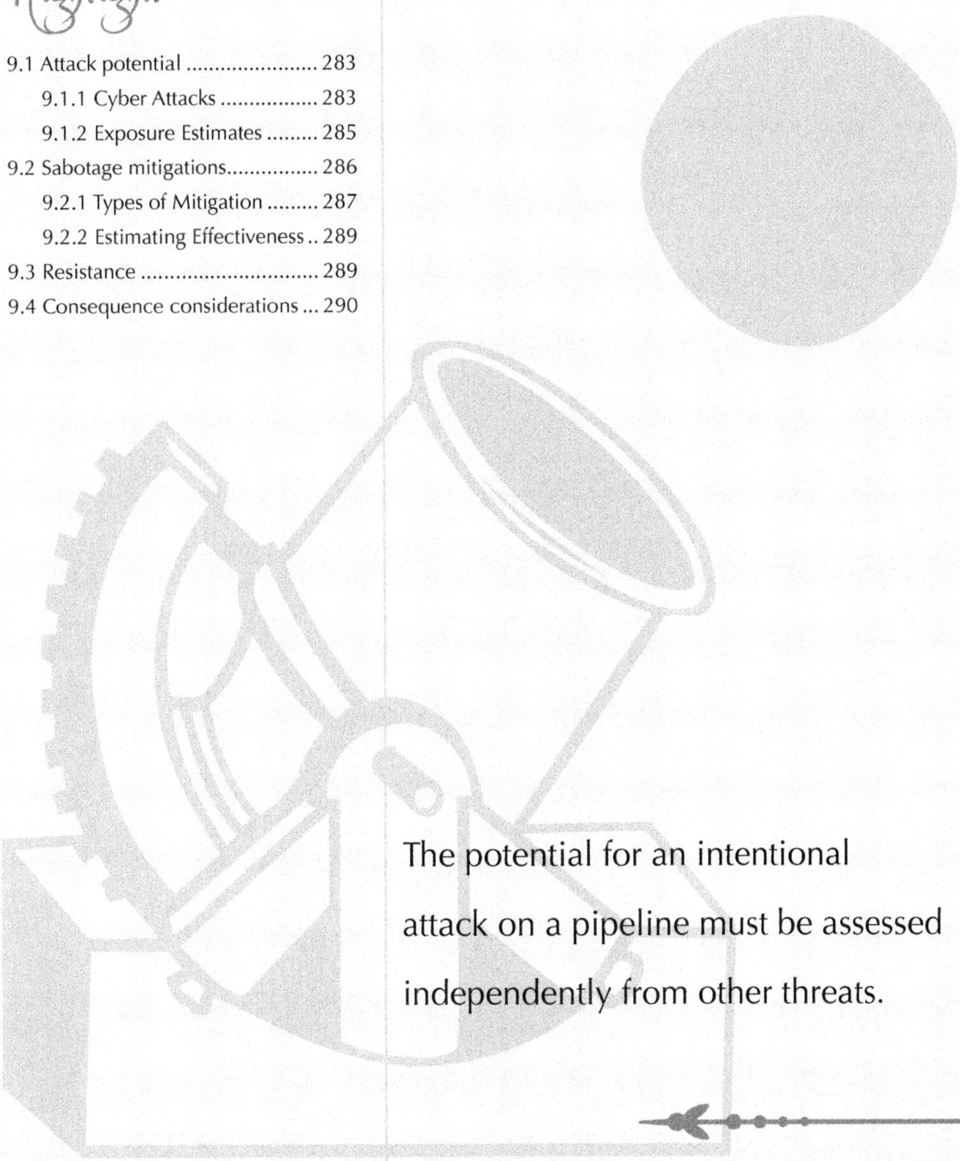

The potential for an intentional attack on a pipeline must be assessed independently from other threats.

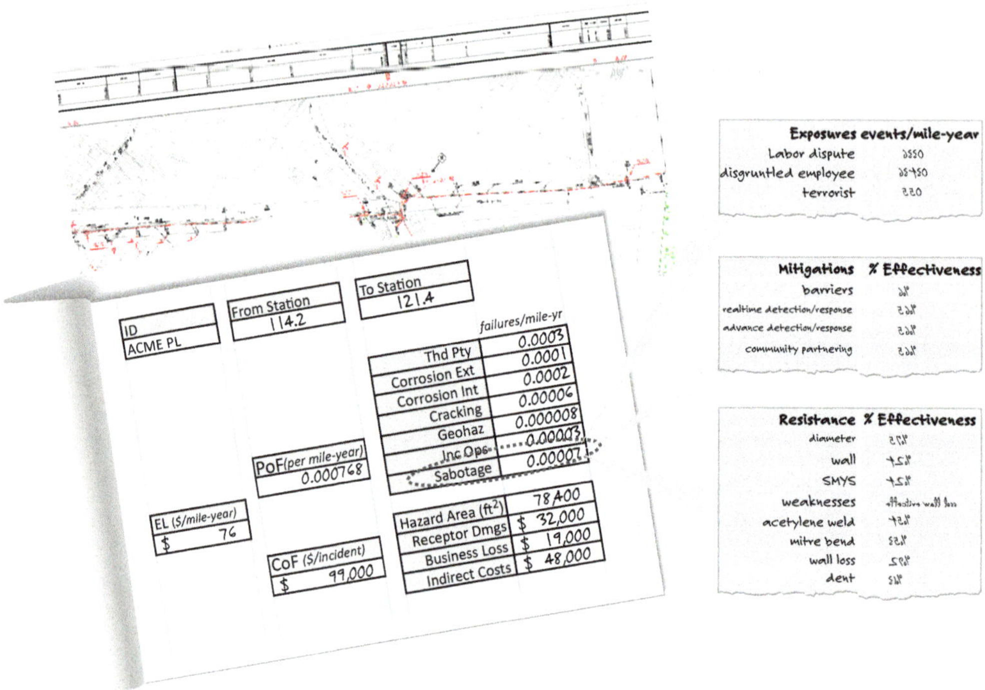

The risk of sabotage is difficult to fully assess because such threats are so situation specific and subject to rapid change. The assessment is usually subject to a great deal of uncertainty. Nonetheless, the potential exists for most pipeline systems and should not be ignored. It is recommended that the sabotage threat be included as a stand-alone assessment. As an intentional, rather than accidental-, event, it represents a unique type of threat that is independent and additive to other threats.

The likelihood of a pipeline system becoming a target of sabotage is a function of many variables, including the relationship of the pipeline owner with the community and with its own employees or former employees. Vulnerability to attack is another aspect. In general, the pipeline system is not thought to be more vulnerable than other municipal systems. The motivation behind a potential sabotage episode would, to a great extent, determine whether or not this pipeline is targeted. Reaction to a specific threat would therefore be very situation specific. Note, that some already-discussed risk variables and possible risk reduction measures overlap the variables and measures that are normally examined in dealing with sabotage threats. These include security measures, accessibility issues, training, safety systems, and patrol.

The exposure level to a sabotage event can first be assessed based on the current socio-political environment in the area of the pipeline as well as inside the pipeline company itself. Then a damage potential can be estimated, based on the presence of mitigating measures. Finally, the ability of the component to resist the attack is estimated.

Guidance documents concerning vulnerability assessments for municipal water systems are available and provide insights into the threat.

9.1 ATTACK POTENTIAL

To assess the attack potential, the definition of 'failure' used for the risk assessment must first be reviewed. If the risk assessment is strictly leak/rupture based, then exposure events are clear—they must threaten integrity. Attacks unrelated to integrity issues can be included in the risk assessment, but must be acknowledged in the 'failure' definition, in order that exposure, mitigation, and resistance values can be assigned. For example, if an event of interest is a cyber attack intended to steal company-sensitive information, (perhaps to give competitive advantage to the thief) that type of event can be included in the definition of 'failure'.

When failure also includes service interruption, then identifying exposure events becomes more challenging. Although there is much overlap, the focus in this chapter will generally be on the former—threats related to leak/rupture potential. See Chapter 12 Service Interruption Risk for a discussion of the latter.

Sabotage can be thought of as intentional third-party damage event. Sabotage often has complex socio-political underpinnings. As such, the likelihood of incidents is usually difficult to judge. Even under higher likelihood situations, mitigative actions, both direct and indirect, are possible.

Vandalism can be considered a type of sabotage. However, defacing (for example, spray painting) or minor theft of materials are exposures that are readily resisted by most pipeline components. If the sabotage exposure count includes vandalism events, then resistance estimates must consider the fraction of exposure events that are vandalism spray-paint-type events and therefore 100% resisted by the component. With the possible exception of instrumentation or control systems, pipeline components are generally more resistive to vandalism than to sabotage. Again, the definition of 'failure' governs how events are included into the risk assessment.

9.1.1 Cyber Attacks

Cyber security is a more recent consideration for pipelines. Historically, pipeline electronic systems were thought to be relatively immune to such attack for several reasons:

- Most critical operations such as valve open/close, pump start, etc, required human physical interaction.
- Control systems were isolated; in particular, they were separate from the Internet.
- Redundancies in control and safety devices prevented malicious threats to integrity, if not also to continuous operation (ie, no flow interruptions).
- The control systems were difficult to understand by outsiders.
- Little damage potential beyond nuisance data interruptions were foreseen.

Today, remote sensing, automation, and interconnectivity is prevalent among control systems. Vulnerability, as well as availability and value of information moving through cyber systems are all much higher than in years past.

Pipeline equipment commonly used and vulnerable, to varying degrees, to cyber attack include components of systems with labels such as:

- PLC (programmable logic controller)
- DCS (distributed control systems)
- SCADA (supervisory control and data acquisition)
- PCS (process control system)
- ICS (industrial control system).

Related to both cyber security and service interruption is the potential use of directed energy weapons, including electromagnetic pulse devices that can destroy electronic components. Such pulses are also naturally occurring (see Chapter 7 Geohazards). When weaponized, a small, perhaps briefcase- sized device, can be placed in proximity (perhaps outside a fenceline) to a surface facility and, when 'detonated' cause significant damages. Some older analog style electronics are relatively immune and more vulnerable components can be 'hardened' to defend against such attacks.

A sometimes complex chain of events needs to be identified and scrutinized to fully understand certain failure scenarios involving failures of electronic components. Most pipeline facilities employ 'failsafe' protocols whereby single or even multiple instrumentation failures may interrupt service but do not threaten integrity.

The ability to orchestrate a failure (by whatever definition of 'failure' is being used in the risk assessment) through a component of such cyber-systems should be identified. This may require a special group of SME's using thorough scenario-generation techniques such as HAZOPS. Susceptible components must then be linked to portions of the pipeline system since the origination of the sabotage event may be different from the point of failure on the pipeline. For example, an attack on a SCADA system's central computer may trigger a valve closure impacting a specific portion of a certain pipeline system.

Once susceptible components are identified and associated with pipeline system failure points, the frequency of potential attacks should be estimated. Several types of potential cyber attackers and their possible motivations are identified [1010]:

- Garden variety hacker: hobby, notoriety, nuisance
- Hactivist: support cause, disrupt or delay project, discredit company, personal agenda
- Cyber-criminal: financial or competitive gain, business disruption, market impact, service for hire, sales of information
- Nation state: intellectual property theft, political agenda, economic gain, disrupt, degrade, or destroy systems.

To the extent that they are consistent with the definition of 'failure' guiding the risk assessment, the contribution from each of these should be included in the sabotage exposure estimate. Even if thought to be 'insignificant', a value—reflecting best estimate of future frequency of events—should still be included in the risk assessment.

9.1.2 Exposure Estimates

In the absence of strong, quantitative data, qualitative descriptors could be linked to exposure frequencies as a starting point in the risk assessment. PRMM provides a sample of such qualitative descriptors. A sample of a quantitative range estimate—future event frequencies—is associated with those descriptors as follows:

- **Low attack probability** P90 exposure frequency is less than 0.001 events per km-yr on buried portions; perhaps 10 to 100 times higher for surface facilities. Indications of impending threats are nonexistent or very minimal. The intent or resources of possible perpetrators are such that real damage to facilities is only a very remote possibility. No attacks other than random (not company or industry specific) mischief have occurred in recent history. Simple vandalism such as spray painting and occasional theft of non-strategic items (building materials, hand tools, chains, etc.) may also warrant this exposure level.
- **Medium probability** P90 exposure frequency = 0.01 to 0.1 events per km-yr on buried portions; perhaps 10 to 100 times higher for surface facilities. A credible threat exists. Attacks on this company or similar operations have occurred in the past few years and/or conditions exist that could cause a flare-up of attacks at any time. Attacks may tend to be propagated by individuals rather than organizations or otherwise lack the full measure of resources that a well-organized and resourced saboteur may have.
- **High probability** P90 exposure frequency = 0.1 to 10 events per km-yr on buried portions; perhaps 10 to 100 times higher for surface facilities. Threat is known and significant. Attacks are an ongoing concern. There is a clear and present danger to facilities or personnel. Conditions under which attacks occur continue to exist (no successful negotiations, no alleviation of grievances that are prompting the hostility). Attacks are seen to be the work of organized guerrilla groups or other well-organized, resourced, and experienced saboteurs.

These are samples only. In any specific situation, actual values may be orders of magnitude higher or lower. Actual situations will always be more complex than what is listed in these much generalized probability descriptions. A more rigorous assessment examine location specific aspects of attack potential.

A less obvious, less newsworthy (at least less 'headlines-grabbing'), but potentially dramatically consequential attack potential lies in sabotage to a corrosion control system. As discussed in the corrosion threat assessment, CP systems are commonly used to protect buried structures from corrosion. These systems are readily converted into damage-causing rather than damage-preventing systems. Simply reversing the polarity on a rectifier can convert the previously protected metal into an anode, causing rapid corrosion. Since thousands of miles of pipe, tanks, foundations, and other critical infrastructure are protected by CP systems, there is great vulnerability. Being hidden from sight, the damage would typically not become apparent until leaks began, at which time extensive and widespread damage may have occurred. Sensitivity to this potential is the first opportunity for prevention. Continuous monitoring via SCADA, additional oversight, and device security are among defense options.

9.2 SABOTAGE MITIGATIONS

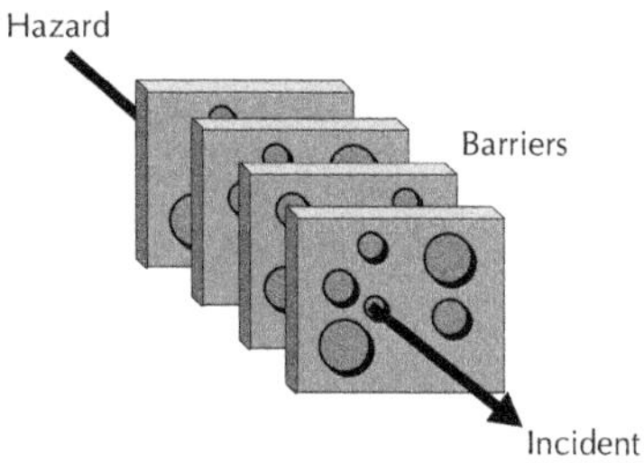

As the potential for an attack increases, preventive measures become more important. However, any mitigating measure can be overcome by determined saboteurs. Therefore, the probability can only be reduced by mitigation, rarely eliminated. Most anti-sabotage measures will be highly situation specific. The designer of the threat assessment should assign values based on experience, judgment, and data, when available. Evaluating the potential for sabotage will often also assesses the host country's ability to assist in preventing damage. Sabotage reduction measures are generally available to the pipeline owner/operator in addition to any support provided by the host country.

Some mitigation measures are specifically designed and installed to prevent sabotage while others are measures that happen to help prevent sabotage while performing another function. Considerations for happenstance mitigative benefits from barriers, detection, and others may also be appropriate. For example:

- Patrolling—A high visibility patrol may act as a deterrent to a casual aggressor; a low-visibility patrol might catch an act in progress.
- Station visits—Regular visits by employees who can quickly spot irregularities such as forced entry, tampering with equipment, etc., can be a deterrent.
- Varying the times of patrol and inspection can make observation more difficult to avoid.
- Monitoring equipment including motion sensors, infrared video, sound detectors, and others.
- Depth of cover—Perhaps a deterrent in extreme cases—ie, >10' of cover—but a few more inches of cover will probably not dissuade a serious perpetrator.
- ROW condition—Clear ROW makes spotting of potential trouble easier, but also makes the pipeline a target that is easier to find and access.

Sabotage prevention benefits from third-party access barriers, including railings, 6-ft chain-link fence, barbed wire, walls, ditches, chains, locks, and others. Also available are various station security detection systems and equipment, including gas/flame detectors, motion detectors, audio/video surveillance,and station lighting systems, including security and perimeter systems covering equipment and working areas.

Beyond mitigation measures designed for an operating facility, other sabotage prevention measures are available to the operating company. For instance, during construction:

- Materials and equipment are secured; extra inspection is employed.
- 24-hour-per-day guarding and inspection
- Employment of several trained, trustworthy inspectors
- Screened, loyal workforce—perhaps brought in from another location
- System of checks for material handling
- Otherwise careful attention to security through thorough planning of all job aspects.

An opportunity to combat sabotage also exists in the training of company employees. Alerting them to common sabotage methods, possible situations that can lead to attacks (disgruntled present and former employees, recruitment activities by saboteurs, etc.), and suspicious activities in general will improve the vigilance. Other human resources opportunities for threat mitigation include the installation of deterrents.

A number of obstacles to internal sabotage can be considered mitigation measures against attacks that may otherwise occur. Common deterrents include:

- Thorough screening of new employees
- Limiting access to the most sensitive areas
- Identification badges
- Training of all employees to be alert to suspicious activities.

9.2.1 Types of Mitigation

Several potential sabotage-specific mitigating measures are discussed in PRMM. These include:

1. Community Partnering
2. Intelligence
3. Security Forces
4. Resolve
5. Industry Cooperation
6. Facility Accessibility (barrier preventions, detection preventions).

9.2.1.1 Community partnering

Supporting communities near to the pipeline by building roads, schools, hospitals, etc. is can change the dynamics of a company's relationship to the local population. This is done not only to become a good neighbor and dissuade some would-be attackers, but also enlist allies—adding to the eyes and ears interested in preserving the assets. See PRMM.

Similarly, efforts to avoid disgruntled employees or former employees is an analogous mitigation.

While some might view such activities as a change in exposure, rather than a mitigation, consider that the attack potential is the starting point and is normally the result of local geopolitical history. The community partnering program intervenes in this attack potential and therefore can be viewed as a mitigation. In some cases, this variable could command a relatively high percentage of possible mitigation benefits—perhaps 20–70%.

9.2.1.2 Intelligence Gathering

Gathering of intelligence regarding potential attacks is commonplace among some corporate security departments. See PRMM.

Effectiveness of intelligence gathering is difficult to measure and can change quickly as fleeting and time-sensitive sources of information appear and disappear. To the extent that the company is able to reliably and regularly obtain information that is applicable in preventing or reducing acts of sabotage, real risk mitigation occurs.

In a preliminary assessment of this mitigation measure, a simple ratio can be used:

Number of acts interrupted through intelligence gathering efforts ÷ number of acts attempted

For example, if it is believed that three acts were avoided (due to forewarning) and eight acts occurred (even if unsuccessful, they should be counted), then 3/11 = 27% may be an appropriate mitigation effectiveness value.

9.2.1.3 Security

Security can take many forms including barriers and accessibility issues, as discussed elsewhere. A security force is another potential mitigation measure. The effectiveness of security measures will be situation specific.

9.2.1.4 Resolve

As discussed in PRMM, a well-publicized intention to protect the company's facilities may be a deterrent and hence can be included as a mitigation measure in a risk assessment.

9.2.1.5 Industry cooperation

As noted in PRMM, sharing of intelligence, training employees to watch neighboring facilities (and, hence, multiplying the patrol effectiveness), sharing of special patrols or guards, sharing of detection devices, etc., are benefits derived from cooperation between companies.

9.2.1.6 Facility accessibility

PRMM describes numerous aspects of accessibility that influence sabotage potential. Attacks will often occur at the readily accessible (most visible and often more vulnerable) targets which are often surface facilities. While a buried pipeline is indeed relatively inaccessible, one common component is a possible exceptions portions of a buried pipeline that are encased in a casing pipe can be more vulnerable to sabotage than directly buried pipe. The vulnerability arises from the common use of vent pipes attached to the casing that provide a route to the carrier pipe from the surface.

Casing vent pipes have historically been used by would-be saboteurs as opportunities to access a carrier pipe. An explosive charge, dropped into a vent pipe, can then detonate against the carrier pipe. Some companies employ design features to prevent intentional and unintentional objects from moving down a vent line to the carrier pipe.

9.2.2 Estimating Effectiveness

As with the estimate of exposure, estimating mitigation effectiveness will necessarily be quite judgmental in many cases. In all assignments of effectiveness, the assessment should carefully consider the "real-world" effectiveness of the anti-sabotage measure. Factors such as training and professionalism of personnel, maintenance and sensitivity of devices, and response time to situations are all critical to the usefulness of most mitigation measures.

The exposures can be offset in the assessment by compiling the effectiveness of all mitigative conditions within the conservatism of the PXX chosen. Preventive measures at each facility can sometimes bring the damage potential nearly to the point of having no such facilities. This is consistent with the idea that "no exposure" will have less risk than "mitigated exposure," regardless of the robustness of the mitigation measures. From a practical standpoint, this allows the pipeline owner to minimize the risk in a number of ways because several means are available to achieve the highest level of preventive measures to offset the exposure level for the surface facility. However, it also shows that even with many preventions in place, the hazard has not been completely removed.

9.3 RESISTANCE

Some sabotage attacks will be unsuccessful not through mitigation—preventing the attack from reaching the component—but rather through the component's resistance. Paralleling the resistance to other external damage mechanisms such as impacts and earth movement, components more able to absorb forces from sabotage attacks will fail less often when damaged.

Earlier, a distinction was made between vandalism and sabotage. The former often includes defacing, theft, and other activities that are not normally direct threats to integrity or even service continuity. Such acts are more readily resisted by the normal designed strength of most components. The 'sabotage' term is reserved for the actions more focused on causing at least service interruption if not also leak/rupture. With a more deliberate attempt to cause significant damage, the ability to resist damages is less certain. It is often conservatively assumed that a determined attacker will eventually be able to inflict damage on a system as difficult to protect as a long pipeline.

9.4 CONSEQUENCE CONSIDERATIONS

The probability of more severe consequences may be increased by an intentional and possibly orchestrated release of pipeline contents. The integrity breach may be more likely to cause a rupture rather than a leak and the timing and subsequent chain of events may be influenced by human interaction seeking to exacerbate the scenario, an attacker could time an event for maximum occupancies in surrounding areas or for more problematic emergency response or he could even directly interfere with emergency response in numerous ways.

Fortunately, it is difficult to orchestrate worst-case pipeline failure events via sabotage, unless significant outside force (weaponry) is deployed against a visible component. Even if, despite numerous safeguards, an integrity breach is created, it would be difficult to maximize the ensuing consequences—ie, ensuring ignition at an optimum time, with receptor proximity, etc.

Nonetheless, it is often prudent to conservatively assume, that in the case of sabotage, there is a greater likelihood of the consequences being more severe. Worst case scenarios possibly occurring more frequently under the threat of sabotage is a conservative and reasonable assumption.

Consider also the less dramatic but highly costly sabotage scenarios. Leaks, below detection limits, continuing for long periods of time, may cause extensive environment damage and costly or impossible remediation. Interference with corrosion control systems could cause widespread, difficult to detect damages that, if allowed to accumulate over time, may cause widespread environmental damages and require extensive infrastructure replacements.

Planning and preparation for repair and replacement, can minimize the impact of attacks. This strategy concentrates on reducing consequences—service interruption—rather than PoF reduction through defensive means. The demonstrated ability to recover quickly and efficiently from any possible damages done by an attack may reduce the incentive of potential saboteurs. There are real examples of this approach. After years of attempting to protect a long pipeline, one owner changed strategies and instead assembled spare parts and rapid response capabilities. These costs were offset by the savings from reduced attempts to protect all locations. With a maximum outage period of two days for even the most successful attacks, the damage to company business was minimized and sabotage events dropped significantly. This strategy will have the added benefit of reducing consequences from any other type of failure mechanisms and is assessed in the cost of service interruption.

Example: 9.1 Sabotage Assessment:

The following example begins with a scenario proposed in PRMM and adds more quantifications, consistent with a newer risk assessment methodology.

The pipeline system for this example has experienced episodes of spray painting on facilities in urban areas and rifle shooting of pipeline markers in rural areas. The community in general seems to be accepting of, or at least indifferent to, the presence of the pipeline. There are no labor disputes or workforce reductions occurring in the company. There are no visible protests against the company in general or the pipeline facilities specifically. The evaluator sees no serious ongoing threat from sabotage or serious vandalism. The painting and shooting are seen as random acts, not targeted attempts to disrupt the pipeline.

Nonetheless, the P99 risk assessment includes the following threat and consequence analyses:

- An estimated near term exposure of 0.5 events per year at an aboveground location and an estimated 20% mitigation effectiveness is assigned. The associated damage probability is assessed to be 0.5 x (1 – 20%) = 0.4 events per year. A resistance value of 50% is used, yielding a PoF = 0.2 failures/year, or a failure every 5 years.
- Consequences, including service interruption costs, are estimated to be $32K per incident based on a collection of P99 scenarios of damage potential. This leads to a near term expected loss of 0.2 events/year x $32K/event = $6.4K/year. This value is carried to risk management meetings to determine appropriate reactions to this conservatively estimated short term risk.

As part of the risk management discussion prompted by this assessment, a related decision is made to address the potential for sabotages during future construction. These are to be addressed primarily via additional inspection and monitoring during installation and a robust post-installation ILI.

10 RESISTANCE MODELING

Highlights

As the third piece of the PoF assessment triad, resistance is a measure of the component's ability to absorb forces and damages without failure—a key determinant of failure probability.

SECTION THUMBNAIL

A basic understanding of a component's strength can be converted into a value that captures its ability to resist failure mechanisms. Stress-carrying capacity is a key determinant.

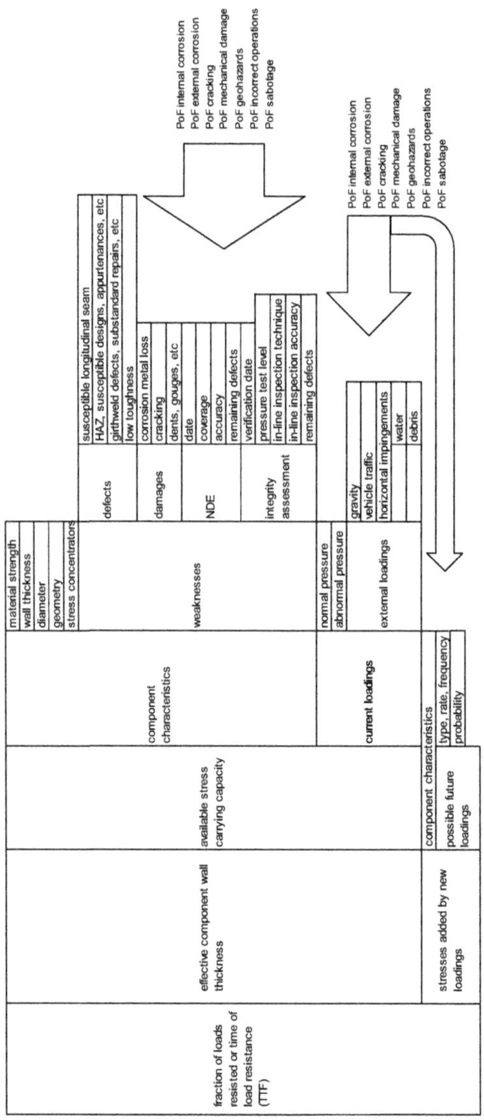

Figure 10.1 Sample of inputs and estimations

The modeling approach of exposure, mitigation, and resistance is an appropriate representation of how actual failure probability manifests and is a complete and efficient way to assess each PoF mechanism. A probability of damage is first produced by assessing the first two terms for all plausible failure mechanisms. Previous chapters have discussed and demonstrated how useful and defensible estimates of damage potential can be generated.

Then, the resistance component is added to discriminate between damage and failure. The ability of the pipeline to withstand failure mechanisms—absorb forces or damages—distinguishes between damage and failure. This resistance to failure will play a significant role in risk calculations involving both time independent failure mechanisms and time-dependent failure mechanisms.

Measuring resistance independently from exposure and mitigation also informs risk management. The simple equation for PoF shows two ways to reduce PoF—either increase mitigation—blocking the failure mechanism—or increase resistance—making the structure stronger to absorb more forces.

As the last piece of the PoF puzzle, an estimate of the component's resistance against all failure mechanisms is sought. This includes a myriad of issues including manufacturing and construction practices, in-service damage rates, and inspection frequencies and capabilities. The need for a formal process is readily apparent upon brief contemplation of the possible combinations of strength issues. For example, it is not possible to intuit the risk prioritization among the following resistance-driven scenarios that will be familiar to many experienced pipeline professionals:

Table 10.1
Scenarios Implying Reduced Component Strength

Potential Strength Issue(s)	Discussion
LF ERW with no known features, MFL ILI 2 years ago	not all LF ERW is problematic, MFL ILI gives only slight evidence of no features—no assurance
LF ERW with minor, known features, specialized ILI (crack tool) conducted last year, suspected low toughness	probability of weaknesses is high, some assurance of integrity via ILI, but not conclusive
high count of laminations discovered in recent ILI, H2 sources, high stress level, wrinkle bends possible	normally benign laminations could be made injurious by the H2, plus wrinkle bends as independent issue
high count of dents found during random excavations, no ILI available, moderate pressure cycling regime	concern of fatigue
low stress, possible miter joints and acetylene girth welds, proposed thermal cycling	concerns of very low resistance to axial forces if these features are present

These sample scenarios are rather complex and difficult to compare to one another. A formal process is required to assimilate all available information and all possible strength issues. This resistance discussion focuses on failure as leak/rupture, but resistance is also an element of the PoF assessment that uses a broader definition: failure = service interruption.

10.1 INTRODUCTION

Resistance calculations in this risk model estimate structural integrity against all anticipated loads—internal, external, time-dependent, and random. This chapter provides guidance on evaluating the component's or pipeline's ability to resist, without failing, all loads.

Varying levels of rigor are available to the risk assessment designer. The underlying engineering, physics, and material science concepts can be complex. However, approximations often provide sufficient accuracy and will be appropriate for many types of risk assessments. When more precision in pipe resistance estimation is desired, pairings of specific weaknesses with specific potential loadings can be analyzed using solutions up to robust finite element analyses.

Whether a more robust or more modest assessment is desired, the general process is the same. The overall strength of the pipeline segment or component and its stress levels are considered. This includes an assessment of foreseeable loads, stresses, and component strengths. Known and suspected weaknesses due to previous damage or questionable manufacturing/construction processes are considered next.

The resistance estimation is akin to calculating a safety factor or a margin of safety, comparing what the pipeline can do (design) versus what it is currently being asked to do (operations). The margin provides protection when unanticipated loads or defects appear. This discussion focuses on steel pipe but concepts apply to any component of any material.

The evaluation process involves an evaluation of loadings and associated stresses, commonly:

- Internal pressure
- External loadings
- Special loadings

System strength (resistance to loadings) is next evaluated:

- First, in the absence of weaknesses, considering
    - material strength
    - structural strength, especially wall thickness
- Next, in consideration of known and possible weaknesses
    - From manufacture
    - From installation
- From damages since installation.

In the interest of completeness, we must cover some basics of material science and stress-strain concepts before adopting a model to capture resistance in the risk assessment. The coverage here is only very rudimentary. The topic warrants much deeper examination, if not a full technical education in the subject area, by the owner of the risk assessment model.

10.1.1 Component resistance determination

As fully discussed herein, resistance can be efficiently measured by modeling a pressure-containing component's effective wall thickness. Wall thickness is a very strong determinant of strength and therefore is a useful surrogate for all other strength-influencing factors. Weaknesses can be efficiently modeled in terms of equivalent reduction in wall thickness.

Increasing forces or defect severities will each reduce effective wall thickness and, hence, the ability to resist additional forces. More reductions in effective pipe wall thickness is the same as forecasting increasing failure rates under assumed loading scenarios. This takes into account the probabilities of various weaknesses coinciding with various loading scenarios.

An assumption here is that wall thickness is a critical determinant of a component's resistance to all failure mechanisms and can be used as a reasonable surrogate for a robust strength analyses. Effective wall then is the basis for modeling resistance to loads. As wall thickness is reduced, implications for component strength include:

- Less capacity for pressure containment
- Faster TTF for degradation mechanisms
- Higher D/t leading to reduced buckling capacity
- Lowered resistance to external forces including localized (puncture) and uniform (subsea hydrostatic pressure).

A probability component is a practical necessity in this part of the assessment since the loads and resistances each involve a spectrum of possibilities—loads and resistance are difficult if not impossible to directly and continuously measure at all points along each pipeline. The risk assessment attempts to accurately represent, at each location, all the possible loading scenarios with all possible weaknesses to estimate how often the two will overlap in a way that causes a failure.

The loads are captured by estimates of exposure and mitigation at all locations. This includes both degradation and random failure mechanisms. The role of defects is similarly represented by probability distributions of severity and likelihood. Either point estimates—representing underlying distributions—can be used, or the distributions themselves can be integrated in the risk calculations.

After loads/stresses are understood, defect potential is the second key ingredient. Each potential defect has both a probability of occurrence and a level of weakness, if it is present. The former can be inferred from an estimated frequency (per mile, for instance) and the latter can be expressed in % loss of wall thickness—an equivalent wall thickness reduction. The probability of occurrence—chances that the weakness is really present—is estimated and used in subsequent steps to determine the probability that the resistance weakness is coincident with the force applied. The weakness estimate resulting from the potential presence of defects is used to predict changes in failure fraction (under certain loads) based first on the severity of the defects.

10.1.2 Including Defect Potential in Risk Assessment

Potential defects and their impact on component resistance must be understood before a model can be developed to efficiently use this understanding in a risk assessment. A robust assessment must consider:

The spectrum of defect types, sizes, and orientations that might be present including those that might have materialized since previous inspections/assessments. Sources of defects include:

- From component manufacture
- From installation
- From operations history.

Knowledge of defects comes from an understanding of possible sources as well as:

- All inspections and integrity assessments that have been performed
- The ability of each inspection/assessment to detect various defects
- The age of each inspection/assessment.

The central question to be answered is: what has been lost, due to the presence of this defect? For instance, how many overpressure events, longitudinal stress loadings, etc. can now no longer be resisted? As a modeling simplification, a probability-weighted summation of potential weaknesses can be used to characterize a component's possible collection of weaknesses. This is an approximation that captures the differences between components with low incidences and/or severities of weaknesses versus those with higher incidences and compounding effects of multiple types of weaknesses co-existing. It captures the frequency and severity of potential weaknesses into a single value while avoiding the intensive approach of a probability distribution of strength reduction versus frequency of occurrence for all possible combinations of pipe weaknesses. Even in a simplified form, this approach also ensures that the intersection of low-probability, high-severity weaknesses with 'sufficient' load scenarios to cause failures, are considered in the risk assessment.

10.1.3 Getting Quick Answers

Since this discussion does not purport to be a full treatment of structural analyses but rather a presentation of a risk assessment methodology, the risk assessor, already familiar with the technical underpinnings, may wish to move directly into the risk assessment methodology—how to embody structural and material science concepts into an efficient risk assessment. Chapter 10.4 Resistance Modeling.

A general technical background discussion follows, for readers seeking more background in material science and structural analyses concepts. See also PRMM for more background discussion.

10.2 BACKGROUND

10.2.1 Material Failure

We now briefly examine the materials science principles that allow estimation of loads and resistance values. Recall the need for a definition of 'failure' in risk assessment. As with the general risk assessment, 'failure' can have any of several meanings in the resistance assessment. Yield strength and ultimate strength are two characteristics typically used to define material failure.

Structural failure can be defined (one of several possible definitions) as the point at which the material changes shape under stress and does not return to its original form when the stress is removed. When this "inelastic" limit is reached, the material has been structurally altered from its original form and its remaining strength might have changed as a result. The structure's ability to resist inelastic deformation is one important measure of its strength.

Resistance can be viewed as the ability to avoid plastic collapse which is related to the difference between applied stresses and material yield point or ultimate strength point. For most pipeline applications, the potential for leaks, unrelated to excess stress, must also be included.

A degradation mechanism active in a pressurized component will require both leak criteria and rupture criteria be considered in concert, ie, as degradation advances through the material, either the pressure-containing capacity (rupture resistance) or the fluid containment capacity (leak resistance) will be lost first. Either results in loss of integrity.

Failure mechanisms/modes include:

- External pressure
- Internal pressure
- Longitudinal bending (longitudinal buckling)
- Axial tension
- Axial compression (axial buckling)
- Lateral compression (crushing)
- Shear
- Cracking (fatigue, etc)
- And various combinations of these.

Concepts from limit state design can be useful here. A limit state is a threshold beyond which a design requirement is no longer satisfied. [9988] Typical limit states include 'ultimate'—corresponding to a rupture or large leak—'leakage', and 'serviceability'. A limit state can be stress-based or strain-based (deformation-controlled).

Changes in material properties over time should be considered. There does not appear to be any evidence that steel strength properties diminish over time. Some researchers even cite minor increases in strength parameters in aged steels. Therefore, the mechanisms resulting in diminished resistance in steel are related to damages suffered, not time-induced changes in metallurgical characteristics. Damages are accounted for as failure mechanisms such as corrosion, cracking, and external forces.

For other pipe and component materials, such as certain types of plastics, degradation mechanisms are expected and should be included in resistance determinations.

SECTION THUMBNAIL

- Pipelines can be built from a variety of materials.
- Different materials have different abilities to resist failure.
- All resistance issues can be efficiently modeled using the same approach.

10.2.2 Toughness

Toughness is a material property playing an important strength role in many types of loadings, sometimes being the difference between failure and not failing and often between rupture and leak.

Material toughness plays a large role in crack failure potential. Crack initiation, activation, and propagation are all influenced. Materials that have little fracture toughness do not offer much resistance to brittle failure. Even small defects can reduce strength dramatically when toughness is low. Rapid crack propagation, perhaps brought on by corrosion and stress, is more likely in these materials, resulting in more violent ruptures.

A common method used to assess material toughness is the Charpy V-notch impact test. Toughness-equivalent considerations for non-steel components—plastics (PVC, PE, etc), cast iron, copper will be required.

The challenge of gauging the likelihood of a more catastrophic failure mode is further complicated by the fact that some materials may change over time. Given the right conditions, a ductile material can become more brittle.

See PRMM for additional discussion.

10.2.3 Pipe materials, joining, and rehabilitation

A basic understanding of common pipe materials is important in assessing the ability to resist failure. Although transmission pipelines are overwhelmingly constructed of carbon steel, distribution lines have historically been built from a variety of materials. The material's behavior under stress is often critical to the evaluation. A more brittle material typically has less impact resistance. Impact resistance is particularly important in reducing the severity of outside force loadings. In regions of unstable ground, materials with higher toughness and more flexible structures will better resist the stresses of earth movements. Traffic loads and pipe handling activities are other stress inducers that must be withstood by properties such as the pipe material's fatigue (cracking) and bending (tensile) strengths. Stresses resulting from earth movements and/or temperature changes may be more significant for certain pipe materials. In certain regions, a primary ground movement is caused by the seasonal freeze/thaw cycle. One study shows that in some pipe materials, as temperature decreases, pipe breaks tend to increase exponentially [51]. Break rates for rigid pipes such as cast iron are found to be several times higher than for welded steel pipelines. Mechanical fittings add rigidity and are common points of failure when external forces are applied.

All of the pipe materials discussed here have viable applications, but not all materials will perform equally well in a given service. Although all pipelines can be inspected to some extent by direct observation and remotely controlled video cameras, larger steel pipelines benefit from maturing technologies employing electromagnetic and ultrasonic inspection devices.

Because there is no "miracle" material, the material selection step of the design process is partly a process of maximizing the desirable properties while minimizing the undesirable properties. The initial cost of the material is not an insignificant property to be considered. However, the long-term "cost of ownership" is a better view of the economics of a particular material selection. The cost of ownership would include ongoing maintenance costs and replacement costs after the design life has expired. This presents a more realistic measure with which to select a material and ultimately impacts the risk picture more directly.

The pipe designs should include appropriate consideration of all loadings and correctly model pipe behavior under load. Design calculations must always allow for the pipe response in determining allowable stresses. Pipe materials can be placed into two general response classes: flexible and rigid. This distinction is a necessary one for purposes of design calculations because in general, a rigid pipe requires more wall thickness to support a given load than a flexible pipe does. This is due to the ability of the flexible pipe to take advantage of the surrounding soil to help carry the load. A small deflection in a flexible pipe does not appreciably add to the pipe stress and allows the soil beneath and to the sides to carry some of the load. This pipe–soil structure is thus a system of high effective strength for flexible pipes [60] but less so for rigid pipes.

Materials with a lack of ductility also have reduced toughness. This makes the material more prone to fatigue and temperature-related failures and also increases the chances for brittle failures. Brittle failures can be more consequential than ductile failures since the potential exists for larger product releases and increased projectile loadings. The potential for catastrophic tank failure should be considered, including shell and seam construction and membrane stress levels for susceptibility to brittle fracture.

Especially in distribution systems, the evaluator must take into account material differences when determining resistance. When the type of material limits its ability to provide 'extra' resistance, the appropriate adjustment to effective wall thickness should be made.

Separation of mechanical fittings can result in large releases. Provisions for mechanical coupling equivalent weaknesses will also be needed, when evaluated systems containing such fittings.

Some common pipe materials, many found only in distribution pipeline systems, are discussed PRMM. That is not an exhaustive list and also does not include multi-material systems. Pipe-in-pipe systems (cased pipe) and plastic-wrapped-in-steel, perhaps also with additional armoring sheaths of various types, are examples of multi-material systems. Resistance calculations may become more challenging with some designs, but are still efficiently modeled using basic principles of material science and physics.

10.2.3.1 Flexible pipe

Steel pipe manufacturing processes have evolved over many years. Processes include furnace butt-welding, continuous butt-welding, lap welding, hammer welding, low frequency electric resistance welding (ERW), flash welding, single submerged arc welding, variations on seamless pipe manufacture, high-frequency ERW, double submerged arc welding (DSAW) either straight or spiral seam. Of these, continuous butt-weld seamless, HFERW, and DSAW processes remain in widespread use today and have since early 1970 whereas the others were phased out around 1970 or before [1020]. Some of these processes, even when meeting the quality standards at the time, had a propensity to introduce weaknesses. LF ERW, lap welding, flash welding, and others have been highlighted as steel manufacturing processes that produced, in some pipe mills, pipe with increased vulnerabilities to failure mechanisms such as selective seam corrosion and cracking. This pipe is often a focus of integrity management, since these weakness features can be difficult to detect and failure modes can be dramatic.

Quality control and inspection of the manufacturing processes have also evolved over the years and impact the types and quantities of weaknesses that might have been introduced. Similarly, construction practices for steel pipelines have evolved. Earlier practices created mechanical couplings, wrinkle bends, acetylene girth welds, and other components that are today considered more susceptible to failure than their more modern counterparts. Linkages between possible weaknesses and resistance related to steel pipe manufacture and construction are examined later in this chapter.

PE failure potential is strongly influenced by stress and temperature. Slow crack propagation is a common long-term failure mechanism that should be considered in risk assessments. Field-performed heat fusions of fittings and joints are similarly susceptible. Secondary loads such as from overburden, bending, and rock impingements should also be included in the assessment.

See the related discussions in PRMM.

10.2.4 Defects and Weaknesses

The detection of weaknesses begins with identifying potential anomalies. The first idea that comes to mind when hearing ‘anomaly’ may involve ‘defect’. All defects are anomalies but not all anomalies are defects. An anomaly is a deviation in some property of the manufactured product. A defect is considered to be any anomaly, such as a crack, gouge, dent, or metal loss, which reduces the component’s capacity to carry a load. Some anomalies—shallow dents, smooth, shallow gouges, minor metal loss, and even some cracks—will not affect the strength or service life of a pipeline. Hence the statement 'not all anomalies are defects'. An anomaly becomes a defect when it introduces a weakness.

As used here, 'defects' include flaws or damages to components from original manufacture, construction, or time-independent mechanisms (not degradations). Examples include dents, gouges, girth weld defects, lack of fusion in welded seams, and others as detailed later.

Besides defects, there are other types of location specific weaknesses, many of which arise through stress concentrators or due to components with inherently less strength than neighboring pipe, such as:

- Wrinkle bends
- Acetylene welds
- Mechanical couplings
- Substandard repairs
- Older and currently-avoided appurtenances

There are also deficiencies in material properties that are efficiently modeled as weaknesses. These deficiencies can be created from inferior or incorrect construction practice. Examples include introduction of hard spots (potential crack initiation sites) and residual stresses. Deficiencies may also be present from undetected errors in original manufacture or from unrecognized issues at the time of manufacture (for example, LF ERW).

The potential for weaknesses introduced in manufacturing and construction is discussed in later in this chapter as well as in Chapter 8 Incorrect Operations.

Finally, weaknesses occur through degradation mechanisms. This aspect is most efficiently captured as part of the degradation mechanism assessment. As corrosion metal loss potential and cracking phenomena are assessed, a degradation rate (or rates) naturally emerges. The rate multiplied by the amount of time the rate could have been active yields a remaining wall thickness. This value is adjusted by the non-degradation potential weaknesses assessed as discussed here.

10.2.4.1 Weakness Identification/Characterization

Defect Types

As noted, some anomalies originate from manufacturing processes, such as laminations, hard spots, inclusions, and seam weaknesses associated with low-frequency ERW and electric flash welded pipe. Others such as girth weld defects, dents, and arc burns occur during installation or repair. Finally, anomalies arise during operations: dents or gouges from excavation damage or other external forces, corrosion wall losses, and cracks. Anomalies introduced during repair/replacement operations are also possible.

API 579 [1021] provides a more extensive listing of causes of types and origins of manufacturing and construction defects in structures. Such listings serve as checklists for designers of risk assessments, helping to ensure that all plausible defects are considered in the assessment.

Anomaly prioritization is often governed by industry standards if not regulations, as described in PRMM.

Probability of original defects

The types of pre-service deficiencies that can be present before equipment enters service are:

a. Material Production Flaws – Flaws which occur during production including laminations and laps in wrought products, and voids, segregation, shrinks, cracks, and bursts in cast products.
b. Welding Related Flaws – Flaws which occur as a result of the welding process including lack of penetration, lack of fusion, delayed hydrogen cracking, porosity, slag, undercut, weld cracking, and hot shortness.
c. Fabrication Related Flaws – Imperfections associated with fabrication including out-of-roundness, forming cracks, grinding cracks and marks, dents, gouges, dent-gouge combinations, and lamellar tearing.
d. Heat Treatment Related Flaws or Embrittlement – Flaws associated with heat treatment including reheat cracking, quench cracking, sensitization, and embrittlement. Similar flaws are also associated with in-service elevated temperature exposure.
e. Wrong Material of Construction – Due to either faulty materials selection, poor choice of a specification break (i.e. a location in a component where a change in material specification is designated), or due to the inadvertent substitution of a different alloy or heat treatment condition due to a lack of positive material identification, the installed component does not have the expected resistance or needed properties to the service or loading.

In most instances, one or more of these pre-service deficiencies do not lead to an immediate failure. Usually, only gross errors cause a failure, normally identified during a pre-service pressure test.

Residual stress

Lack of toughness should also be considered in the resistance assessment. Lower toughness makes crack initiation, activation, and propagation more probable and rupture more likely. At higher stress levels, more toughness is required to arrest a running brittle fracture. Larger diameter or thinner wall pipes require proportionally higher toughness to prevent running brittle fracture. Hole size, also a function of toughness, is discussed in Chapter 11 Consequence of Failure.

Manufacturing/Construction Weaknesses

A list of common manufacturing and construction weaknesses found in onshore steel pipelines over many decades has been compiled in several references (including [1020, 1022, 1035]). The following information is extracted from such references:

Table 10.2

Feature	Comments on source	Impact on resistance
hook crack	older ERW, both LF and HF	fatigue cracking
cold weld; pinhole	inadequate bonding in LF and DC-welded ERW	small leaks
Penetrator	inadequate bonding in flash-welded or HF ERW	small leaks
mismatched skelp edges	DSAW, ERW, flash-welds	fatigue cracking
off seam weld; incomplete penetration; incomplete fusion; centerline crack; toe crack	DSAW and/or SSAW	fatigue cracking
excessively hard HAZ	late 40's early 50's X-grade Youngstown pipe mill	increased crack probability, especially with H exposure
unbonded or partially bonded seam	lap-weld pipe	fatigue cracking
burned metal	crack-like voids in lap-welded pipe	loss of effective wall
Lamination	common in pre 1980 seamless pipe	blister formation if H exposure
hard spot	arc burns are one cause of hard spots	increased crack probability, especially with H exposure
defective weld		leaks; rupture when external forces applied
acetylene girth weld	pre WW II	little strain resistance; rupture when external force applied
mechanical coupling	pre WW II	low resistance to axial and lateral forces
wrinkle bends	pre WW II	cold-working reduces toughness; increased crack potential
transportation fatigue cracks	cracks produced during transportation, more common on pipe with D/t>70 produced prior to 1970 and shipped by rail	fatigue cracking
high levels of impurities and non-metallic inclusions		increased crack probability, especially with H exposure
Toughness		fatigue cracking

Some source references cite incident statistics linked to these features, sometimes tracing back to specific steel mills and dates. This information can be very useful in assigning probabilities of defects to pipeline segments. It can also provide inferential information on strength-reduction magnitudes of certain defects. However, without a full understanding of the incidents underlying these statistics, caution in their use is recommended. Recall that these defects, normally having survived pressure tests, inspections, and on-going service loads for many years, fail only when additional loads are introduced or after degradation has occurred. Without knowledge of the degradation and/or additional loads, the knowledge provided by the statistics alone is incomplete.

Manufacturing issues

It is commonly accepted that older manufacturing and construction methods do not match today's standards for rigor of specifications nor quality control. Nonetheless, many very old systems have successfully and admirably withstood the test of time—decades of service in sometimes challenging environments, with no reduction in strength.

All other things equal however, it is reasonable to assume superior product quality in modern manufacturing. Technological and quality-control advances have improved quality and consistency of both manufactured components and construction techniques. These improvements have varying degrees of importance in a risk assessment. In a more extreme case, depending on the method and age of manufacture, the assumption of uniform material may not be valid. If this is the case, the maximum allowable strength value should reflect the true strength of the material.

Purchasing specifications now cover strength properties such as minimum yield strength (SMYS) and toughness, all of which are certified by the manufacturer. The risk assessment should consider the probabilities that the specifications were correct, were followed, and were applicable to the pipe or component in question.

A pattern of failures connected to a particular manufacturer or process should lead the risk evaluator to question the strength of any components produced in that way. Materials from steel mills whose pipe has been known to have higher rates of weakness should be penalized in the risk assessment where appropriate.

Some weaknesses are actually an increased susceptibility to later damages such as from corrosion and cracking. Preferential corrosion (selective corrosion, seam corrosion, etc) is a possibility for several types of steel pipe. It is commonly associated with variable quality LF ERW or flash weld seams or non-heat treated HF ERW seams. Certain steel pipe manufacture dates and locations (pipe mill) can be correlated with increased occurrence rates [1035]. This information can be efficiently modeled as reduced wall thickness in the resistance estimation.

Hard spots created during pipe manufacture or construction (for example, arc burns, girth weld HAZ) can support cracking, especially in the presence of hydrogen. Hard spots can be large—covering the full circumference of the pipe over several inches of length [1020]. H_2 stress cracking (HSC) occurs at a hard spot when sources of hydrogen are present and sufficient stress exists. H_2 sources include sour service (H_2S), higher CP (cathodic protection applied for external corrosion control) levels, and in association with higher microbiological activity (swamps, MIC, etc). Susceptibility factors include sufficient hardness, hydrogen availability, and sufficient stress level.

Increasing crack susceptibility can be assumed when:

- H_2 charging of steel could have occurred
- There may be or have been temperature effects on toughness
- hard spot, arc burns could be present.

When no inspection information is available, increasing susceptibility to cracking can be modeled to occur in pipe manufactured before 1960 and/or with higher CP levels (perhaps a threshold of -1.2volts pipe-to-soil, $CuCuSO_4$ reference electrode) and with increasing stress and with higher potential H_2 availability. [1020]

Construction issues

Similar to the evolution of pipe manufacturing techniques, the methods for construction practices such as welding pipe joints have improved over the years. See PRMM for a relevant discussion on girth weld defects.

A wrinkle bend is a type of buckle, often an artifact of an intentional bending process used in early pipeline installations. Wrinkle bends are known locations of stress concentrations, with the severity of the effect increasing with decreasing D/t and severity of the wrinkle (height and width of wrinkle). Axial stress cycles, combined with the stress concentration effect, reduces the fatigue life of a component with a wrinkle bend. Depending on material properties, a doubling of stress due to a stress concentrator can shorten life by a factor of 16 or more. [1023]

As an artifact of a past, discontinued practice, a wrinkle bend is today considered by most to be an anomaly, sometimes requiring repair or replacement. In the risk assessment, the anomaly could be modeled as a resistance vulnerability.

Date of construction provides evidence of the existence of older features of concern, when inspection data is not available. For example, mechanical couplings were used from 1890's until about 1940, acetylene welding was employed from about 1915 to 1940, miter bends are found in pipelines built prior to 1940, and wrinkle bends pre-1955 [1020]. Prior to the introduction and adoption of engineering standards and regulations, all repair practices may be suspect.

All such features should be considered in evaluating the strength of the system. Buried bends, girth welds, substandard repairs, and couplings are not normally highly loaded during normal service [1020] and hence, these features may enjoy long service lives. However, when abnormal loadings—including external forces and pressure or thermal cycling—occur, they will often be the points of failure.

With all this in mind, the fact of a pipeline's long-term reliable operation can to some extent offset these concerns and be a "plus" in the overall evaluation. This is the "withstood the test of time" argument for evidence of low probability of failure. See discussion in Chapter 2.8.6 The Test of Time Estimation of Exposure.

Damages during operations is the final opportunity for weaknesses to be introduced. PRMM describes common mechanical damages to pipelines. The Pipeline Research Council International & Institute (PRCI) provides useful insights into these mechanical damages on pipelines [1036]:

Mechanical damage can cause changes to:

1. The shape of the pipeline's cross section, as for example where the line sits on a rock ledge, and
2. The wall thickness or its properties, as for example where earthmoving equipment scrapes along the pipeline displacing, or cold-working, and/or tearing the wall as it passes.

Mechanical damage also can involve combinations of (1) and (2).

The consequences of mechanical damage fall into one of four categories, depending on the nature of the outside force, the pipeline's design and operating conditions, and the line-pipe properties. These consequence categories are:

- immediate failure due to plastic collapse or cracking on the inside diameter (ID) during contact
- immediate failure due to plastic collapse or OD cracking during re-rounding in the wake of the contact
- delayed failure due to in-service cracking, and
- no threat for failure for the current service or possible upset conditions.

Other important observations are that re-rounding of dents has been shown to cause crack initiation and that damages incurred prior to pressurization are more benign than those post pressurization. Damage inflicted at zero pressure is not as severe as that inflicted under pressure, all else being equal. This occurs because the unpressurized pipe changes shape over much of its cross section and consequently avoids the localized deformation that leads to puncture or cracking. In contrast, pressure in the pipeline keeps the pipe round except where outside forces contact, which leads to localized deformation and possibly severe damage. Thus, while a severity criterion for damage done at zero pressure could prove useful, such a criterion would be nonconservative for applications involving damage done at pressure.

Although the pipeline is subjected to a pre-service pressure test, it is unlikely that existing damage would be detected, except for areas pierced as a result of the damage incident. Data in published literature indicate that very severe damage involving gouges in dents with depths greater than 15 percent of the diameter seldom leads to failure in full-scale testing after just one major pressure cycle. For this reason, line pipe damaged at zero pressure probably survives the pre-service pressure and thus may exist in operating pipelines, or possibly lead to delayed failure.

Repairs and Reinforcements

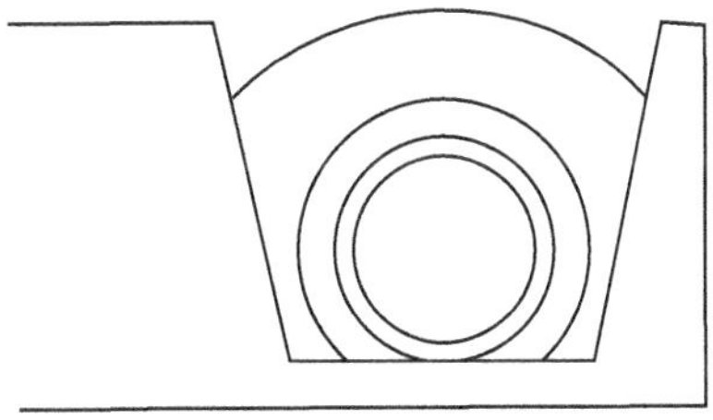

As with general construction practice, repair practice has evolved over the years. Some previously acceptable repair methods would no longer be considered by most modern operators. Examples include deposition of weld metal to fill in corrosion damages; use of metal patches or complex shaped shells installed over leaks; converting temporary clamps to permanent installations; and even the use of wooden plugs driven into holes in low pressure steel and cast iron pipelines. Repairs that, by today's standards, are judged to be inferior, may contribute weaknesses. Their likelihood of existence must be estimated, especially when inspection cannot reliably provide identification and characterization. This is discussed in a later section.

A full encirclement sleeve serves to carry some of the stresses otherwise carried by the pipeline, thereby providing increased resistance to new loads. It also provides increased impact resistance and, made from a composite material, corrosion protection equivalent or superior to a coating system. If pressure-containing, it increases TTF from degradation mechanisms by effectively increasing the amount of material that must be degraded before leak or rupture. The sleeve also provides benefit as a crack arrestor, potentially reducing consequence potential by limiting hole size.

Composite sleeve materials are popular repair choices. The underlying concept of composite materials is very old. Straw and mud bricks and concrete (cement and aggregate) take advantage of the best properties of multiple materials to provide a stronger final product. Modern pipeline repair wraps or sleeves are layered systems of solid fibers, such as carbon or fiberglass, and a bonding resin such as urethane or epoxy, installed around a short section of pipeline containing a defect. The characteristics of the applied and cured repair wrap, such as flexibility, yield strength, UV resistance, and others, will determine the ability of these types of repairs to not only restore the component's strength, but also provide additional resistance, perhaps beyond original capabilities. [1036, 1037]

Modeling of Repairs in Risk Assessment

Modern repairs will reduce risk, sometimes far beyond their role in offsetting weakness caused by a defect. Repairs often act as reinforcement, mitigation, and consequence reduction in addition to restoration of desired strength. These normally cover a small portion of a system, but a detailed risk assessment can recognize that risk is significantly reduced at these short locations. This is often in stark contrast to the risk immediately prior to the repair.

Repairs, especially when made with full encirclement sleeves, can be modeled as providing a general increased resistance, perhaps using a simple factor to increase effective wall thickness by some amount. Alternatively, a repair's role in specific risk reduction can be modeled in a detailed way, quantifying its independent contributions to:

- Increased impact resistance
- Increased stress carrying capability
- Corrosion mitigation (when non-corrosive sleeve material is used)
- Increased effective wall thickness for TTF estimates
- Crack arresting is modeled as consequence hole size reduction
- Clamps and non-pressure-containing repairs often provide less resistance. Given the typically short length of repairs, detailed modeling may not be warranted and a simple factor, scaling up resistance at the repair location, will be sufficient.

Older repair techniques, no longer allowed in current industry recommended practice, may cause unintended weaknesses such as stress concentration points and brittleness at welds. Even acceptable repairs may have unintended consequences as was noted in the example of hydrogen permeation into the annular space between a repair sleeve and the carrier pipe, eventually causing buckling of the carrier pipe [1001]. In some of these cases, the repair actually causes a new exposure to be included in the risk assessment.

The evaluation of resistance will also include non-pipe components since they will typically be included in the risk assessment. These include flanges, valve bodies, fittings, filters, pumps, compressors, flow measurement devices, pressure vessels, and others. Each will be acted upon by various exposures, have mitigations to protect it, and will have varying amounts of resistance to failure.

Characterizing Potential Weaknesses

A risk assessment that examines available pipe strength should probably treat anomalies (identified defects whose severity has not yet been evaluated) as evidence of reduced strength and possible active failure mechanisms.

A complete assessment of remaining pipe strength in consideration of an anomaly requires accurate characterization of the anomaly—its dimensions and shape. In the absence of detailed remaining strength calculations, the evaluator can reduce pipe strength by a percentage based on the severity of the anomaly.

Increased crack susceptibility is a common concern for all of these features. This is efficiently modeled as reduced wall thickness and/or increased probability of crack initiation/activation/propagation, both used in the cracking PoF estimation. Some features may also impact the ability to resist other loadings including internal pressure and external forces.

10.2.5 Loads and Forces

Loads and forces, and their resulting stresses, obviously play a large role in failure potential. The design process first considers the loads and forces that are to be resisted.

This discussion is an examination of how the pipeline's design characteristics impact its ability to resist forces/damages. Certain design concepts are presented to

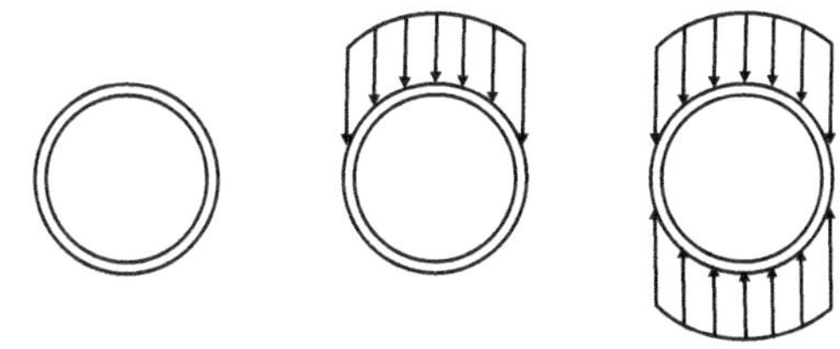

give the evaluator who is not already familiar with pipeline design methods a feel for some of the considerations. This obviously does not replace a design manual or design methodology. Used with the corresponding risk evaluation sections, this section can assist one unfamiliar with design concepts in understanding strength/resistance aspects of the pipeline being examined.

Design of any structure involves examinations of loads and forces. Load is a general term meaning a force applied to a structure. Internal pressure, gravity (or weight), and temperature-induced strains are examples of loads typically experienced by pipelines.

Loads have effects on structures—pipeline components in this case. Those effects include stresses, strains, and deformations. Resistance can be measured in terms of any of these—the ability to withstand a stress, strain, or deformation. Even a pinhole leak from an unpressurized component conceptually falls into this model. The leak will only occur with some driving force, if only gravity or a tiny amount of hydrostatic fluid head. This tiny driving force is no longer resisted if the pinhole has penetrated the entire component wall.

In general, any influence that tries to change the shape of the pipe will cause a stress. Pipe stress can originate from loads that cause or exacerbate:

- Internal pressure
- External pressure
- Longitudinal bending (longitudinal buckling)
- Axial tension
- Axial compression (axial buckling)
- Lateral compression (crushing)
- Thermal expansion/contraction
- Shear
- Cracking (fatigue, etc.)
- And various combinations of these.

Defects in component walls will heavily influence resistance. That will be considered separately from the defect-free analysis.

As a pressure containment system, internal pressure obviously plays a key role in many pipeline strength determinations. While often the dominant load, internal pressure is not the only loading on a typical component. External forces also add stress to the pipe. Loads causing external stresses include the weight of the soil over a buried line, the weight of the pipe itself when it is unsupported, temperature changes, etc. Some of these stresses are additive to the stresses caused by internal pressure. As such, they must be allowed for in the design pressure calculations. Hence, care must be taken to ensure that the pipeline will never be subjected to any combination of internal pressures and external forces that will cause the pipe material to be overstressed.

Tolerable loads are set by maximum stress-carrying capacity. The design phase includes consideration of all loadings to which the pipeline will be subjected. Pipeline loadings typically include internal pressure and physical weights such as soil and traffic over the line. A typical analysis of anticipated basic loads for a buried pipeline would include provisions for:

- static internal pressure
- dynamic internal pressures such as surge pressures
- overburden (Soil loadings, including soil movements).

Additional criteria are considered in detailed design and for special installation circumstances such as drilled crossing and spans. These criteria include provisions for:

- Bending Stresses
- Tensile Loads l Buoyancy.
- Span loadings including gravity and lateral forces
- Traffic loadings
- Strain induced loadings such as from temperature changes.

For each loading combination, all stresses and failure modes must be identified. Failure is often defined as permanent deformation of the material. After permanent deformation, the component may no longer be suitable for the service intended. Permanent deformation occurs through failure modes such as bending, buckling, crushing, rupture, bulging, and tearing. In engineering terms, these relate to stresses of shear, compression, torsion, and tension. These stresses are further defined by the directions in which they act; axial, radial, circumferential, tangential, hoop, and longitudinal are common terms used to refer to stress direction. Some of these stress direction terms are used interchangeably.

As discussed in the previous sections, pipeline component materials have different properties and different abilities to resist loads. Ductility, tensile strength, impact toughness, and a host of other material properties will determine the weakest aspect of the material. If the pipe is considered to be flexible (will deflect at least 2% without excessive stress), the failure mode will likely be different from a rigid pipe. The highest level of stress directed in the pipe material's weakest direction will normally be the critical failure mode. The exception may be buckling, which is more dependent on the geometry of the pipe and the forces applied.

The critical failure mode for each loading will be the one that fails under the lowest stress level.

10.2.5.1 Load Types

A useful listing of load types can be found in [9988] as part of the limit state discussion. Limit states included are 'ultimate' (ULS), 'leakage' (LLS), and 'serviceability' (SLS). These limits may be established based on stress or strain or both. This particular reference categorizes loads based on their potential appearance in the system's life cycle. It also assigns a time dependency to each combination of loads and limit states, as well as a cross reference to potentially interacting load cases.

When loss of integrity is the focus of the risk assessment, limit states dealing with ruptures and leaks are the focus. Some of the pertinent loads are further discussed below.

10.2.5.2 Pressure containment

The most commonly used measure of a pipeline's strength will normally be the documented design pressure—the maximum internal pressure that can be withstood without damage (including permanent deformation). Design pressure is determined from stress calculations, with internal pressure normally causing the largest stresses in the wall of the pipe. Material stress limits are theoretical values, confirmed (or at least evidenced) by testing, that predict the point at which the material will fail when subjected to high stress.

Several key aspects of risk are directly linked to the amount of internal pressure in the line. Pressure levels may vary widely along a pipeline or at a single location over time. The pressure to which a component will be subjected is needed to calculate stress levels and other risk factors in the risk assessment. The assessment may choose any of several commonly cited pressure levels: the maximum tolerable design pressures, the maximum allowable pressure (including safety factors), the maximum working pressure, the normal operating pressures, and others. The terms maximum operating pressure (MOP), maximum allowable operating pressure (MAOP), maximum permissible pressure, and design pressure have specific definitions in some regulatory and industry guidance documents. However, they are often used interchangeably. They all imply an internal pressure level that comports with design intent and certain safety considerations—whether the latter stem from regulatory requirements, industry standards, or a company's internal policies. In this risk assessment discussion, the term 'design pressure' is used for the maximum internal pressure that can be sustained by the component without permanent deformation or other harm to the material.

For purposes of this discussion, design pressure will be used to describe the pressure to which the defect-free component can be subjected without failure (such as yielding). By this definition, design pressure should exclude all safety factors that are mandated by government regulations or chosen by the designer. It should also exclude engineering safety factors that reflect the uncertainty and variability of material strengths and the simplifying assumptions of design formulas since these are technically based limitations on operating pressure. These include safety factors for temperature, joint types, and other considerations. Safety factors that usually allow for errors and omissions, deterioration of facilities, and provide extra 'cushioning' between actual conditions and tolerable limits. Such allowances are certainly needed, but can be confusing if they are included in the risk assessment. There is always an actual margin of safety between the maximum stress level caused by the highest pressure and the stress tolerance of the pipeline. Measuring this directly without including the confounding influences of a regulated stress level and stress tolerance, makes the assessment more intuitive and useful, especially when differing regulatory requirements make comparisons more complicated. Regulatory safety factors are therefore omitted from the design pressure calculations for risk assessment purposes.

The design or other 'maximum allowable' pressure is appropriate for characterizing the maximum stress levels to which all portions of the pipeline might be subjected, even if the normal operating pressures for most of the pipe are far below this level. This avoids the potential criticism that the assessment is not appropriately conservative.

Although the design pressure could be conservatively used here, this would not differentiate between the upstream sections (often higher pressures) and the downstream sections (usually lower pressures). The alternative of using normal operating pressures, provides a more realistic view of actual stress levels along the pipeline. Pipeline segments immediately downstream of pumps or compressors would routinely see higher pressures, and downstream segments might never see pressures even close to the maximum limits. One approach would be to create a hypothetical pressure profile of the entire line and, from this, identify normal maximum pressures in the section being evaluated.

This approach might be more appropriate for operational risk assessments where actual differences along the pipeline are of most interest. A challenge in using 'normal' pressures will be the time period implied: ie, the highest pressure seen in last year? 5 years? The average or median pressure seen in the last 12 months? Etc.

Provisions for surge (water hammer) or other temporary pressures should be independent of design pressure determination. Potential for pressure levels in excess of system tolerances should be considered separately as exposures. Surge potential is discussed in Chapter 8 Incorrect Operations.

Pipe wall damages or suspected weaknesses—anomalies—may impact pipe strength and hence allowable pressures or safety margins. Formal reductions of maximum operating pressure resulting from pipeline anomalies are normally based on approaches described in industry standards[1]. If a new pressure limit is determined based on calculations of remaining strength after a detected weakness, then that should be the new design pressure used in the risk assessment of the component. In this case, it may be hard to determine how much conservatism in the form of extra safety margin has been added in the treatment of some anomalies. If the assessment is able to ascertain the true pressure limit, free from any safety factor, that is the better value to use as design pressure.

The design pressure also plays a role in probability of damage estimates. For instance, in the incorrect operations assessment, there is an important distinction made between a safety-system-protected component and one that is impossible to overpres-

1 Such as ASME/ ANSI B31G, Manual for Determining the Remaining Strength of Corroded Pipelines, or AGA Pipeline Research Committee Project PR–3–805, A Modified Criterion for Evaluating the Remaining Strength of Corroded Pipe

sure due to the absence of adequate pressure production—where it is physically impossible to exceed the design pressure because there is no pressure source (including static head and temperature effects) that can cause an exceedance.

Note also that pressure, from the standpoint of a small leak, can mean the tiny driving force created by hydrostatic head or gravity.

The degree of pressure cycling is another factor to take into account in the evaluation since this can also contribute to failure probability as discussed in Chapter 6.8 Cracking.

10.2.5.3 Load Estimations

Both continuous and intermittent loads are appropriately included in risk assessments.

Normal, continuous loads are addressed in the design phase. Normal, intermittent loads should also be addressed during design, but may not receive the same amount of rigor or they may be compromised over time by changes in system characteristics during its life cycle. Fatigue loadings are an example. Even if considered during design, changes in use over time may change the originally planned number and magnitude of pressure cycles and changes in environment may add new sources of external fatigue cycles.

Intermittent loads, especially when both abnormal and intermittent, require both a categorization of intensity or damage potential and an estimate of frequency. Frequencies may already have been partially captured in exposure estimates for the various time-independent forces—excavator hits, vehicle impacts, landslides, surge pressures, anchor strikes, etc.

Normal loads can often be estimated from design documents, as previously discussed, and can produce a baseline level of resistance.

10.2.5.4 Special External loadings

Normal external loadings listed in PRMM include the weight of the soil over a buried component, the loadings caused by moving traffic, possible soil movements (settling, faults, etc.), external pressures and buoyancy forces for submerged lines, temperature effects, lateral forces due to water flow and debris impacts, and component weight. See discussion of these in PRMM.

As a special case of 'failure', infiltration of a component and subsequent product contamination can occur. For example, groundwater infiltration into a distribution system. This is a form of integrity loss since, for infiltration to occur, the outside pressure exceeds the internal pressure and the components ability to resist. There would presumably also be an integrity loss when groundwater pressures are lower and the component's internal pressure produces the driving force to create a leak.

Overburden
This is a measure of the weight of soil, objects and anything else over the pipeline. In an offshore environment, this would also include the pressure due to water depth. Uncased pipe under roadways may require additional wall thickness to handle the increased loads from vehicles. The speed and weight of the vehicles, as well as depth of cover, cover type, and other factors will be important determinants of how much stress is transferred to the buried component.

Spans
Similar to the forces of gravity on an onshore spanning component, the stresses from lateral forces of moving waters, debris accumulations, should be considered for offshore susceptible components. Spans are a unique feature in a risk assessment, as discussed in Chapter 2 Definitions and Concepts.

Buckling
Pipelines under compressive forces from pressure or thermal forces, can buckle if the axial compression goes beyond a certain level. Buckling can also occur under excessive external force.

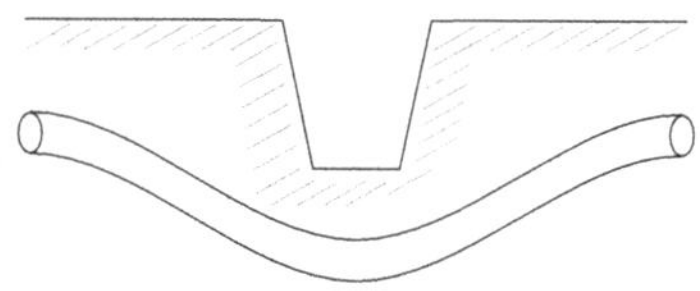

Buckling is more common concern with pipelines in deep water. Some offshore designs incorporate controlled lateral buckling as a means to dissipate pressure and thermal expansion induced forces on a long pipeline.

However, buckling as a failure mode can manifest at other, unexpected conditions, far from common external pressure sources. In one operator's experience, hydrogen permeation through steel repair sleeves caused numerous buckles to the pipe beneath. The source of hydrogen was generated from high CP levels external to the sleeve. An annular space pressure of around 300 psig was sufficient to cause the buckling. [1001]

Accounting for unspecified external loads
Especially for preliminary or screening type risk assessments, it may be appropriate to simply use a factor to account for unknown or unquantified loads. The factor can be set according to the desired level of conservatism in the risk assessment. See also PRMM.

10.2.5.5 Design Factors & Safety Margin

Actual designs almost always provide for component strengths beyond what is required for actual loads.

Designs are based on calculations that must, for practical reasons, incorporate conservative assumptions. These assumptions deal with the variable material strengths and potential stresses over the life of the pipeline—usually involving many miles over many decades. Safety factors and conservativeness in design help to ensure long term system reliability. They are assigned by regulations, industry standards, or by choice of a design engineer or corporate mandate. The real safety margin—the difference between likely loadings and the component's load-carrying capacity—is most important in risk assessment. The pre-assigned safety or design factors cloud the view of the actual safety margin and should be avoided in risk assessments. This drives the previous recommendation to use strength estimates free from safety factors. This is in the interest of simplicity and clarity, a risk assessment can be viewed as a means of quantifying the safety margin in a system at any point in its life, regardless of what the original safety margin intent was. The safety margin can be re-set, as discussed in the role of integrity assessments in a load-resistance model. See more in PRMM.

Discrimination between intended safety margin and actual safety margin is an important deliverable of a good risk assessment.

10.2.6 Stress calculations

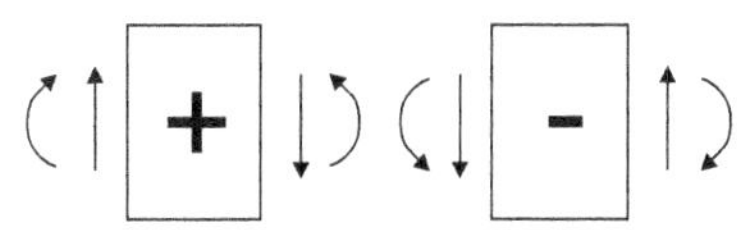

Once loads are identified and quantified, the accompanying stresses can be examined. Of particular interest here, is the relationship between stress and component wall thickness. This establishes the risk modeling opportunity to represent resistance in terms of 'effective' wall thickness.

The moment capacity for metallic pipes is a frequently used measure of their strength and is a function of many parameters. The most common are:

- Diameter to wall thickness ratio
- Material stress-strain relationship
- Material imperfections
- Welds (Longitudinal as well as circumferential)
- Initial out-of-roundness
- Reduction in wall thickness due to e.g. corrosion
- Cracks (in pipe and/or welds)
- Local stress concentrations due to e.g. corrosion damage or dents
- Additional loads and their amplitude
- Temperature.

In any failure mode, pipe wall thickness and strength will be key determinants of resistance to loads. The D/t ratio is seen in many expressions of resistance to external force damage.

The strength of a thin walled container such as pipe, from both an internal pressure and an external loading standpoint, is related to the pipe's wall thickness and diameter. In general, can contain more pressure and larger diameter and thicker walled pipes have stronger load-bearing capacities and should be more resistive to external loadings. A thinner wall thickness and smaller diameter will logically increase a pipe's susceptibility to external force failure [48].

The D/t ratio is seen in many expressions of pipe strength. Some risk evaluators have used D/t as a variable for both resistance against external loadings and as a susceptibility-to-cracking indicator. As D/t gets larger, stress levels increase—increasing failure potential and risk [1039]:

Under pure bending load:

For low D/t, the failure will be initiated on the tensile side of the pipe due to stresses at the outer fibers exceeding the limiting longitudinal stress. For D/t higher than approximately 30-35, the hoop strength of the pipe will be so low compared to the tensile strength that the failure mode will be an inward buckling on the compressive side of the pipe.

Under external load:

For low D/t ratios, material softening will occur at these points and the points will behave as a kind of hinge at collapse. The average hoop stress at failure due to external pressure changes with the D/t ratio. For small D/t ratios, the failure is governed by yielding of the cross section, while for larger D/t ratios it is governed by elastic buckling. By elastic buckling is meant that the collapse occurs before the average hoop stress over the cross section has reached the yield stress. At D/t ratios in-between, the failure is a combination of yielding and elastic collapse.

Under combined loads:

In general, the ultimate strength interaction between longitudinal force and bending may be expressed by the fully plastic interaction curve for tubular cross-sections. However, if D/t is higher than 35, local buckling may occur at the compressive side, leading to a failure slightly inside the fully plastic interaction curve.

Either stress criteria or strain criteria can be used. Discussion here is on stress.

10.2.6.1 Stress Equations

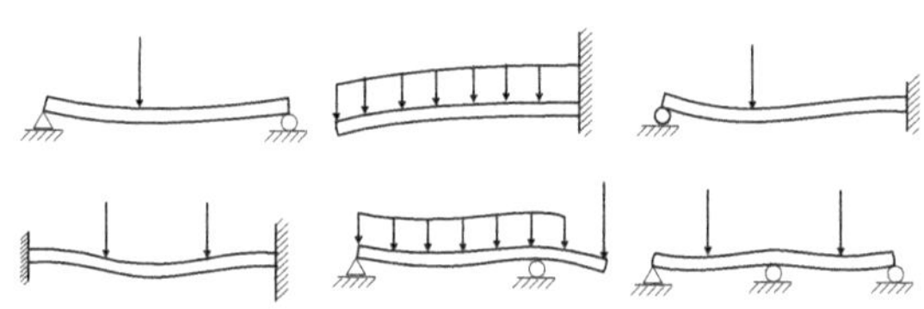

Resistance estimates will ideally involve combined stress formulae such as Tresca, Von Mises, and others, plus additional consideration of certain highly localized stresses, plus degradation/damage mechanisms. Whatever stress carrying capacity is not already 'used up' by existing loads (internal pressure, spans, overburden, etc) is available to resist additional loads.

Pipelines are normally designed to operate at a stress well below the yield strength of the component material. The principal stresses in a pipeline are the hoop stress due to internal pressure and the longitudinal stress, which is a function of internal pressure (axial), external force, weight of the pipe between spans (bending), etc. Yielding can occur as a result of either of these stresses, or under combination loading.

Yield, as a criteria for 'failure', is often conservative, even for older components. Ref [1020] says vintage pipe fails at UTS which is typically about 25% higher than SMYS. With a typical maximum allowable stress (per many regulations and standards) of 72% SMYS (1.39 safety factor), this implies a total safety factor for defect-free line pipe of about 1.74.

Formulae for calculating individual stresses are well known. Barlow's calculation is a commonly used equation for relating internal pressure to stress in a pipe. Alternative calculations may be available for pressure-stress relationships in non-pipe components or manufacturers' information may need to be used for more complex components.

External loadings are also related to stresses via well-documented equations. Understanding effects of external forces involves complex calculations both in determining actual loadings and the pipe responses to those loadings. Longitudinal stresses and buckling due to external pressure are common considerations for pipelines.

Residual stresses play an important role in some failure mechanisms. These are stresses that remain in a component after their source load is no longer active. Manufacturing processes and mechanical 'working' of materials are common causes. Residual stresses can have effects on material strength similar to conventional stresses, but their presence is more difficult to calculate. Some measurement tools to quantify residual stress are available but may not be readily applicable to most pipelines.

10.2.6.2 SRA

Structural Reliability Analysis is an analysis technique designed to improve upon the traditional use of safety factors that typically rely on a high level of conservatism in dealing with uncertainty. Compounding conservatisms in the traditional approach can produce unnecessarily conservative (and expensive) designs.

In the use of fixed, pre-determined safety factors, the true margin of safety or probability of failure is not quantified. As a 'one size fits all' design practice, this naturally leads to costly over-protection in some areas and perhaps under-protection in others. On the other hand, it avoids the potential errors and bias that may occur when more situation-specific safety margins are calculated.

Limit state threshold identification and calculations comparing actual conditions with these thresholds normally underpin SRA.

10.3 INSPECTIONS AND INTEGRITY VERIFICATIONS

SECTION THUMBNAIL
Inspections and integrity verifications provide direct input into estimations of remaining strength—the ability to resist failure.

Pipeline integrity is ensured by two main efforts: (1) the detection and removal of any integrity-threatening anomalies and (2) the avoidance of future threats to the integrity (protecting the asset). The latter is addressed by the many risk mitigation measures commonly employed by a pipeline operator, as discussed in Chapters 5 through 9.

The former effort involves inspection[2] and testing and is fundamental to ensuring pipeline integrity, given the uncertainty surrounding the protection efforts. The purpose of integrity assessment inspection and testing is to validate the structural integrity of the pipeline and its ability to sustain the operating pressures and other anticipated loads. Recall the load-resistance curve discussion in PRMM where, after conservatively assuming a shifting resistance distribution, an integrity assessment can re-set the clock, verifying available resistance to loads. Inspections serve as intervention opportunities. They interrupt a sequence of events that would have otherwise resulted in a failure. Their success in this depends on the timing and robustness of inspection compared to the degradation mechanisms possible active.

Conservatism in verifying pipeline integrity assumes that defects are present and growing. Inspection and testing at defined intervals allow for intervention so that their growth can be interrupted before they become serious threats. In theory, a defect will be largest immediately before the next verification. Uncertainty in measurements and calculations relates the estimated size of the defect, just prior to re-inspection, to probability of failure. The inspection or re-verification interval therefore establishes the maximum failure probability for each mode of failure.

Inspections and integrity verifications serve to 're-set the clock', overriding conservatively *assumed* appearance of new weaknesses since the last verification. They also provide evidence for refinement of exposure and mitigation estimates—calibration of previous estimates.

The goal is to test and inspect the pipeline system at frequent enough intervals to ensure pipeline integrity and maintain the margin of safety. The risk assessment's resistance estimate is improved by removal of any damages present or confirmation that no injurious defects exist. A pipeline segment that is partially replaced or repaired

2 See also discussion of inspections related to pipeline support condition, coatings, changes in immediate environment, etc. Here, 'inspection' refers to the identification of damages on the pipeline component itself.

will show an improvement under this protocol since either the anomaly count/severity will have been reduced via repairs or defect-free components have been installed. If a root cause analysis of the detected anomalies concludes that active mechanisms are not present, then only the resistance estimate is affected. For example, the root cause analysis might use sequential inspection results to demonstrate that corrosion damage is old and corrosion has been halted. In the absence of such findings, the risk assessment's previous estimates of exposure and mitigation may need to be modified based on the inspection results.

Inspection and integrity verifications are methods employed to find weaknesses in a component. Prior to assigning a label of 'weakness' or 'defect' to an anomalous feature of a component, its presence and characteristics as an anomaly are identified or posited. Once identified (or posited) and sized, an anomaly's role, if any, in resistance can be determined. For metal loss from corrosion, the failure potential for purposes of probability calculations is normally determined by two criteria: (1) the depth of the anomaly and (2) a calculated remaining pressure-containing capacity of the defect configuration. Both are required to account for the two failure modes of leak versus rupture. For crack-like defects, fracture mechanics and estimates of stress cycles (frequency and magnitude) are required to fully understand resistance implications.

As noted previously, a defect is considered to be any undesirable pipe anomaly, such as a crack, gouge, dent, or metal loss, that could lead to a and not all anomalies are defects. Possible defects include seam weaknesses associated with low-frequency ERW and electric flash welded pipe, dents or gouges from past excavation damage or other external forces, external corrosion wall losses, internal corrosion wall losses, laminations, pipe body cracks, and circumferential weld defects and hard spots.

The absence of any defect of sufficient size to compromise the integrity of the pipeline is most commonly proven through pressure testing and/or ILI, the two most comprehensive integrity validation techniques used in the hydrocarbon transmission pipeline industry today. Integrity is also sometimes inferred through absence of leaks and verifications of protective systems. For instance, CP counteracts external corrosion of steel pipe and its potential effectiveness is determined through pipe-to-soil voltage surveys along the length of the pipeline, as described in Chapter 6 Time-Dependent Failure Mechanisms. All of these measurement-based inspections and tests are occasionally supported by visual inspections of the system. Each of these components of inspection and testing of the pipeline can be—and usually should be—a part of the risk assessment.

Common methods of pipeline survey, inspection, and testing are listed in PRMM. Pipe wall inspections include non-destructive examination (NDE) techniques such as ultrasonic, magnetic particle, dye penetrant, etc., to find pipe wall flaws that are difficult or impossible to detect with the naked eye.

Offshore inspection is usually more expensive and can be less accurate due to challenging conditions. Inspections by divers or from submersible vessels will not normally generate the same level of confidence as their onshore integrity verifications due to numerous issues including reduced visibility, inability to use many of the NDE techniques, and the presence of concrete coatings often used offshore. Offshore inspection can also include side-scan sonar and ROV.

Recall the early discussion in this book regarding the use of measurements versus estimates in a risk assessment. Inspections and integrity verifications are measurements that typically override conservative estimates of component wall weakness. In a conservative risk assessment, they demonstrate that damages did not actually occur and 're-set the clock'.

10.3.1 Inspections

Similarly, formal in-ditch assessments of coating or pipe condition should be integrated into the risk assessment. The inspection information from other activities and analyses such as corrosion control surveys, effectiveness of coating and cathodic protection systems, and even leak detection surveys are relevant. Inspection results inform many aspects of the risk assessment—often providing evidence of exposure, mitigation, and resistance simultaneously. Types of inspections common to the pipeline industry are listed and discussed in PRMM. The use of inspection results is discussed here and in previous chapters.

10.3.2 Visual and NDE Inspections

Nondestructive examination (NDE) refers to numerous specific inspection and examination techniques. Usually done in conjunction with visual inspection, an NDE is used to find wall flaws that are hard to detect visually. NDE can involve various forms of ultrasonic wave analyses, magnetic particle, dye penetrant, etc. ILI is a type of NDE conducted remotely with subsequent visual examination sampling or confirmations. Integrity assessment can also include NDT (and 'destructive' testing) for assessing component strength or coating properties such as thickness, adhesion, strength, numbers of holidays, etc.

A visual and NDE inspection of an internal or external component surface may be triggered by an ILI anomaly investigation, a leak, a pressure test, or routine maintenance. For risk assessment purposes, a visual inspection can be extrapolated, ie, assumed to reflect conditions for some length of pipe beyond the portions actually viewed. A conservative zone some distance either side of the damage location can be assumed. This should reflect the degree of belief and desired level of conservatism. For instance, if poor coating condition is observed in one site, then poor coating condition should be assumed for as far as those conditions (coating type and age, soil conditions, etc.) might extend.

10.3.3 Integrity Verifications

As special types of inspection, the integrity verification processes of pressure testing and ILI are further discussed in following sections.

Pressure test

Pressure testing is a long-used method to ensure integrity. By stressing components to levels above what they will see during their service lives, integrity is verified and a margin of safety is established. However, the higher stress levels during the test may also cause damages—growing some defects that might otherwise not grow. This leads to some controversy in the use of pressure testing. See PRMM for further discussion.

In-line inspection (ILI)

See PRMM for a background discussion on the evolution and application of in-line inspection. ILI has been compared to medical diagnostic devices, where the doctors' interpretation of the inspection data is at least as critical as the data itself. Ref [1024] notes a typical ILI vendor's sequence of events:

- The ILI tool runs at 9 mph, capturing 1.2M measurements per second.
- Automatic data analyses algorithms identify over 1 million areas of interest in an ILI run.
- The human analyst spends 75% of his time scrutinizing every one of these, perhaps prioritizing down to 100,000 possible defects.
- Subsequent analyses utilizes knowledge of the kinds of defects that could emerge from the subject pipe's manufacture, construction, and operational history to produce categorizations of anomalies.

These steps would also ideally consider any and all excursions from ideal inspection conditions—tool travel speed, magnetization level, sensor failures, etc.—that potentially impact the inspection results.

The operator's direct examination of selected anomalies finalizes the process by linking the often more exact field NDE measurements with the ILI measurements to gain a sense of the accuracy of the entire inspection.

Not all pipelines can be internally inspected with conventional ILI tools. Certain geometries and/or flow conditions make ILI difficult or impossible. Even the best ILI tools have difficulty detecting certain kinds of anomalies, and a combination of tools may be needed. ILI can be costly, too, requiring pre-cleaning, service interruptions in some cases, challenging excavations, etc. The ILI process originally involved trade-offs between more sensitive tools (and the accompanying more expensive analyses) requiring fewer excavation verifications and less expensive tools that generate less accurate results and hence require more excavation verifications. While less accurate tool types are generally no longer used, a similar trade-off may still exist in choosing the optimum level of post-ILI analyses.

ILI and pressure testing detect damage that has already occurred and therefore provide lagging indicators of damage potential. They must be done at appropriate intervals to ensure severe defects are found and remediated before they become critical. In ILI, exceptions exist when pre-cursors to failure (other than damages) can be found. Examples include laminations, hard spots, and inferior manufacture/construction features, all of which may, under certain conditions, lead to increased failure potential even though they are not the result of damages.

Anomaly categories that can be detected to varying degrees by ILI include:

- Geometric anomalies (ovality, dents, wrinkles)
- Volumetric anomalies (metal loss from gouging and general, pitting, and channeling corrosion)
- Crack-like indications (cracks, narrow axial corrosion, certain laminations.

In every case, the size and orientation of the smallest detectable anomaly is dependent on several general and inspection-run-specific factors. Tethered or self-propelled inspection devices are also available for special applications.

10.3.3.1 Evaluating the integrity assessment

> FOCUS POINT
> Normalizing inspection and integrity assessment data in terms of age and accuracy, allows newer and more accurate information to override older, less accurate information.

Inspection and integrity verifications are powerful tools in weakness assessments. Inspection is a critical aspect of many maintenance activities and should be a key part of any risk assessment. But these should also not be relied upon as the sole driver of a PoF estimate.

For example, a practitioner of pipeline risk assessment had done this, basing his time-dependent PoF estimates solely on results of ILI. The obvious flaw in this approach is that it fails to include other valuable evidence. For instance, corrosion control was managed by a different group in this company and information between them and the ILI/risk assessors was not routinely exchanged. Results of corrosion control surveys, which tend to provide more forward-looking evidence than does ILI, were not included in the PoF determination. So, while a 3-year old ILI may have shown no metal loss and active corrosion was not suggested, last month's overline surveys show inadequate CP and coating in corrosive soils, suggesting corrosion is imminent. As an extreme example of this error of not including all available information, an operator could stop CP, scratch the coating off, add corrosive contaminants to the soil, and an ILI-based risk assessment would not report any change in external corrosion PoF until actual metal loss was occurring and detected by a subsequent ILI.

Inspection and integrity verifications are also not the final answer in resistance determinations. Their inability to detect or correctly characterize certain defects, as well as their time-sensitive nature, requires that their results be supplemented with other information.

In the risk assessment, inspection and test results are best used as confirmations of or contraindications to previously-estimated feature frequencies rather than complete assessments of feature frequencies.

For purposes of risk assessment, the age and robustness of the integrity verifications should be included in a risk assessment. With appropriate consideration, the most recent inspection is not always providing the best information. An older but more robust inspection may still provide better information than a more recent but less robust inspection. That is why both age and accuracy must be considered. The results from the best combination of the two should override the older, less accurate results. When the inspection or test is more accurate and more recent, it overrides previous estimates more completely. When only less accurate and/or older inspection/test information is available (for example, a 20 year old pressure test), estimates based on other information may dominate in the risk assessment.

A defect or theoretical defect must be characterized in order to calculate its role in resistance and/or a time to failure when subjected to degradation. With knowledge of maximum surviving defect size after the previous integrity assessments, defect rate of appearance/growth, and defect failure size, all of the ingredients are available to establish (or evaluate) an optimum integrity verification schedule. Unfortunately, most of these parameters are difficult to estimate to a high degree of confidence and resulting re-assessment schedules will also be rather uncertain.

Age of verification

Information deterioration refers to the diminishing usefulness of past data to determine current pipe condition. See related discussions in Chapter 2.8.5 Age as a Risk Variablee and Chapter 2.14 Measurements and Estimates. The past data should be used to characterize the current effective wall thickness only with considerations for what might have happened since the data was obtained and only until better information replaces it.

A re-inspection or integrity reassessment interval is best established on the basis of three factors: (1) the largest defect that could have survived or been undetected in the last test or inspection (2) the types and rates at which new anomalies are introduced into the component and (3) an assumed anomaly growth rate, all since the last assessment.

Robustness of integrity assessment

Integrity verifications vary in terms of their accuracy and ability to detect all types of potential integrity threats. Regardless of the inspection or integrity assessment technique, an inspection efficiency or robustness should be included. This includes the probability of detection and the accuracy of the anomaly dimension/orientation measurements. Building upon the matrix of possible defects created earlier, the robustness of inspection can now be added.

Robustness is a measure of the quality of the inspection or integrity assessment. The robustness consideration for a pressure test can simply be the pressure level above the maximum operating pressure. This establishes the largest theoretical surviving defect. Inspection-type assessment also involve a largest theoretical surviving—undetected—defect.

Evaluation of the effectiveness of NDE for identifying weaknesses such as metal loss, cracking, and dents is based on the NDE performance criteria used, number and location of inspection points (coverage), frequency of inspection point readings, variance of readings from criteria, equipment used and its PPM, equipment operator skill, weather/environment at time of inspection, component cleanliness, accessibility, time available to inspect, and others.

Further complicating this evaluation is the fact that inspections have varying sensitivities to anomaly types, sizes, orientations, and configurations. A separate set of capabilities will be required for at least several classes of anomalies.

The approach used in the more rigorous risk assessments is to characterize the ILI program—tool accuracy, data interpretation accuracy, excavation verification protocol—against all possible defect types under both ideal and as-inspected conditions. The performance of a series of inspections where results can be overlaid so trends and more minor changes detected, is even more valuable.

Much has been written on the subject of inspection capability and efficiency and industry standards are available for certain inspection techniques. A recommendation here is to consider separately, the inspection capabilities under ideal conditions and then under actual conditions on the day of inspection. This allows the risk assessment to 'value' separately, an improved inspection technique or an improvement to the conditions under which the inspection occurs. This can be especially important for expensive inspections such as ILI, where pipe cleanliness, configuration, flow control, and other inspection day parameters are important and also potentially costly to manage.

The two-part inspection capability assessment can be represented as follows:

PoI = probability of identifying a potentially injurious defect; probability per inspection = PoI1 x PoI2

Injurious defect = one of size, orientation, etc, (characteristic set of 'N') that, under at least one plausible scenario, reduces pipe resistance to one or more failure mechanisms

PoI1= based on tools and process designed to and capable of, under ideal conditions, finding defects of N or larger

PoI2= considers the amount of deviation from ideal conditions, expressed as a reduced PoI, compared to ideal.

Both require consideration of all steps in the process, especially tasks whose accuracies are susceptible to human error.

Assessing the ILI process

ILI results provide direct evidence of damages and, by inference, they also provide evidence of damage potential. ILI results provide evidence about possibly active failure mechanisms. Such evidence should be included in a risk assessment. The specific use of direct evidence in evaluating risk variables is explored in specific failure mechanism discussions.

The ILI PoI is improved through follow-up direct inspections. The capabilities of both (1) the ILI tool and data interpretation accuracies and (2) the excavation verification program should be considered. These two capabilities combine to show how much inaccuracy may be associated with a particular pipeline segment's assessment. The largest theoretical surviving defect best characterizes the robustness of any integrity assessment.

An excursion during an inspection is a deviation from intended or specified inspection characteristic that could lead to data collection inaccuracies. Various types of excursions during a specific ILI are common. These have varying effects on detection and sizing of anomalies. Excursions include:

- Loss of carrier signals on one or more channels.
- Velocity range exceeded—accuracy is lost when the tool travels at speeds outside its design parameters.
- Reduction in magnetization— accuracy is lost when the pipe's magnetization level falls outside its design parameters.

It is often necessary to supplement the ILI vendors' stated tool tolerances—which are typically stated for ideal run conditions—with the run-specific effects of excursions.

Another challenge often faced by risk evaluators is the array of inspection results from different tools, which may have varying capabilities and accuracies. This may require establishing equivalences between indications from different tools at different times, perhaps involving vendor-reported tool accuracies and statistical analysis of anomaly measurements, considering all run-specific characteristics and capabilities of the post-run data interpretations.

Integrity assessment and component strength

Defects left uncorrected should reduce calculated resistance in a risk assessment, in accordance with reductions in stress-carrying capacity Where inspection occurs and no defects are detected, uncertainty has been reduced, usually with a corresponding reduction in previously (and conservatively) assumed degradation and/or damage rates. In this way, the role of the integrity assessment in risk reduction can be quantified.

Such extrapolation should, of course, carry increased uncertainty. This provides the means to quantify the benefits of the inspection actually applied versus inspection results that have been extrapolated.

ILI Summarizations

The previously described direct consideration of ILI results presumes that specific anomalies have been mapped to specific locations and that anomalies are considered individually. There is rarely a justification for anything other than this complete, anomaly-by-anomaly analysis of ILI data in a permanent risk assessment. The cost of data storage and computer processing is so low that lesser solutions are unwarranted. However, for temporary risk assessments—preliminary or very approximate—or special applications, a summarization approach may be an alternative.

If this is the case, ILI results can be used to generally characterize the current integrity condition of longer stretches. Fewer segments are created under this approach. Even though each anomaly still contributes to the characterization of a pipe segment, the avoidance of a new dynamic segment for each anomaly saves some subsequent processing and analyses time. This is intended to be an approximate and rapidly deployable solution to the more correct anomaly-by-anomaly characterization. It can be done either as a preliminary step pending full anomaly-specific investigations or as stand-alone input into some special types of risk assessment.

Under such a summarization approach, pipeline segments could be generally characterized in terms of anomaly indications that might reduce pipe strength and indicate possibly active failure mechanisms.

10.3.3.2 NOP as Pressure 'Test'

With an assurance of leak free condition, a normal operating pressure can serve as an on-going pressure test. The fact that a component withstands a certain amount of pressure provides some evidence of resistance. Higher NOP provides evidence of higher resistance. A highest recent pressure to which the component has been exposed serves the same role. All other things equal, a component successfully containing 2,000 psig of internal pressure shows evidence of higher strength than does a component holding 200 psig. While higher pressures and stresses cause 'penalties' in most parts of a risk assessment, the pressure as evidence of resistance, plays the opposite role by suggesting more strength. This evidence is admittedly weak—a severe defect could exist at normal pressure. It may, however, be the only data available upon which to base an estimate of wall thickness.

This is often the default for the effective wall estimate when inspection and integrity assessments are too old or too inaccurate to provide better evidence of resistance. The wall thickness implied by leak-free operation at normal operating pressure (NOP) or a recent high pressure can be calculated by simply using a hoop stress calculation to infer a minimum wall thickness.

With assumptions therefore, a wall thickness based solely on operating leak-free at NOP, pipe_wall_NOP, can be inferred as with a pressure test: ie, using the Barlow formula for stress in the extreme fiber of a cylinder under internal pressure as follows:

pipe_wall_NOP = ([NOP]*[Diameter]/(2*[SMYS])

This simple analysis does not account for defects that are present but are small enough that they do not impact pressure containment capability at NOP. Since defects can be present but not failing due to internal pressure, a value for "max depth of defect surviving NOP" can also be assumed and included in the calculation for more conservatism. The depth of defect that can survive at any pressure is a function of the defect's overall geometry. Since countless defect geometries are possible, assumptions are required as discussed next.

Effective pipe strength can be estimated by adjusting the NOP-based wall thickness estimate for an assumed population of possible defects. There is some precedent in using 80% to 90% of the Barlow-calculated wall thickness to allow for non-critical defects that might soon grow critical. The analysis could be made even more robust by incorporating a matrix of defect types and sizes that could be present even though the pipe has integrity at NOP. An appropriate value can be selected knowing, for example, that a pressure test at 100% SMYS on 16", 0.312, X52 pipe could leave anomalies that range from 90% deep 0.6" long to 20% deep, 12" long. All combinations of geometries having deeper and/or longer dimensions would fail. Curves showing failure envelopes can be developed for any pipe.

Of course, the estimate of wall thickness based on NOP pre-supposes that the portion of pipe being evaluated is indeed not leaking and is exposed to the assumed NOP.

10.3.3.3 Inspection Used in Calibration of Risk Assessment

Inspection and integrity assessment results provide powerful evidence to be used in a risk assessment, especially for resistance estimates. They also can play a large role in assigning inputs into many exposure and mitigation variables for many failure mechanisms, as detailed in earlier chapters. The parallel paths between estimates relying only on inferred knowledge of underlying failure mechanisms versus those estimates also benefiting from inspection (estimates versus measurements), is also discussed in Chapter 2.14 Measurements and Estimates.

It is normally conservatively assumed that some deterioration mechanisms are active in any pipeline (even though this is certainly not the case in many systems). As time passes, these mechanisms have an opportunity to reduce the pipe integrity. A good risk assessment model will show this possibility as increased failure probability over time. An assumed deterioration rate is confirmed or revised by inspection in hydrocarbon transmission pipelines and often by the presence of leaks in other systems. An effective inspection has the effect of "resetting the clock" in terms of assumed events since it can show whether the forecasted count has indeed not occurred.

Leaks sometimes replace inspection as the early warning mechanism in some systems. Integrity is sometimes not thought to be compromised unless or until leaks are seen to be increasing over time. Only an unacceptably high and/or increasing leak rate, above permissible original installation leak rates, would be an indication of loss of integrity. As already noted, distribution system leakage is normally more tolerable with some amount of leakage acceptable even for some newly installed systems. Careful monitoring of leaks also confirms or refutes assumed deterioration. So, leak detection surveys can be credited as a type of integrity verification when results are intelligently and appropriately used to assess integrity.

10.4 RESISTANCE MODELING

The final step in resistance assessment is to capture knowledge about loads, stresses, damages, and defects into a resistance estimate. The goal of the resistance assessment is to estimate the ability of the component to resist failure, given that forces are active.

The stress carrying capacity is the measure of resistance and is efficiently expressed as an effective wall thickness. It captures the ability to resist new loads, given the need to resist existing loads and the possible presence of any weaknesses. The effective wall thickness requires two things: 1) the best estimate of the current wall thickness and 2) the impacts of known or possible weaknesses.

In this modeling approach, interaction of failure mechanisms with resistance issues happens automatically. As more mechanisms or stronger mechanisms overlay more weaknesses or more severe weaknesses, failure potential increases.

For a risk assessment with a definition of failure as leak/rupture, the resistance estimate must respond to two general types of failure mechanisms:

- applied loads: the fraction of applied loads that are successfully resisted without loss of containment.
- degradation: the amount of material available to be degraded before loss of containment.

These are related since, at some point in the degradation process, the load carrying capacity is compromised. All failures can essentially be understood in terms of load-resistance pairings. Degradation caused failures can also be viewed as a subset of applied loads since it is ultimately the load that generates the leak/rupture. As previously noted, even a minor leak requires a load—some driving force, if only hydrostatic pressure or gravity—to precipitate loss of containment. The degradation focus is preserved for clarity here since it is a separate branch of the PoF estimation methodology.

Essential in the resistance assessment is an understanding of:

- Component characteristics, including possible defects
- Loads applied and the corresponding stresses created in the component

The full, robust solution involves structural analyses techniques for loads and defects combinations. Textbooks and college post-graduate curriculae are dedicated to the study of stresses. Fitness for service and finite element analyses are formal methodologies to apply structural theory to specific components. While these detailed analyses certainly play a role in RA, they are beyond the scope of this text. Rather, the results that would emerge from these detailed analyses are envisioned; placeholders established, and simpler values inserted. Risk assessment model slots are available for the more robust solutions when/if warranted but the simpler estimates will often suffice.

A normalized PoF considers length effects. This 'rate' of failure probability per mile is used to establish the failure probability for any length of pipeline or collection of components. The resistance aspect of PoF—also normalized to a length of pipe—'follows along' in this calculation. Both key parts of the resistance estimate are usually 'area of opportunity'-based and hence, length-based. A current wall thickness estimation uses per-unit-length information for corrosion and cracking—for example, active corrosion points per mile; coating holidays per square foot of coated pipe; etc. The estimation of weakness potential also uses a per-unit-length approach—for example, dents per mile.

10.4.1 Resistance to Degradation

For metal loss and cracking, pressure containing capacity is generally proportional to wall thickness. A reduction in wall thickness effectively reduces the TTF from these time-dependent mechanisms. In simplest terms, a wall loss can be modeled as only a reduction in time-to-leak. Adding to this some considerations for wall loss leading to rupture failure improves upon this. The role of wall loss in loads other than internal pressure, such as those causing longitudinal stresses, can also be included. The first two considerations lead to a defensible TTF for each degradation mechanism. The third is included as a weakness in all stress-carrying capacity analyses.

10.4.2 Resistance as a Function of Failure Fraction

Resistance for time-independent failure mechanisms is often more complicated. The key is in modeling resistance as a reduction in failure fraction. Failure fraction is the number of loadings that are ***not*** resisted compared to the total number of loadings.

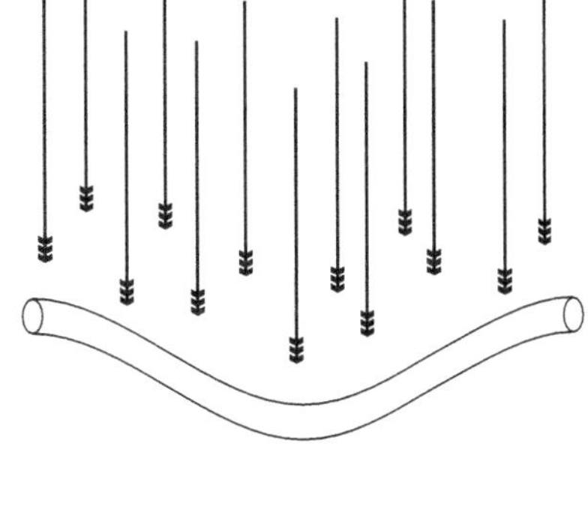

Full understanding of resistance requires examination of two strength aspects:

1. Defect-free stress-carrying capability
2. Stress-carrying capability, adjusted for defect potential

Estimation of the failure fraction under an assumed set of loadings and where there are no defects or weaknesses present is the first step. This failure fraction may be close to zero—resistance = 100%—when a defect-free component easily carries all the stresses created by even the extreme ranges of all normal loadings. Both normal and abnormal loadings should be captured as in damage rate estimates for the failure mechanisms assessed.

There are countless possible combinations of loads, stresses, and weaknesses. A cumulative probability distribution shows the probability of various combinations of stress carrying capacities and loads at any point along the pipeline. This distribution is comprised of separate distributions for the loads and resistances. Ideally, a cumulative probability distributions of all possible stress carrying capacities—considering all possible weaknesses—would intersect the distribution of possible loads in order to see how many scenarios result in damage and/or loss of integrity. That would obviously be a complex undertaking for pipelines since conditions are constantly changing along their length and full inspection is not practical.

While known weaknesses can and should trigger very specific assessments of resistance, unknown, suspected, possible weaknesses must be treated differently. A superior modeling approach offers an assessment solution that can be rapidly deployed over hundreds of miles of pipeline. It should simultaneously include detailed analyses on individual anomalies when available. The more detailed analysis will also be useful for FFS, incident investigations, and other anomaly-specific applications.

For longitudinal overstress, excessive hoop stress, buckling, and other failure modes, a reduction in wall thickness has the effect of increasing failure potential under applied loads. This increases the estimates of failure counts arising from damage scenarios. Many pipe failure mode estimates use D/t as a prime factor in predicting failure potential. D/t can therefore also be a focus for resistance reduction by effective wall thickness reductions.

This modeling approach of reducing effective pipe wall thickness based on weakness potential has the effect of increasing D/t. Higher D/t changes the failure mode under some loading scenarios and reduces pipe resistance in most.

10.4.2.1 Resistance Estimation Process

The detailed assessment of resistance involves the following steps:

1. Estimate the defect-free stress carrying capacity available to resist loads
   a. Identify normal loadings applied to the component
   b. Identify stresses generated by the loadings—ie, general structural stresses
   c. Compare to maximum tolerable stress capacity of component.
2. Adjust stress carrying capacity, based on role of known and suspected defects
   a. Estimate the probability of potential defects in the component
   b. Determine the effect of potential defects—ie, highly localized stresses.
3. Estimate the ability of the component to resist additional loads or failure mechanisms

   a. Estimate the amount of stress carrying capacity 'used up' by normal loads and potential presence of defects
   b. Express the remaining stress carrying capacity as effective wall thickness.
4. Estimate the spectrum of future abnormal loads that may be experienced (for example, from PoD estimates).
5. Estimate the fraction of future loads resisted by the stress carrying capacity implied by the effective wall thickness.

Assessing resistance in this way also ensures that interaction among all threat issues is fully included. Most of the modeled weaknesses are additive, as are most loading scenarios, and all loads and weaknesses should be included. Similarly, delayed failure potential is fully included since weaknesses remain (until repaired) and continue to interact with modeled future loadings, including external forces, pressure surges, corrosion, cracking, etc.

Later in this section is a discussion of practical modeling considerations, including how this aspect of risk assessment can be modeled in a very robust way or a very simple way, depending on the needs of the assessment.

10.4.3 Effective Wall Thickness Concept

With an understanding of loads, stresses, and potential weaknesses, the next step in estimating resistance is the bridge between this information and a resistance value to be applied to each component in the risk assessment. This too can be a complex step unless some simplifying assumptions are made. An effective wall thickness can be an efficient intermediate step or at least a conceptual framework for this final assignment.

Next to internal pressure capacity, a component's wall thickness is probably the most referenced characteristic used in strength and safety margin determinations of pipeline components. Minimum required wall thicknesses are determined from material properties and the amount of stress that the component must withstand. As a pressure containment system, the importance of wall thickness is intuitive. The role of increased wall thickness in risk reduction is also intuitive and verified by experimental work. Component wall thickness, above what is needed for internal pressure and known loadings, provides a margin of safety against unanticipated loads as well as an increased survival time when corrosion or cracking mechanisms are active. Certain wall thicknesses are also thought to substantially reduce the chances of failure from external forces such as from excavating equipment. Some wall thickness–internal pressure combinations provide enough strength (safety margin) that most conventional excavating equipment cannot puncture them. Of course, material type must be considered along with the component dimensions. Even among steels, the material strength, often reported as SMYS, can vary greatly.[3]

3 SMYS is only one aspect of material strength but is often used to generally characterize a steel.

However, experience also indicates that increased wall thickness is not a cure-all. Increased brittleness, greater difficulties in detecting material defects, and installation challenges are cited as factors that might partially offset the desired increase in damage resistance [58].

Furthermore, avoidance of immediate failure is only part of the threat reduction—nonlethal damages can still precipitate future failures through fatigue and/or corrosion mechanisms. Nonetheless, increased wall thickness provides failure protection in most failure scenarios.

Defects can also be modeled as equivalent reductions in wall thickness. An effective wall thickness—actual thickness less some amount of wall loss to account for defects—can be estimated. Effective wall thickness then is an efficient basis for modeling pipe resistance to loads. As wall thickness is reduced, implications for component strength include:

- Less capacity for pressure containment
- Faster TTF for degradation mechanisms
- Higher D/t leading to reduced buckling capacity
- Lowered resistance to external forces including localized (puncture) and uniform (subsea hydrostatic pressure).

With a modeling assumption that all potential weaknesses can be effectively treated as reductions in pipe wall thickness, an 'effective' or 'equivalent' wall thickness can be used to represent resistance. The term 'effective' is added to the wall thickness label to capture the idea of equivalencies. It provides a common denominator by which all stress-carrying capacity reductions can be captured in similar units. When evaluating a variety of pipe materials, distinctions in material strengths and toughness will be needed when assessing the role of component wall thickness. With respect to resisting many types of loadings, a tenth of an inch of steel offers more than does a tenth of an inch of fiberglass. When evaluating defects, some will have a more profound effect on strength than others.

As a measure of strength, or stress-carrying capacity, wall thickness is a useful surrogate for the whole suite of factors to be considered in a full strength assessment. The evaluation of stress levels in the component will focus on wall thickness, enabling a risk assessment methodology to similarly focus on 'effective' wall thickness as the modeled resistance. The concept of effective wall thickness is therefore efficiently used in risk assessment.

10.4.3.1 Nominal Wall Thickness

Effective wall thickness estimation begins with actual wall thickness. In the absence of recent measurements of wall thickness, the actual wall thickness may need to be derived from the originally specified or nominal wall thickness.

General stress calculations assume a uniform pipe wall, free from any defect that might reduce the material strength. It discounts possible reductions in actual or effective wall thickness caused by defects such as cracks, laminations, hard spots, gouges, etc. A specific stress calculation on a component requires consideration of such features. Finding or positing all differences between specified and actual and effective wall thickness is essential to risk assessment. Pipeline integrity assessments are designed to identify areas of weaknesses, in the form of wall thinning or in-wall defects, which might have originated from any of several causes. Other inspection may also reveal areas of actual or a high-probability of wall loss, pinhole corrosion, graphitization (in the case of cast iron), and leaks.

Most pipeline systems have incorporated some "extra" wall thickness—beyond that required for anticipated loads, and hence have extra strength. This is often because of the availability of standard manufactured pipe and appurtenance wall thicknesses. Such "off-the-shelf" purchases are normally more economical than special designs even though they may involve more material than may be required for the intended service. This extra thickness will provide some additional protection against corrosion, external damage, and most other failure mechanisms.

When actual wall thickness and wall condition measurements are not available, the nominal wall thickness can be the starting point for estimating current wall thickness. The difference between nominal or "specified" wall thickness and actual wall thickness is a key aspect of resistance determination in this risk assessment. Especially in a conservative risk assessment, the nominal value as a estimate of current, must be adjusted for all variances pertinent to the estimation of the strength provided by likely (or worst case) actual wall thickness.

Differences between nominal and effective wall thickness include:

- Allowable manufacturing tolerances—the actual wall thickness can be some percentage thicker or thinner than specified and still be within acceptable specification.
- Manufacturing defects including material inclusions, voids, and laminations.
- Installation/construction damages or errors such as during joining (welding, fusion, coupling, etc.) processes
- Damages suffered since manufacture: ie, during transportation, installation, and operation, including corrosion and cracking.

Some of these adjustments are actual reductions in thickness while others are reductions in effective strength, ie, features such as cracks, girth weld defects, hard spots, etc are not measured in terms of thinning but rather by some other loss of stress-carrying capacity.

10.4.3.2 Current Wall Thickness

As used here, the current wall thickness is not always a direct measurement of the component's wall by UT, caliper, or other means. It also includes inferential indications of current wall thickness that often must be made in the absence of the direct measurement. Actual or current wall thickness values emerge from whichever of the following provides the strongest evidence:

- Direct measurements, with considerations for age and accuracies of all readings, as well as other uncertainties, such as if a measurement at one location is to be extrapolated to another location. These measurements include those taken by NDE examinations including direct-measurement ILI, UT, etc.
- Thickness inferred by ILI techniques designed to find changes in wall rather than measure thickness, external only indications such as visual and pit depth gauge
- Thickness inferred by pressure test
- Thickness inferred by normal or recent high pressure levels (see NOP as Pressure Test discussion)
- Specified or nominal thickness, with previously described adjustments (manufacturing tolerances and error rates, damage rates during and since installation, etc.)

All of these possible information sources will grow more uncertain over time except for wall thickness implied by a current operating pressure (which carries its own significant uncertainties).

It is not unusual to have data from several or all of these information types available at the same location but with widely varying accuracies and age. For instance, one or more ILI's, multiple excavations, and at least a post-installation pressure test, will each offer one or more pieces of information in each category, for an operating pipeline. The risk assessment will need to efficiently filter through the disparate information to determine the best indicator of today's thickness. This mirrors the process the SME would also have to use when faced with the same information set and the need to determine the single best estimate.

With a consistent application of conservatism in uncertainty estimates, the more optimistic value—the information suggesting the best wall thickness after adjustments for age and accuracy—will usually govern, as discussed early in this text. Refer to early discussion of measurements versus estimates—the general approach for efficiently integrating many disparate pieces of evidence into the risk assessment. See Chapter 2.14 Measurements and Estimates.

10.4.3.3 Effective Wall Thickness Estimation

Beginning with the best available estimate of current wall thickness, we now assess for any weaknesses that may detract from the strength implied by this wall thickness. A weakness will reduce the current, actual wall thickness into an 'effective wall thicknesses'.

Since resistance, as it is being modeled here, is proportional to available stress-carrying capacity it is also generally proportional to material thickness. Wall thickness is often the single most important component characteristic in most loadings of components. Use of wall thickness to represent resistance is intuitive for degradation from corrosion and withstanding internal pressure, longitudinal loads, and puncture. It is less intuitive, for example, in assessing cracking.

Nonetheless, it is still efficient, as previously discussed. Cracking can be modeled as follows: defects that increase crack initiation potential and/or stress intensifications and/or lower toughness, are modeled as either reductions in pipe wall or increases in cracking rates. Technically, lower toughness does not directly cause faster cracking but rather allows smaller defects to initiate/activate a crack.

Either results in increased failure probability as the probability or severity of defects increases. General insights from structural theory can be incorporated into a risk analysis. Component wall thickness is usually proportional to structural strength—greater wall thickness leads to greater structural strength (not always linearly)—with the accompanying assumption of uniform material properties and absence of defects.

Defects modeled as reductions in effective pipe wall thickness is a simplification of the complex analysis that would require consideration for each possible anomaly under every possible loading scenario. In a robust solution, for each anomaly's characteristics such as:

- length, width, depth;
- location in wall;
- clock position on circumference; and
- orientation relative to axes (axial, radial, circumferential)

loads would be applied, stresses calculated, and ability to survive under various scenarios assessed. The simplification is intended to represent this spectrum of scenarios with an equivalent wall thickness: defect X causes an equivalent loss of strength as does a reduction of wall thickness by Y%.

Knowledge or suspicion of potential weaknesses arises from:

- discovery via NDE
- era of manufacture including manufacture specifications used
- construction practices including construction specifications used
- experience on current component or with similar (relevant) collections of components
- defect-introduction mechanisms possibly active
- includes benefits from sleeves and other repairs

The ratio of effective pipe wall thickness to required wall thickness is another way to view the resistance concept. A ratio greater than one means that extra wall thickness (above design requirements) exists. For instance, a ratio of 1.1 means that there is 10% more pipe wall material than is required by design and 1.25 means 25% more material. If this ratio of effective wall thickness to required wall thickness is less than one, the pipe does not meet the design criteria—there is less actual wall thickness than is required by design calculations. The pipeline system has not failed either because it has not yet been exposed to the maximum design conditions, there is excess conservatism in the calculation, or some error in the calculations or associated assumptions has been made.

This ratio concept is used in some inspections. Certain NDE, especially ILI, often reports wall loss not only in terms of length, width, and depth, but also as implications in pressure containing capacity. Estimated Repair Factor (ERF) and Rupture Pressure Ratio (RPR) are common types of ratios reported by ILI. These reported ratios based on theoretical rupture pressure versus MAOP are readily converted into equivalent wall thicknesses.

10.4.3.4 Resistance and Effective Wall Thickness

All resistance estimates can use 'effective wall thickness' as an efficient foundation. While applied loads produce stress in different ways, wall thickness is a key strength determinant in most loading scenarios of thin-walled structures (most pipeline components are modeled as shell type structures). Degradation resistance considers potential wall loss by corrosion as well as fatigue life reduction—'wall loss' by cracking [4]. Reduced wall thickness leads to reduced load carrying capacity. So, wall thickness as a measure of load-carrying capacity, when coupled with degradation rate (mpy or mm per year), leads to an estimate of time before degradation advances to point of containment loss (or yield).

10.4.4 Resistance Baseline

Since resistance is to be measured as a simple percentage, a starting point or baseline is required. This involves basic definitions of exposure, as discussed in Chapter 2.8.12 Nuances of Exposure, Mitigation, Resistance. The resistance baseline could be the remnant stress carrying capacity after normal loads are applied. Then the percentage resistance shows the fraction of additional loads that could be resisted, given that the existing loads are 'using up' some of the stress-carrying capacity. The resistance baseline could also be essentially zero, whereas the percentage shows the fraction of all loads resisted. The 'aluminum can' analogy represents this scenario.

Recall that exposure events were quantified by imagining that there is no mitigation nor resistance. An aluminum drink container—a can, crushable between two fingers—is the right mental image for lack of resistance from outside force. So, the image of a soda can lying atop the ground, is the correct image to estimate exposure event frequencies. If such a can be broken by an event, then that event should be counted as an exposure.

The resistance estimate using the 'aluminum can' analogy is therefore capturing the ability to withstand forces beyond those that would fail a component with virtually no ability to resist. Resistance values will therefore be very high. Most unflawed steel pipe components will have, for example, 100% resistance to pedestrian traffic.

---

4 Cracking is modeled as effective wall loss even though there may be no actual loss of material associated with some forms of cracking

10.4.5 Logic and Mathematics Proof

The use of resistance as the third component in the PoF calculation warrants more conceptual discussion. Specific equations are proposed as an efficient means of including all weakness issues into the risk assessment. These equations, as applied to a pipeline segment or other component with potentially multiple weaknesses, is discussed here.

Weakness

A weakness is any structural feature that reduces a component's stress carrying capacity. That reduction increases failure potential by causing one or more of the following:

- Less capacity for pressure containment
- Faster TTF for degradation mechanisms
- Lowered resistance to external forces including localized (puncture) and uniform (subsea hydrostatic pressure).

The role of the weakness is intuitive, directly proportional, and well documented for the first two of these. Converting all weaknesses into equivalent wall thinning is an effective approach to show changes in resistance. The third, also efficiently modeled as equivalent wall thinning, involves more complexity, as described below.

Weakness Equals 'Increased Failure Fraction'

A weakness, as used here in modeling time-independent failure mechanisms, actually represents a failure fraction, not necessarily a direct reduction in strength. Assigning an equivalent wall thinning to each weakness is a useful intermediate step, but its role in failure fraction must still be estimated.

Failure fraction implies a probabilistic aspect. This warrants examination. Assume, in some length of pipe, there is a 10% probability of a weakness that introduces a 60% loss of strength. Can this be modeled as a 0.1 x 0.6 = 6% weakness? Probably not. The probability-adjusted weakness estimate should not be used in direct comparison to an absolute level of strength that triggers failure. A 10% chance of 60% weakness may predict occasional failure while a 6% weakness may suggest that no failure is possible. For instance, if nothing less than a 10% weakness allows failure under a certain loading condition, then using 6% weaknesses shows that the pipe always survives even though there is a chance of a serious weakness being present. In reality, we expect that 10% of the time, an applied load will involve the weakness and the pipe will fail. 90% of the time, the weakness is not involved and the pipe survives.

However, as used in this risk assessment, the 60% weakness actually represents a 60% increase in failure potential. If the 10% probability of weakness is 'per mile', then after about 10 miles, we would be fairly certain of the weakness occurring at least once; 10%/mile x 10 miles = 100% (taking some liberties with probability theory). So, let's say that we have a 100% chance of at least one 60% weakness somewhere in the 10 miles. Under certain assumptions, that is mathematically the same as the 6% probability of a weakness per mile in the assessment equations used here. The key is that the 6% weakness is actually modeled as a 6% increase in failure potential. Each mile has a relatively low chance of failing from the weakness—6%. The aggregation of all ten miles, however, shows a high chance of a failure point—10 miles x 6%/mile = 60%. Due to the possible presence of a weakness, each mile carries a 6% increase in failure probability and the whole ten miles carries an 60% increase. We expect a failure somewhere but do not know in which mile it will occur.

Multiple weaknesses increases the failure fraction as is discussed in a following section.

Resistance vs Failure Fraction

The recommendation to measure resistance rather than failure fraction in the top level PoF equation arises from the notion that increased resistance is intuitively a good thing. Adding more resistance, just as adding more mitigation, reduces PoF. Numerically increasing either will prevent failures. This is useful in communications of risk. Discussing a reduction in failure fraction, rather than an increase in resistance, is less convenient in the everyday conversations that will hopefully emerge in risk management, based on the assessment.

Multiple Weaknesses

While it only takes one weakness to coincide with a sufficient load to precipitate a failure, the number of potential weaknesses logically increased the opportunity for the unfavorable load-resistance overlap. The potential density of weaknesses is captured in the probability estimate. This discriminates between components with potentially few or none and those with many weaknesses.

When there are potentially multiple weaknesses, all combinations should be considered—portions of the segment with no weakness, one weakness, two weaknesses, etc. The full analyses to include potential weaknesses in the PoF involves combining all possibilities for each portion of the segment being assessed: PoF_with weakness1 + PoF_without_weakness1 + PoF_with weakness2 + PoF_without_weakness2 + PoF_with both weaknesses1-2 + PoF_with neither weaknesses1-2 + etc for each potential weakness and weakness combination. Each pairing sums to be 1.0 or 100% since all possibilities are being considered, ie with and without the weakness or weakness combination.

This equation would be repeated for each combination of failure mechanism (PoF) and resistance (weakness or collection of weaknesses) scenario.

Examining the mathematics involved

Using the standard form for PoF estimation:

time independent: exp*(1-mit)*(1-res) = PoD*(1-res)

time-dependent: exp*(1-mit)/(res) PoF=1/TTF = 1/(res/(exp*(1-mit)))= exp*(1-mit)/res = PoD/res

PoD is probability of damage. Once that is determined, then resistance is added to the equation to predict failure potential.

RES is %—fraction of damage events that do not immediately lead to failure (loss of integrity, in these examples).

(1-RES) is the fraction of failures after damage occurs. For example, 80% resistance means that one out of every five loadings—20% of the damage-causing events—will result in immediate failure while in four out of the five events (80% of the time), damage may occur but failure will be successfully resisted. Failure fraction is 20% and RES is 80%.

Implicit in the estimate of PoD is the existence of one or more 'damage' scenarios that could result in failure. But the frequency/probability of ***damages*** is always equal to or less than the frequency/probability of ***failures.*** We can't have more failure scenarios than damage scenarios [5]. So, RES is <1.0 and approaching 1.0 if a high fraction of damages result in immediate failures.

Since (1 – resistance) is indeed the failure fraction for time-independent failure mechanisms, FailFrac can be used in the original PoF relationship to make this proof more transparent:

PoF = PoD x FailFrac

If a weakness exists, RES is reduced and FailFrac increases. Since we often don't know for sure where/if weaknesses exist, a probability consideration is added.

Pr = probability that weakness RES exists and generates the corresponding failure fraction.

(1-RES)*Pr = FailFrac *Pr = probability of the failure fraction occurring = FailFrac given weakness

Recall that PoF = PoD*(FailFrac if weakness exists) + PoD*(FailFrac if no weakness)

5 At least not with a leak/rupture type risk assessment where damage is a prerequisite for failure.

Assume (FailFrac if no weakness) = 0, so the second term can be ignored.

PoF = PoD(1-RES1) Pr1+PoD(1-RES2) Pr2+PoD(1-RES3) Pr3.... PoD(1-RESn) Prn with no coincident occurrences

n = count of weakness scenarios (ie, girth weld defect, hard spot, low freq seam, wrinkle bend, etc and various combinations of these)
RESn = resistance scenario—fraction not failing if weakness exists, at least one RES scenario must exist; the sum of all RES scenarios is <= 1.0

Prn = prob of weakness scenario n existing; sum of Prn represents all possible scenarios, ie, (Pr1 + Pr2 + ... Prn) = 1.0

The above equation simplifies to:

PoF = PoD[(1-RES1)Pr1+(1-RES2)Pr2+(1-RES3)Pr3....(1-RESn)Prn]

so, PoF = PoD[(Pr1+Pr2+Pr3...Prn)-(RES1(Pr1)+RES2(Pr2)+RES3(Pr3)... RESn(Prn))]

where FailFrac = [(1-RES1)Pr1+(1-RES2)Pr2+(1-RES3)Pr3....(1-RESn)Prn]

using our initial example with a simple one-weakness resistance scenario, a 60% weakness means 40% resistance and there is a 10% chance of the weakness and 90% chance of no weakness (resistance = 100%), so:

PoF = PoD[(1-40%)]10% + PoD[(1-100%)]90% = PoD(6% + 0%) = PoD*6%

PoF should never exceed PoD, so sum of Prn's should equal 1.0, all scenarios. In this equation, it is necessary to include all combinations in Pr—ie, all combinations of weaknesses where more than one weakness exists. Alternatively, an OR gate can be used (discussed in next section) to aggregate possible scenarios of weaknesses, including coincidences.

10.4.5.1 Using the OR gate math:

The OR gate math approach to combining probabilistic elements also supports the modeling of probabilistic failure fractions as proposed. By a similar logic as previously shown, resistance scenarios, can be combined as follows:

PoF = PoD[(1-RES1)Pr1 OR (1-RES2)Pr2 OR (1-RES3)Pr3....(1-RESn)Prn]

The OR gate method of summation does not require that the 'no weakness' scenarios are included. Since not all possible scenarios are included here (only the 'with weakness' scenarios, not the 'without weakness' scenarios) summations to 100% probability are not expected. Any potential resistance scenario is added to the others via the OR gate. This makes modeling much easier.

The OR gate applies for both combining multiple weaknesses in the same component and aggregating the resistance of a collection of components. The latter is of less interest since each component will have its own failure probability. Aggregating failure probabilities from a collection of components has many applications but aggregating their resistance values serves no apparent purpose other than perhaps a point of interest.

10.4.5.2 Resistance in Time-dependent Failure Mechanisms

This same justification enables the use of probabilistic pipe weaknesses PoF calculations for time-dependent failure mechanisms. This includes delayed failure potential, where a defect is introduced, does not precipitate immediate failure, but contributes to a later failure.

Resistance in time-dependent failure mechanisms is efficiently measured as effective reductions in wall thickness, as discussed in this section. This is illustrated in the following example:

> Say there is a 10% probability of one or more defects per mile is present and that each defect results in 50% effective pipe wall. Some miles will have one or more defects while others will have no defects (100% effective pipe wall). Miles with no defects will have a leak-based TTF1 = wall1/mpy. Miles with one or more defects that are coincident with the degradation rate will have TTF2 = wall2/mpy = (0.5 x wall1)/mpy = 0.5TTF1. Under certain assumptions, we expect 10% of the miles to have TTF2 = 1/2TTF1 and 90% of the miles to have TTF1. If PoF is modeled to be 1/TTF, then any random mile will have a 10% chance of PoF2 and a 90% chance of PoF1. To obtain a point estimate of the potential pipe weakness in the mile, we use a probability-weighted value calculated as:

10% x 50% + 90% x 100% = 95%

So, a 90% chance of PoF1 and 10% chance of 2XPoF1 is modeled as 1.05%PoF1. The three values that arise from this reasoning are as follows:

TTFprobable = pipe wall/mpy = PoF1

TTFmodel = 95% pipe wall / mpy = 1.05% PoF1

TTFworst = 50% pipe wall/mpy = 2 X PoF1

The modeled TTF uses both the most probable and the worst case TTF, in a two-part relationship converting TTF to PoF, as is discussed in Chapter 2.8.4 From TTF to PoF.

10.4.6 Modeling of Weaknesses

SECTION THUMBNAIL

- List any and all types of weaknesses that could be present
- For each weakness, estimate 1) its probability—rate of occurrence—along each pipeline and 2) the equivalent amount of wall loss if the weakness is present

Recall the advice to begin with the robust solution before contemplating any shortcuts. In the case of resistance estimation, the robust solution entails analyses of every combination of load, stress, and potential defect. The first two have been discussed already, so the last remains. Here, the role of defects—weaknesses—is considered in the risk assessment.

10.4.6.1 Process

Given all the types of potential weaknesses, the varying abilities to detect each, and the role each may play in component strength, there are countless combinations to consider in an assessment. This seemingly daunting task can be made manageable via the establishment of a matrix. This takes some initial effort, but then is simple to maintain and adjust as needed. More specifically, some parts of this process require an initial set up but then only very infrequent maintenance and updates. Other parts will be location specific and sensitive to inspection results, therefore requiring sometimes frequent updating.

In outline form, the following ingredients will be needed for the matrix:

1. List of all possible defects/weaknesses: any that could appear anywhere on any pipeline component
2. Estimation of representative size/configuration of defect populations, covering at least two possibilities:
    a. Noteworthy Defects: The size/configuration combination that first results in a measurable strength reduction—ie, the smallest size/configuration that noticeably reduces strength under design loads. This sets the lower threshold for what types of features should be included in the assessment. Non-injurious features can usually be disregarded.
    b. Worst-case defects that could be undetected: The combination that yields the worst-case strength reduction AND is undetectable (just below detection limits) by integrity assessment or

inspection methods. This establishes the largest defects that could remain undetected by an inspection or integrity assessment.

3. Inspection and integrity assessment capability evaluations. The probability of detection of each size/configuration combination using each type of anticipated integrity assessment technique.
4. Assignments of effective wall thickness reductions to each defect
5. Conversions of wall thickness reductions into increased failure fractions for time-independent failure mechanisms

The above set of estimates can be established for all possible pipeline systems to be included in the risk assessment. Having initially set this up, tested it with real-world applications, and gaining the acceptance of SME's, it should only infrequently require maintenance.

Then, the location-specific elements are added for each segment under evaluation. That is, each length of pipe or individual component, requires a current estimate of:

- The failure fraction under an assumed set of loadings when there are no defects or weaknesses present. This failure fraction may be close to zero, when a defect-free component easily carries all the stresses created by even the extreme ranges of all normal loadings[6]. Note that both normal and abnormal loadings should be captured as exposure estimates for the failure mechanisms assessed.
- The probability of each size/configuration existing in the subject segment prior to the integrity assessment. In the absence of better information, this may have to be a rate per mile, broadcast along many miles of apparently-similar pipeline.
- The probability of each size/configuration existing in the subject segment immediately after the integrity assessment. This uses the general inspection capability analyses generated above. But it adds the location- and application-specific nuances of each inspection—ie, the accuracy of that particular inspection, considering weather, cleanliness, ILI excursions of speed, magnetization, etc, operator skills, and others.
- The rate of re-emergence of each size/configuration. This may be zero for many anomalies such as those associated with original manufacture or construction and not possible to introduce during modern repair.

This second list will require more maintenance, given its role in measuring changing conditions at specific locations and the situation-specific nature of many inspections.

After applying this exercise, each component will have an effective wall thickness estimate. This will lead to a resistance estimate to be used in all PoF calculations

6 When the resistance baseline is the remnant stress carrying capacity after normal loads are applied. The resistance baseline could also be essentially zero, if the 'aluminum can' analogy is used.

For defects whose contribution to increased failure potential is primarily through stress concentration, the defect can be treated either a decrease in effective wall thickness or an increase in crack growth rate. To keep the association between the anomaly and the effect on failure potential, the former is usually the more efficient modeling choice. Each anomaly can be treated as a reduction in effective wall thickness, resulting in reduced TTF and increased PoF, compared to anomaly-free components.

10.4.6.2 Listing of Potential Weaknesses

A general listing of potential weaknesses may be appropriate for early phase risk assessment. For instance, Table 10.3 shows a sample of general categories of potential weaknesses, where SME's have assigned an effective wall loss reduction to each category.

Type	Sub-Type	Effective Wall Loss (%)
Dent	>6% of diameter	
	Dent with gouge	
	Dent with re-rounding	
	<6% of diameter	
Mechanical coupling	Flange	
	Screwed	
	Dresser style	
Stress concentrator	Wrinkle bend	
	Miter joint	
	Substandard appurtenance	

Such generalizations require many assumptions and will not result in the most accurate assessments. More detailed listings will provide more ability to discriminate weaknesses. Dents and gouges will often need to be better characterized in terms of their dimensions and orientations in order to assess their realistic impact on resistance. Furthermore, their existence in regimes of higher stress and/or more pressure or thermal cycling will be important. Creating more extensive lists to capture more characteristics and/or combinations of effects will often be appropriate.

Type	Sub-Type	Effective Wall Loss (%)
Dent	>6% of diameter	5%
	Dent with gouge	10%
	Dent with re-rounding	10%
	<6% of diameter	2%
Mechanical coupling	Flange	5%
	Screwed	10%
	Dresser style	[illegible]
Stress concentrator	Wrinkle bend	
	Miter joint	
	Substandard appurtenance	

Considerations either beyond or more general than defects can also be included here. Characteristics such as toughness, old repair methods, and certain appurtenances are not defects but may impact resistance. Nuances such as laminations vs laminations plus source of hydrogen can also be considered in the matrix. So, rather than a focus on specific defect types, a more generalized list of locations that may harbor a resistance issue can be used instead or can supplement. For example:

> Note that features caused by mechanisms such as metal loss (from corrosion) and cracking do not appear on these sample lists of weaknesses. This is due to their role as independent failure mechanisms, modeled elsewhere in the risk assessment. The assessments of corrosion and cracking yield estimates of effective wall thickness. To these estimates, the potential for additional weaknesses will be considered, further reducing the effective wall thickness in many cases. This ensures appropriate consideration of interaction of all degradation mechanisms (as well as random failure mechanisms) with all potential weaknesses.

This listing of potential weaknesses is a generalized part of the matrix that will not often change. Only with new or improved inspection or new knowledge of structural resistance will changes be needed.

10.4.6.3 Estimating Strength Reductions

The amount of strength reduction that should be attributed to each potential weakness is well understood for some weaknesses, such as corrosion metal loss, but less so for others such as stress concentrators. In other cases such as modeling for crack progression, the amount can be calculated, but only after acquiring additional costly information or making highly uncertain assumptions.

Some anomalies are only a defect under certain loadings and/or sufficient stress. A stress concentrator may lead to crack initiation, leading to increased crack susceptibility (modeled as more rapid crack propagation), but only when sufficient stress exists. Below this threshold stress level, no crack activity occurs. To handle these situations in a risk assessment, it is usually prudent to include the strength reduction effect as if the sufficient loadings were present. When this is paired with a low probability of that 'sufficient loading' scenario, the effect on risk is appropriately quantified.

Some strength reduction estimates can be derived from design standards. ASME B31 series standards report stress intensification factors for various pipeline components. While not directly designated for this use, an extrapolation of such factors into strength reduction values is logical. Similarly, design factors of various types often imply the amount of strength reduction accompanying design or construction features. These too may be valuable as guidance for assigning weakness values in the risk assessment. Others may be derived from published research. Of course, a finite element analysis for a specific component with specific defects or stress concentrators will be the full and most accurate guidance on resistance.

By first modeling each feature as an effective wall thickness reduction, quantitative assignments of strength reductions can then be made. In some cases, the effective wall thickness is simply inserted into stress calculations, replacing the nominal or measured wall thickness that would otherwise be used. However, the quantifications of strength reduction will also require assumptions and modeling shortcuts to make it manageable for most practical applications.

10.4.6.4 Probability of Weakness

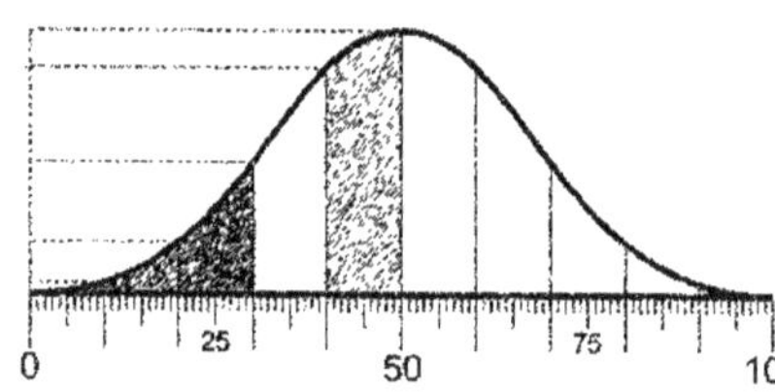

Once weakness potential is understood, the probability of each weakness (or of each category) is estimated for each stretch of pipeline or each component. This is an input data set. Whenever the rate of occurrence changes along the pipeline, a new dynamic segment is warranted. Changes in rate of occurrence are often linked to characteristics such as:

- Era of manufacture
- Manufacturing process and plant
- Construction/installation process
- Construction challenges
- Outside force changes
- Pipe specification
- Surface type—pavement, water, agriculture, urban, etc
- Burial depth
- Inspection/test history
- Etc.

Consistent with other parts of this risk assessment, it is advantageous to have parallel branches in the model for estimates and measurements. Estimates are 'best guesses' of how often a weakness may appear. They may have to be deduced from era of manufacture/construction knowledge or experience with similar systems. Measurements are the results of surveys or inspections that more directly identify weaknesses. Estimates override older and less accurate measurements while newer, more accurate measurements override older, less accurate measurements and estimates. This way, the absence of a measurement (no inspection) is penalized (shows as higher risk) when conservative estimates are used.

All of these potentially impact previous frequency and severity estimates. For instance, the discovery of an old metallurgy report noting steel toughness may warrant a change in that 'weakness'. Other examples include the ILI discovery of old fittings or appurtenances or wrinkle bends; the occurrence of aggressive MIC activity; etc.

For some pipeline segments, some potential issues will be immediately dismissible—for example, no low frequency ERW seam issues where ERW pipe does not exist. Even in these cases, the matrix serves a valuable function. It documents that 1) the potential issue is considered and 2) that it plays no role in risk at the subject location.

The rate of appearance of new defects depends firstly on the origin of the component. New defects originating from manufacturing and construction processes would not be expected unless new components had been added or existing component modified. The additions or modifications would not be expected to harbor defects of the kind associated with older practices now known to be inferior, unless errors (for example, use of improper material) or sabotage are suspected. Otherwise, only errors in the pertinent manufacture/construction processes could introduce new defects of those types.

Defects also appear in the operations and maintenance phase of the pipeline's life cycle. New anomalies can be introduced by unintentional contacts with excavation or agricultural equipment, earth movements, and others. Anomalies can transition into defects under the influence of degradation mechanisms or new stresses.

SECTION THUMBNAIL

An estimate of future weaknesses will be needed. The PoD's from threats assessed will inform such estimates.

All possible defect origination scenarios should be included in the resistance assessment. Each should be estimated for each component in the risk assessment. Since there are myriad types of anomalies that can arise, sometimes from multiple causes, and grow under countless scenarios, this is not a trivial task. But it is reflective of the real world and must at least be understood and approximated before the risk assessment can be accurate.

The defect rates of growth and appearance can often be better estimated after successive integrity evaluations. Care must be taken to separate temporary aberrations from trends. Third party construction associated with a housing subdivision under development may have led to multiple dents and gouges but, once completed, will no longer be a source.

Defect rates may also be based on previously assessed rates of underlying degradation mechanisms (corrosion or cracking, normally) and rates of time-independent damage (ie, PoD's from impacts, excavations, geohazards, etc). Since exposures and mitigation effectivenesses for future rates have already been quantified in the risk assessment, those values can be used directly for estimating future defect rates and indirectly (ie, modified based on changes over time) for past defect rates.

To make the assessment more manageable, it may be appropriate to group some defect types and origin causes. In the following example, two estimates are made for potential weakness category; one for frequency of each weakness at the time of installation or last measurement (integrity assessment) and one for frequency of introduction, ie the rate at which the weaknesses are being created. The latter is estimated as a rate per mile-year that might have been introduced since the most recent inspection or assessment that should have detected the anomaly. A defect from a manufacturing process would have a zero rate of introduction unless replacements are being made.

A relevant integrity assessment 'resets the clock' to some extent, establishing the number and severity of defects existing at the time of inspection. The interest is then in the defects that escaped detection plus any new defects occurring since that inspection. If no such inspection or assessment had been done, the pipe installation date is used with a PXX plausible rate of introduction of defects—for example, 1 mile of pipeline, 20 years old = 20 mile years area of opportunity; 20 mile-years x 0.2 defects introduced per mile per year = 4 defects dispersed along the mile of pipe.

Again, note that some resistance issues are assigned a zero rate of emergence since they are associated with outdated manufacturing and construction practices that could not have occurred since the last assessment. The emergence rate also takes into account improvements in inspection and quality control during actions on the pipeline.

For instance, in this example, the assessor assigns the rate of new substandard girthwelds to be once every 5 miles (per year of new welds being produced), while the older portions are assigned a rate of once every mile. Discounting 'per year' implications (ignoring, for the moment, any defective girth welds introduced during repairs) and with an average girthweld spacing of every 40 ft, this implies error rates of one in every 132 welds on the older portions and one in every 660 welds for the newer.

Paired with the probability of each feature existing on a hypothetical pipeline segment yields listings such as the following example table.

Estimates are first captured as frequencies rather than probabilities, since the assessment may need to discriminate between high counts—ie, multiple features per unit length—rather than "100% probability of one or more". In other words, a frequency of 7 per mile is different from 12 per mile, but in both cases, an associated estimate of 'probability of one or more per mile' will largely mask this difference. Only one occurrence is sufficient to generate the weakness. Multiple occurrences results in higher probability of a weakness being coincident with a damaging load, but does not increase the amount of weakness.

Table 10.3
Sample Defect Rates and Rates of Introduction

Resistance Issue or location	Rate of Defects being introduced (count/mile-yr)	Current Number of Defects (count/mile)
substandard appurt	0.001	0.01
substandard repairs	0.001	0.01
Pre 1960 repair	0	0.01
Girthweld anomaly	0.2	1
lamination	0	0.01
wrinklebend	0	10
transportation fatigue crack	0	0.
hard spot; arc burn	0.02	0.
Acetylene weld	0	0
dent	0.15	
Mechanical coupling	0	
gouge	0.1	
low toughness from manufacturing	0.008	
low toughness from in-svc phenomena	0.02	

Using units such as 'per mile' for rates of features can help in visualization by an SME. At some point in the process, the frequencies can be converted to probabilities using a reasonable distribution assumption, such as the exponential distribution.

Defect frequencies should include all available evidence including all NDE (for example ILI) indications; history on similar lines; recent research; knowledge of construction and manufacture processes, etc. The estimates are then adjusted based on evidence from subsequent integrity assessments including all NDE and press test. Adjustments should consider the strength of the evidence. Higher PoI is achieved by more robust NDE or higher pressure testing. There is reduced PoI with sub-optimal NDE technique, application, follow-up, etc or lower pressure testing.

Table 10.4
Sample Matrix for Detectability

Defect type	defect size/ configuration	Detectability by Integrity Assessment Method			
		ILI, type 1	ILI type 2	Pressure Test	DA
External metal loss	Category 1: depth a, length b, width c	99%	0.95	99.9%	50%
External metal loss	Category 2: depth x, length y, width z	80%	0.9	5%	
Crack, circumferential	Depth a, length b	60%	80%	99%	
Crack, circumferential	Depth x, length y	20%	75%	2%	
Crack, axial	Depth a, length b	70%	50%	5	
Crack, axial	Depth x, length y	70%	50%		
Dent	type 1	90%	80%		
Dent	type 2	70%	50%		

The modeler chooses the number of defect categories as well as the number of differentiating characteristics of the integrity assessment method. Recall the earlier discussions on detectability sensitivity to specific inspection/assessment characteristics such as conditions and level of expertise. The more robust risk assessments will include all of the inspection accuracy determinants previously discussed. This includes, for ILI, reductions in detectability/characterization of defects are assigned for losses in ILI carrier signals, magnetization, and speed excursions.

While the weakness listings and assignments of effective wall loss are relatively unchanging in a resistance assessment, the 'probability of weakness' is the more variable part of the analyses. It will change routinely, by changes in:

- Integrity assessment and inspection results
- Overline survey results (for example, coating holidays may indicate increased external force incident rates)
- Excavation results, confirming or refuting previous estimates of defects
- Risk assessments—changes in exposure and/or mitigation estimates (for example, new sources of dents)
- New information regarding design, manufacture, or installation

Fortunately, maintenance of this analysis structure is straightforward. Only a few inspection-specific characteristics must be added for each new inspection or integrity assessment. Then, the rate of defect introduction must be reviewed and updated as necessary. With a spreadsheet-based calculation routine, the impacts of these updates carry through the analyses and thereby provide the new estimates of resistance.

10.4.6.5 Effects of Weaknesses

Defects have varying effects on stress-carrying capacity. The equivalent stress at any location depends on component geometry, defect type and size, including damages (metal loss due to corrosion, dents, buckles, etc), support condition, all existing stresses including residual stresses, and knowledge of the original design state. A detailed finite element analysis will best determine the stress state in a component. However, some basic assumptions can be made to allow for a simplified calculation without the use of finite element modeling. The result is less accurate, but is more convenient, reasonably conservative, and of sufficient accuracy for many risk assessment applications.

For example, corrosion damages, metal losses, obviously impact a component's stress carrying capacity including its leak resistance. Internal corrosion is typically very localized and therefore does not typically affect the stress state. In fact, most leaks due to internal corrosion result from 100% wall thickness penetration by metal loss from corrosion (for example, the leak is independent of the stress). By contrast, leaks due to external corrosion, especially corrosion under coatings in buried components and under insulation on aboveground components, typically result from ~80% wall thickness penetration by metal loss, and then the large, thin area fails in tensile overload. In contrast to internal corrosion, the stress state is often affected by the often-larger area of metal loss that results from external corrosion.

Rather than perform finite element analysis for each possible case, it is possible to estimate the worst-case longitudinal bending stress by assuming a large external corrosion metal loss network centered at the 6 o'clock position of the pipe that wraps 1/3 of the circumference and has a uniform metal loss equal to the maximum metal loss. This is a very conservative assumption, because in reality the maximum metal loss is very localized and gradually tapers off toward the edges of the damaged area. Similar worst-case assumptions can be used for how the metal loss network affects the axial stress and the hoop stress. The equivalent stress can be calculated using both the longitudinal stress (axial plus bending) in the corroded condition and the hoop stress in the corroded condition.

10.4.6.6 Assigning Wall Thickness Reductions to Defects

This step is required for time-dependent failure mechanism evaluation and is intended to simplify the understanding and processing of resistance estimates for time-independent failure mechanisms. To some, it may instead be an unnecessary additional step for time-independent forces. If so, it can be eliminated from that part of the risk assessment. We first examine the incentive to include the step for all resistance estimates.

The effective wall thickness is used directly in PoF calculations for degradation mechanisms. It also serves to establish equivalencies among the multitude of possible defect types, sizes, and configurations. The role of a 2% dent with a gouge versus an acetylene girth weld is captured in the risk assessment by assigning an amount of equivalent wall thinning to each. These equivalencies can be used in a very detailed way—actually using the effective wall thickness values in subsequent stress calculations—or in a relative way—using the effective wall thickness values to help assign the general effect on stress carrying capacity and failure fraction. If nothing else, it helps to ground an SME's assignment of final values: 'Mr SME, in general, if we have X % wall loss at this location, how many failures of type Y will now NOT be resisted?' The difference between the damaged and undamaged will be the estimate of resistance. See discussion in next section.

On the other hand, this step is not always needed in time-independent failure mechanisms analyses when the risk assessment directly links weaknesses with changes in failure fraction without the intermediate steps of evaluating the details of which stresses are more impacted and to what extent. To some, a direct estimation of increased failure fraction caused by the 2% dent or the acetylene girth weld is preferable to first producing an equivalent wall thinning. In this case, the intermediate assignment of an equivalent wall thickness reduction is not necessary for the time-independent part of the risk assessment.

Note however that an effective wall thickness is always required to complete the modeling of degradation (time-dependent) failure mechanisms. This may provide incentive to prepare the estimate for all failure mechanisms in the interest of consistency.

When assigning an effective wall thickness estimate, the task need not be a complex, academia-style undertaking. In the absence of publications or specific calculations, it is not unreasonable and often within the accuracy tolerances of a risk assessment for a knowledgeable expert to assign equivalent wall thinning to various weaknesses. The question to be answered is: "what is the equivalent reduction in wall thickness caused by this defect?" In the absence of a full set of calculations, the SME is challenged to estimate that "defect X is equivalent to a Y% reduction in wall thickness". For increased accuracy, he may discriminate among load types, when the defect has significantly differing effects on different loadings. For example, a girth weld defect generally contributes more weakness (increased wall reduction) under an external force loading such as landslide, than it does under the loading from internal pressure. So the effective thinning for external loadings is different than for internal pressure. Assigning different effective wall thinning when exposed to external forces compared to internal pressures allows this discrimination to appear in the assessment.

To account for the varying effects on resistance without a detailed assessment of every possible combination of defect and load/stress, some grouping can be done without excessive loss of accuracy. Short cuts:

- Group defect types—for example, metal loss, crack-like, geometry.
- Reduce number of size/configuration combinations included.

The range of potential weakness scenarios—the various combinations of many factors noted previously—at least somewhat justifies the use of groupings and other simplifications to make the modeling more manageable.

For example, perhaps three categories of load/stress will sufficiently model all possible combinations:

- Resistance Issue or location.
- Potential Strength Reduction (effective wall loss %).

10.4.6.7 Assigning Strength Reduction to Wall Thickness Thinning

The amount of stress carrying capacity available depends on the component's properties and the amount already committed to normal loads. This requires analyses of all loading combinations and resulting primary and secondary stresses created.

Having estimated the equivalent wall thinning caused by potential defects, that thinning effect can be related to increased failure potential. For time-dependent failure mechanisms, the use is intuitive—thinning wall leads to shorter TTF and higher PoF. For time-independent failure mechanisms, it is less intuitive. The full solution is to insert into stress calculations the effective wall thickness, replacing the nominal or measured wall thickness that would otherwise be used. To make this step more manageable, groupings of loads or stresses can be made. The general effect of wall thinning on each grouping can be estimated.

This will then be used with load exposure estimates (PoD estimates, actually) to model changes in failure fraction.

10.4.6.8 Assigning Failure Fraction to Changes in Strength

These strength-reduction values are used with previously estimated PoD values. Each has an assumed distribution of loads—how often loads of various magnitudes are expected. Based on these distributions, the reductions in resistance are modeled to have changes in failure fraction—some loads that could be resisted if there were no weakness will now cause failure, due to the weakness.

In the absence of specific calculations, it is not unreasonable and often within the accuracy tolerances of a risk assessment for a knowledgeable expert to assign general strength reduction values to the previously generated wall thinning estimates. The question to be answered is: "what is the increase in failure fraction caused by this wall thickness reduction, when acted upon by the spectrum of loadings in the exposure estimate?"

Failure fraction is the needed measure of weakness and is equal to 1–resistance. In the absence of a full set of calculations, the SME is challenged to estimate that "a wall thinning of X% results in a Y% increase in failure fraction for the range of loadings expected at this location".

Example: 10.1

The simpler approach is illustrated in the following example:

- Weaknesses are suspected or conservatively assumed.
- An equivalent wall thinning of 20% is estimated based on the frequency and severity of defects known or suspected.

This is assumed to have the following effects on three primary resistance types:

- 20% reduction is hoop stress carrying capacity.
- 10% reduction in longitudinal stress carrying capacity.
- 10% reduction in puncture resistance.

These values are used with previously estimated PoD values for surge and vehicle impact. Surges are resisted by hoop stress capacity and vehicle impacts are modeled to be resisted by longitudinal stress capacity and puncture resistance.

Each has an assumed distribution of loads—how often loads of various magnitudes are expected. Based on these distributions, the reductions in resistance are modeled to have changes in failure fraction—some loads that could be resisted if there were no weakness will now cause failure, due to the weakness.

Table 10.1
Example Assignment of Resistance Changes

Load	Surge	Vehicle Impact
failure fraction if no weakness	0.1/yr	0.05/yr
failure fraction with weakness of type x	0.15/yr	0.07/yr

In this example, some important steps are not detailed here, notably: 1) setting the relationship between wall thinning and loss of stress carrying capability and 2) setting the relationship between reduced stress carrying capacity and increased failure fraction. The example shows that the 20% reduction in hoop stress capacity results in an increase of 0.15 – 0.1 = 0.05 failures/yr. This implies that 0.05 events are of such magnitude that they can no longer be resisted when the 20% hoop stress capacity is lost. Similarly, 0.02 additional failures per year are expected from vehicle impacts due to the loss of 10% in longitudinal stress carrying capacity and 10% loss in puncture resistance.

As will be discussed, these relationships can be very robust and more defensible or, at the other extreme, simply based on SME judgments in order to quickly obtain preliminary risk estimates.

Using the failure fraction with and without the weakness, allows the estimation of a cost benefit calculation of removing the weaknesses. For instance:

- Assume an incident cost at this location: $67K per failure event
- Weakness-induced increase in PoF: 0.07 failure events per year
- Increased risk due to weaknesses 0.07 x $67K = $4,690/yr

- The cost of removal of weaknesses can now be compared to this annual loss exposure. Note that removal of the weaknesses does not change the PoD, only the PoF.

10.5 MANAGEABLE RESISTANCE MODELING

SECTION THUMBNAIL
Changes in stress-carrying-capacity can be modeled in very simple ways or very robust ways, depending on the needs of the risk assessment.

Resistance is a critical aspect of PoF estimation. It is also an inherently complex aspect of the real-world failure potential. The objective is to capture knowledge about loads, stresses, damages, and defects into a resistance estimate using a manageable process. The idea of a 'manageable process' will mean different things to different risk assessors. The trade-offs between modeling complexity and accuracy/defensibility will not appear the same to all. The goal of this section, therefore, is to present modeling options that are all grounded in the same underlying principles, but vary in their level of technical rigor (and, hence, complexity).

Even when available stress carrying capacity can be confidently estimated, perhaps captured in an effective wall thickness, the types of loadings often cannot. For instance, we can know the force required to puncture a certain component but must still estimate the frequencies of scenarios that can cause that amount of force (ie, equipment power, angle of contact, operator reaction, etc.)

As an example of different levels of analysis rigor, consider the common problem of accidental third party mechanical contact with buried pipelines. The simplest approach would be to assume a percentage resistance reflecting an average or worst case (depending on PXX) fraction of damages that would not result in immediate failure. The extremely simple version would use the same fraction for all type of components. This could be improved by adjusting the fraction based on the component's material type, diameter, and wall thickness. It could be further improved by linking the fraction to component characteristics AND several groupings of loadings—for example, excavator hits versus agricultural equipment hits versus vehicle impacts. Vehicle impacts could be further categorized into land vehicles, aircraft, and offshore, including potential ship wrecks impacting a submerged pipeline. Types and speeds of the vehicles will be important also. Additional considerations could continue to be added until, at the other end of the technical robustness spectrum, the result is a FEA type analyses specific to each component, at every location, interacting with all plausible loading (impacts) scenarios.

This choice in modeling rigor should be driven by the intended use of the risk assessment. Initial, higher level risk assessment applications will often have sufficient rigor when driven by simple yet appropriate assumptions. Applications requiring the most technically defensible status will migrate towards the FEA end of the spectrum.

Both the simple and rigorous solutions utilize the same framework so nothing is lost by beginning with the simpler approach (gaining immediately useful answers) and then migrating to increasingly more detailed analyses later.

Following are some examples, illustrating the range of modeling possibilities.

10.5.1 Simple Resistance Approximations

Often a simpler, more approximate solution is sufficient— some loss of rigor is acknowledged and is acceptable. As an initial risk estimate, SME's can assign equivalent wall thicknesses to abnormal loadings and defects using their judgment and experience. These, coupled with similarly estimated corresponding increases in failure fraction, provides the necessary ingredients to perform a preliminary assessment.

The SME will be able to readily distinguish between, say a feature causing a 1% effective wall reduction (negligible impact) and a 50% effective wall reduction (large impact) under a certain set of scenarios. Given the often wide range of possible loadings and other variables, the level of discrimination available solely from SME judgments may be sufficient. Far more discrimination will be available for some defects than others—metal loss is more readily and accurately linked to equivalent wall reduction than a dent with gouge.

Provisions for multiple weakness types and coincident occurrences of weaknesses can also be included, at least conceptually, by SME approximations.

The resulting approximate solutions are immediately valuable since they use all of the important factors, at least approximately, to arrive at risk conclusions. The process will correctly show that a higher incidence rate of more severe defects leads to the higher PoF values and that all combinations of quantities and severities, sometimes of multiple types of defects in close proximity, are included in the assessment. Furthermore, the simplified approach does not encumber attempts to later make the assessment more robust.

10.5.1.1 Time-Dependent

Two time-dependent failure mechanisms are normally included in a risk assessment: corrosion and cracking. As detailed in earlier chapters, each is included in assessments which produce 'wall thickness available' values, after considerations of degradation rates through the component's life, inspection accuracies and timing, and remaining strength calculations for both leak and rupture criteria. As part of this resistance estimation, each 'available wall thickness' is adjusted for possible weaknesses. This adjustment for weakness can be called the wall-adjustment-factor and, when applied to the best estimate of current wall thickness, converts that value into the effective wall thickness. The adjustment factor should reflect the desired level of uncertainty and can be approximated in a simple way, as previously discussed, or in a more rigorous way, as shown in the following section.

The final step in PoF assessment is simple and intuitive for time-dependent failure potential, once the effective wall thickness (including adjustments for weaknesses) is available. The effective wall thickness is directly used with the future degradation rate estimates—mpy internal and external corrosion and mpy cracking—to yield a TTF or remaining life estimate. TTF is then used to generate the PoF estimate.

10.5.1.2 Time-Independent

As in the time-dependent estimation, an 'available wall thickness' can be[7] adjusted for possible weaknesses in the time-independent (random) analysis. This adjustment for weakness, when applied to the best estimate of current wall thickness, converts that value into the effective wall thickness. The effective wall thickness is now used to estimate the ability to resist possible future loads. This relates to the fraction of failures that are avoided due to the strength of the component.

Multiple time independent, random force, failure mechanisms are recognized as load-producing events here. They can be grouped into exposures, for example:

- loads creating hoop stress
- loads creating longitudinal stress
- loads causing puncture
- loads causing buckling.

Each is estimated in terms of events per mile-year. To be considered an event, the load must be sufficient to break the hypothetical component imagined to have no resistance (ie, the beverage can analogy). These estimates are recognized to be point estimates of underlying probability density functions which suggest the range of loadings possible along the pipeline or over time at a single location.

As with time-dependent failure mechanisms, the adjustment factor to capture these possible weaknesses and their effects on PoF should reflect the desired level of uncertainty of the risk assessment. They can be approximated in a simple way, as previously discussed, or in a more rigorous way, as shown in the following section.

7 Recall the earlier observation that an estimate of effective wall thickness is not necessarily an essential step in estimating failure fraction for many time-independent phenomena.

10.5.2 More Detailed Resistance Valuation

In seeking the most defensible and accurate estimate of resistance, the robust risk assessment will embody more resolution and more accurate modeling of resistance. As earlier noted, much of this deeper analysis involves a one-time 'set up' of general relationships that can be universally applied to all pipeline locations. This takes initial effort but then very little on-going maintenance. Only the location-specific data will need to be routinely refreshed for subsequent risk assessments.

Even after a more sophisticated analyses of available resistance, the future frequency of loads with sufficient force to cause failure must still be estimated. As described in previous chapters, these values can arise from a variety of estimation approaches, ranging from simple SME judgments to detailed studies of local equipment availability and planned third party excavation projects, speed and volume of various landslide events and vehicle impacts, moving waters with debris forces, and many others. Some level of uncertainty will remain, even with the most detailed analyses

The following example application illustrates the use of more analyses to better model the interactions of component characteristics and potential weaknesses with specific loadings. A guiding principle of this approach is that, in order to understand the fraction of loads that can be resisted, the load-carrying capacities under various loads must first be quantified.

10.5.2.1 Examples of Weakness Estimations

In a more robust resistance assessment, a more detailed analysis is performed. This analysis includes multiple steps to more fully model resistance.

To begin, each exposure has mitigations identified and estimated in terms of effectiveness. The pairings of exposures with corresponding mitigations yields PoD's for each. These are used with the resistance estimations (about to be detailed) to yield the PoF's for all time-independent failure mechanisms.

The spreadsheet lists the PoD for each threat. It then adds resistance analysis components as follows:

- Listing of weakness types or categories that are possible. Some weaknesses may not significantly impact some threats.
- Quantification of what is known from recent integrity assessments
    - o Measured defect rates (pressure test and ILI results adjusted for age and accuracy to desired PXX level)
        - PoI and age of last pressure test
        - PoI and age of last ILI
        - PoI reductions due to ILI run-specific characteristics
        - Rates of new damages possibly introduced since the assessment
- Quantification of what can be inferred in the absence of integrity assessment information

  - o Estimated defect rates (based on pipe manufacture type and age, era of construction practice, rates of new damages, etc to desired PXX level)
- Today's best estimate of probability of weakness, considering all of the above
- Selection of representative stress types. For example, three stresses may be considered—hoop, longitudinal, and shear—neglecting the influence of axial stresses. Most buried pipelines have few significant stresses beyond those created by internal pressure, but for this example, an unsupported span is being modeled to illustrate a case where additional loads play a role. Standard hoop stress and beam stress formula are used. Yield stress is used as the limiting factor, but ultimate stress would be less conservative (more realistic).
- Baseline estimates of available resistance. This step provides the defect-free stress carrying capacity available for additional loading. This is the difference between stresses already being carried due to normal loadings and the full stress carrying capacity of the component. This answers the question: " after resisting internal pressure, external forces, and any other stresses, how much strength remains for abnormal loads? "
- Selection of representative loading scenarios such as pressure surges, outside force by excavator or landslide, puncture by excavator, etc.
- For each loading type, a resistance is estimated. These estimates are made by first comparing the load-carrying capacity available with the loads that would result in failure. The load-carrying capacity is derived from the maximum stress carrying capacity. The conversion from stress to loads is made in order to more easily estimate the fraction of exceedances that might occur. See further discussion below. Once the available load-carrying capacity is known, the fraction of loads that will exceed this capacity can be estimated.
- Amount of strength loss, per stress type, if weakness is present.
- Probability-adjusted amount of wall loss, for each stress type. The modeler chooses how many stress types to include. Hoop stress and longitudinal would commonly be chosen; puncture, buckling, and others will be included in the more detailed analyses. These values are next used in estimations of failure fractions.
- Amount of increased failure fraction due to the strength loss caused by the presence of the weakness. An increased failure fraction is generated for each pairing of weakness with each stress type.
- These failure fractions are converted into resistance estimates, where failure fraction = (1-resistance), and used to complete the PoF assessment for each threat. Each loading scenario's resistance can now be paired with the previously estimated PoD—exposure x (1 – mitigation). This provides a PoF for each loading produced by each threat.

10.5.2.2 Load-Resistance Estimations

The above process of obtaining 'fractions of failures avoided' as estimates of resistance, warrants further discussion.

Estimating the number of loads that could result in failure can arise from analyses ranging from a robust, technically complete study to simple estimate from engineering judgment. For example, once it is known that the component can withstand a maximum of, say, 544,000 kN force from an excavator, the analyst can research availability of equipment that can produce this type loading and its frequency of use in the area, to estimate the fraction. Or he can simply use his field experience and perhaps a cursory review of published equipment capabilities, to make an initial estimate. Again, different levels of analyses rigor will be warranted depending on the intended use of the risk assessment.

As another illustration, in the case of hoop stress, suppose that the pipe specification used in the Barlow calculation shows that an additional hoop stress of about 14K psi can be tolerated by a component. This is derived from a comparison of the combined existing stresses (created from normal loads) with the maximum yield stress (ultimate stress could alternatively be used). So, an additional load corresponding to a hoop stress of 14K psi can be tolerated. For a certain component configuration, suppose that this equates to an additional internal pressure of 535 psig that can be resisted. An estimate of how often the internal pressure can exceed this value—ie, how often the mitigated exposure will result in 535 psig or more additional internal pressure—yields the fraction of events that will be resisted. HAZOPS, PHA, and other techniques, coupled with physics equations for stresses, are available to quantify the frequency and magnitude of accidental overpressure events, ie, how many events > 535 psig are plausible. Potential scenarios include surges, thermal overpressure (for example, from blocked in above ground portions subjected to daytime heating), and malfunctioning control/safety systems. Again, in the absence of the full HAZOPS type study, an experienced SME can usually produce a reasonable estimate simply based on his knowledge of the system hydraulics.

10.6 HOLE SIZE

Many of the same determinants of resistance also inform the potential hole size created with any load/stress scenario. With an assumption that most risk assessments will be measuring failure as any loss of integrity, hole size becomes an aspect of consequence potential. See discussion in Chapter 11 Consequence of Failure.

11 CONSEQUENCE OF FAILURE

Highlights

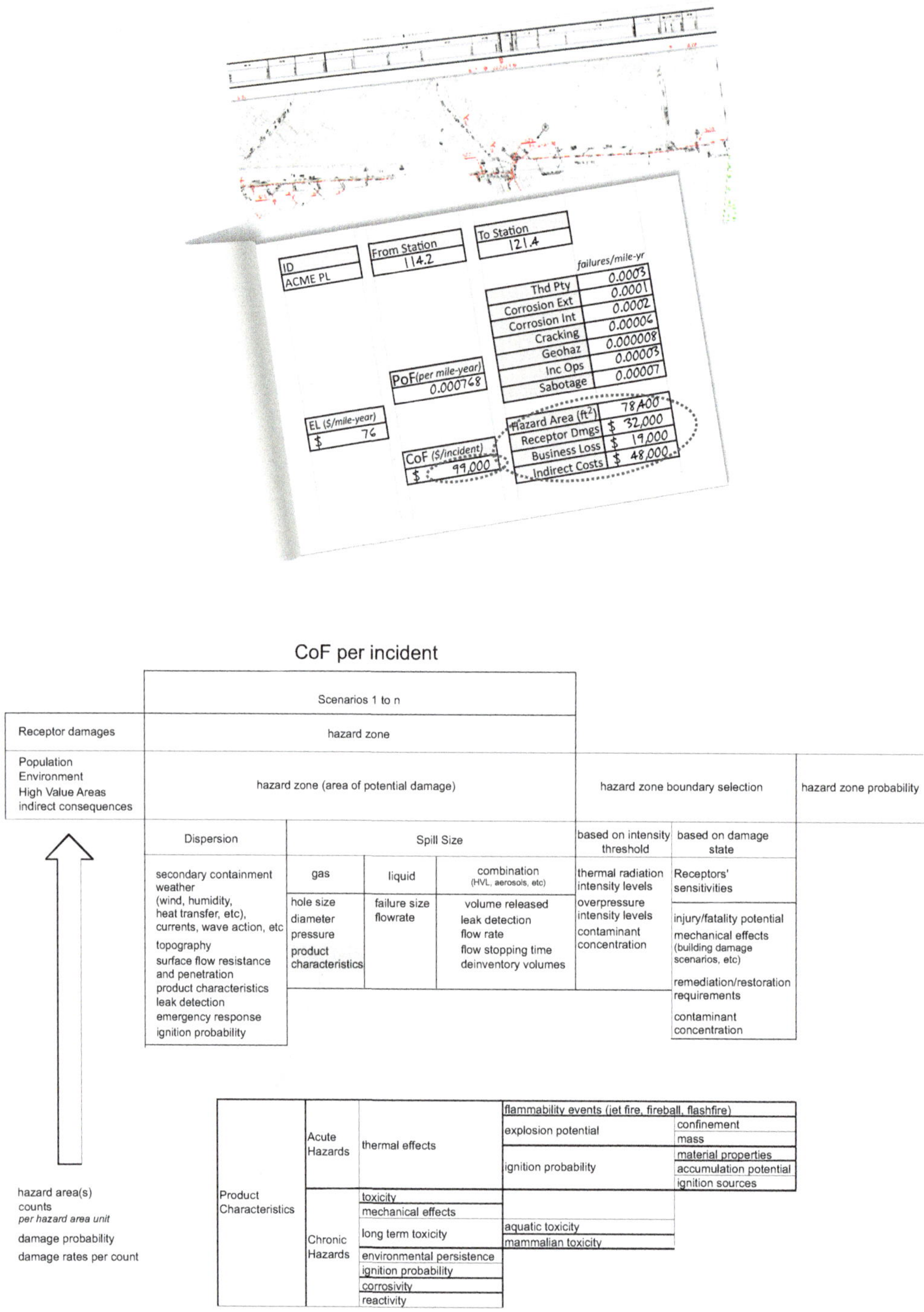

Figure 11.1 Assessing Potential Consequences

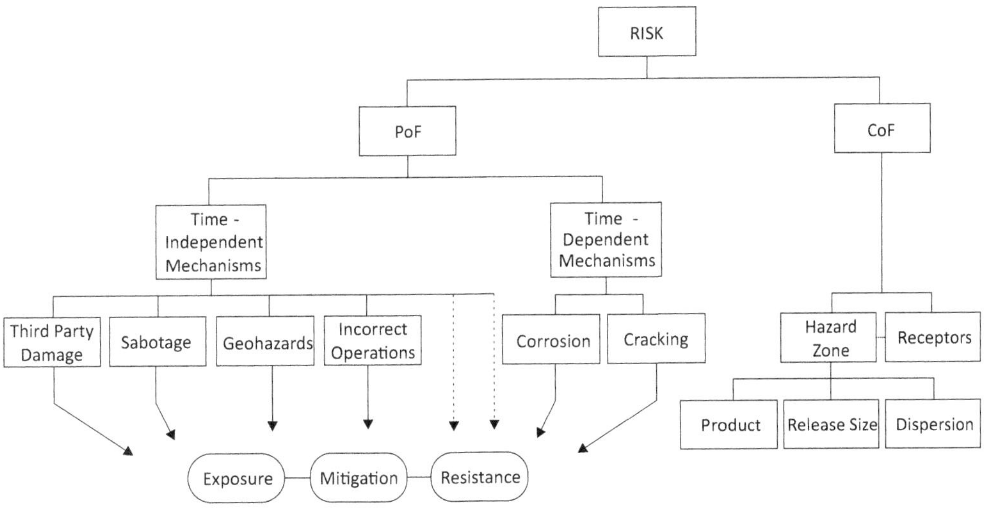

Figure 11.2 Modeling of Pipeline Risk

"There are in nature neither rewards nor punishments — there are consequences."

Robert G. Ingersoll, The Christian Religion An Enquiry

11.1 INTRODUCTION

Risk assessment measures the frequency and/or impacts of some consequence created by some failure. The definition of failure determines the measurement units for consequence.

Once we understand what can go wrong and how likely it is for something to go wrong, the next logical question is 'how bad can this event be?' More specifically: What can be harmed by this pipeline failure? And how badly are 'receptors' likely to be harmed? and other various forms of the question "What are the consequences?" are answered by estimating damages that may occur. When failure is defined as loss of integrity, then the complex and variable interaction between the product transported and the pipeline's environment must be evaluated in terms of damage potential. For example, topography, soil types, vegetation cover, populations nearby, weather conditions etc., are often variable and unpredictable. When they interact with the countless possible leak/rupture scenarios, the problem becomes reasonably solvable only by making assumptions and approximations. Consequences associated with broader definitions of 'failure' add even more complexity since they add to the leak/rupture scenarios.

In a risk assessment, potential consequence estimates are combined with the PoF estimates to arrive at final risk estimates. With failure defined as a leak/rupture (loss of integrity), this full risk assessment approach requires estimates, all along each pipeline, of the following:

1. Probabilities of various spill sizes and dispersion scenarios.
2. Consequences associated with each spill at each possible location
   a. Estimates of hazard zone distances associated with each spill size
   b. Characterization of receptors at various distances from the release
   c. Counts or valuations associated with potential damages to the various receptors.

When estimates from these are combined, the results will represent probability and magnitude of consequences. While this task list is short, producing estimates for each item can be very challenging. Initial chapters of this book focused on the failure potential and this chapter addresses the consequence estimation step.

As with PoF, the designer of the CoF assessment model must strike a balance between complexity and utility—using enough information to capture all meaningful nuances (and satisfy data requirements of all regulatory oversight) but not insisting upon information that adds little value to the analysis. By identifying more critical variables and taking advantage of some modeling conveniences, a methodology structure is offered here as a possible assessment approach that is both manageable and robust enough to be a complete decision-support tool. Initial applications can be completed quickly, although some accuracy will usually be sacrificed with 'short cut' approaches. More robust and more defensible iterations can be subsequently completed by eliminating the short cuts and assumptions initially employed. In other words, the assessment can improve over time, with no change in methodology required.

The recommendations here parallel the robust consequence assessments seen in many QRA's and improve upon assessments typically associated with older scoring or indexing risk assessments. The main enhancements are:

1. Use of hazard zones and their associated probabilities of occurring, as a key ingredient in the assessment.
2. Characterization of receptors and their potential damage rates within hazard zones.
3. Recognition of the range of consequence scenarios, including their respective probabilities of occurrence, rather than basing the assessment solely on a point estimate like 'worst case'.

11.1.1 Terminology

To quantify consequence, a choice of some measurable level of harm or damage is first required. Fatalities or monetized values are common measures. Alternatively, one could choose a generic incident count, for example 'leak', 'failure', etc, or some general effect such as thermal radiation level or overpressure level which in turn implies a certain possible range of damages. This is discussed in Chapter 11.1.5 Measuring Consequence and Chapter 11.2.2 Hazard Area Boundary.

Most pipeline risk assessments will examine the potential for unintended release of the pipeline's contents, even if an expanded definition of 'failure' also brings in other scenarios. In discussion of these events, the terms leak, release, spill, and others are used interchangeably and apply to both liquid and gaseous release events.

11.1.2 Facility Types

The same risk assessment methodology can be used for any pipeline component on any type of pipeline system. Each component can create its own hazard area, even if that area is due solely to a short-distance event such as rapid depressurization. There will generally be more leak/rupture sources in a more complex facility, but also more control, safety, and consequence minimization aspects. Unlike PoF, a larger or more complex facility does not necessarily add to consequence potential. The maximum or average or most likely consequence scenario is usually the most meaningful comparison between facilities (collections of components) so a small, simple facility may have the higher consequence potential. Secondary or sympathetic reactions—one component's failure results in a nearby component's damage or failure—are, however, logically more likely in more complex and larger facilities, adding to those consequence scenarios.

11.1.3 Segmentation/Aggregation

CoF variables are used to generate dynamic segments, just as with the PoF variables. This creates changing CoF values whenever any aspect of CoF changes, from the more obvious changes such as population density, to the less obvious, such as vapor confinement potential. CoF values are typically generated per potential spill/release location. Aggregating risk or failure probabilities for a collection of components, such as 'trap to trap' or all components of a compressor station or tank farm, has many applications. Aggregating consequence values is not generally useful although the maximums and the average or most likely per-incident consequences will be.

11.1.4 A Guiding Equation

The focus here will initially be on integrity—failure as leak/rupture. This is also the initial focus for most pipeline risk assessments: 'failure' as 'loss of integrity', ie an unintentional release of pipeline contents and the possible associated consequences to public health, property, and the environment. Consequences associated with expanded definitions of 'failure' are discussed in the assessment of service interruption.

A leak impact emerges from an analysis of the nature of the product released—its potential hazard(s)—the size of the release, the release dispersion, and the receptor sensitivities.

An interesting high-level view of the leak impact analysis is a simple mathematical formula. The product of four variables essentially determines the magnitude of the impact:

$$RI = PH \times RQ \times D \times R$$

Where

LI	=	Release impact
PH	=	product hazard (toxicity, flammability, etc)
RQ	=	release quantity (quantity of the liquid or vapor release)
D	=	dispersion (spread or range of the release)
R	=	receptors (all things that could be damaged by contact with the release).

While not a unitized and directly employable equation to fully quantify consequences in a modern risk assessment, this is a useful underlying equation to guide the analyses. Since each variable is multiplied by all others, each can independently and radically impact the final consequence. This represents real-world situations. For instance, as noted in PRMM, this equation shows that if any one of the four components is zero, then the consequence (and the risk) is zero. Therefore, if the product is absolutely nonhazardous (including depressurization effects), there is no consequence, and no risk. If the leak volume or dispersion is zero, either because there is no leak or because some type of secondary containment is used, then again there is no risk. Similarly, if there are no receptors (human or environmental or property values) to be endangered from a leak, then there is no risk. Likewise, as each aspect gets higher, the consequence and overall risks will usually also increase.

To reduce consequence potential, any single component can be reduced. While some exceptions can be identified (see later discussions), any directional changes—higher or lower—in any of these four variables will generally forecast the change in consequence potential.

As in the modeling of PoF, this reductionist approach to CoF modeling —breaking the issue to be assessed into its key components—is critical to understanding and managing risk.

A consequence assessment sequence will normally follow these steps for each scenario (or representative set of scenarios):

1. Identify release scenarios
2. Determine damage states of interest
3. Calculate hazard distances associated with damage states of interest
4. Estimate hazard areas based on hazard distances,source (burning pools, vapor cloud centroid, etc.), and location-specific characteristics
5. Characterize receptor vulnerabilities within the hazard areas

Limited modeling resources often requires some short cuts to this process—leading to the use of screening simplifications and detailed analyses at only critical points. Such simplifications and the use of conservative assumptions for modeling convenience are common.

11.1.5 Measuring Consequence

As earlier noted, a unit of measurement for consequence must be chosen. Common choices include:

- Release events, where any unintentional release of product is the consequence and any leak scenario is an event to be counted (or predicted)
- Leaks/ruptures that specifically involve loss of integrity
- Leak size, sometimes categorized by volume of releases so that only leaks of a certain size produce consequence and larger leaks produce greater consequences
- Incidents, with pre-determined definitions, sometimes categorized by type, example: major, significant, minor
- Fatalities
- Injuries
- Costs

Some of these units are based on direct indications of damage, others use implied damages. To say the consequence being measured is 'leak' implies that damage occurs from the leak, even if only loss of product. Categorizations of events goes a step farther in linking incidents to damages. For instance, in the US, PHMSA tracks 'reportable incidents', as a measure of consequence, with a further discrimination into 'serious' and 'significant'. The definitions of 'reportable', as of this writing, include aspects such as:

- Involves death or personal injury requiring hospitalization; or
- Involves fire or explosion; or
- Is 5 barrels or more; or
- Has property damage greater than $50,000; or
- Results in pollution of a body of water; or
- In the judgment of the operator was significant even though it did not meet these criteria.

Consequences are driven by damages to receptors. Quantifying potential damages on a common scale can be challenging. Using a measure such as cost—the monetized loss associated with the damages—forces some difficult judgments to be made among various receptor damages. For example, not only must a value be assigned to human life, but also to various injury types, environmental damage, damage to or extinction of a threatened and endangered species, historical sites, pristine areas, irreparable contamination of a recreational or drinking water source, and any other potential consequence. Some of these valuations involve socio-political and moral/ethical considerations that vary greatly among different cultures, decision-makers, and even over time. Monetizing all potential loss is obviously controversial. However, the ability to express risk in monetary terms is a great advantage in many applications. It is a universally understood 'common denominator' of all loss potential and its use as a measure of risk is quite compelling.

Valuations assigned to certain receptors are discussed in subsequent sections.

11.1.6 Scenarios

A release of pipeline contents can impact a very specific area, determined by a host of pipeline and site characteristics. The size of that impacted area is the subject of this portion of the consequence assessment discussion.

The range of hazard scenarios from loss of integrity of any operating pipeline includes the following:

- Mechanical effects—debris, erosion, washouts, projectiles, etc. and even boat instability offshore, from actions of escaping product.
- Toxicity/asphyxiation—contact toxicity or exclusion of air.
- Contamination pollution—acute and chronic damage to property, flora, fauna, drinking waters, etc. can cause soil, groundwater, surface water, and environmental damages due to spilled product
- Fire/ignition scenarios:
    a. Flame jets—an ignited stream of material leaving a pressurized container creating a long flame .direct flame impingement and/or radiant heat damages are commonly associated with this scenario.
    b. Vapor cloud fire, flash fire; fireball —a cloud of released flammable material encounters an ignition source and causes the entire cloud to combust as air and fuel are drawn together. Where a gaseous fluid is released from a high-pressure vessel engulfed in flames, a special type event is possible. This scenario potentially supports the creation of a large fireball that can arise from boiling liquid expanding vapor explosion (BLEVE) episodes. A BLEVE fireball, while not thought to be a potential event for subsurface pipeline facilities, is normally caused by episodes in which an aboveground vessel, usually engulfed in flames, violently explodes, creating a large fireball (but not blast effects) with the generation of intense radiant heat.
    c. Vapor cloud explosion—occurs when an ignited flammable vapor cloud combusts in a way that leads to detonation and the generation of blast waves. This scenario potentially occurs as a vapor cloud combusts in such a rapid manner that a blast wave is generated. The transition from normal burning in a cloud to a rapid, explosive event is not fully understood. Deflagration—a steady burning of the flammable material-- is the more common event, with flamefront speeds through the cloud not supporting detonation. Under certain conditions, however, the flamefront can accelerate, reaching speeds that support detonation. Confinement is a key de-

terminant of the transition from burning to explosion. A confined vapor cloud explosion is more common than unconfined, but note that even in an atmospheric release, the mixing dynamics of the material in the air, as well as physical barriers such as trees, buildings, terrain, etc., can create partial confinement conditions. An explosive event can generate pressure wave effects as well as associated missiles and high-velocity debris. The damage potentials from vapor cloud explosions have been dramatically demonstrated, but are very difficult to accurately model.

d. Liquid pool fires—an ignited pool of liquid flammable material burns and creates radiant heat hazards.

Naturally, not all of these hazards accompany all pipeline releases. The product being transported is the single largest determinant of hazard type. A water pipeline will often have only the hazard of "mechanical effects." A gasoline pipeline, on the other hand, may carry several of the above hazards.

There is a range of possible outcomes—consequences—associated with these release scenarios. This range can be seen as a distribution of possible consequences; from a minor nuisance leak to a catastrophic event. Even at a single location along a pipeline, the potential scenarios can vary widely. At least a set of representative scenarios must be analyzed in order to understand the possibilities.

Table 11.1 shows some common pipeline products and how the consequences can be modeled. Each of the modeling types are discussed in this chapter.

Table 11.1
Common pipeline products and modeling of consequences

Product	Dominant hazard models
Pressurized, flammable gas (methane, etc.)	Jet fire; thermal radiation, mechanical effects
Toxic gas (chlorine, H_2S, etc.)	Vapor cloud dispersion modeling
Highly volatile liquids (propane, butane, ethylene, etc.)	Vapor cloud dispersion modeling; jet fire; overpressure (blast) event, mechanical effects
Flammable liquid (gasoline, etc.)	Pool fire; contamination, mechanical effects
Relatively nonflammable liquid (diesel, fuel oil, etc.)	Contamination, mechanical effects
Water	Mechanical effects

Additional scenarios are certainly possible. Consider an offshore gas pipeline. A rupture or even leak could threaten a nearby platform or ship's stability as large quantities of escaping gas reach the water surface. With ignition, the scenario is akin to onshore scenarios but perhaps more consequential due to population density and reduced escape potential for the offshore populations (for example, ships, boats, platforms, etc.)

Example: 11.1 Example Scenario for Toxicity and Thermal Effects

The following is an excerpt from a risk assessment conducted on a sour gas (H_2S in natural gas) production well and pipeline network. This excerpt covers only the general description of potential scenarios and an initial basis for frequency estimations (prior to the full risk assessment).

> Accidental releases of sour natural gas from the well/ pipeline network could create potentially life-threatening hazards to persons near the location of the release. Due to the presence of hydrogen sulfide in the natural gas, the vapor cloud created by a release of gas to the atmosphere would be toxic as well as flammable. Persons inhaling air containing toxic hydrogen sulfide vapor could be fatally injured if the combination of hydrogen sulfide concentration and time of exposure exceeds the lethality threshold. If the cloud is ignited, persons in or very near the flammable vapor cloud could be fatally injured by the heat energy released by the fire.
>
> An initial frequency of occurrence of a potential pipeline accident was estimated from historical pipeline failure rate data gathered by the U.S. Department of Transportation. Event trees were then used to estimate the percentage of releases of various sizes that would create a toxic or fire hazard. For example, it was estimated that 50 percent of moderate-sized releases of sour natural gas from the pipeline do not ignite but do create a toxic cloud; 10 percent ignite immediately on release and create a torch fire; and 40 percent ignite after some delay, thus creating a toxic cloud followed by a torch fire.
>
> The frequency of sour gas well blowouts was derived from sour gas well historical data. The largest documented database covers wells in the Province of Alberta, Canada. According to the data, an uncontrolled sour gas well blowout occurs with a frequency of 3.55E–06 blowouts per well per year. This failure rate is for wells equipped with subsurface safety valves.
>
> Computerized consequence models were used to calculate the extent of potentially lethal hazard zones for toxic vapor clouds and/or gas fires created by each potential accident identified. Calculations were repeated for numerous combinations of wind speed and atmospheric stability conditions in order to account for the effects of local weather data.
>
> For each pipeline section or well site, one particular accident will create the largest potentially lethal hazard zone for that section. As an example, one accident is a full rupture of the pipeline without ignition of the flammable cloud, thus resulting in a possible toxic exposure downwind of the release. Under worst case atmospheric conditions, the toxic hazard zone extends 2,600 feet

> from the point of release. Under the worst case conditions, it takes about 11 minutes for the cloud to reach its maximum extent. The hazard "footprint" associated with this event is illustrated in two ways. One method presents the footprint as a "hazard corridor" that extends 2,600 feet on both sides of the pipeline for the entire length. This presentation is misleading since everyone within this corridor cannot be simultaneously exposed to potentially lethal hazards from any single accident. A more realistic illustration of the maximum potential hazard zone along the pipeline is the hazard footprint that would be expected IF a full rupture of the pipeline were to occur, AND the wind is blowing perpendicular to the pipeline at a low speed, AND "worst case" atmospheric conditions exist, AND the vapor cloud does not ignite. The probability of the simultaneous occurrence of these conditions is about 1.87E–07 occurrences/pipeline mile-year, or approximately once in 5,330,000 years for a particular mile of pipeline.
>
> The highest risk along this section of the pipeline network is to persons located immediately above the pipeline. The maximum risk posed by this portion of pipeline is about 5.0E–6 chances of fatality per year. This is for an individual located directly above the pipeline 24 hours per day for 365 days. In other words, an individual in this area of the pipeline network would have one chance in 200,000 of being fatally injured by some release from the pipeline for an entire year, if this individual remained directly above the pipeline for an entire year. An individual in this same area, but located 50 meters from the pipeline, would have about one chance in one million of being fatally injured by a release from the pipeline, if the individual were present at that location for the entire year.

This example excerpt illustrates the types of conclusions often sought by pipeline risk assessments. The risk posed to the population within the appropriate "hazard corridor" for the pipeline/well network can also be presented in the form of graphical tools such as FN curves.

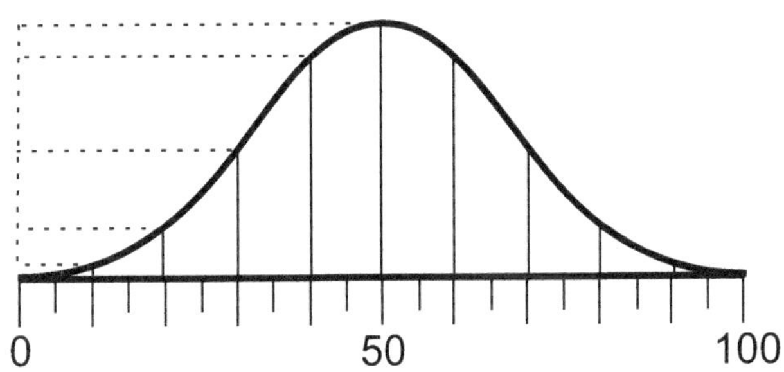

Figure 11.3 Normal Distribution

11.1.7 Distributions Showing Probability of Consequence

As is evident from the previous example, elements of scenario probability must be considered in CoF evaluations. This is a probability aspect beyond those already included in estimating failure event likelihoods.

The variables that are needed to assess consequence potential include specifics of and interactions among receptors, product, spill, and dispersion. Since there are an infinite number of combinations of receptors interacting with an infinite number of spill scenarios, the range of possibilities is literally infinite. So, all consequence estimations will include some simplifications and assumptions in order to make the solution process manageable. Lower level models tend to model only worst case scenarios. Point estimates of the more severe potential consequences are often used as a surrogate for the full distribution of scenario possibilities, downplaying the normally very low probability of such scenarios actually occurring. In reality, the vast majority of possible failure and consequence scenarios do not nearly approach the magnitude of the worst case. The worst case scenario certainly must be understood, and using it, no matter how improbable, as the entire basis of the estimate, may be useful for certain types of risk assessments, but does not convey full understanding of risk.

Higher level models will characterize the range of possibilities, perhaps even producing a distribution to represent all possible CoF scenarios. The full range of possibilities is best viewed as a frequency or probability distribution—distribution graphs show the range of possibilities. Unfortunately, distributions can be cumbersome to work with, especially since these distributions must be understood at all potential spill locations along a pipeline. Since there are innumerable potential spill points along a typical pipeline, this is an impractical approach.

The underlying distributions are more readily assimilated into decision-making when they are approximated by point estimates that capture the range of potential scenarios. If done properly, this simulation of real probability distributions will bound all plausible scenarios and provide better understanding of all events within those bounds. The most useful analysis acknowledges the high-consequence-extremely-improbable scenarios; the low-consequence-higher-probability scenarios, and all variations between. It does this without overstating the influence of either end of the range of possibilities. The use of probabilities ensures that the influences of certain scenarios are not over- or under-impacting the results. All scenarios are considered with appropriate 'weight' for more objective decision support.

11.2 HAZARD ZONES

SECTION THUMBNAIL

The hazard zone approach--estimate the areas potentially impacted and then estimate receptor impacts within—is a critical part of modern consequence assessment.

Seek a 'broadcast' application to efficiently model many miles of pipeline:

Gas release hazard zones can be more generalized, but liquid spills almost always require consideration of local conditions (topography, surface flow resistance, etc)

A modern pipeline risk assessment uses hazard zones in the estimation of consequence potential from leak/rupture[1]. A hazard zone is a geographical area in which certain spill/leak effects are expected. They are often based on the "stress" such as a thermal radiation level or blast overpressure level created by the leak/rupture. Hazard zones will vary in size depending on the scenario (product type, hole size, pressure, etc.) and the environmental conditions (wind, temperature, topography, soil infiltration, etc.).

The simple formula presented earlier is our guideline for conceptualizing hazard zones.

$$RI = PH \times RQ \times D \times R$$

All components are combined to determine consequence and also hazard areas, even though the last term, receptors, initially appears to be independent from hazard areas. Let's examine that premise. Higher intensity from the product hazard, greater release volume, greater dispersion of released product, or increased receptor counts or sensitivities are each able to independently increase consequence potential. If the hazard zone is based on a threshold intensity, then only three of the four factors is needed. The presence of receptors only impacts a hazard zone if the threshold is contingent upon some damage level to a receptor, For example, when a receptor is harmed by a lower airborne concentration of a product, the hazard distance is usually longer. However, receptor damage potential is the reason we define a hazard area, so receptors are never completely de-coupled from hazard area estimates.

The probability of a given hazard area occurring is a function of the probability of the associated scenario occurring. The scenario probability is dependent upon the

1 A risk assessment not focused on leak/rupture may not require hazard area estimations.

probabilities of failure, leak size, product dispersion, ignition, and others. The potential consequences from each scenario are dependent upon the receptors exposed.

A hazard area requires the definition of a hazard extent—at what distance will harm be realized. The effects that define the boundary of a hazard area can be expressed as a level of damage to a receptor—number of fatalities or injuries; fatality rate; dollar damages to property; remediation costs to sensitive environment, etc—or as an effect—overpressure level; thermal radiation; direct flame impingement, etc. These are linked, as is discussed in a following section on hazard zone boundaries. Hazard areas are formed by both acute and chronic releases or by their components within a single release event (see discussion of product hazard). An example of a damage threshold is a thermal radiation (heat flux) level that causes injury or fatality in a certain percentage of humans exposed for a specified period of time. Another example is the overpressure level that causes human injury or specific damage levels to certain kinds of structures.

It is the interaction between the product hazard and the receptor that creates the hazard zone. Recall that a receptor is anything that might be harmed by contact with the release or the effects of the release. Receptors within the defined hazard area must be characterized. All exposure pathways to potential receptors should be considered. Population densities, both permanent and transient (vehicle traffic, time-of-day, day-of-week, and seasonal considerations, etc.); environmental sensitivities; property types; land use; and groundwater are some of the receptors typically characterized. The receptor's vulnerability will often be a function of exposure time, which is a function of the receptor's mobility—that is, its ability to escape the area.

Receptors falling within the hazard zones are considered to be vulnerable to damage from a pipeline release. In the case of a gas release, receptors that lie between the release point and the lower flammable concentration boundary of the cloud may be considered to be susceptible to direct contact with a flame. Receptors that lie between the release point and the explosive damage boundary may additionally be at risk from direct overpressure effects. Receptors within the hazard zone would also be at risk from thermal radiation effects—but not direct contact with a flame—from a jet fire as well as from any secondary fires resulting from the ignition event. In the case of liquid spills, migration of spilled product, thermal radiation from a potential pool fire, and potential contamination could define the hazard zone.

This analysis is efficiently applied to any component in any type of pipeline system. Variations in components' pressure, volume, flowrate, failure mechanism likelihood, etc are expected and appropriately included in the assessment of hazard zone potential.

11.2.1 Conservatism

Because an infinite number of release scenarios—and subsequent hazard zones—are possible, some simplifying assumptions are required. A very unlikely combination of events is often chosen to represent maximum hazard zone distances. The assumptions underlying such event combinations produce very conservative (highly unlikely) scenarios that typically overestimate the actual hazard zone distances. This is done intentionally in order to ensure that hazard zones encompass the vast majority of possible pipeline release scenarios. A further benefit of such conservatism is the increased ability of such estimations to weather close scrutiny and criticism from outside reviewers.

As an example of a conservative hazard zone estimation, the calculations might be based on the distance at which a full pipeline rupture, at maximum operating pressure with subsequent ignition, and with unfavorable weather conditions (ie, promoting increased consequence), could expose receptors to significant thermal damages, plus the additional distance at which blast (overpressure) injuries could occur in the event of a subsequent vapor cloud explosion. The resulting hazard zone would then represent the distances at which damages could occur, but would exceed the actual distances that the vast majority of pipeline release scenarios would impact.

More specifically, the calculations could be first based on conservative assumptions generating distances to the LFL boundary, but then doubling this distance to account for inconsistent mixing, and adding the overpressure distance for a scenario where the ignition and epicenter of the blast occur at the farthest point.

Conservatism in a risk assessment is useful for a number of reasons, as discussed in an early chapter. However, conservatism may also be excessive, leading to inefficient and costly repercussions—in the case of land-use decisions, for example. To supplement the worst case, but normally very rare, release consequence scenario analyses, the more likely scenarios should also be understood. Just as with PoF, a PXX approach to selecting levels of conservatism for CoF estimation are appropriate.

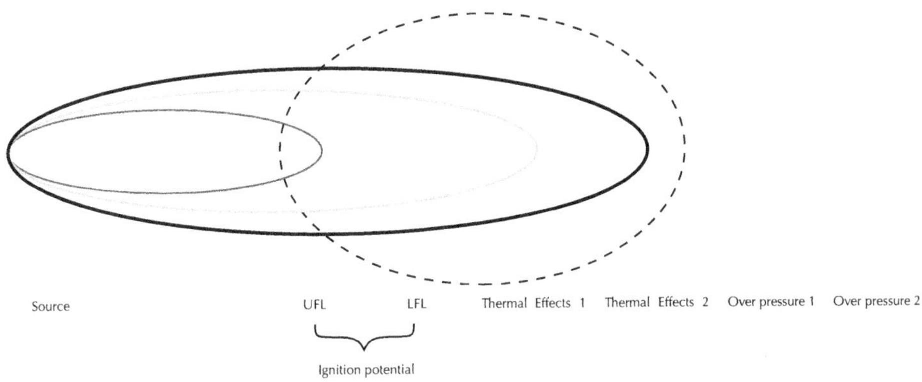

Figure 11.4 Hazard Zones

11.2.2 Hazard Area Boundary

The boundaries of a hazard area must be defined. A boundary can be defined in two general ways: by the intensity of the damaging phenomena or by the effect on the receptor. Each requires the definition of threshold.

11.2.2.1 Thresholds

The intensity of an exposure—heat flux level in the case of thermal events, overpressure level in the case of explosions, concentration or dose in the case of toxicity—can be viewed as a threshold. Similarly, the resulting damage state from intensity of exposure can also be viewed as a threshold. As used here, a threshold is a decision point, a point of interest, a point above which some certain impact is expected or some action will be taken. It is important to recognize that a hazard zone requires an associated threshold—thresholds define hazard boundaries which in turn set hazard zones. A threshold can either directly define the hazard zone—distance to a certain effect—or it can imply a damage state on which the hazard zone is based—10% mortality, if people are present. Speaking of a hazard zone without knowing what threshold is expected at that distance, is not meaningful. The hazard zone's boundary definition must be stated.

11.2.2.2 Intensity Boundary

The most common intensity measures for pipeline failures are concentration levels (contamination,toxicity), thermal radiation (fires), and overpressures levels (blasts). These values are measured/estimated at various distances from a defined source and then used to generate the corresponding hazard areas. The distances are themselves a function of many factors including release rate, release volume, flammability limits, threshold levels of thermal/overpressure effects, product characteristics, and weather conditions.

For example, under a certain set of assumptions, an ignited rupture of a natural gas pipeline might generate a vertical torch fire producing 3 kW/hr/m2 thermal radiation at a distance of 235 ft from the fire (at the rupture location). Perhaps this thermal radiation level is identified as the extent of a certain type of hazard area. Under an assumption of circular effect, the 235 ft becomes a radius generating a hazard area of about 173,500 square feet.

Secondary effects may also define a hazard zone boundary. This includes fires ignited and/or spreading by autoignition from heat flux; delayed explosions such as BLEVE's; soot and ash fallout; pollution; additional hazard effects caused by sympathetic failures/ignitions of nearby equipment, etc.

11.2.2.3 Receptor Impact Boundary

In an alternative approach to threshold definition, the hazard zone boundary can be linked to the specific type of damage, eg 1% fatality rate; third degree burns likely; auto-ignition point for wooden structures, glass shattering, etc. A hazard zone might also be based on potential liquid contamination thresholds that render water sources unfit for consumption or cause defined levels of damage to other sensitive environments.

Defining the hazard zone by the type of harm normally uses the previous intensity estimate. Beginning with that value, an additional step is taken by equating an intensity to the amount of damage a certain receptor will experience when exposed for a certain amount of time. Using the example above, 3 kW-hr/m^2 can cause various levels of harm to human populations exposed for several minutes. So, depending on the level chosen, the previous 173,500 square foot hazard zone can be called, for example, the "second degree burn" hazard zone or the "0.5% mortality rate" hazard zone.

11.2.2.4 PIR Hazard Area Thresholds

As an example of the creation and use of a threshold, consider the equation for natural gas "potential impact radius" (PIR) described in ref [83]. This has been adopted by US regulations and is a mandatory consideration for determining HCA's for US natural gas transmission pipelines. Since countless gas pipeline release scenarios are possible and various types of damage can occur, some choices were made in determining this hazard distance. In ref [83], some of the implicit assumptions used to estimate the PIR include the following:

- Full, guillotine rupture, leak is fed by both open ends of pipe;
- No vapor cloud explosion potential;
- Trench fire (horizontal jet fire) is dominant effect;
- Rapid ignition of escaping gas;
- Effective release rate as a multiple of the peak initial release rate; and
- Heat intensity of 5000 BTU/(hr-ft^2) as the appropriate threshold.

The chosen heat intensity level corresponds to a level below which wooden structures would probably not burn and sheltered persons are not injured. Unsheltered persons would be exposed to a 1% chance of fatality as they seek shelter or distance from the heat.

According to this reference, a level of 5,000 BTU/(hr-ft^2) "...establishes the sustained heat intensity level above which the effects on people and property are consistent with the definition of a high consequence area. Note that in the context of this study, an HCA is defined as the area within which the extent of property damage and the chance of serious or fatal injury would be expected to be significant in the event of a rupture failure" [83]. These assumptions and choices have been deemed appropriate for US gas pipelines by US legislators and regulators.

This illustrates the use of threshold intensities—5,000 BTU/(ft^2-hr)—to establish a damage state based threshold—1% chance of fatality. The threshold intensity is relevant in terms of its expected damage potential and can be used to set geographical boundaries around any pipeline component. Damage requires the presence of receptors. The 1% fatality rate in the above example occurs IF the assumed population is present and exposed as assumed. So, following the setting of the geographical boundaries of the hazard area, receptor counts and characterizations can be made.

Similar PIR hazard zone boundary equations are available as shown in refs [1040] and [1041].

As an example of a similar application, but with expanded consideration of other mortality rates and receptor characteristics, consider the following, excerpted from a risk assessment study:

> The principal hazard range criterion for people exposed to thermal radiation has been taken as the distance from the fire from which there is a Significant Likelihood of Death (SLOD), equivalent to 1800 thermal dose units (tdu) [kW/m^2]4/3s or 12.8 kW/m^2 (3800 Btu/ft^{2}h) exposure for 1 minute. This dose is considered by the UK Health and Safety Executive (HSE) to be equivalent to 50% lethality for normal populations. In calculating the 'escape distance', a lower threshold of 1 kW/m2 (320 Btu/ft^{2}h) was used, to which it is assumed that a person can be exposed to an indefinite period of time without injury. It was further assumed that people who are not inside buildings are able to escape the effects of the fire at a speed of 2.5 m/s (8.2 ft/s). (For "sensitive" populations such as schools, hospitals etc., a more onerous 1% lethality criterion is used with reduced escape speed of 0.7m/s (2.3 ft/s)). The reduced escape speed of 0.7 m/s (2.3 ft/s) is also used for adults at a location where a sensitive population is present as they are assumed to assist the sensitive population to escape.

Here we see additional thresholds used and justified, plus a focus on escape potential. Shielding by clothing, buildings, structures, and specific population demographics are a few of many aspects that could be added as yet additional focus areas. In other similar applications, the PIR is scaled to produce property damage rates. Table 11.2 and Table 11.3 list some PIR formulae for hazard zones based on torch fires and overpressure (explosion) scenarios, respectively. In both listings, r is the distance (in ft for the first table, and miles for the second) from the ignition point where 'significant' damages likely occur; d is the pipe diameter in inches; and p is the pressure in psig.

11.2.2.5 Combined Boundaries

The distinction between the types of thresholds can become blurred as a modeler will often associate a heat-, overpressure-, or toxicity-based intensity threshold with a level of damage to a receptor, and then use the threshold definitions interchangeably. For instance, a heat intensity of X units will result in an estimated Y% mortality of exposed, unshielded populations. When chosen as a threshold, the X units of heat intensity may be referred to as the "1% mortality" threshold. However, preserving the "X units of heat intensity" definition is important since the alternate definition implies that receptors are always present and have certain characteristics regarding shielding clothing, mobility, etc. Losing the original exposure intensity of interest may result in modeling confusion as probabilities of thresholds are integrated with varying receptor characteristics.

Most hazard zone estimates and receptor characterizations are closely intertwined. The former usually embed some assumptions about potential receptors as well as a choice of a damage level for the receptor of interest. The level of damage chosen—1% fatality rate, for instance—sets the effect of interest—thermal radiation level, for instance—which in turn determines the distance to the edge of the hazard zone. All are based on numerous assumptions. Atmospheric conditions, orientation of flame, mobility of populations, shielding, are but a few of the required assumptions for the mortality criteria exampled.

A hazard zone that is to be expressed as a distance from a point on a pipeline is most easily based on some threshold intensity effect, independent of possible receptors. It could alternatively be based directly upon some damage level such as 90% chance of at least one fatality or 50% chance of more than $100K in property damage or any of countless other damage states. However, this would make the distance dependent upon the nearby receptors rather than upon the pipeline alone. Granted, the thresholds are themselves based upon some possible damage state, but keeping that basis indirect allows the threshold to be a function solely of pipeline properties. This makes modeling easier.

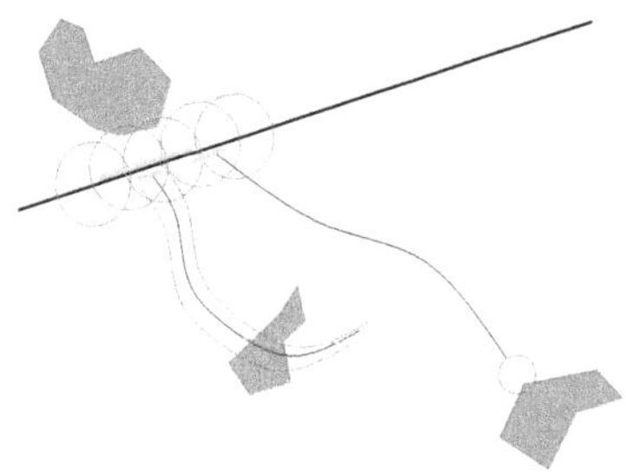

More detailed assessments will use multiple thresholds for each type of impact. For instance, thermal effect thresholds corresponding to third degree burns, first degree burns, and autoignition of wood could be used to set three different hazard distances. Overpressure (blast) levels corresponding to window breakage only, heavy structural damage to wood frame buildings, ear drum rupture, and serious internal injuries could be used to establish yet more. In the case of toxicity, multiple exposure-effect levels (dose) might also be of interest, as noted in the discussion of probits.

Table 11.2
Summary of Potential Impact Radius Formulae [1040]

Product	PIR Formula
Ethylene	$r = 1.04 \cdot \sqrt{p \cdot d^2}$
Hydrogen	$r = 0.47 \cdot \sqrt{p \cdot d^2}$
Natural Gas (Lean)	$r = 0.69 \cdot \sqrt{p \cdot d^2}$
Natural Gas (Rich)	$r = 0.73 \cdot \sqrt{p \cdot d^2}$
Syngas	$r = 0.49 \cdot \sqrt{p \cdot d^2}$ Note 1

Table 11.3
Summary of PIR Formulae [1041]

Product		PIR Formula
Acetylene	1 psi Overpressure	$r = 0.021 \cdot (d^2 \cdot p)^{1/3}$
Anhydrous Ammonia (Liquefied under pressure)	1 psi Overpressure	$r = 0.014 \cdot (d^2 \cdot p)^{1/3}$
	Rural Conditions	$r = 0.08 \cdot (d^2 \cdot p)^{0.48}$
	Urban Conditions	$r = 0.07 \cdot (d^2 \cdot p)^{0.45}$
Carbon Monoxide	1 psi Overpressure	$r = 0.012 \cdot (d^2 \cdot p)^{1/3}$
	Rural Conditions	$r = 0.04 \cdot (d^2 \cdot p)^{0.5}$
	Urban Conditions	$r = 0.03 \cdot (d^2 \cdot p)^{0.45}$
Chlorine	Rural Conditions	$r = 0.38 \cdot (d^2 \cdot p)^{0.49}$
	Urban Conditions	$r = 0.16 \cdot (d^2 \cdot p)^{0.5}$
Ethylene	1 psi Overpressure	$r = 0.021 \cdot (d^2 \cdot p)^{1/3}$
Hydrogen Sulfide	1 psi Overpressure	$r = 0.015 \cdot (d^2 \cdot p)^{1/3}$
	Rural Conditions	$r = 0.37 \cdot (d^2 \cdot p)^{0.45}$
	Urban Conditions	$r = 0.27 \cdot (d^2 \cdot p)^{0.46}$
Methane	1 psi Overpressure	$r = 0.019 \cdot (d^2 \cdot p)^{1/3}$
Rich Gas	1 psi Overpressure	$r = 0.020 \cdot (d^2 \cdot p)^{1/3}$

11.3 PRODUCT HAZARD

One of the primary factors in determining the consequences from a release is the characteristics of the product being transported in the pipeline. It is the product that determines the nature of the hazard.

In studying the impact of a leak, it is useful to make a distinction between acute and chronic hazards. Acute, as used here, means sudden onset, or demanding urgent attention, or of short duration. Hazards such as fire, explosion, or contact toxicity are considered to be acute hazards. They are immediate threats caused by a release.

Chronic means marked by a long duration. A time variable is therefore implied. Hazards such as groundwater contamination, carcinogenicity, and other long-term health effects are considered to be chronic hazards. Many releases to the environment are chronic hazards because they can cause long-term damages perhaps worsening with the passage of time.

The primary difference between acute and chronic hazards is the time element. An immediate hazard, created instantly upon initiation of an event, growing to its worst case level within a few minutes and then improving, is an acute hazard. The hazard that potentially manifests slowly or grows worse with the passage of time is a chronic hazard.

A natural gas release poses mostly an acute hazard. The largest possible gas cloud normally forms immediately (unless confinement occurs), creating a fire/explosion hazard, and then begins to shrink as pipeline pressure decreases. If the cloud does not find an ignition source, the hazard is reduced as the release quickly dissipates and the vapor cloud shrinks. If the natural gas vapors can accumulate inside a building, the hazard may become more severe as time passes—it then becomes a chronic hazard.

The spill of crude oil is more chronic in nature because the potential for ignition and accompanying thermal effects is more remote, but environmental damages are likely, slowing killing plants and and contaminating ever increasing areas.

A gasoline spill contains both chronic and acute hazard characteristics. It is easily ignited, leading to acute thermal damage scenarios, and it is also has the potential to cause short- and long-term environmental damages.

Many products will have some acute hazard characteristics and some chronic hazard characteristics. A product's hazard nature depends on several key aspects such as ignitability, how readily it disperses (the persistence), and its energy content. Some product hazards are almost purely acute in nature, such as natural gas. Others, such as brine, may pose little immediate (acute) threat, but cause environmental harm as a chronic hazard.

Acute ◄·············► Chronic

Immediate threat — Long-term threat

A normally chronic hazard can take on acute consequences. For instance, a leaking hydrocarbon liquid can accumulate in buildings, beneath pavement, etc. and have its flammable vapors confined, concentrated, and ignited—ie, the scenario has worsened with the passage of time.

Many hydrocarbons have both an acute and chronic component to their hazard zone potential. A gasoline and a fuel oil spill of the same quantity may have equivalent contamination potential but the gasoline potentially produces more thermal effects due to its propensity to readily ignite. Determining the release behavior of the type of product transported is a first step in characterizing scenarios. The release categories of liquid, gas, and HVL are useful here.

Gas

- Hazardous vapor releases from products or constituents typically transported in pipelines include

- Natural gas (95%+ methane)
- Ethane
- O_2
- Hydrogen H_2
- Ammonia
- CO_2
- Cl_2
- H_2S
- hydrocarbons

Liquids
- water, potable, non-potable
- brine
- hydrocarbons

Hazardous liquid pipelines typically transport hydrocarbons of various types:
a. Crude oil
b. Refined products
c. Highly volatile liquids

Refined products are liquids such as:
a. Gasoline
b. Diesel
c. Fuel oil
d. Jet fuel
e. Kerosene

Refined products are liquids inside the pipeline and usually remain liquids when released from the pipeline.

Styrene, toluene, benzene

HVL's

Highly volatile fluids are in a liquid state inside the pipeline and gaseous state when outside the pipeline at ambient conditions. Common Highly volatile liquids include:
a. Liquefied petroleum gas (LPG)
b. Natural gas liquid (NGL)
c. Anhydrous ammonia
d. Ethane
e. Propane
f. Butane
g. ISO-butane
h. Ethylene
i. Propylene
j. Butylene
k. Mixtures

LPG is a term used mostly for mixtures of ethane, propane, and/or butane, behaving as an HVL—liquid while pressurized, gaseous when released at ambient conditions.
NGL is a term used mostly for mixtures of ethane, propane, butanes and higher order saturated hydrocarbons that mostly remain in liquid state when released at ambient conditions. [1011]

11.3.1 Acute hazards

A very serious threat from a pipeline is the potential loss of life directly caused by a release of the pipeline contents. This is usually considered to be an acute, immediate hazard. Both gaseous and liquid products pipelines should be assessed in terms of their potential flammability, reactivity (including pressurization, mechanical effects), and toxicity impacts on receptors. This assessment should conclude with a list of acute damages that would potentially be experienced by the receptors of interest. Ultimately, probability-weighted distances will be associated with each damage state of interest.

Toxic, thermal, and mechanical (erosion, debris, and projectiles from violent depressurization or deinventorying) are typical acute hazards. Each of these hazards has a potential to cause varying levels of damage at various distances from the leak/rupture. These damage level-distance combinations are the bases of hazard areas—geographical areas within which certain damage levels could occur.

Damage distances for releases of acutely toxic pipeline contents are most often linked to airborne concentrations causing certain health consequences. Thermal events—fire and explosion-- are normally of prime interest for the hydrocarbon products typically moved by pipelines. The intensity of a thermal event is related to the energy content of the product which is a function of product characteristics like specific heat, heat of combustion Hc (BTU/lb) and boiling point. The boiling point is a readily available property that correlates reasonably well with specific heat ratios and hence burning velocity. This allows relative consequence comparisons since burning velocity is related to fire size, duration, and radiant heat levels (emissive power), for both pool fires and torches. The likelihood of an ignition source is a function of the nearby environment including density of flame sources, likelihood of spark generation, and the type of product.

In the case of water systems, the main product hazard will be related to the more mechanical effects of escaping water. This includes flood, erosion, undermining of structures, and so on. The potential for people to drown as a result of escaping water is another consideration. Oxygen and nitrogen pipelines may similarly only create mechanical hazards. Mechanical impacts will also be important for large storage tanks. Catastrophic failure of a liquid-full, large atmospheric storage tank can cause much damage, even without ignition.

PRMM suggested the use of NFPA ratings for relative assessment of acute hazards. From this acute leak impact consequences model, we could rank the immediate hazard from fire and explosion for the flammable products transported by pipeline and from direct contact for the toxic materials. While the scoring (assignment of points) methodology is no longer appropriate for most of today's risk assessment applications, this analyses provides insight into product behavior upon release.

The acute damage states—the types of receptor harm—potentially created by the pipeline will be used to initially determine the boundary of the hazard area at each potential release point along the pipeline. When the release scenario has a chronic component, a similar exercise of determining potential chronic damage states will also be used in establishing hazard areas.

11.3.1.1 Thermal effects

The possibility of thermal effects—flame and explosion scenarios--from a flammable product released from a pipeline is an important part of most hazard scenarios for hydrocarbon pipelines. Ignition followed by product burning is usually thought to increase consequences, but can also theoretically reduce them. A scenario where immediate ignition causes no damage to receptors but eliminates a contamination potential (preventing groundwater contamination or shoreline damage from an offshore spill, for example) is such a case.

In this section, thermal effects caused by ignited pipeline releases are examined. Terminology, as used in these discussions, is as follows:

- Auto-ignition temperature: A fixed temperature above which material may not require an external ignition source for combustion.
- Flash point: Lowest temperature at which liquid gives enough vapor to form a flammable mixture.
- Fire point: Lowest temperature at which liquid generates enough vapor to maintain a continuous flame.
- Flammability limit: Range of vapor concentration which, when coming in contact with an ignition source, would cause combustion. There are two limits LFL and UFL.
- Explosions: a rapid release of energy causing development of pressure or shock wave.
- Shock wave: An abrupt pressure wave (energy front) generated due to sudden release of energy.
- Blast wave: A shock wave in open air generally followed by strong wind, the combined shock and wind is called blast wave.
- Overpressure: The pressure on an object as a result of an impacting shock wave.
- Deflagration: An rapid combustion in which the flame front moves at a speed less than the speed of the sound in the medium.
- Detonation: An explosion in which the reaction front (energy front) exceeds the speed of the sound in the medium.
- Confined vapor cloud explosion: An explosion in vessel or building. It may be caused due to release of high pressure or chemical energy.

- Vapor cloud explosion: An explosion caused by the instantaneous burning of vapor cloud formed in air due to release of flammable chemical.
- Boiling liquid expanding vapor explosion: Explosion caused due to instantaneous release of large amount of vapor through narrow opening under pressurized conditions.

Direct measurement of thermal acute hazards

Acute hazards general involve fire and explosion effects when contact toxicity is not an issue. In fire scenarios, possible damages extend beyond the actual flame impingement area,. Heat intensity is normally measured as thermal radiation (or heat flux or radiant heat) and is expressed in units of Btu/ft^2-hr or kW/m^2. Certain doses—intensity and duration of exposure—of thermal radiation can cause fatality, injury, and/or property damage, depending on the vulnerability of the receptor.

Explosion intensity is normally characterized by the blast wave, measured as overpressure and expressed in pressure units of psig or kPa.

The level of harm to receptors potentially caused by any form of thermal hazard depends on the distance, shielding, and time of exposure of the receptors.

Ignition probabilities

Ignition is a prerequisite for a thermal event. The consequences of ignition range from a jet or pool fire to a large fireball and detonation.

Ignition probability is, of course, very situation specific. Countless sourcing and timing of ignition scenarios are possible. Ignition can occur at either the source or a location some distance away—a delayed ignition. The source of ignition may be from numerous nearby sources or related to the loss of containment event itself, such as sparks generated by involved excavation machinery or by the release of energy, including static electricity arcing (created from high dry gas velocities), contact sparking from flying debris (e.g., metal to metal, rock to rock, rock to metal), or electric shorts (e.g., movement of overhead power lines).

Common sources of ignition include:

- Vehicles or equipment operating nearby
- Grinding and welding
- Residential pilot lights or other open flames
- External lighting or decorative fixtures (gas or electric).
- Cigarettes
- Engines
- Open flames of any kind

One source cites the following ignition source of major fires.

Source	%	Source	%
Electric	23%	Hot surfaces	7%
Smoking	18%	Flames	7%
Friction	10%	Sparks	5%
Overheated material	8%	Other	22%

These may be relevant to certain scenarios of pipeline leaks/ruptures.

A release that covers a larger area logically has an increased chance of encountering a source of ignition. Ignition can only occur within a susceptible air/fuel mixture, typically found at the edge[2] of a vapor cloud or close to the surface of a pool of flammable liquid. See more discussion of this under Vapor Cloud Ignition.

A buoyant gas such as hydrogen or natural gas will rise rapidly on release and limits the formation of a flammable gas cloud in open space. With the assumption that most ignition sources are at or near ground level, this reduces the probability of remote ignition for these lighter gases. Vapor release orientations other than vertical, accumulation and/or containment, and increasing gas density generally increase the probability of ignition. Higher vapor generation from spilled liquids also lead higher ignition probabilities. The role of gas density in vapor cloud formation supports the presumption that a heavier gas leads to a more cohesive cloud (less dispersion) leading to a higher ignition probability. Confinement of a vapor cloud (caused by topography, proximity to structures, entry into enclosed spaces, etc) also leads to less dispersion and greater opportunity for accumulations within the flammability range, also implying higher ignition probabilities.

Estimates of ignition probabilities can be generated from company experience, pipeline failure databases, or obtained via literature searches. One well-regarded source shows ignition probabilities of natural gas transmission pipelines to be related to pipe diameter by the formula:

$$0.0125 \times \text{diameter}$$

The following empirical formula is recommended for use in quantitative risk assessments for gas pipelines in Australia [67]:

$$\text{Ignition probability} = 0.0156(\text{release rate in kg/s})^{0.642}$$

PRMM discusses several ignition probabilities from various studies, including the use of 12% as the ignition probability of NGL (natural gas liquids, referring to highly volatile liquids such as propane) based on U.S. data. [43]

A conclusion that the overall ignition probability for natural gas pipeline accidents is about 3.2% [95]. nominal natural gas leak ignition probabilities ranging from 3.1 to

2 Of course, the 'edge' is defined by some chosen criteria and tends to grow from the point of origin

7.2% depending on accumulation potential and proximity to structures (confinement), the ignition probabilities for natural gas ruptures ranging from about 4 to 15%

For buried gasoline pipeline leaks/ruptures, ignition probabilities ranging from <1% (rural leak) to >6% (urban rupture) are commonly reported.

Thermal radiation damage levels

Flames from an ignited release of a gas or liquid will normally occur at all points of the spill footprint where the fuel-oxygen mixture promotes combustion. Due to mixing and entrainment of oxygen, this is generally the entire footprint area. Flames are therefore expected initially at a distance equal to the physical extent of the product release—the edge of the cloud or pool.

Adding to this 'direct flame impingement' distance, is the potentially harmful thermal radiation distances arising from the burning. Thermal radiation at any point away from the flame is related to the emissivity and transmissivity.

A US regulatory agency published a guidebook on acceptable separation distances of government housing from explosive and flammable hazards. The guidebook presents a method for calculating a level ground separation distance from pool fires, based on simplified radiation heat flux modeling. Some useful information from this guidebook includes that agency's use of certain thresholds and underlying assumptions:[3]

Ref [83] recommends the use of 5000Btu/hr-ft^2 as a heat intensity threshold for defining a "high consequence area." It is chosen because it corresponds to a level below which:

- Property, as represented by a typical wooden structure would not be expected to burn,
- People located indoors at the time of failure would likely be afforded indefinite protection, and
- People located outdoors at the time of failure would be exposed to a finite but low chance of fatality.

Note that these thermal radiation intensity levels only imply damage states. Actual damages are dependent on the quantity and types of receptors that are potentially exposed to these levels. A preliminary assessment of structures has been performed,

3 U.S. Department of Housing and Urban Development (HUD) published a guidebook in 1987 titled Siting of HUD-Assisted Projects Near Hazardous Facilities: Acceptable Separation Distances from Explosive and Flammable Hazards. The guidebook was developed specifically for implementing the technical requirements of 24 CFR Part 51, Subpart C, of the Code of Federal Regulations. The guidebook presents a method for calculating a level ground separation distance (ASD) from pool fires that is based on simplified radiation heat flux modeling. The ASD is determined using nomographs relating the area of the fire to the following levels of thermal radiation flux

identifying the types of buildings and distances from the pipeline. This information is not yet included in these calculations but will be used in emergency planning.

Jet fire

Direct flame impingement or thermal radiation from a sustained jet or torch fire, is a primary hazard to people,property, and other receptors in the immediate vicinity of a gas pipeline failure.

This scenario is often used as the most likely event in the unlikely case of ignition.

Paradoxically, a long-running brittle pipe failure may produce less thermal consequences under certain circumstances. If the long rupture causes the release to behave more like two or more release points rather than a single, guillotine type release, the differences in fuel source proximities may produce less concentrated thermal damages.

Vapor cloud ignition

A vapor cloud, formed from a pipeline leak or rupture, will be flammable within a specific fuel-to-air ratio range, the vapor cloud will be flammable.

Although ignition is normally not the most probable event there is often a reasonable probability of ignition due to the typically large number of possible ignition sources. Upon ignition, a flame entrains surrounding air and fuel and propagates through the cloud. A fireball and possibly a detonation can occur, generating thermal radiation and shock waves.

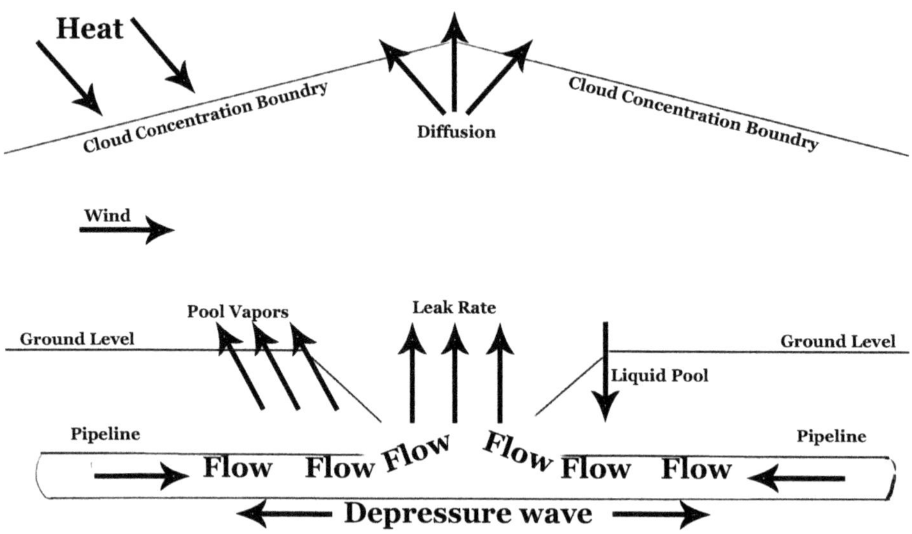

Figure 11.5 Gaseous Release Thermodynamics

Detonation

In rare cases, a vapor cloud ignition can lead to an explosion. This is possible in either a gas pipeline release or liquid pipeline release. In the latter, sufficient vapor generation must occur. In both cases, confinement of the vapor increases the chance of explosion. An explosion involves a detonation and the generation of blast waves.

An vapor cloud explosion occurs when a cloud is ignited and the flame front travels through the cloud quickly enough to generate a shock wave—detonation. This deflagration to detonation transition is possible only under certain conditions. It rarely occurs when the weight of airborne vapor is less than 1000 pounds [83] or when there is no confinement of the vapors.

Expected damages from various levels of overpressure are shown in PRMM

The possibility of vapor cloud explosions is enhanced by any type of confinement, including not only enclosed areas, but also partial enclosures created by topography, trees, buildings, or even weather phenomena. While a confined cloud is more likely to explode, confinement is difficult to accurately model for an open-terrain release. In an atmospheric release trees, buildings, topography, and weather can all add to confinement effects.

Mechanical Effects

The energy contained in pressurized pipeline components can cause damages even when no thermal (ignition) event is involved. This includes debris and pipeline fragments that could become projectiles in the event of a violent pipeline failure. Other mechanical effects associated with violent releases of compressed fluids and gases include product impingements, shock waves, and erosion. Violent depressurization or deinventorying, including tank collapse and pressurized vessel rupture, are typical generators.

Large fragments of ruptured pipelines have not only been unearthed by the force of a rupture, but have subsequently been propelled hundreds of feet from the rupture site. Directional jets and rapid deinventorying can cause erosion, undermining support of nearby structures. Public safety is threatened, as with thermal effects. Environmental and property damages are also potentially involved, but generally in more localized effects compared to thermal events, ie, damage from a single projectile impact, rather than a wide burn radius.

A compressed gas will normally have much more potential energy and hence the greater chance to do debris-related damage, compared to an incompressible fluid. The increased hazard area due solely to the mechanical effects is thought to be usually more limited for a buried pipeline and more extensive for above-ground components.

11.3.1.2 Acute Hazard Minimization

Few mitigative actions are able to reliably and substantially reduce acute hazards as a pipeline leak/rupture event is unfolding. To be effective, a mitigative action must change the characteristics of the emerging hazard zone itself. Secondary containments, fire suppression, quenching a vapor release instantly or otherwise preventing the formation of a hazardous cloud are examples of hazard zone reductions. Subsequent effects associated with acute releases—secondary fires, for instance—can often be reduced. See also the discussions of leak detection, emergency response, secondary containment, and other general CoF reduction opportunities.

11.3.2 Chronic hazard

Another potentially serious leak consequence is the contamination of the environment due to the release of the pipeline contents.

Chronic damage states are often efficiently estimated by concentration levels and associated restoration costs--damage compensations, clean-up/remediation, etc. Concentrations of interest are readily obtained in published materials on mammalian and aquatic toxicities, environmental persistence, and other considerations. Facilitating the practical use of these concentration-to-level-of-harm linkages are environmental regulations which have integrated the available dose-response information and made determinations regarding unacceptable concentration levels under various circumstances. The use of 'reportable quantities' (RQ) in US regulations demonstrates the establishment of unacceptable spill amounts *regulatory references*. RQ's, as supplemented by the addition of hydrocarbonsPPRM, can improve understanding of chronic harm potential.

By-products of a release, and potentially a subsequent thermal event, may include aerosol sprays, soot and ash fallout, or other pollution. Damage payments associated with these are not uncommon. These effects can be considered in either the hazard zone determination or otherwise as a cost of potential consequence scenarios.

The input ultimately sought by the risk assessment will be the actionable contamination extents of the pipeline release and the associated costs of clean-up and remediations for various contamination levels. The contamination extents form the basis of the hazard area or add to the previously estimated acute hazard areas. The hazard area is then used with the clean-up/remediation costs to estimate the consequence potential of the pipeline release.

11.3.2.1 Contamination potential

Estimations of contamination effects are complex, involving many difficult to estimate factors. It is generally not required that all of these parameters be individually linked to levels of harm for purposes of a risk assessment. Such considerations have normally already been synthesized into regulatory definitions of unacceptable concentrations—contamination levels—and mandated amounts of remediations when contamination occurs. It is useful, however, to understand the complexities underlying determinations of 'what constitutes unacceptable levels of contamination'.

The following excerpt from ref [1029] illustrates one approach:

For many substances, the effect of concentration is magnified and, for concentration C and exposure time t, the relevant dose A is given by:

$$A = C^n t$$

Note that the exponent n is not necessarily an integer.

In its regulatory work the UK HSE uses two values of A:

- SLOT (Specified Level Of Toxicity) Dangerous Toxic Load: the dose that results in highly susceptible people being killed and a substantial portion of the exposed population requiring medical attention and severe distress to the remainder exposed. It represents the dose that will result in the onset of fatality for an exposed population (commonly referred to as LD1 or LD1-5)
- SLOD (Significant Likelihood Of Death): is defined as the dose to typically result in 50% fatality (LD50) of an exposed population and is the value typically used for group risk of death calculation onshore.

Values of the SLOT and SLOD for selected materials are shown below. As can be seen in the final column, values of "n" for these materials range from 1 to 4.

Table 11.4
SLOT & SLOD Values for Selected Materials

Substance	SLOT	SLOD	"n"
Ammonia	3.78 × 108	1.09 × 109	2
Carbon monoxide	40125	57000	1
Chlorine	1.08 × 105	4.84 × 105	2
Hydrogen sulphide	2.0 × 1012	1.5 × 1013	4
Sulphur dioxide	4.66 × 106	7.45 × 107	2
Hydrogen fluoride	12000	41000	1
Oxides of nitrogen	96000	6.24 × 105	2

Note: these values are based on concentration in ppm, time in minutes.

Related to this is the use of probit equations to better model dose-response behaviors of exposed populations. This is discussed in a later section.

11.4 LEAK VOLUME

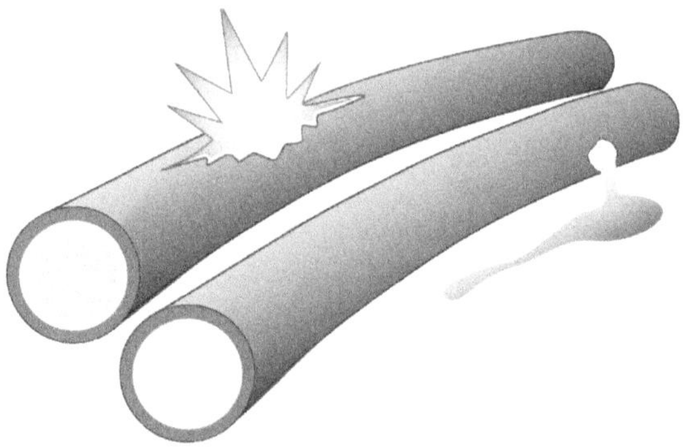

Figure 11.6 Rupture vs Leak

A normal supposition in risk assessment is that larger spill quantities create larger consequences. This will generally be true, but a robust risk assessment will also capture the unusual scenarios where this is not the case. For instance, a smaller total volume and/or small leak rate, contaminating a difficult-to-radiate receptor such as a subterranean aquifer, or accumulating in the basement of a multi-family dwelling, could be far more consequential than many large volume release scenarios.

The most costly small leaks occur below detection levels for long periods of time. Larger leak rates tend to occur under catastrophic failures such as external force (equipment impact, earthquake, etc.), avalanche crack failures, and with shocks to brittle materials, such as graphitized cast iron pipes.

11.4.1 Spill size

A spill or release size in any scenario is a function of many factors such as the failure mechanism, operating conditions, product characteristics, and leak rate. Smaller leak rates can occur due to corrosion (pinholes) or in mechanical connections. The most damaging leaks may be small leaks persisting below detection levels for long periods of time. Larger leak rates tend to occur under catastrophic failures such as external force (for example, equipment impact, ground movement) and avalanche crack failures. Up to the maximum component volume being instantaneously released, almost any size leak is possible.

Potential spill volume is estimated from potential leak rates and leak times.

11.4.2 Hole size

As a worst-case scenario, as well as a means to easily incorporate the intuitive belief that large diameter can mean higher consequence, pipe failures can be modeled as having opening (hole) areas equal to the cross-sectional area of the pipe—a guillotine rupture. This provides a consistent way to compare the maximum hazard zones from equipment of varying sizes and operating pressures. However, a rupture is a very rare event and can lead to over-conservatism and associated misunderstandings of true risk. It will also not recognize the differences that influence hole size and therefore will not 'reward' those components less susceptible to large hole size and 'punish' those that are more susceptible.

Including, in the assessment, the various potential hole sizes adds more robustness and realism to the analysis. The leak size probabilities—derived from hole size and other factors—can offset a consequence potential that would otherwise be modeled as being higher. For example, a smaller diameter line that is more prone to rupture can exceed the consequence potential of a larger line that is vulnerable only to small leaks. So, the larger line may actually carry less consequence potential.

The hole size is related to the failure mode, which in turn is a function of pipe material, stress conditions, and the failure mechanism. Failure modes can be categorized in different ways, such as: pinhole, large holes, ruptures; tearing, cracking, etc. Interrelationships among many factors determine the likely type of pipeline leak/rupture (hole size) for any failure scenario.

One intent of including hole size in estimating consequence potential is to identify components more likely to fail in a catastrophic fashion. Material toughness, including the implications of joints which may have greatly reduced toughness equivalents, is a key determinant of catastrophic failure potential in some scenarios. Where pipe material toughness is constant, changing pipe stress levels or initiating mechanisms will discriminate components more susceptible.

As an extreme example of catastrophic failure mode, an avalanche failure is characterized by rapid crack propagation, sometimes for thousands of feet along a pipeline, which completely opens the pipe, sometimes violently launching fragments into the air. (See discussion under Cracking). A crack will move at the speed of sound through a material. If the crack speed is higher than the depressurization wave—where pressure is the driving force creating the failure stress—then cracking continues. When the depressurization wave passes the cracking location, the driving force is lost and cracking halts.

Product compressibility and the level of pressurization play a role in crack length. Less compressible products can have relatively fast depressurization speeds. In other words, on initiation of the leak, the pipeline depressures quickly with an incompressible fluid. This means that usually insufficient energy is remaining at the failure point to support continued crack propagation.

A compressed gas, due to the higher energy potential of the compressible fluid, can promote significantly larger crack growth and, consequently, leak size. This is because the stored energy in a compressed fluid is relatively slow to release, allowing continued pressure on a crack that is opening.

Material toughness and thickness can each reduce crack speed. Crack arrestors take advantage of this. A crack arrestor is designed to slow the crack propagation sufficiently to allow the depressurization wave to pass. Once past the crack area, the reduced pressure can no longer drive crack growth. A more ductile or thicker material (stress levels are reduced as wall thickness increases), sometimes used intermittently along a pipeline, can act as a crack arrestor.

Given this model of crack growth, main contributing factors to an avalanche failure include low material toughness (a more brittle material that allows crack formation and growth), high stress level in the pipe wall (especially when at the base of a crack), and an energy source that can sustain rapid crack growth (usually a gas compressed under high pressure).

A hole size probability distribution can be generalized from research and/or an examination of past releases. This provides insight into what hole sizes have more often been associated with what types of failure mechanisms and pipeline characteristics—ie, incident frequencies typically show corrosion causing smaller holes and mechanical damage causing larger.

While useful as a calibration tool for populations of components, care should be taken to ensure that a statistical analysis does not introduce an inappropriate bias into assessing the spill size for a specific scenario. The subject pipeline being assessed may behave in ways drastically different from the population underlying the summary statistics.

11.4.2.1 Component Materials

Material types and their various failure modes are important aspects of a risk analysis and contribute to the PoF (exposure, mitigation, resistance) and CoF assessments. While especially important in addressing the widely different materials often encountered in an older distribution systems, for example, it is also useful in addressing more subtle differences in pipelines of basically the same material but operated under different conditions. For example, a higher strength steel pipeline may have slightly less ductility than Grade B steel and, when combined with factors such as changing stress levels and crack initiators, this raises the likelihood of an avalanche-type line break.

An important difference lies in materials that are inherently prone to more consequential failure modes. A large leak area is often created by the action of a crack in the pipe wall. A crack is more likely to activate in a higher stress environment and is more able to propagate in a brittle material; that is, a brittle pipe material is more likely to fail in a fashion that creates a large leak area—equal to or greater than the pipe cross-sectional area.

11.4.2.2 Stresses

Material stress levels in a component are a main determinant in the probability of a larger hole size. Stress is often expressed as a fraction of SMYS. For many years, 30% SMYS has been used as a discrimination point between leak and rupture. This level changes as defect size increases, with large defects susceptible to generating large failure areas at low stress levels. This is not a hard rule however. While rare, ruptures at lower stress levels have also been documented.

11.4.2.3 Initiating mechanisms

The role of initiating mechanisms in failure potential is discussed in Chapter 10 Resistance Modeling. Their role in influencing hole size is briefly noted here.

Shorter defects under less stress tend to fail as leaks. As defects get longer and stresses increase, rupture becomes more likely. Weld seam anomalies, which can be relatively long, often fail as ruptures.

Damage type is another consideration a failure mechanism such as corrosion is often characterized by a slow removal of metal and is often modeled as producing smaller leak sites, whereas cracking and third-party damage initiators often have a relatively higher chance of leading to large opening.

11.4.3 Release models

Having determined the failure hole sizes to be used in the risk assessment, release scenarios can now be modeled. Several hazard area mechanisms—the underlying processes that create the dispersion or hazard zone area—have been identified. The hazard area for a gas release is established through either a jet fire or a vapor cloud. The hazard zone for a liquid release arises from either a pool fire or a contamination scenario. HVL hazard zones can arise from a combination of these mechanisms.

As noted, some leak/rupture scenarios are more sensitive to release rate, while others are more sensitive to total volume released. The rate of release is the dominant mechanism for most short-term thermal damage potential scenarios, whereas the volume of release is the dominant mechanism for many contamination-potential scenarios. Based on the expected potential hazards, consequences from gas releases are more often leak rate dependent. In a liquid spill, hazards are pool fire and contamination potential, so the spill volume is the critical determinant. Differences between and among these types of scenarios determines the potential consequences.

Potential leak rate and volume is dependent upon factors such as product characteristics, pressure, flowrate, hole size, system hydraulics, and the reliability and reaction times of safety equipment and pipeline personnel.

Leaks of gaseous products are driven primarily by hole size, pressure, and gas density.

Liquid leaks are more influenced by hole size, flowrate, and gravity effects. Because the release of a relatively small volume of an incompressible liquid can depressure the pipeline quickly, the longer term driving force to feed the leak may be pumping equipment or gravity and siphoning effects. A leak in a low-lying area may be fed for some time by the draining of the rest of the pipeline, so the evaluator should find the worst case leak location for the section being assessed. The leak rate should include product flow from pumping equipment. Reliability of pump shutdown following a pipeline failure is considered elsewhere.

There are more opportunities for consequence mitigation in V2 dominated scenarios, as is discussed in a later section. While actually consequence mitigation measures, leak detection and component isolation are inextricably linked to spill volumes are therefore covered here and again under mitigation. But first, an examination of hole size as a key determinant of leak rate.

Flow halt time and drain volume are often the determining factors for liquid releases and orifice flow to atmosphere (sonic velocity) determines vapor release rates. In simplest terms, low spots on large-diameter, high–flow-rate pipelines can be the sites of largest potential spills and larger diameter, higher pressure gas pipeline mains can generally cause greater releases.

Leak rates (V2) are typically determined via well established orifice flow equations. Leak volume (V1) determinations use these leak rates (V2), plus time to halt flow and deinventorying volumes.

11.5 DISPERSION

Dispersion is often the initial determining factor of a hazard zone. As noted, however, hazard area can extend beyond the physical movement of leaked product when thermal and explosion effects are included. Toxic and asphyxiate characteristics of some clouds will be pertinent to most risk assessments. Flammability is the more common hazard associated with pipelined gases and HVL's.

In most modern risk assessments, some type of release and dispersion modeling will need to be performed to understand distances at which possible intensities occur. This can be as simple as the application of an equation with only two variables, such as that for PIR of natural gas pipelines (only diameter and pressure are needed) or as rigorous as a vapor cloud dispersion or particle trace analysis requiring dozens of inputs at each potential spill location.

Software solutions range from simple calculations to assist first responders, up to extremely sophisticated and expensive models.

11.5.1 Hazardous vapor releases

The flammable and toxic limits of interest generally define the gas cloud boundaries. Upon ignition of a flammable cloud, the thermal effects may initially extend beyond the cloud boundaries—a fireball—and then retreat back to the source as the cloud is consumed, finally becoming a jet fire.

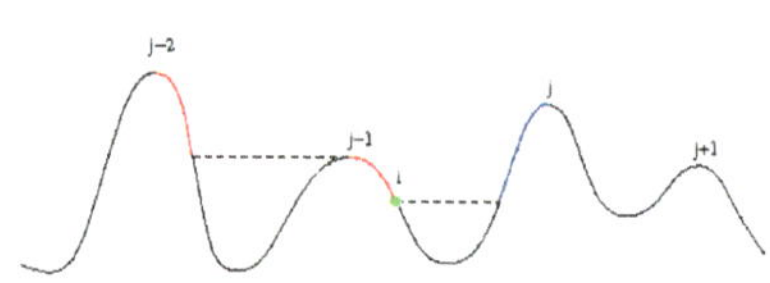

An accepted approach to modeling jet fire releases simplifies the calculation complexities associated with estimating release quantities by using pressure and diameter as proxies for the release quantities. Using a fixed damage threshold, it has been demonstrated that the extent of the threat from a burning release of gas can be modeled to be proportional to pressure and diameter [83]. Therefore, pressure and diameter are suitable variables for assessing at least one critical aspect of the potential consequences from a gas release.

Because the immediate hazards from vapor releases are mostly influenced by leak rate, leak detection will not normally play a large role in risk reduction. One notable exception is a scenario where leak detection could minimize vapor accumulation in a confined space.

11.5.1.1 Vapor cloud size

The release of a gaseous pipeline product creates a vapor cloud. The extent and cohesiveness of a vapor cloud are critical parameters in determining possible threats from that cloud. A vapor cloud that envelopes more near-ground surface area has a greater area of opportunity to find an ignition source or to harm living creatures. This should be reflected in the risk assessment. The cloud boundary is typically defined by some concentration of the vapor mixed with air.

A flammable limit is often chosen as a cloud boundary threshold for hydrocarbon gases. The use of the lower flammability limit—the minimum concentration of gas that will support combustion—is the most common cloud boundary. It conservatively represents the maximum distance from the leak site where ignition could occur. Sometimes 1/2 of the LFL is used to allow for uneven mixing and the effects of random cloud movements. This lower concentration creates a larger cloud.

In the case of a toxic gas, the cloud boundary must be defined in terms of toxic concentrations. These might exceed thermal hazard distances. For instance, unignited sour gas (hydrogen sulfide, H_2S) releases have been estimated to cause potential hazard zones 4 to 17 times greater than from an ignited release [95].

Sophisticated dispersion studies have revealed a few simplifying truths that can be used to better understand cloud size. In general, the rate of vapor generation rather than the total volume of released vapor is a more important determinant of the cloud size. Due to a cloud reaching an equilibrius with the atmosphere, release duration—total release volume—is not as critical in estimating maximum cloud size as is release rate. Released product balances the product dispersing at the cloud boundaries, resulting in a relatively stable cloud size. The release rate will normally diminish quickly as the pipeline rapidly depressures under a pipeline rupture scenario, which is normally the more interesting cloud-generating event.

Cloud stability and hence, size are significantly influenced by meteorological conditions. Conditions that favor mixing and more rapid dispersion minimize cloud size while more stable atmospheric conditions support a more stable and larger cloud. Meteorological conditions are often categorized into stability classes for purposes of dispersion modeling. Each stability class represents some fraction of possible weather type days for a specific location in any year. Under very favorable conditions, unignited cloud drift may lead to extended hazard zone distances.

11.5.2 Liquid spill dispersion

11.5.2.1 Physical extent of spill

The physical extent of a liquid spill is highly variable and on the spill rate and duration including drain effects, the type of product spilled, and the characteristics of the spill site. The first two of these are known or readily estimable at all locations along a pipeline. The third, spill site characteristics, will usually be the information most challenging to obtain and integrate into the risk assessment.

Pipeline pressure is not a main determinant in liquid spill volume since the product is assumed to be relatively incompressible. Except for a scenario involving spray of liquids, the potential damage area is not thought to be very dependent on pressure in any other regard.

Some form of at least rudimentary site-specific analyses will be required to properly assess liquid spill characteristics. A range of options in analysis rigor is available, GIS based models that generate spill footprints along a pipeline are commonly used tools given the increased availability of powerful computing environments and information (for example, soils, topography, surface resistance, groundwater depth, etc) in electronic databases. Such models vary in complexity, with the more robust taking into account all of the spill-determining characteristics noted previously.

11.5.2.2 Spills onto Soil

References detailed in PRMM can then be used to assess the soil permeability for liquid spills into soil. This implies that more or faster liquid movements into the soil increase the range of the spill. Of course, greater soil penetration will decrease surface flows and vice versa. Either surface or subsurface flow might be the main determinant of contamination area, depending on site-specific conditions. When groundwater contamination is the greater perceived threat, the risk assessment should show greater consequences with increasing soil permeability.

The soil permeability is normally used with an accompanying assumption that larger volumes, spilled in a higher permeability soil, lead to proportionally greater consequence areas. A low-penetration soil promotes a wider spill-surface area and hence places additional laterally-located receptors at risk. A spill of a more acutely hazardous product might generate less consequence if accompanied by greater soil penetration and reduced lateral spread and/or ignition probability.

Table 11.5
Soil permeability

Description	Permeability (cm/sec)
Impervious barrier	0
Clay, compact till, unfractured rock	$<10^{-7}$
Silt, silty clay, loess, clay loams, sandstone	10^{-5}–10^{-7}
Fine sand, silty sand, moderately fractured rock	10^{-3}–10^{-5}
Gravel, sand, highly fractured rock	$>10^{-3}$

Ultimately, an assessment of the spilled substance's hazards and persistence (considering biodegradation, hydrolysis, and photolysis) will be needed in evaluating the consequences of a liquid spill.

Subsurface water contact is also an important aspect of liquid spills.

One source [1014] notes a simple equation for determining spill pool diameter. The US HUD guidelines [1012] and the SFPE Handbook [1013] discuss methods of estimating the diameter of an unconfined spill fire.

A simple method of obtaining a spill diameter is:

D 10sqrt V

Where D is in meters and V is in cubic meters 1

This equation asserts that the liquid will continue to spread until it is about 1 cm in depth.

In using any simplified approach, assumptions must be made regarding rate of penetration into the soil, evaporation, and other considerations.

11.5.2.3 Spills on Water

Spills into water should take into account the miscibility of the substance with water and the water movement. A spill of immiscible material into stagnant water would be the equivalent of a spill on flat terrain with impermeable soil. A highly miscible material spilled into a flowing stream results in widespread dispersion.

For the more persistent liquid spills, including oil, mixing and transport phenomena should be considered.

For subsea gas releases, a common assumption is that the diameter of the plume at the sea surface is 20% of the water depth at the release point, regardless of the gas flow rate. This diameter together with the gas flow rate can then be used as input to a plume model.

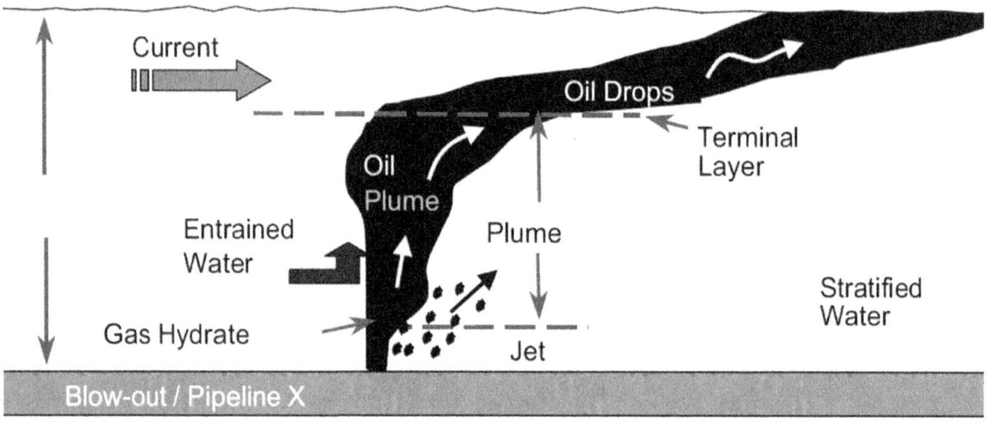

Figure 11.7 Oil Release Into Water

11.5.3 Highly volatile liquid releases

HVL releases involve characteristics of both gas and liquid releases. Since multiphase fluids are involved. Material released under flashing conditions is a complex nonlinear, non-equilibrium process that is difficult to model.

As with liquids, the initial release rate will usually be the highest rate of the event, and then rapidly decrease. Inside the pipe, the depressurization wave from a rupture moves from the rupture site and pressures inside the pipeline quickly drop to the product's vapor pressure. At vapor pressure, the pipeline contents will vaporize (boil), generating quantities of vapor that emerge from the leak.

Outside the pipe, flashing liquids will initially emerge and a gas cloud will be formed, including immediately flashing material, the vapor generation from a liquid pool, and the evaporation of airborne droplets from any aerosol phase release components.

After the immediate depressurization from the leak event, the scenario will unfold as a dense gas release. Release characteristics are then similar in many respects to pure vapor release scenario.

11.5.4 Distance From Leak Site

A hazard area may ***originate*** some distance from the point of pipeline failure. Envision a sloped topography where the spilled liquid will accumulate some distance from the leak site or the accumulation of natural gas into a basement, following migration through the soil from the source of a minor leak.

In the case of delayed or no ignition, the product will usually have migrated some distance prior to ignition (unless the ignition source moves into the leak source area). This moves the origination point for the thermal effects. The cloud centroid or liquid pool center then become the point from which the thermal hazard zone extends. The thermal effect can also move back towards the leak site as the 'trail' of combustible spilled product is consumed. This creates a hazard zone along the 'trail'.

A receptor can be very close to a leak site and not suffer any damages, depending on variables such as wind strength and direction, topography, or the presence of barriers, while areas farther away are damaged. Scenarios envisioned include a liquid spill where a ditch or sewer catches and moves the spilled product away from the leak; or an HVL 'puff' release where the cloud, fully decoupled from any other vapors escaping from the pipeline, drifts some distance before finding an ignition source. These scenarios are challenging to model and require location-specific analyses. Including the migration possibility without the decoupling-from-the-source possibility produces larger (more conservative) hazard zones.

Making a distinction between the path and the event centroid is useful. Centroid is used to refer to the center from which thermal or overpressure effects are emerging. In the absence of some type of dispersion modeling, the path is often set to zero distance, making the centroid coincident with the spill site (on top of the pipe). This is a convenient way to model, but will miss-characterize damage potential when, for instance, scenarios like those described above occur.

For general consequence assessment, the recommendation is to simply add the migration distances to the hazard zone distances. While this inflates the hazard zone distances for many scenarios, it also captures the scenarios where the hazard zone is actually enlarged by the migration path of material that can combust or contaminate.

In the case of liquid spills, the distance estimate should consider topography, surface flow resistance, permeability, and other factors making these scenarios more location-specific and difficult to model. Where the topography is relatively consistent, some 'rules' can be developed to facilitate assessment, adjusting estimates only when certain changes are encountered. For example, a hazard area can be based on a predominant topography—say, 'prairie' or 'level pasture'—and, where the pipeline crosses a ditch or stream of certain characteristics, a different set of assumptions creates a different hazard zone.

In the case of HVL's and gas releases, the hazard zone should also consider meteorology. This is generally stable over long stretches of pipeline, but conceivably can cause modeling complications in scenarios where weather patterns change over short distances. Examples include canyons, intermittent forest cover, buildings, coastal regions, and perhaps even shielded (from wind) versus unshielded locations where 'confinement' increases the ignition and/or explosion potential of a vapor cloud.

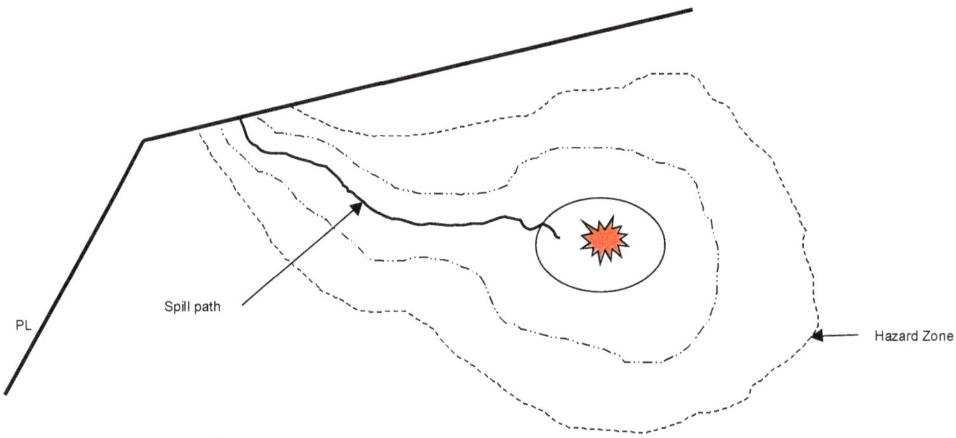

Figure 11.8 Spill Migration with Subsequent Ignition

11.5.5 Accumulation and Confinement

As noted previously, confinement and accumulation of release flammable products generally increases the potential for both ignition and explosion. In an urban environment, the confinement/accumulation potential is greater because product can migrate for long distances under pavement, route through adjacent conduits (sewer, water lines, etc.), permeable soils, or find other pathways to enter buildings intended for human occupancy.

Similar scenarios may also emerge in rural areas, involving gas or HVL transmission pipelines.

Gas cloud confinement potential in both urban and rural areas was previously noted and can dramatically increase damage potential when detonation events are triggered.

11.6 HAZARD ZONE ESTIMATION

We again turn to our simple, summary equation of consequence estimation:

Release Impact (RI) = product hazard (PH) x Release Quantity (RQ) x dispersion (D) x receptors (R)

Hazard zones based on threshold intensities such as heat, overpressure, and toxicity/contamination are a function of the first three factors, which can be grouped into just two general sets of release conditions:

- Pipeline / product characteristics
- Dispersion potential

o Topography effects if liquid release
o Meteorology effects if gaseous release

Product characteristics are grouped with pipeline characteristics since the operating conditions—pressure, temperature, flowrate—will influence how the product behaves when released.

As previously noted, thresholds based on a receptor effect or damage state, such as fatality, injury, property damage, environmental harm, require the above plus another:

• Receptor proximities and characteristics

A countless number of hazard distances can be created from possible failure scenarios of most hydrocarbon pipelines. The range of scenarios used to evaluate hazard zones is narrower when the receptor characterizations are separated from the threshold definitions. For instance, initially avoiding the complexity of approximating population density, shielding, mobility, and potential exposure times reduces the number of permutations required to estimate a hazard zone. Hazard zone estimation can therefore efficiently begin using only the factors that establish threshold intensity distances. These are primarily the pipeline and product characteristics and dispersion potential. Then, receptor characterizations can be later added to the analysis.

One modeling objective is to establish hazard zone distances in a way that the same distance can apply to large stretches of pipeline. This allows for efficient and consistent characterization of receptors within hazard zones.

Three aspects of hazard zones should be considered in building a simplifying model: distance from event; the threshold of interest; and probability of the threshold appearing at a certain distance. The goal is to model a manageable number of scenarios ***while ensuring that*** the chosen scenarios represent the full range of possibilities.

Hazard zones should represent reasonable assumptions and capture the logical premise that damage severity—thresholds—will normally decrease as distance from the event increases. When establishing threshold zones, the modeler should keep in mind that actual intensities of thermal events—normally the events of most interest—are in fact usually proportional to the *square* of the distance. Therefore, potential damages will normally drop very dramatically with increasing distance. See transmissivity / emissivity discussions. Contamination potential can often be assumed to decrease with increasing distance since dilution, absorption, evaporation, etc. have more opportunity to reduce contaminant levels after the spill has moved some distance overland. The rate of drop in damage potential with increasing distance might be receptor- or threshold-dependent.

As a further simplifying opportunity, expressing a hazard zone threshold as a fraction of the theoretical maximum hazard distance might improve modeling efficiency. The underlying assumption is that a certain percentage of the maximum hazard zone produces a certain threshold. For instance, the first 10% of the maximum hazard zone may be assumed to produce a high probability of fatalities and 100% property destruction; between 10% and 60% of maximum hazard zone produces no fatalities—injuries only, and 50% property destruction; etc.

The probability of the hazard distance and the probability of various damages states are both captured in the probability number assigned to the distance. So, a hazard zone distance of 1000 ft with a 1% probability embodies the belief that there is only a 1% chance of a threshold extending this far, and, if it does reach this distance, damages will only be 1% of what they would be immediately adjacent to the centroid.

In this suggested approach, some liberties with measurement units are taken. Probabilities of occurrence are combined with possible distances to thresholds and expressed as distance. Probabilities can be represent either the chance of a hazard zone occurring or the probability of a certain damage state, given the manifestation of the hazard zone. Mathematically, the two are treated as identical. Given the high levels of uncertainty and variability in possibilities, such liberties and simultaneous representations are not unreasonable.

It may be assumed that contamination areas are encompassed by the thermal effects or, alternatively, a separate contamination assessment can be performed.

Mechanical effects hazard zones can be estimated via analyses of underlying phenomena such as product release forces, impingement forces, projectile trajectories, and submerged gas releases (ship instability due to offshore gas pipeline rupture).

11.6.1 Hazard zone calculations

With an understanding of the potential hazards generated by a product plus the dispersion characteristics of product release scenarios, hazard zones can now be estimated in order to characterize the receptors that might be vulnerable to a pipeline release. The hazard zone, as previously defined, is the physical area in which receptor damage is possible.

As noted earlier, thermal, toxic, and mechanical hazards are potentially produced from unintended releases of products typically transported in pipelines. Thermal effects are the dominant threat in many hydrocarbon releases. Thermal radiation is generated from flames jets (or torch fires), fireballs, or pools of burning liquids. Overpressure events are potentially generated if a flammable vapor cloud is detonated.

Each of these scenarios has its own probability of occurrence and generates its own hazard distance. In some consequence assessments, each will need to be individually analyzed and included.

Most damage state or hazard zone calculations result in an estimated threat distance from a source, such as the center of a burning liquid pool or a vapor cloud centroid. It is important to recognize that the source of a thermal event might not be at the pipeline failure location. The source can actually be some distance from the leak site and this must be considered when assessing potential receptor impacts. Note also that a receptor can be very close to a leak site and not suffer any damages, depending on variables such as wind direction, topography, or the presence of barriers.

11.6.1.1 Air Dispersion

Vapor dispersion estimates will govern scenarios of toxic gas releases as well as fireballs and flashfires that predominantly involve gases, and vapor cloud explosions. These phenomena were discussed in the previous section. While there are few, if any, short cut estimation solutions for vapor cloud modeling, there are widely available models for first responders, air pollution, and hazard area calculations.

11.6.1.2 Jet fire modeling

The potential consequences from a pipeline release will depend on the failure mode (uch as leak versus rupture), discharge configuration (such as vertical versus inclined jet, obstructed versus unobstructed), and the time to ignite (such as immediate versus delayed). For natural gas pipelines, the possibility of a significant flash fire or vapor cloud explosion resulting from delayed remote ignition is low due to the buoyant nature of gas, which prevents the formation of a persistent flammable vapor cloud near ignition sources.

Ref [83] "Model of Sizing High Consequence Areas (HCAs) Associated with Natural Gas Pipelines" is commonly used to determine the point of 'significant' potential pipeline natural gas jet fire impacts on surrounding people and property. The Gas Research Institute (GRI) funded the development of this model for U.S. gas transmission lines in 2000, in association with the U.S. Office of Pipeline Safety (OPS), to help define and size HCAs as part of new integrity management regulations. This model uses a conservative and simple equation that calculates the size of the affected worst case failure release area based on the pipeline's diameter and operating pressure. Evaluating the potential consequences from a natural gas release is often based on the hazard zone generated by a jet fire from such a release.

A jet fire is a common result of an ignited release from a flammable gas pipeline. With some reasonable assumptions, the associated hazard zone can be modeled with some readily available data and efficiently applied to long stretches of pipeline. The most well known model, GRI PIR, was highlighted in a previous discussion of hazard zone thresholds. That model illustrated the identification and use of intensity levels and damage levels (ie, human mortality) in a hazard zone determination. That same model is also relevant as a tool for efficiently establishing hazard zones for releases that behave under the assumptions used in the model development.

The GRI model first seeks to characterize the heat intensity associated with ignited gas releases from high-pressure natural gas pipelines. Escaping gas is assumed to feed a fire that ignites shortly after pipe failure. The affected ground area can be estimated by quantifying the radiant heat intensity associated with a sustained jet fire.

A relationship is proposed and described in PRMM that uses a simple equation to calculate the potential size of 'significant' damage from a natural gas pipeline failure based on the pipeline's diameter and operating pressure.

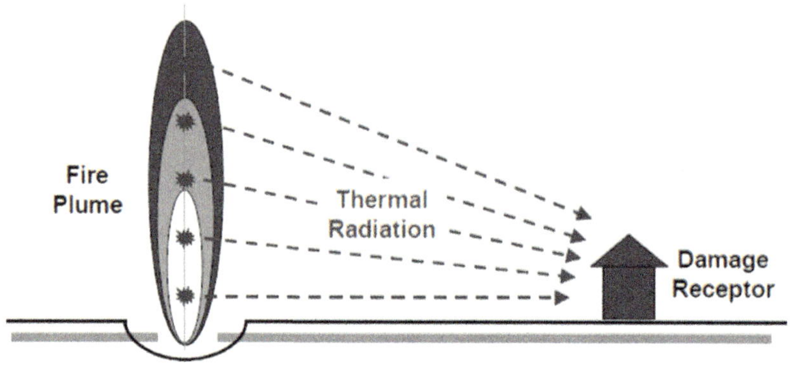

Figure 11.9 From ref [83]

Other models are available, but this GRI model has gained a level of acceptance worldwide and is unrivaled in its ease of application. A related set of equations, by the same authors, can be used to calculate distances for other damage states, ie, other than the 1% mortality used here. Alternative threshold values for thermal radiation intensity can also be used in the above equations to calculate hazard areas for other types of damage such as property damage, secondary fires, injuries, etc. This is important since a robust risk assessment will seek to characterize all consequence potential, not just the worst case scenario. This requires estimation of various levels of harm to various receptor types.

Similar equations are available for other gases but not all gases nor all scenarios. When a model is needed to evaluate risks from a variety of flammable gases, then additional variables are needed to distinguish among potential hazard zones. Density might be appropriate when the consequences are thought to be more sensitive to release rate. MW or heat of combustion might be more appropriate for consequences more sensitive to thermal radiation. If a gas to be included is thought to have the potential for an unconfined vapor cloud explosion, then the model should also include overpressure (explosion) effects as discussed for HVL scenarios.

The previous thermal radiation relationship [83] along with a supposition that dispersion, thermal radiation, and vapor cloud explosive potential are proportional to MW could lead to a modified equation to capture differences among gases for which there is no deterministic equation.

Even when a simple model such as this appears to be pertinent to the scenario being assessed, caution is in order. To reduce the complex real-world phenomena into such a simple equation involving only two inputs, requires numerous assumptions. Some of these assumptions may not be appropriate for scenarios being evaluated.

11.6.1.3 Pool fire modeling

For damage potential from pool fires, the pool diameter and the flammable material's heat of combustion are the most critical factors in most calculation procedures. Factors such as release rate, topography, and soil permeability are needed to estimate pool size. To broadcast a pool size estimate along long distances of pipeline, a pool depth can be assumed and the radius calculated according to the volume of leaked product at each spill point. This will not take into account location specific characteristics and should not be considered a robust approach.

Once the pool size has been estimated, thermal radiation damage distances can be added. Models such as the one shown earlier in the product hazard discussion, work well to show distances beyond the pool edges where damages can be expected.

One source predicts distance to a certain thermal radiation intensity with an equation based on factors for estimating the distance to a heat radiation level that could cause second degree burns from a 40-second exposure. This heat radiation level was calculated to be 5,000 watts per square meter. The equation for estimating the distance from pool fires of flammable liquids with boiling points above ambient temperature is:

$$X = H_c \sqrt{\frac{0.0001A}{5000\,\Pi\,(H_v + C_p\,(T_B - T_A))}}$$

Where:

X	=	distance to the 5 kilowatt per square meter endpoint (m)
HC	=	heat of combustion of the flammable liquid (joules/kg)
HV	=	heat of vaporization of the flammable liquid (joules/kg)
A	=	pool area (m2)
CP	=	liquid heat capacity (joules/kg-°K)
TB	=	boiling temperature of the liquid (°K)
TA	=	ambient temperature (°K)

EPA's RMP Off-Site Consequence Analysis Guidance (May 24, 1996)

One source presents maximum separation distances from a fire beyond which the thermal radiation flux impinging on a structure or person is less than the acceptable separation distance (ASD) threshold values regardless of the fire size. Table 11.6 lists these maximum values for the different fuels considered. The values are obtained by extrapolating the ASD from a simplified chart solution for extremely large fire diameters. These maximum ASD values can be used as "screening" values because distances greater than the "Screen ASD" meet the criteria for thermal radiation flux regardless of fire size.

Table 11.6

Liquid	Mass Burning Rate, m''	Heat of Combustion	HRR Per Unit Area, q''f	Screen ASD	
				Struct.	People
	kg/m2/s	kJ/kg	kW/m2	m	m
Acetic Acid	0.033	13,100	400	10	90
Acetone	0.041	25,800	1,100	10	250
Acrylonitrile	0.052	31,900	1,700	15	390
Amyl Acetate	0.102	32,400	3,300	30	750
Amyl Alcohol	0.069	34,500	2,400	20	550
Benzene	0.048	44,700	2,100	20	480
Butyl Acetate	0.100	37,700	3,800	35	860
Butyl Alcohol	0.054	35,900	1,900	15	430
m-Cresol	0.082	32,600	2,700	25	620
Crude Oil	0.045	42,600	1,900	15	430
Cumene	0.132	41,200	5,400	50	1220
Cyclohexane	0.122	43,500	5,300	45	1200
No. 2 Diesel Fuel	0.035	39,700	1,400	12	320
Ethyl Acetate	0.064	23,400	1,500	15	340
Ethyl Acrylate	0.089	25,700	2,300	20	530
Ethyl Alcohol	0.015	26,800	400	10	90
Ethyl Benzene	0.121	40,900	4,900	40	1100
Ethyl Ether	0.094	33,800	3,200	30	730
Gasoline	0.055	43,700	2,400	20	550
Hexane	0.074	44,700	3,300	30	750
Heptane	0.101	44,600	4,500	40	1000
Isobutyl Alcohol	0.054	35,900	1,900	15	430
Isopropyl Acetate	0.073	27,200	2,000	20	460
Isopropyl Alcohol	0.046	30,500	1,400	15	320
JP-4	0.051	43,500	2,200	20	500
JP-5	0.054	43,000	2,300	20	530
Kerosene	0.039	43,200	1,700	15	400
Methyl Alcohol	0.017	20,000	340	10	80
Methyl Ethyl Ketone	0.072	31,500	2,300	20	530
Pentane	0.126	45,000	5,700	50	1300
Toluene	0.112	40,500	4,500	40	1000
Vinyl Acetate	0.136	22,700	3,100	25	700
Xylene	0.090	40,800	3,700	30	850

[pool fire U802.pdf] NISTIR 6546 Thermal Radiation from Large Pool Fires, Kevin B. McGrattan, Howard R. Baum, Anthony Hamins; Fire Safety Engineering Division

[Building and Fire Research Laboratory, November 2000, National Institute of Standards and Technology, U.S. Department of Commerce]

While these distances are conservative and fixed to pre-determined threshold effects, they are useful, perhaps particularly so in examining the relative differences in safe distances for various types of hazardous liquids.

11.6.1.4 Highly volatile liquids

HVL releases are complex, nonlinear processes, as previously discussed. Hazards associated with the release of an HVL include several flammability scenarios, explosion potential, and the more rare scenario of spilled material displacing air and asphyxiating creatures in the oxygen-free space created. The flammability scenarios of concern include the following (previously described):

- Flame jets
- Vapor cloud fire, flashfire, fireball
- Liquid pool fires
- Vapor cloud explosion

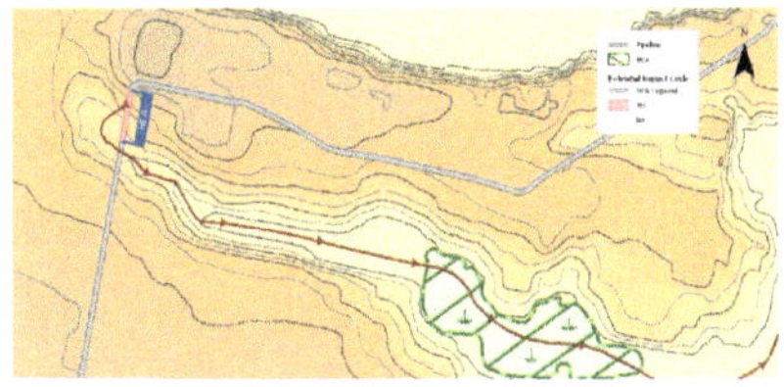

Because precise modeling is so difficult, many assumptions are often employed. Use of conservative assumptions helps to avoid unpleasant surprises and to ensure acceptability of the calculations, should they come under outside scrutiny. A conservative hazard zone distance adopted for an HVL pipeline release, for example, should be based upon a compilation of calculation results generally corresponding to the distance at which a full pipeline rupture, at maximum operating pressure, with subsequent ignition, could expose receptors to significant thermal damages, plus the additional distance at which blast (overpressure) injuries could occur in the event of a subsequent vapor cloud explosion. Some sources of conservatism that can be introduced into HVL hazard zone calculations include:

- Overestimation of probable pipe hole size (can use full-bore rupture as an unlikely, but worst case release)
- Overestimation of probable pipeline pressure at release (assume maximum pressures)
- Stable atmospheric weather conditions at time of release
- Ground-level release event.
- Maximum cloud size occurring prior to ignition
- Extremely rare unconfined vapor cloud explosion scenario with overpressure limits set at minimal damage levels
- Overpressure effects distance added to ignition distance (assume explosion epicenter is at farthest point from release).

These conservative parameters would ensure that actual damage areas are well within the hazard zones for the vast majority of pipeline release scenarios. Additional parameters that could be adjusted in terms of conservatism include mass of cloud involved in explosion event, overpressure damage thresholds, effects of mixing on LFL distance, weather parameters that might promote more cohesive cloud conditions and/or cloud drift, release scenarios that do not rapidly depressurize the pipeline, possibility for sympathetic failures of adjacent pipelines or plant facilities, ground-level versus atmospheric events, and the potential for a high-velocity jet release of vapor and liquid in a downwind direction.

Available models and modeling services for HVL releases are numerous. They range from public domain (free) software designed for first responders, to extremely sophisticated models run only by specialists.

An example calculation, based on equations from the EPA's *RMP Off-Site Consequence Analysis Guidance (May 24, 1996)* is as follows:

> For vapor cloud explosion, the total quantity of flammable substance is assumed to form a vapor cloud. The entire cloud is assumed to be within the flammability limits, and the cloud is assumed to explode. Ten percent of the flammable vapor in the cloud is assumed to participate in the explosion. The distance to the one pound per square inch overpressure level is determined using equation C-1.

$$X = 17\left(0.1\ W_f\ \frac{H_{cf}}{H_{CTNT}}\right)^{1/3}$$

Where:

X	= distance to overpressure of 1 psi (meters)
Wf	= weight of flammable substance (kg)
HCf	= heat of combustion of flammable substance (joules/kg)
HCTNT	= heat of combustion of trinitrotoluene (4.68 E+06 joules/kg)

11.6.1.5 Secondary Fire Effects

One Canadian study concludes that there are on average about two pipeline-related fires in Canada each year, compared to 70,000 other fires and 9,000 forest fires. Their conclusion is that gas pipelines generally pose little threat to the environment based on the low incident of fires initiated by gas pipelines [95]. This conclusion is consistent with the generally accepted low risk of environmental harm from most gas releases.

Nonetheless, when thermal effects from ignition of any released pipeline product do occur, secondary fires are commonly seen. Post incident aerial photos clearly show this.

There will normally be much uncertainty in estimating the potential spread of a fire, given the multitude of variables impacting the spread, including heat flux, emissivity, transmissivity, types of combustibles, wind, humidity, recent rainfall, emergency response, and others.

Thermal radiation threshold levels for non-piloted ignition of wood products and aerial photographs from incidents in similar environments can be used to inform the selection of a distance for secondary fires to be added to the hazard zone.

11.6.2 Hazard zone examples

A hazard zone for a pipeline could be based on generalized distances from specific receptors representing "distances of concern", based on receptor vulnerability or other damage distances from a pipeline release. Some examples are noted here.

One high profile assessment uses a default 1250-ft radius around an 18-in. gasoline pipeline as a hazard zone, but allows for farther distances where modeling around specific receptors has shown that the topography supports a larger potential spill-impact radius.

In cases of HVL pipeline modeling, conservative (near worst case) distances of 1000 to 2500ft are commonly used, depending on pipeline diameter, pressure, and product characteristics. HVL releases cases are very sensitive to weather conditions and carry the potential for unconfined vapor cloud explosions, each of which can greatly extend impact zones to more than a mile.

Regulatory set back distances also provide insight into hazard zones determined by others. A draft Michigan regulatory document suggests setback distances for buried high-pressure gas pipelines based on the HUD guideline thermal radiation criteria. The proposed setback distances are tabulated for pipeline diameters (from 4 to 26in.) and pressures (from 400 to 1800 psig in 100-psig increments). It is not known if these distances will be codified into regulations. In some cases, the larger distances might cause repercussions regarding alternative land uses for existing pipelines. Land use regulations can have significant social, political, and economic ramifications. (See also the discussion on land-use issues in a following section for thoughts on setback distances that are logically related to hazard zones.)

The U.S. Coast Guard (USCG) provides guidance on the safe distance for people and wooden buildings from the edge of a burning spill in their *Hazard Assessment Handbook, Commandant Instruction Manual M 16465.13*. Safe distances range widely depending on the size of the burning area, which is assumed to be on open water. For people, the distances vary from 150 to 10,100ft, whereas for buildings the distances vary from 32 to 1900ft for the same size spill. The spill radii for these distances range between 10 and 2000ft [1025].

A summary of setback distances was published in a consultant report and is shown in Table 14.36 of PRMM.

Any time default hazard zone distances replace situation-specific calculations, the defaults should be validated by actual calculations to ensure that they encompass most, if not all, possible release scenarios for the pipeline systems being evaluated.

11.6.3 Using a Fixed Hazard Zone Distance

Based on sound analyses, hazard zones for groups of similar pipelines—same product, diameter, pressure range, etc—could be set at some consistent nominal distance. A fixed hazard zone distance sacrifices some resolution since the distance must be based on a set of parameters that will not be exactly correct for every portion along a long pipeline.

Fixed hazard buffers may be more appropriate for vapor releases—gas and HVL—since those releases are often less sensitive to minor changes in location-specific characteristics. In contrast, a liquid spill is often heavily influenced by minor location-specific changes such as drainage ditches, storm sewers, surface flow resistance and permeability, topography, etc. therefore, the use of a fixed buffer could carry an acceptable loss of accuracy in assessing a gas or HVL pipeline, but will often not suffice when assessing liquid pipeline risks.

Depending on the desired level of conservatism, the selected hazard zone will often represent the distances at which damages *could* occur, but are thought to exceed the actual distances that the vast majority of pipeline release scenarios would impact. For many practical applications of a risk assessment, such conservatism will be warranted.

11.6.4 Characterizing Hazard Zone Potential Using Scenarios

Since an infinite range of hazard distances (areas, zones) are possible, a methodology to efficiently characterize this range without undue loss of accuracy is desired. A good choice is to select a sufficient number of scenarios to represent all possible scenarios and their relative frequency of occurrence. Using a dozen or less scenarios to represent the thousands that are possible, will often generate sufficient resolution for the risk assessment. The selected scenarios should certainly represent both the most common and slight variations on the most common, as well as the worst case.

Estimate hazard distances (threshold distances) for representative pairings of leak size and ignition scenarios. For example, using hole size as a surrogate for leak size, holes sizes of "rupture", "leak", and "pinhole" could be paired with ignition scenarios of "immediate", "delayed", and "no ignition", resulting in 9 combinations, as is shown in the following example. Hole size probabilities could be linked directly to failure mechanism, material toughness, and other pertinent factors.

As another example, Table 11.7, which coincidentally also uses nine scenarios to represent all possible scenarios, is offered. This table is created in a different way from the previous. Here, various combinations of hole size (up to full rupture of the 16" pipe being modeled) and pressure (up to maximum operating pressure) are selected. They encompass the full range of larger sized releases, ignoring smaller, <0.5" diameter holes.

Table 11.7
Establishing Hazard Zone Distances and Probabilities

Product	Hole Size	Probability of Hole	Ignition Scenario	Probability of ignition scenario	Distance from source (ft)	Threshold Distances (ft): Thermal impact	Overpress impact	Contamination Impact	Maximum Distance (ft)	Probability of Maximum Distance
propane	rupture	8%	immediate	60%	0	400	0	0	400	4.8%
			delayed	20%	300	400	800	0	1500	1.6%
			no ignition	20%	300	0	0	0	300	1.6%
	medium	12%	immediate	15%	0	300	0	0	300	1.8%
			delayed	15%	100	300	200	0	600	1.8%
			no ignition	70%	100	0	0	0	100	8.4%
	small	80%	immediate	10%	0	50	0	0	50	8.0%
			delayed	10%	30	50	0	0	80	8.0%
			no ignition	80%	30	0	0	0	30	64.0%
		100%								100.0%

Each pairing is assigned conservative probabilities of the hole size and pressure occurring, as well as ignition subsequently happening; hole probability x pressure probability x ignition probability = scenario probability. This is thought to fairly represent the range of plausible large hazard zone generating scenarios.

When multiple hazardous liquid and vapor releases are to be assessed, some comparisons can be useful. Equivalences are challenging, though, given the different types of hazards and potential damages (thermal versus overpressure versus contamination damages, for example). For instance, 10,000 square feet of contaminated soil or groundwater is a different damage state than a 10,000-square-foot burn radius. When consequences are to be monetized, equivalences will emerge—the costs of the incidents is the common denominator to make comparisons meaningful.

For instance, using some very specific assumptions, some human fatality and serious injury distances involving multiple products, diameters, pressures, and flow rates were calculated to generate Table 7.11 in PRMM.

11.7 CONSEQUENCE MITIGATION MEASURES

Consequence reduction measures are opportunities to reduce the potential losses from an event already in progress. The pipeline operator's ability to seize these opportunities should be included in the risk assessment.

The simple formula of consequence factors is again a useful summary of the CoF ingredients and shows, in a more structured way, the opportunities to reduce potential consequences.

Release Impact (RI) = product hazard (PH) x Release Quantity (RQ) x dispersion (D) x receptors (R)

Reduction to any factor or combination of factors will reduce consequence potential. Reductions to some will not often be practical changing product or permanently moving receptors, for instance. In the interest of completeness, however, such options should be acknowledged. Other options are usually viable—reduce spill volumes and/or dispersion of released product.

Discounting business consequences, consequence-reducing actions must do at least one of two things:

1. Limit the damage area.
2. Limit damages to receptors within the damage area.

Given a release, associated damage/hazard areas are reduced by limiting the amount of product spilled by isolating the pipeline quickly or changing some transport parameter (pressure, flowrate, type of product, etc), by preventing ignition, and/or by limiting the extent of the spill. If a reduction measure can reduce the size of the hazard zone, then fewer receptors may be exposed and consequences will be lower.

Additionally, the potential damage rate within the hazard zone can be limited by protecting or removing vulnerable receptors. Additional actions to limit receptor damages include prompt medical attention, quick containment, avoidance of secondary damages, and rapid cleanup of the spill.

Chronic hazards have a time factor implied: events tend to worsen with the passage of time. Actions that can influence what occurs during the time period of the spill will impact the consequences. Therefore, there are more opportunities to reduce hazard areas associated with chronic events. If a small release is detected before a spill plume can become larger or migrate to additional sensitive receptors, the hazard zone may be reduced by flow halting, secondary containment, and others. In chronic hazard scenarios, emergency response actions such as evacuation, blockades, and rapid pipeline shutoff are effective in reducing the hazard area.

Most acute events offer less intervention opportunities since the largest hazard zones tend to occur immediately after release and then improve over time. The more probable leak scenarios involving acute hazards show that the consequences would not increase over time because the driving force (pressure) is being reduced immediately after the leak event begins and dispersion of spilled product occurs rapidly. This means that reaction times swift enough to impact the immediate degree of hazard are not very likely. The emphasis here is on 'immediate' so as not to downplay the importance of emergency response. Emergency response can indeed influence the final outcome of an acute event in terms of loss of life, injuries, property damage, and other potential losses.

In many scenarios, reaction to a liquid spill plays a larger role in consequence minimization than does reaction to a gas release.

Additional opportunities, less common for pipelines, include fire suppression systems higher-volume containment does not always warrant more risk mitigation than smaller containments. The larger containment component or facility has a greater potential leak volume due to its larger stored volume, but either can produce a smaller, but consequential leak.

11.7.1 Mitigation of CoF vs PoF

The first determination for the risk implications of a mitigation measure is whether it plays a role mostly in terms of failure avoidance or consequence minimization. For example, it can be argued that leak detection should be assessed only in the consequence analyses because it acts as a consequence-limiting activity—the leak has already occurred and early detection can reduce the potential consequences of the leak. However, leak detection can also play a role in leak size—sometimes allowing intervention before a larger leak manifests. Depending on the definition of 'failure', this scenario may reduce failure probability in addition to consequence potential.

Distribution systems are a good example of this nuance. Distribution systems tend to have a higher incidence of leaks compared to transmission systems. This is due to differences in the age, materials, construction techniques, and operating environment between the two types of pipelines. Leakage in these low pressure systems is more routine and leak detection and repair is a normal aspect of operations. Some leaks are not actionable except for perhaps inclusion on a 'monitoring' list. Before some threshold leak rate (or leak circumstance) is reached, the leak is not a 'failure'. Furthermore, leaks often provide early warning of deteriorating system integrity. The number of leak locations is often used as a forecaster of 'failures', with failure being a leak of actionable size.

Therefore, there may be overlap where a mitigation measure such as leak detection plays a role in both PoF and CoF estimations. This is not an obstacle for the risk assessment approach recommended here—any and all measures reducing either can be readily included in the assessment. When a measure such as leak detection is thought to play a significant role in failure rates—by some definition of 'failure'--it is readily incorporated into the exposure, mitigation, and resistance modeling of PoF. It will often be best modeled as an inspection, playing a similar role as other inspections such as ILI. It first provides some indications of resistance—where damage has already occurred. It then provides inferential evidence of both exposure—failure rates may be higher when the leak suggest system deterioration—and mitigation—the leak, having occurred despite mitigation, informs the assessment of mitigation effectiveness.

11.7.2 Sympathetic Failures

Note that CoF mitigation plays a measurable role in PoF reduction through avoidance of secondary damages. That is, reducing the hazard area from event 1 prevents event 2, 3, 4, etc. where subsequent events are avoided by fire suppression systems, depressurizations, secondary containment, blast walls, etc. Especially in complex facilities, each component's PoF will include its neighbor's PoF scenarios that can generate sufficient forces, including thermal effects, to cause sympathetic failures.

11.7.3 Measuring CoF Mitigation

Much discussion on consequence mitigation is offered in the following sections. Note however, that the assessment of such capabilities is straightforward. This is done by measuring (quantifying) and including in the risk assessment, the ability to reliably minimize the area of exposure or exposure time.

In other words, the assessor accounts for the abilities of the mitigation measures to reduce the hazard zone itself or to minimize damages to receptors within the hazard zone. Specifically, this involves the quantification of one or more of the following aspects:

- Reduction in spill volume
- Reduction is release dispersion
- Fewer receptors harmed
- Less harm to exposed receptors.

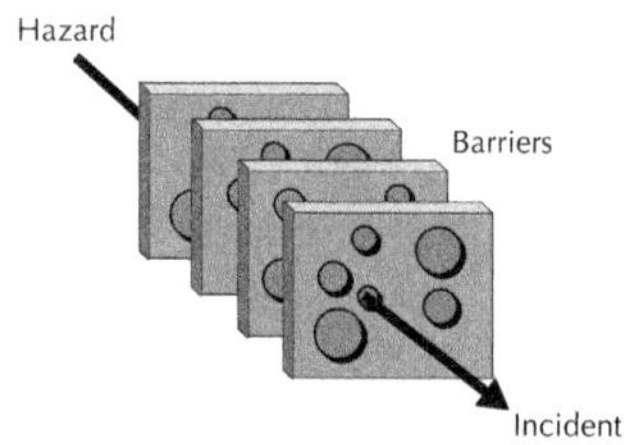

Especially for the first two, the quantification can be based on robust calculations. For example, the role of extra valves in reducing draining-by-gravity volumes can be calculated and the leak detection/reaction capabilities can be assessed at all points along the pipeline as a function of instrumentation, ability to stop flows, and abilities to mobilize and execute loss-minimizing reactions. In other cases, only assumptions and judgments may be available. Realistically, the assessor will sometimes have to simply estimate a percentage reduction, based on the perceived effectiveness and reliability of the mitigation. For example, if emergency response is thought to reduce receptor damages that would otherwise occur, the quantification may be the result of examinations of scenarios to estimate amount of receptor protections afforded by actions such as evacuation, rapid boom deployment, removal of ignition sources, etc. these actions will be very much location- and incident-specific, making general estimates especially uncertain.

Even if the quantification is imprecise, the estimation exercise is important. The quantification puts a value on the emergency response, leak detection, secondary containment, etc, thereby providing the 'benefit' portion of cost/benefit analyses for these measures. Different mitigation measures will have different benefits (and costs) at various potential spill locations along a pipeline. The cost/benefit all along a pipeline guides decision-makers in risk management. Even when imprecise, the quantifications demonstrate a defensible, process-based approach to understanding and therefore managing risk.

Reduction measures are valued in the same way as mitigation measures in PoF. Two questions are asked and answered in performing the valuation—'how effective can the measure be if it is done as well as can be imagined'? and then, 'how well is it being done in the situation being assessed?' in measuring the effectiveness, 'probability of success' will need to be considered, since many measures. The reduction may be expressed as a reduced damage state—a fraction of the damage that would otherwise occur.

As with PoF measurements, it is most efficient to compartmentalize events (exposures) from mitigations. This means that the hazard zone associated with the unmitigated event should first be estimated. Then, that theoretical hazard zone may be reduced by mitigation measures. For instance, the spill footprint is first estimated as if no temporary spill containment measurements occur. Then, the reductions in area due to emergency response, secondary containment, etc are estimated. (An exception is leak detection and isolation time capabilities which are, for practical reasons, normally a part of the initial spill size determination rather than an imagined scenario of infinite leak rate and duration.)

Similarly, the receptor damages should be first estimated as if no protections were in place. Then, reductions to the theoretical receptor damages may be afforded by protections. Shielding and reduction in exposure time (perhaps by enhanced escape opportunities through early warning and/or rapid evacuation) are examples of protection opportunities for human receptors.

If the hazard zone is created directly from a threshold intensity—thermal radiation or overpressure level, for example—then receptor protection can be evaluated separately and used to reduce the modeled hazard zone that would otherwise occur.

Note the partial overlap in emergency response actions. This is due to the fact that, in some cases, the same action may reduce the hazard area while in other cases, the hazard area is unaffected but the receptor damage potential within the area is reduced. The distinction is somewhat esoteric since loss limiting actions mostly reduce the receptor exposure duration and the hazard area boundary already implicitly includes duration of exposure (thermal or toxic) considerations.

The following consequence-reducing opportunities are common:

- Hazard Area Limiting Actions
    - o Secondary containment
    - o Suppression systems
    - o Detection (leak, fire, concentrations, etc.)
    - o Emergency response (temporary secondary containment, shielding, removal of ignition sources, intentional ignition, dilution, suppression, etc.)
- Loss limiting actions
    - o Detection
    - o Emergency Response (evacuations, removal of ignition sources, intentional ignition, other exposure duration reductions)

11.7.4 Spill volume/dispersion limiting actions

Reductions in spill size are made by reducing the product containment volume in the case of volume-dependent spills, and by reducing the source rate (e.g., pressure, density, hole, time-to-detect) in the case of rate-dependent spills. Smaller volumes that can potentially be released, for example, smaller vessel, or smaller leak rates, for example, lower pressure, smaller holes, reduce spill sizes. Note that improvements in leak detection also effectively reduce the source, in the leak-rate dependent case.

Secondary containment and emergency response, especially leak detection/reaction, are considered to be risk mitigation measures that minimize potential consequences by minimizing product leak volumes and/or dispersion. The effectiveness of each varies depending on the type of system being evaluated.

This opportunity for consequence reduction includes leak detection/reaction and is often the most realistic way for the operator to reduce the consequences of a pipeline failure. Some common approaches to limiting spill volumes are discussed below.

11.7.5 Pipeline Isolation Protocols

The ability to quickly isolate leaks and reduce volume to a leak location are logically important consequence minimizations. Sequencing of pipeline isolation can be important to spill size estimations. To minimize release volumes in the event of a leak, the pressure at the leak location must be minimized as quickly as possible. This is accomplished by halting all sources of pressure and allowing the leak location to depressure as rapidly as possible. Providing an alternative flow path—other than through the integrity breach—assists in the depressurization. In many leak scenarios, therefore, maintaining an alternate flow path away from the leak minimizes consequences.

Elevation profiles and hole size also play an important role in isolation protocols for liquid pipelines. Leaks in low spots or in the rare scenario where the pipe is completely separated may be worsened by attempts to maintain an open flow path away from the leak.

It may therefore be difficult to quickly ascertain the optimal action to take. While larger and more rapid changes in monitored points such as flow and pressure are associated with large leak events, the guillotine rupture type of event—where all flow paths should be quickly closed—is largely indistinguishable—from a remote control center or even from the leak site itself—from a larger leak where maintaining the alternative flow path is beneficial.

A downstream flow meter (or manual observation) accurately indicating that no flow is passing the leak site would be the most compelling evidence that full isolation is appropriate.

Isolation must also consider surge potential. In certain circumstances, damages could be caused to other parts of the pipeline while trying to minimize the consequences of a leak in progress. This is readily avoided by commonly used surge prevention equipment. See full discussion of surge potential as a contributor to pipeline PoF in Chapter 8.7.3 Surge potential.

11.7.6 Valving

This is especially true for incompressible fluids transported in pipelines.

Two key components of a release volume from a liquid line are (1) the continued pumping that occurs before the line can be shut down and (2) the liquid that drains from the pipe after the line has been shut down. The former is only minimally impacted by additional isolation capability—perhaps only helping to stop momentum effects from pumping if a valve is rapidly closed (but potentially generating threatening pressure waves). The main role of additional isolation capabilities, therefore, seems to be in reducing drain volumes. Because a pipeline is a closed system, hydraulic head and/or a displacement gas is needed to affect line drainage. Hilly terrain can create natural check valves that limit hydraulic head and gas displacement of pipeline liquids.

Faster response scenarios may include valves that automatically isolate a leaking pipeline section based on continuously monitored parameters that indicate a leak. However, in real applications, the value of such valves and the practicality of such automation is often uncertain. The use of valves as spill limiting equipment are discussed below:

A. **Automatic and/or remotely operated valves.** Automatic valves are often triggered on low pressure, high pressure, high flow, rate of change of pressure or flow, or more complex combinations of these. This includes automatic shutoffs of pumps, wells, and other pressure sources. Regular maintenance is required to ensure proper operation. Experience warns that this type of equipment is often plagued by false trips from transient conditions, nearby electrical storms, and other system or environment causes. Such valve actuations may create additional stresses such as surge pressures in addition to unnecessary supply interruptions. Avoidance of false triggers is sometimes accomplished by setting relatively insensitive response trigger points, thereby reducing the automation reaction time and the benefits sought.
   Check valves are another form of automatic valves and play an important spill-reducing role in some systems. A check valve might be especially useful for liquid lines with elevation changes. Strategically placed check valves may reduce the draining or siphoning to a spill at a lower elevation.
B. **Valve spacing**. Closer valve spacing logically provides a benefit in reducing the spill amount in many scenarios. Spacing benefits must be coupled with the most probable reaction time in closing those valves since valves may be near to a leak sites but lack a quick activation time (for example, manual valves that are difficult to access or slow to operate). Many countries' regulations require valves be placed within certain distances, sometimes related to receptors such as population densities (US natural gas transmission pipeline valve maximum permissible spacings are a function of population density) or water bodies (US hazardous liquid pipelines). Regulations also commonly require situation-specific analyses to determine when additional valves or improvements in valve

swiftness of operation is warranted. Regulations using ALARP have such considerations implicitly required.

Concerns with the use of additional block valves include costs and increased system vulnerabilities from malfunctioning components and/or accidental closures, especially where automatic or remote capabilities are included. For unidirectional pipelines, check valves (preventing backflow) can provide some consequence minimization benefits. Check valves respond almost immediately to reverse flow and are not subject to most of the incremental risks associated with block valves since they have less chance of accidental closure due to human error or, in the case of automatic/remote valves, failure due to system malfunctions. Their failure rate (failure as unwanted closure or failure to close when needed) can be considered against benefits provided.

Studies of possible benefits of shorter distances between valves of any type produce mixed conclusions. Evaluations of previous accidents can provide insight into possible benefits of closer valve spacing in reducing consequences of specific scenarios. By one study of 336 liquid pipeline accidents, such valves could, at best, have provided a 37% reduction in damage [76]. Offsetting potential benefits is the often substantial costs of additional valves and the increased potential for equipment malfunction, which may increase certain risks (surge potential, customer interruption, etc.). Rusin and Savvides-Gellerson [76] calculate that the costs (installation and ongoing maintenance) of additional valves would far outweigh the possible benefits, and also imply that such valves may actually introduce new hazards.

More recent work presents findings that also might be useful to the risk assessor. A 2012 study [1015] focusing on full ruptures with subsequent ignition (of transmission pipeline carrying natural gas and using propane as the worst-case hazardous liquid scenario) plus a spill scenario of unignited crude oil, concluded the following:

Natural Gas

- "... block valves have no influence on the volume of natural gas released during the detection phase..."
- "Fire damage to buildings and personal property located in Class 1, Class 2, Class 3, and Class 4 HCAs resulting from natural gas combustion immediately following guillotine-type breaks in natural gas pipelines is considered potentially severe for all areas within 1.5 to 1.7 times the PIR."
- "Without fire fighter intervention, the swiftness of block valve closure has no effect on mitigating potential fire damage to buildings and personal property in Class 1, Class 2, Class 3, and Class 4 HCAs resulting from natural gas pipeline releases."
- "Block valve closure swiftness also has no effect on reducing building and personal property damage costs."

- "The benefit in terms of cost avoidance is based on the ability of fire fighters to mitigate fire damage to buildings and personal property located within a distance of approximately 1.5 times the PIR by conducting fire fighting activities as soon as possible upon arrival at the scene."
- "The study results further show that for natural gas release scenarios, block valve closure within 8 minutes after the break can result in a potential cost avoidance of at least $2,000,000 for 12-in nominal diameter natural gas pipelines and $8,000,000 for 42-in nominal diameter natural gas pipelines depending on the configuration of buildings within the Class 3 HCA."
- "Delaying block valve closure by an additional 5 minutes can reduce the cost avoidance by approximately 50%."

Hazardous Liquids[4] with Ignition

- "The effectiveness of block valve closure swiftness on limiting the spill volume of a release is influenced by the location of the block valves relative to the location of the break, the pipeline elevation profile between adjacent block valves, and the time required to close the block valves after the break is detected and the pumps are shut down."
- "Fire damage to buildings and personal property in a HCA resulting from liquid propane combustion immediately following guillotine-type breaks in hazardous liquid pipelines is considered potentially severe for a radius up to 2.6 times the equilibrium diameter."[5] "These conclusions are based on computed heat flux versus time data for liquid propane pipelines with nominal diameters ranging from 8 to 30 in. and operating pressures ranging from 400 psig to 1,480 psig."
- "The benefit in terms of cost avoidance for damage to buildings and personal property attributed to block valve closure swiftness increases as the duration of the block valve shutdown phase decreases. Risk analysis results for a hypothetical 30-in. nominal diameter hazardous liquid pipeline release of liquid propane show that the estimated avoided cost of moderate building and property damage resulting from block valve closure in 13 rather than 70 minutes is over $300,000,000."

4 As defined in US regulations

5 pool diameters, the study produced 'equilibrium diameters' of around 300 ft for 8" pipelines and 1,900 ft for 30" pipelines. The report does not discuss how such pools can be formed by propane under atmospheric conditions (ie, the expected HVL behavior is not explained)

Hazardous Liquids without Ignition

- "The swiftness of block valve closure has a significant effect on mitigating potential socioeconomic and environmental damage to the human and natural environments resulting from hazardous liquid pipeline releases because damage costs increase as the spill size increases. The benefit in terms of cost avoidance for damage to the human and natural environments attributed to block valve closure swiftness increases as the duration of the block valve shutdown phase decreases."
- "The damage cost for crude oil released in the Enbridge Line 6B pipeline rupture in Marshall, Michigan in 2010 was approximately $38,000 per barrel. "
- "It is also important that inadvertent block valve closure does not occur. It is undesirable to disrupt service to critical customers, and also sudden block valve closure that occurs inadvertently may cause a pressure surge that could damage equipment."

Note that cost and benefit conclusions are incomplete here since benefits expressed on a per incident basis do not provide the complete story. The frequency of incidents is also needed before meaningful conclusions can be drawn. While the conclusions and analyses in this study are interesting, they have used many assumptions that may not be appropriate in many applications. The emphasized point that location-specific characteristics can readily invalidate the underlying assumptions, points to the need to consider more than these conclusions in a risk assessment.

11.7.7 Sensing devices.

Part of response time is the first opportunity to take action. This opportunity depends on the sensitivity of the leak detection. All leak detection will have an element of uncertainty, from the possibility of crank phone calls to the false alarms generated by instrumentation failures or instrument reactions to pipeline transients. This uncertainty must also be included in reaction times.

11.7.8 Reaction times

If a human intervention is required to initiate the proper response, this intervention must be assessed in terms of timeliness and appropriateness. A control room operator must often diagnose the leak based on instrument readings transmitted to him. How quickly he can make this diagnosis depends on his training, his experience, and the level of instrumentation that is supporting his diagnosis. Probable reaction times can be judged from mock emergency drill records when available. If the control room can remotely operate equipment to reduce the spill size, the reaction time is improved. Travel time by first responders must otherwise be factored in. If the pipeline operator has provided enough training and communications to public emergency response personnel so that they may operate pipeline equipment, response time may be improved, but possibly at the expense of increased human error potential. Public emergency response personnel are probably not able to devote much training time to a rare event such as a pipeline failure. If the reaction is automatic (computer-generated valve closure, for instance) a sensitivity is necessarily built in to eliminate false alarms.

11.7.9 Secondary containment

Hazard area is reduced when secondary containment is present. The greater the leak or receptor isolation offered by secondary containment, the footprint within which damages can occur. Secondary containment benefit is usually proportional to the size of the effective area it protects.

Opportunities to contain or limit the spread of a release can be considered here. These opportunities include:

- Natural barriers or accumulation points
- Casing pipe, pipe-in-pipe designs
- Tunnels
- Lined trench
- Berms or levees
- Containment systems
- Impervious/ Semipervious liner
- Immediate fill indication
- Overflow alarms
- Double-walled tanks
- Reducing Receptor Contact Times
    - o Fire suppression system Deluge system foam systems, water curtains
    - o Depressurization systems (for example, flares) dump/blowdowns.

Limited secondary containments such as pump seal vessels and sumps are designed to capture specific leaks. As such they provide risk reduction for a limited range of scenarios.

Many secondary containment opportunities apply only to liquid releases and are found at stations. The presence of secondary containment can be considered as an opportunity to reduce (or eliminate) the "area of opportunity" for consequences to occur—fewer exposed receptors.

Secondary containment can be evaluated in terms of its ability to:

- Contain the majority of all foreseeable spills scenarios.
- Contain 100% of a potential spill plus firewater, debris, or other volume reducers that might compete for containment space—largest tank contents plus 30 minutes of maximum firewater flow is sometimes used [26].
- Contain spilled volumes safely—not exposing additional equipment to hazards.
- Contain spills until removal can be effected—no leaks.

Note that ease of cleanup of the containment area is a secondary consideration (business risk).

Within station limits, the drainage of spills away from other equipment is important. A slope of at least 2% (1% on hard surfaces) to a safe impoundment area of sufficient volume is seen as adequate. [26]

Some secondary containment designs provide a great deal of additional risk reduction benefits, beyond their role in preventing dispersion of releases. Pipe-in-pipe designs and installations in tunnels often support continuous and improved leak detection, improved inspectability, reduced threats from external forces and corrosion, etc, all in addition to the important secondary containment benefits. They are not, however, free from practical challenges including very high initial costs and additional maintenance requirements.

Where man-made secondary containment exists, or it is recognized that special natural containment exists, the evaluator can adjust the hazard area accordingly.

11.7.10 Leak detection

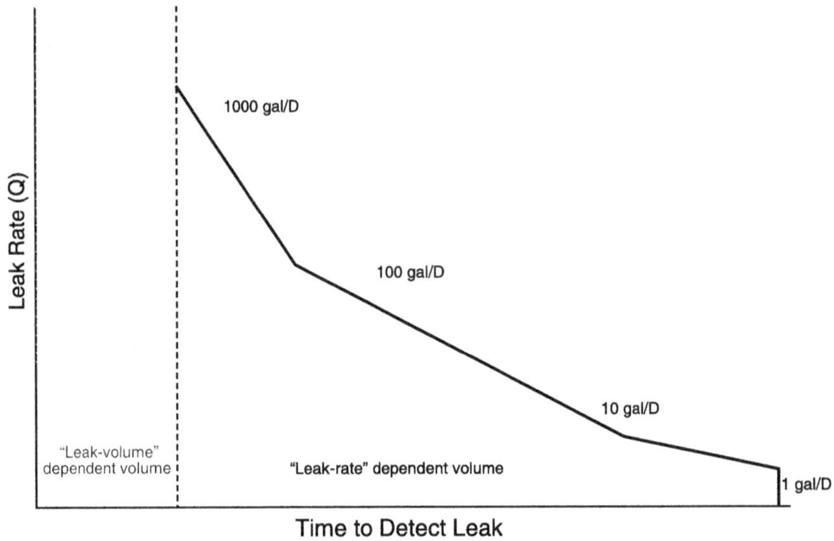

Figure 11.10 Leak detection capabilities

Including leak detection and emergency response considerations impacts the volumes released and adds an important level of resolution to any risk analysis. Their inclusion also provides a way to assign values to this largely-discretionary risk reduction measures. By quantifying the avoided potential losses (expected loss valuations), the costs of new systems or enhancements to existing systems can be justified.

It is especially important to consider leak detection capabilities for scenarios involving toxic or environmentally persistent products. In those cases, a full line rupture might not be the worst case scenario. Slow leaks gone undetected for long periods can be more damaging than massive leaks that are quickly detected and addressed.

The ability to detect smaller leaks is important since the smaller leaks tend to be more prevalent and can also be very consequential. The negative impact of smaller leaks often far exceed the scale predicted by a simple proportion to leak rate. For example, a 1gal/day leak detected after 100 days is often far worse than a 100gal/day leak rate detected in 1 day, even though the same amount of product is spilled in either case. Unknown and complex interactions between small spills, subsurface transport, and groundwater contamination, as well as the increased ground transport opportunity, account for increased chronic hazard in many scenarios.

11.7.10.1 Leak detection and vapor dispersion

Leak detection plays a relatively minor role in minimizing consequence in most scenarios of gas pipelines large leaks or ruptures. Therefore, many large gas release scenarios will not be significantly impacted by any assumptions relative to leak detection capabilities. This is especially true when defined damage states use short exposure times to thermal radiation, as is often warranted.

Gas pipeline release hazards depend on release rates which in turn are governed by pressure and hole size. In the case of larger releases, the pressure diminishes quickly—more quickly than would be affected by any actions that could be taken by a control center. In the case of smaller leaks, pressures decline more slowly but ignition probability is much lower and hazard areas are much smaller. In general, there are few opportunities to evacuate a pressurized gas pipeline more rapidly than occurs through the leak process itself, when the leak rate is significant. A notable exception to this case is that of possible gas accumulation in confined spaces. This is a common hazard associated with urban gas distribution systems.

Another exception would be a scenario involving the ignition of a small leak that causes immediate localized damages and then more widespread damages as more combustible surroundings are ignited over time as the fire spreads. In that scenario, leak detection might be more useful in minimizing potential impacts to the public.

11.7.10.2 Leak detection and liquid dispersion

Leak detection capabilities play a larger role in liquid spills compared to gas releases. Long after a leak has occurred, liquid products can be detected because they have more opportunities for accumulation and are usually more persistent in the environment. A small, difficult-to-detect leak that is allowed to continue for a long period of time can cause widespread contamination damages, especially to aquifers. Therefore, the ability to quickly locate and identify even small leaks is critical for some liquid pipelines.

A leak detection capability curve can be used to establish the largest potential volume release.

11.7.10.3 Leak Detection and CoF

Leak detection can be seen as a critical part of emergency response. It provides early notification and allows more rapid response. Leak detection is considered a spill reducing opportunity aspect of emergency response.

The role of leak detection is evaluated in the determination of spill size and dispersion.

As discussed previously, leak size is partially dependent on failure mode. Small leak rates tend to occur due to corrosion (pinholes) or some other failure modes. The more damaging of these small leaks occur below detection levels and continue for long periods of time. Larger leak rates tend to occur under catastrophic failure conditions such as external force (e.g., third party, ground movement) and avalanche crack failures.

Larger leaks can be detected more quickly and located more precisely. Smaller leaks may not be found at all by some methods due to the sensitivity limitations. The trade-offs involved between sensitivity and leak size can be expressed in terms of probability of detection over time.

Computational pipeline monitoring (CPM) is a part of most modern transmission pipeline operations and includes leak detection capabilities ranging from rudimentary to extremely sophisticated. The specific method of CPM leak detection chosen depends on a variety of factors including the type of product, flow rates, pressures, the amount of instrumentation available, the instrumentation characteristics, the communications network, the topography, the soil type, and economics. Especially when sophisticated modeling is involved, there is often a trade-off between the sensitivity and the number of false alarms, especially in "noisy" systems with high levels of transients.

As is the case with other aspects of post-incident response, leak detection is thought to normally play a minor role, in reducing the hazard, reducing the probability of the hazard, or reducing the acute consequences. Leak detection can, however, play a larger role in reducing the chronic consequences of a release. As such, its importance in risk management for chronic consequence scenarios is more significant.

This is not to say that leak detection benefits that mitigate acute risks are not possible. One can imagine a scenario in which a smaller leak, rapidly detected and corrected, averted the creation of a larger, more dangerous leak. This would theoretically reduce the acute consequences by preventing the potentially larger leak. We can also imagine the case where rapid leak detection coupled with the fortunate happenstance of pipeline personnel being close by might cause reaction time to be swift enough to reduce the extent of the hazard. This would also impact the acute consequences. These scenarios are obviously limited and it is conservative to assume that leak detection has limited ability to reduce the acute impacts from a pipeline break. Increasing use of leak detection methodology is to be expected as modeling techniques become more refined and instrumentation becomes more accurate. As this happens, leak detection may play an increasingly important role, leak volume and leak rate are both critical determinants of dispersion and hence of hazard zone size. Leak rate is important under the assumption that larger rates cause more spread of hazardous product and higher thermal impacts (more acute impacts), and lower rates impact detectability (more chronic impacts). Leak volume is more important in chronic scenarios such as environmental cleanup. The rate of leakage multiplied by the time the leak continues is often the best estimate of total leak volume. Some potential consequences are more volume sensitive than leak-rate dependent. Spills from catastrophic failures or those occurring at pipeline low points are more volume dependent than leak-rate dependent. Such events are better assessed by leak volumes because the entire volume of a pipeline segment will often be involved, regardless of response actions.

11.7.10.4 Detection methodologies

Common methods of Pipeline leak detection are shown in PRMM. Each method has its strengths and weaknesses and an associated spectrum of capabilities.

Regular leakage surveys are routinely performed on hydrocarbon pipelines, (especially gas) systems in many countries. Hand-carried or vehicle-mounted sensing equipment is available to detect trace amounts of leaking gas in the atmosphere near the ground level. Such overline leak detection by instrumentation (sniffers), vehicle-based systems, or even by trained animals—usually dogs (which reportedly have detection thresholds far below instrument capabilities)--is an available technique. The effectiveness of leak surveys depends partly on environmental actors such as wind, temperature, and the presence of other interfering fumes in the area. Therefore, specific survey conditions and the technology used will make many evaluations situation specific. Pipeline patrolling and surveying can generally be made more capable of detection by adjusting observer training (the observer seeks visual indications of a leak such as dying vegetation, bubbles in water, or sheens on the water or ground surface), speed of survey or patrol, equipment carried (may include detection based on flame ionization detectors (FID), thermal conductivity, infrared sensors, laser-based detection systems, etc.), altitude/speed of air patrol, training of ground personnel, and allowing for specific topography, ROW conditions, product characteristics, weather—both current and, for instance, recent rainfall, etc. Although the capabilities of direct observation techniques are inconsistent, experience shows them to still play a viable role in leak detection, computer-based leak detection methods require instrumentation and computational analysis. A common type of pipeline leak detection employs SCADA-based capabilities of monitoring of pressures, flows, temperatures, equipment status, etc. plus balancing flows in and out of segments. SCADA and control center procedures might call for a leak detection investigation when (1) abnormally low pressures or an abnormal rate of change of pressure is detected; and (2) a flow balance analysis, in which flows into a pipeline section are compared with flows out of the section and discrepancies are detected. SCADA-based alarms can be set to alert the operator of such unusual pressure levels, differences between flow rates, abnormal temperatures, or equipment status (such as unexplained pump/compressor stops).

SCADA-based capabilities are commonly enhanced by computational techniques that use SCADA data in conjunction with mathematical algorithms to analyze pipeline flows and pressures on a real-time basis. Some use only relatively simple mass-balance calculations, perhaps with corrections for linefill. More robust versions add conservation of momentum calculations, conservation of energy calculations, with considerations for fluid properties, instrument performance, using a host of sophisticated equations to characterize flows, including transient flow analyses. The nature of the operations will impact leak detection capabilities, with more less steady flows and more compressible fluids reducing the capabilities.

The more instruments (and the more optimized the instrument locations) that are accurately transmitting data into the SCADA-based leak detection model, the higher the accuracy of the model and the confidence level of leak indications. Ideally, the model would receive data on flows, temperatures, pressures, densities, viscosities, etc., along the entire pipeline length. By tuning the computer model to simulate mathematically all flowing conditions along the entire pipeline and then continuously comparing this simulation to actual data, the model tries to distinguish between instrument errors, normal transients, and leaks. Depending on the system characteristics, relatively small leaks can often be accurately located in a timely fashion. How small a leak and how swift a detection is specific to the situation, given the large numbers of variables to consider. Refs [3] and [4] discuss these leak detection systems and methodologies for evaluating their capabilities.

Another computer-based method is designed to detect pressure waves. A leak will cause a negative pressure wave at the leak site. This wave will travel in both directions from the leak at high speed through the pipeline product (much faster in liquids than in gases). By simply detecting this wave, leak size and location can be estimated. A technique called pressure point analysis (PPA) detects this wave and also statistically analyzes all changes at a single pressure or flow monitoring point. By statistically analyzing all of these data, the technique can reportedly, with a higher degree of confidence, distinguish between leaks and many normal transients as well as identify instrument drift and reading errors. Ultrasonic leak detectors—in which instrumentation is used to detect the sonic energy from an escaping product are used in permanent and pig-based applications.

Another method of leak detection involves various methods of continuous direct detection of leaks immediately adjacent to a pipeline. One variation of this method is the installation of a secondary conduit along the entire pipeline length. This secondary conduit is designed to sense leaks originating from the pipeline. The secondary conduit may take the form of a small-diameter perforated tube, installed parallel to the pipeline, which allows vapor samples to be drawn into a sensor that can detect the product leaks. Variations on this type of system can detect temperature changes or react specifically to certain hydrocarbons, based on electrical conductivity or other characteristics. Floating hydrocarbon sensors used at river crossings and other offshore locations fall into this method. Use of hydrocarbon sensors, 'fire eyes', and other above-ground, atmospheric-based sensing systems are also included here.

Additional leak detection methods include the following:

- Subsurface detector survey—in which atmospheric sampling points are found (or created) near the pipe. Such sampling points include manways, sewers, vaults, other conduits, and holes excavated over the pipeline. This technique may be required when conditions do not allow an adequate surface survey (perhaps high wind or surface coverage by pavement or ice). A sampling pattern is usually designed to optimize this technique.
- Pressure loss test—in which an isolated section of pipeline is closely monitored for loss of pressure, indicating a leak.
- Bubble leakage—used on exposed piping, the bubble leakage test in one in which a bubble-forming solution can be applied and observed for evidence of gas leakage.

In a pipe-in-pipe design, where an exterior pipe totally encloses the product pipeline, the annular space can be continuously monitored for leaks. This emergency response improvement supplements the secondary containment benefits. Furthermore, PoF benefits include enhanced protection from external forces and corrosive environments. Therefore, both Pof and CoF reductions are achieved by such designs. Pipelines in tunnels offer similar advantages, often with the additional benefit of improved inspectability. These systems are much more expensive than conventional designs, can cause a host of logistical problems, and are usually not employed except on short lines. Their impact on risk reduction can be dramatic, however. See discussion under Secondary Containment.

Offshore, a small amount of spilled hydrocarbon is not always easy to visually spot, especially from moving aircraft. A variety of sensing devices have been or are being investigated to facilitate spill detection. Detection methods proposed or in use include infrared, passive microwave, active microwave, laser-thermal propagation, laser acoustic sensors [78], and sonar-based technologies. Some of these technologies offer the opportunity for continuous monitoring with automatic notifications, thereby improving response times.

Gas odorization

As a special leak detection and early warning system for most natural gas and LPG distribution systems, gas odorization warrants further discussion. Methane has very little odor detectable to humans. Natural gas is mostly methane and will therefore be odorless unless an artificial odorant is introduced. It is common practice to inject an odorant so that gas will be detected at levels far below the lower flammable limit of the gas in air mixture—often one-fifth of the flammable limit. This means that accumulations of 5 times the detection level are required before fire or explosion is possible. This allows early warning of a gas pipe leak and reduces the potential for human injury.

A 1937 incident in New London, TX is often cited as the beginnings of the widespread use of odorization (even though it had been used in Germany as early as 1880).

In this incident, a school house filled with undetectable natural gas, ignited, exploded, and resulted in 239 fatalities. In the US, odorization is always required in distribution systems and sometimes for transmission pipelines also [1008].

With the increased opportunity for leaked products to accumulate beneath pavement, in buildings, and in other dangerous locations and with the higher population densities seen in hydrocarbon distribution systems, special risk reduction provisions are warranted. One of the primary means of leak detection for gas distribution is the use of an odorant in the gas to allow people to smell the presence of the gas before flammable concentrations are reached.

Odorization system design

Aspects of optimum system design include selection of the proper odorant chemical, the proper dosage to ensure early detection, the proper equipment to inject the chemical, the proper injection location(s), and the ability to vary injection rates to compensate for varied gas flows. Ideally, the odorant will be persistent enough to maintain required concentrations in the gas even after leakage through soil, water, and other anticipated leak paths. The optimum design will consider gas flow rates and the potential for odor fade to ensure that gas at any point in the piping is properly odorized. Fade can occur through absorption of the odorant in some pipe materials, for example, new steels, especially for larger diameter, longer lengths. When new piping is placed in service, "over-odorizing" for a period of time is sometimes done to ensure adequate odorization. When gas flows change, odorant injection levels must be changed appropriately. Testing should verify odorization at the new flow rates. Odorant removal (de-odorization) possibilities should be minimized, even as gas permeates through soil or water. Odor desensitization and disguise by other environmental odors also impact the odorization program's ability for early alert.

System operation/maintenance

Odorant injection equipment is best inspected and maintained according to well-defined, thorough procedures. Trained personnel should oversee system operation and maintenance. Inspections should be designed to ensure that proper detection levels are seen at all points on the piping network. Provisions are needed to quickly detect and correct any odorization equipment malfunctions.

Performance

Evidence should confirm that odorant concentration is effective (provides early warning to potentially hazardous concentrations) at all points on the system. Odorant levels are often confirmed by tests using human subjects who have not been desensitized to the odor. Gas odorization can be a more powerful leak detection mechanism than many other techniques discussed. While it can be argued that many leak survey methods detect gas leaks at very low levels, proper gas odorization has the undeniable benefits of alerting the right people (those in most danger) at the right time.

Odorization Assessment

The role that a given gas odorization effort plays as a consequence reducer depends on the reliability of the system and the fraction of incidents whose consequence are reduced and by what amount.

High-reliability odorization—99%+ reliability, a segment without effective odorization is extremely rare, occurring perhaps once every 0.001 mile-years. The likelihood of an unodorized segment coinciding with a leak location would therefore be very improbable. Qualitative descriptors associated with a high reliability system would typically include the following:

- A modern or well-maintained, well-designed system exists. There is no evidence of system failures or inadequacies of any kind. Extra steps (above regulatory minimums) are taken to ensure system functionality. Also falling into this category is a consistent, naturally occurring odor in a product stream that allows early detection of a hazardous vapor, if the odor is indeed a reliable, omnipresent factor.
- Reduced reliability may be associated with scenarios such as:
    - o Where an odorization system exists and is minimally maintained (by minimum regulatory standards, perhaps) but the evaluator does not feel that enough extra steps have been taken to make this a high-reliability system, the assessment may show reduced reliability.
- Questionable odorization system may be associated with scenarios such as:
    - o A system exists; however, the evaluator has concerns over its reliability or effectiveness. Inadequate record keeping, inadequate maintenance, lack of knowledge among system operators, and inadequate inspections would all indicate this condition. A history of odorization system failures would be even stronger evidence.
- Absence of odorization means the assessed distribution system is carrying higher potential consequences, compared to otherwise equivalent systems.

A formal event tree or fault tree analysis can be used to estimate the fraction of leaks whose consequence scenarios may be reduced by odorization. Experience has shown that the fraction is fairly high. It may be difficult to, even in an imagineering exercise, separate this factor since it has been a part of most distribution system transportation for so long. Scenarios of un-odorized gas would rely on naturally occurring odors as well as sound and sight indications of nearby leaks. Such indications would not be universally recognized and may even invite investigation, putting persons at increased risk.

Given the location- and situation-specific benefits derived from odorization, as well as the ample margin between detection levels and flammability levels, human injury/fatality reduction estimates of over 90% or even 99% compared to un-odorized systems would not seem unreasonable. Such values suggest that in only one in ten to one in one hundred incidents would exposed populations not be alerted to the danger and subsequently be able to reduce their chance of harm.

Facilities

Hydrocarbon stations often have several levels of monitoring systems (e.g., relief device, tank overfill, tank bottom, seal piping, and sump float sensors/alarms), operations systems (e.g., SCADA, flow-balancing algorithms), secondary containment (e.g., seal leak piping, collection sumps, equipment pad drains, tank berms, stormwater controls), and emergency response actions. Therefore, small liquid station equipment-related leaks are designed to be detected and remedied before they can progress into large leaks. If redundant safety systems fail, larger spills can often be detected quickly and contained within station berms. Where a leaking liquid can accumulate under or be rinsed from station facilities, stormwater (prior to discharge) or groundwater can be gathered and sampled for hydrocarbon contamination, enabling the detection of very small leaks.

Gaseous product pipeline stations often control compressor or pressure relief discharges by venting the gas through a vent stack within the station. In the case of high-pressure/volume releases, large-diameter flare stacks (with a piloted ignition flame) combust vented gases into the atmosphere. Gas facilities are normally leak checked periodically and remotely monitored for equipment or piping leaks.

11.7.10.5 Evaluation of leak detection capabilities

The most suitable method of leak detection depends on a variety of factors including the type of product, flow rates, pressures, the amount of instrumentation available, the instrumentation characteristics, the communications network, the topography, the soil type, and economics. Some systems are designed or calibrated for certain leak rates or spill volumes, with reduced sensitivities for leaks outside of their optimum ranges. Multiple systems, offering redundancy and/or capabilities to detect wider ranges of leak rates, are common. As previously mentioned, when highly sophisticated instruments are employed, a trade-off often takes place between the sensitivity and the number of false alarms, especially in "noisy" systems with high levels of transients.

The operator's use of established procedures to positively locate a leak can be included in this evaluation. Follow-up actions including the use of leak rates to assess system integrity and the criteria and procedures for leak repair should also be considered.

In assessing the potential benefits—consequence mitigation in the form of spill volume reduction—from leak detection, some conclusions from a detailed study are relevant.

1. The pipeline controller/control room identified a release occurred around 17% of the time.
2. Air patrols, operator ground crew and contractors were more likely to identify a release than the pipeline controller/control room.
3. An emergency responder or a member of the public was more likely to identify a release than air patrols, operator ground crew and contractors.

4. A CPM LDS was the leak identifier in 17 (20%) out of 86 releases where a CPM system was functional at the time of the release.
5. SCADA was the leak identifier in 43 (28%) out of 152 releases where a SCADA was functional at the time of the release.
6. For hazardous liquid pipelines, SCADA or CPM systems by themselves did not appear to respond more often than personnel on the ROW or members of the public passing by the release incident.
7. It appeared that procedures may have allowed alarms to be ignored by controllers in several of the larger volume releases or to re-start pumps or open a valve, thus aggravating the size of the release.
8. Large distances between block valves may also have been a contributory factor in the size of the release. (Kiefner)

[1011]

The evaluator should assess the nature of leak detection abilities in the pipeline section he is evaluating. The assessment should include:

- What size leak can be reliably detected
- How long before a leak is positively detected
- How accurately can the leak location be determined.

A leak detection capability can be defined as the relationship between leak rate and time to detect. This relationship encompasses both volume-dependent and leak-rate-dependent scenarios. The former is the dominant consideration as product containment size increases (larger diameter pipe at higher pressures), but the latter becomes dominant as smaller leaks continue for long periods.

As shown in Figure 11.7, this relationship can be displayed as a curve with axes of "Time to Detect Leak" versus "Leak Size." The area under such a curve represents the worst case spill volume, prior to detection. The shape of this curve is logically asymptotic to each axis because some leak rate level is never detectable and an instant release of large volumes approaches an infinite leak rate.

Many leak detection systems perform best for only a certain range of leak sizes and therefore require independent evaluation. Overlapping leak detection capabilities are usually present in a pipeline, often with reliance on equipment and instruments located in stations. In assessing station leak detection capabilities, all opportunities to detect can be considered producing curves for each type of leak detection as well as for the combined capabilities at the station. A leak detection capability curve can be developed by estimating, for each pipeline component, the leak detection capabilities of each available method for a variety of leak rates. A listing of leak rates is first created. For each leak rate, each detection system's time to detect is estimated. When a detection system reacts at a certain spill volume, then various leak rate-duration pairings will result in that system being triggered. For instance, if a detection system responds when 10 gallons of leak volume is present (perhaps a hydrocarbon sensor in a sump), then that system reacts when a 1 gallon/hr leak persists for 10 hrs, or a 0.5 gallon/min leak persists for 20 minutes, etc.

In assessing leak detection capabilities, all opportunities to detect should be considered. Therefore, all leak detection systems available should be evaluated in terms of their respective abilities to detect various leak rates. A matrix can be used for this.

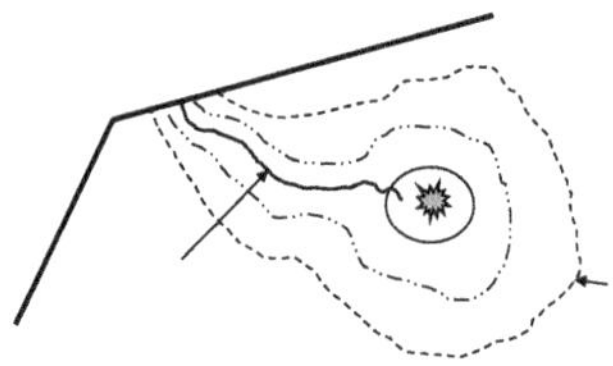

Refs [3] and [4] discuss SCADA-based leak detection systems and offer methodologies for evaluating their capabilities. Other techniques will likely have to be estimated based on time between observations and the time for visual, olfactory, or auditory indications to appear. The latter will be situation dependent and include considerations for spill migration and evidence (soil penetration, dead vegetation, sheen on water, etc.). The total leak time will involve detection, reaction, and isolation time.

As a further evaluation step, an additional column can be added to the matrix for estimates of reaction time for each detection system. This assumes that there are differences in reactions, depending on the source of the leak indication. Reaction time includes estimates of how long it would take to isolate and contain the leak, after detection. This recognizes that some leak detection/reaction opportunities, such as 24–7 staffing of a station, provide for more immediate reactions compared to patrol or off-site SCADA monitoring. A series of SCADA alarms will perhaps generate more immediate reaction than a passerby report that is lacking in details and/or credibility. The former scenario has an additional advantage in reaction, since steps involving telephone or radio communications may not be part of the reaction sequence. Such considerations can be factored into assessments that place values on various leak detection methodologies.

In Germany, the Technical Rule for Pipeline Systems (TRFL) covers:

- Pipelines transporting flammable liquids.
- Pipelines transporting liquids that may contaminate water, and
- Most pipelines transporting gas.

It requires these pipelines to implement an LDS, and this system must at a minimum contain these subsystems:

- Two independent LDS for continually operating leak detection during steady state operation. One of these systems or an additional one must also be able to detect leaks during transient operation, e.g. during start-up of the pipeline.

These two LDS must be based upon different physical principles.

- One LDS for leak detection during shut-in periods.
- One LDS for small, creeping leaks.
- One LDS for fast leak localization.

Most other international regulation is far less specific in demanding these engineering principles. It is very rare in the U.S. for an operator to implement more than one monolithic leak detection system.

Facility Staffing

Staffing, as a means of leak detection, is seen to supplement and partially overlap any other means of leak detection that might be present. As such, the staffing level leak detection can be combined with other types of leak detection. The benefit is normally more of a redundancy rather than an increased sensitivity. This recognizes the benefit of a secondary system that is as good or almost as good as the first line of defense, with diminishing benefit as the secondary system is less effective.

An simple approach to evaluating the staffing level as it adds leak detection capability is to consider the maximum interval in which the station is unmanned, ie, the time that staffing as leak detection is unavailable:

Leak detection capability = maximum interval unobserved

This is basing the capability on the worst case detectability. As an opportunity to detect and react to a leak, the staffing level of a facility can be more fully evaluated by considering the following relationship:

Opportunity to detect = [(inspection hours) + (happenstance detection)]

Where

Inspection hour	=	an inspection that occurs within each hour
Happenstance detection	=	% of manned time per week.

In this relationship, it is assumed that station personnel would have a certain % chance of detecting any size leak while they were on site. This is of course a simplification since some leaks would not be detectable and others (larger in size) would be 100% detectable by sound, sight, or odor. Additional factors that are ignored in the interest of simplicity include training, thoroughness of inspection, and product characteristics that assist in detectability.

The maximum unobserved interval method is simple, but it appears worthwhile to also consider the slightly more complicated "opportunity" method, since the "max interval" method ignores the benefit of actions taken while a station is manned, that is, while performing formal inspections of station equipment—rounds. The "opportunity" method shows benefits that more closely agree with the belief that more directed attention during episodes of occupancy (performing inspection rounds) are valuable.

Various 'staffing of stations' scenarios can be evaluated in terms of their leak detection contributions and those contributions can be a part of the overall risk assessment. A drawback of an incomplete "opportunity" scheme would be the inability to show preference of a 1 hr per day / 5 days per week staffing protocol over a 5 hours / 1 day per week protocol, even though most would intuitively believe the former to be more effective. A 7–24 staffing arrangement, with formal inspection rounds, logically has leak detection capabilities far superior to than a weekly station visit.

Added to the detection time is the reaction time, which is generally defined as the amount of additional time that will probably elapse between the strong leak indication and the isolation of the leaking facility (including drain downtime). Here, consideration can be given to automatic operations, remote operations, proximity of shutdown devices, etc. Benefits of remote and automatic operations as well as staffing levels should be captured in the risk assessment.

11.7.11 Emergency response

Emergency response, as used here, focuses on on-site actions taken during the unfolding of a pipeline release. Leak detection, leak isolation, and automatic/semi-automatic equipment available to reduce hazard areas are included in the assessment elsewhere, as previously discussed.

Emergency response effectiveness in reducing hazard zones and damage rates can be assessed by first recognizing the two different ways that such actions impact consequence scenarios. The first is reducing hazard areas—normally by spill volume reductions—and the second is limiting losses within the hazard zone.

11.7.11.1 Reducing Damage Potential

As noted previously, the area of opportunity can sometimes be limited by protecting or removing vulnerable receptors, by removing possible ignition sources, or by limiting the extent of the spilled product.

A. Evacuation. Under the right conditions, emergency response personnel may be able to safely evacuate receptors (usually people) from the hazard area. To do this, they must be trained in pipeline emergencies. This includes having pipeline maps, knowledge of the product characteristics, communications equipment, and the proper equipment for entering to the danger area (breathing apparatus, fire-retardant clothing, hazardous material clothing, etc.). Obviously, entering a dangerous area in an attempt to evacuate people is a situation-specific action. The evaluator should look for evidence that emergency responders are properly trained and equipped to exercise any reasonable options after the situation has been assessed. Again, the criteria must include the time factor. Damage rates within hazard zones can be assessed to be lower for scenarios where evacuation plays a significant role.

B. Blockades. Another limiting action in this category is to limit the possible ignition sources and the entry of additional receptors. Preventing vehicles from entering the danger zone has the double benefit of reducing human exposure and reducing ignition potential.

C. Containment. Especially in the case of restricting the movement of hazardous materials into sewers, buildings, groundwater, etc, quick containment can reduce the consequences of the spill. To reduce the spreading potential during emergency response, equipment such as booms, absorbents, vacuum trucks, dispersion or neutralizing agents, and others are available. Some of these act as temporary secondary containment. Permanent forms of secondary containment were previously discussed.

D. Shielding. Protecting receptors by the use of thermal or blast walls is an option in some cases. These structures are sometimes used in production facilities, protecting control rooms and other areas of normal human occupancy. They are less common, but nonetheless an available option, for limiting consequences beyond a facilities borders.

E. Pre-emptive actions. Some operators allow responders (company personnel only) to ignite releases in instances where such action would limit damages that may occur by later ignition. Using a flare gun to ignite a vapor cloud, is an example of such a procedure. Such procedures also add an amount of risk. The potential for unintended consequences is relatively high, given the uncertainties and high energy release potentials associated with unconfined vapor cloud explosions and vapor fires. With a limited ability to fully diagnose an unfolding scenario, such actions should be very carefully considered.

11.7.11.2 Loss limiting actions

Prompt and proper medical care of persons affected by releases can reduce losses. Again, product knowledge, proper equipment, proper training, and quick action on the part of the responders are necessary factors.

Other items that play a role in achieving the consequence-limiting benefits include the following:

- Emergency drills
- Emergency plans
- Communications equipment
- Proper maintenance of emergency equipment
- Updated phone numbers readily available
- Extensive training including product characteristics

- Regular contacts and training information provided to fire departments, police, sheriff, highway patrol, hospitals, emergency response teams, government officials.

These can be thought of as characteristics that help to increase the chances of correct and timely responses to pipeline leaks. Perhaps the first item, emergency drills, is the single most important characteristic. It requires the use of many other list items and demonstrates the overall degree of preparedness of the response efforts.

Equipment that may need to be readily available includes:

- Hazardous waste personnel suites
- Breathing apparatus
- Containers to store picked up product
- Vacuum trucks
- Booms
- Absorbent materials
- Surface-washing agents
- Dispersing agents
- Freshwater or a neutralizing agent to rinse contaminants
- Wildlife treatment facilities.

The evaluator/operator should look for evidence that such equipment is properly inventoried, stored, and maintained. Expertise is assessed by the thoroughness of response plans (each product should be addressed), the level of training of response personnel, and the results of the emergency drills. Note that environmental cleanup is often contracted to companies with specialized capabilities.

11.8 RECEPTORS

A receptor is anything that could "receive" damage from a pipeline leak/rupture. It includes all biological life forms, structures, land areas, etc. Some possible receptor types include: people (human fatality; human injury); property; environment; and even service, when 'service interruption' is part of the definition of failure.

The damage potential of various receptors should be based on the vulnerability and consequence potential of each receptor-spill pairing. This includes direct damages and secondary effects such as public outrage.

Understanding the damage threshold leads to a hazard area estimation and the ability to characterize receptor vulnerability within that hazard area. In the earlier discussion of hazard area determination, it was shown that receptor damage potential sets the boundaries for the hazard area. However, the suggestion was made to initially ignore receptors after their role in thresholds was acknowledged, in producing the hazard areas around the pipeline components. The areas are efficiently produces using only the threshold intensity values. Damage threshold levels for thermal radiation and overpressure intensity effects were discussed earlier in this chapter.

After the hazard areas have been 'drawn', then the counting, valuations, and potential damage rates of receptors can be efficiently included in the assessment.

11.8.1 Receptor vulnerabilities

Receptor sensitivities are an aspect that should be considered in the consequence assessment. Receptor damage is dependent upon the nature of the scenario—acute versus chronic—as well as the intensity. Longer duration, higher intensity events generally cause the most damage; low intensity, short duration usually cause the least, and many possibilities exist between the extremes. Included with chronic impacts—consequences that tend to worsen over time—is secondary effects. This includes fires ignited and/or spreading by autoignition from heat flux; explosions such as BLEVE's; soot and ash fallout; pollution; etc. damages from more persistent pipeline releases also include contamination scenarios.

Valuations and sensitivities require certain information, even if only simplified assumptions. For each receptor, such as population, environment, drinking water, waterways, etc., key information needed for valuations includes:

- Receptor characterization (type of people, type of buildings, water flowrates, etc.)
- Receptor density (count per area unit)
- Receptor vulnerabilities (susceptibility to harm at various exposure intensities and durations)
- Shielding, distance, and mobility (ability to escape) of receptors

An "estimate of risk expressed in an absolute terms" requires identification of a hazard zone, a characterization of receptors within that zone, and an estimate of the extent of damages to those receptors. The levels of damage possible and their associated likelihood of occurrence require an understanding of receptor sensitivity to the effect. A dose–response type assessment, as is often seen in medical or epidemiological studies, may be necessary for certain receptors and certain threats. Focusing on possible acute damages to humans, property, and the environment, some simplifying assumptions can be made, as discussed below.

As noted, a robust consequence assessment sequence will generally follow these steps:

1. Determine damage states of interest (see discussions this chapter)
2. Calculate hazard distances associated with damage states of interest
3. Estimate hazard areas based on hazard distances and source (burning pools, vapor cloud centroid, etc.) location
4. Characterize receptor vulnerabilities (damage potential) within the hazard areas

This process is rather essential to absolute risk calculations. Having addressed the first three in earlier sections of this chapter, we now turn our attention to the fourth.

An important benefit of the more complex GIS spill footprint analysis (over the older, buffer zone) approaches is the ability to better characterize the receptors that are potentially exposed to a spill—those that are actually "in harm's way." In many cases, receptors may be relatively close to, but upslope of, the pipeline and hence at much less risk in a liquid spill scenario. Focusing on the locations that are more at risk is obviously an advantage in risk management.

The probability of various damage levels to various receptors requires an understanding of very location-specific factors such as escape potential, shielding and sheltering options, wind direction, and many others. General assumptions are used in many risk analyses including several detailed in PRMM. Listings such as this provide insight into those authors' beliefs about receptor damage potential.

11.8.2 Population

Most pipeline release consequence assessments focus on threats to humans, especially threats to the general public. Risks specific to pipeline operators and pipeline company personnel can be included, often as a separate classification in order to discriminate between voluntary and involuntary risks.

Potential injury and fatality counts relies on an understanding of the population within the potential hazard zone must be characterized. Hazard intensities and durations, coupled with population densities, characteristics, and protections at any point in time, yield injury and fatality potentials. Characterization of a population vulnerability includes estimating:

- Permanent vs Transitory/occasional population density
- Special population (restricted mobility)
- Barriers, shielding, and escape capabilities.

Even within a hazard zone, there are differences in level of harm. In addition to thermal effects being very sensitive to receptor proximities, the potential for ingesting, inhaling, and having dermal contact with contaminants may be higher at some locations if less dilution has occurred and there is less opportunity for detection and remediation before the normal pathways are contaminated. Recall that common pathways for contact with humans is through direct contact (with skin, eyes, etc), or via an ingestion/inhalation pathway: air, drinking water, vegetation, fish, or others.

Especially for acute hazard zones scenarios, a detailed analyses of human health effects is often unnecessary when the pipeline's products are common and epidemiological effects are well known. However, more advanced assessment techniques are available, as is illustrated in the discussion of probit equations. These may be needed to determine cleanup and remediation requirements for more chronic hazard zone scenarios.

In either a simple or advanced assessment, understanding the potential for injury or fatality from thermal effects requires consideration of the time and intensity of exposure. This is discussed in PRMM and methods for quantifying these effects are available. Shielding and ability to evacuate are critical assumptions in such calculations.

11.8.2.1 Population Density

Most risk assessments use the simple and logical premise that risk increases as nearby population density increases. Population density estimates are often already available along a pipeline. Many operators, by choice or regulatory mandate, use published population density scales such as the class locations 1, 2, 3, and 4 (used in US regulations (CFR49 Part 192) These are for rural to urban areas, respectively.

Sometimes landuse data along a pipeline is available and can be used for characterizing population densities categories such as urban, rural, light residential, heavy commercial (shopping center, business complex, etc.), and many others appear in various landuse categorizations. These can be converted into population densities.

The population density, as measured by class location or other categorization, or when based on large geographical areas is an inexact method of estimating the number of people likely to be impacted by a pipeline failure. A thorough analysis will make more accurate counts and characterizations of buildings, roadways, assembly areas, and other indicators of population. It will also necessarily require estimates of people density (instead of building density), people's away-from-home patterns, nearby road traffic, evacuation opportunities, time of day, day of week, and a host of other factors. Several methods can be devised to incorporate at least some of these considerations. An example methodology, from an Australian standard ref [67], illustrates this. According to this ref [67], average population densities per hectare can be determined for a particular land use by applying the following formula:

Population per hectare = [10,000/(area per person)] x (% area utilized) x (% presence)

This reference describes the process of population density estimation as follows:

- Indoor population densities have been based on the number of square meters required per person according to the local building code. Residential dwellings are not covered in this building code, but have been assigned a value of 100 m^2 per person, on the basis of a typical suburban density of 30 persons per hectare and one-third actual dwelling area. For nonresidential use, available floor space has been set at 75% of the actual area, to allow for spaces set aside for elevators, corridors, etc.

- For rural and semirural areas, the outdoors population is generally expected to be greatest on major roads (excluding commercial areas). If an appropriate value for vehicular populations can be determined, then this can be conservatively applied to all outdoor areas. Assuming that a major rural road is 10 m wide, 1 hectare covers a total length of 1km. For rural areas, an average car speed of 100km/hr and an average rate of 1 car per minute has been assumed. Based on this and an average of 1.2 persons per car, an outdoor population density of 1 person per hectare has been determined. Using 60km/hr and a 30-second average separation, a population density of 4 people per hectare is applied to semirural areas.

Other typical population densities from another source [43] are shown in:

Table 11.8
Population density by location class

Class	Average population density (people per hectare)
1	0.04
2	3.3
3	18
4	100

Assessments of occupancies based on time-of-day, day-of-week, and/or season, traffic volumes on roadways, and populations associated with offshore locations or activities (for example, platforms, shipping lanes, anchoring areas, fishing areas, coastal proximity, etc.) will strengthen the risk analyses. Identification of individuals with reduced escape capabilities, such as restricted mobility populations (nursing homes, rehabilitation centers, etc.) and difficult-to-evacuate populations, may be warranted.

Especially for early phase risk assessments, rule sets can be developed to assign exposures. For instance, in the offshore environments, water depths and/or shore proximity can be used to set initial estimates of populations associated with fishing and recreational activities. Shipping lane proximity can influence estimates of transient populations moving near a facility.

11.8.2.2 Probit

PROBIT is a method to take into account the total damage received by the receptor. For consequences requiring an understanding of the dosage influences, this represents an improvement over a fixed limit approach since time of exposure is included in the analysis. A higher intensity of exposure can be safely absorbed if the exposure time is less, so a measure of 'dose' is more representative of actual damages. Probit equations are based on experimental dose-response data. According to probit equations, all combinations of concentration and time that result in an equal dose also result in equal values for the probit and therefore produce equal expected fatality rates for the exposed population. When using a probit equation, the value of the probit (P r) that corresponds to a specific dose must be compared to a statistical table to determine the expected fatality rate.

An example of the use of probits in common pipeline failure consequence effects (thermal and overpressure) is excerpted below:

> The physiological effects of fire on humans depend on the rate at which heat is transferred from the fire to the person, and the time the person is exposed to the fire. Even short-term exposure to high heat flux levels may be fatal. This situation could occur to persons wearing ordinary clothes who are inside a flammable vapor cloud (defined by the lower flammable limit) when it is ignited. In risk analysis studies, it is common practice to make the simplifying assumption that all persons inside a flammable cloud at the time of ignition are killed and those outside the flammable zone are not.

In the event of a torch fire or pool fire, the radiation levels necessary to cause injury to the public must be defined as a function of exposure time. The following probit equation for thermal radiation was developed for the U.S. Coast Guard [1045]:

$$Pr = -36.378 + 2.56 \ln [t (I^{4/3})]$$

Where: t = exposure time, seconds
I = effective radiation intensity, W/m^2

> The physiological effects of explosion overpressures depend on the peak overpressure that reaches the person. Direct exposure to high overpressure levels may be fatal. If the person is far enough from the edge of the exploding cloud, the overpressure is incapable of directly causing fatal injuries, but may indirectly result in a fatality. For example, a blast wave may collapse a structure which falls on a person. The fatality is a result of the explosion even though the overpressure that caused the structure to collapse would not directly result in a fatality if the person were in an open area.
>
> In the event of a vapor cloud explosion, the overpressure levels necessary to cause injury to the public are typically defined as a function of peak overpressure, without regard to exposure time. Persons who are exposed to explosion overpressures have no time to react or take shelter; thus, time does not enter into the relationship. An example probit relationship based on peak overpressure is as follows:
>
> $$Pr = 1.47 + 1.37 \ln (p)$$
>
> Where: p = peak overpressure, psig
>
> The following explosion/lethality relationships have been used.

p = 1 psig	1% mortality
p = 5 psig	50% mortality
p = 7 psig	95% mortality

11.8.2.3 Generalized damage states

Historical data on fatal accidents involving natural gas gathering and transmission pipelines have been compiled by the U.S. Department of Transportation (DOT). During a recent 14.5 year period for which summary data are available, the maximum number of fatalities due to any single accident was six, and two accidents actually caused six fatalities.

Numerous studies and publications are available dealing with the potential extent of injury from exposures to various toxic, thermal, and mechanical effects. These lead to more general assumptions that can be used to set overall damage states. Under a set of assumptions, one study concluded a full rupture of a natural gas transmission pipeline produces a 1% mortality rate at distances corresponding to

$$r = 0.685 \times SQRT(p \times d)^2$$

Where
r = radius from pipe release point for given radiant heat intensity (feet)
p = maximum pipeline pressure (psi)
d = pipeline diameter (inches).

This study [83] used an approximate exposure time of 30 seconds and several other assumptions to set a suggested damage threshold at a thermal radiation (from ignited natural gas release) level 5,000 Btu/ft^2-hr. Distances suggested by this equation have become the hazard zone at which the designation of HCA is applied in US regulations for natural gas transmission pipelines.

In a related study, other mortality rates are linked to distances dependent on pressure and diameter.

Two hazard areas are defined that correspond to the lower and upper heat intensity thresholds associated with fatal injury. The lower and upper thresholds adopted are 12.6 and 31.6 kw/m2 for outdoor exposure, and 15.8 and 31.6 kw/m2 for indoor exposure. The probability of fatality is assumed to be 100% within the area bounded by the upper threshold and 0% outside of the area bounded by the lower threshold. Between these two thresholds, the probability of fatality is assumed to be 50% for outdoor exposure and 25% for indoor exposure. [333]

Another study of thermal radiation impacts from ignited pools of gasoline assumes the following:

- There is a 100% chance of fatality in pools of diameter greater than 5m.
- The fatality rate falls linearly to 0% at a thermal radiation level of 10kW/m^2 [59].

Due to the sensitive nature of fatality rate potential, extra caution in producing such estimates is warranted. A risk model with a conservative bias intended to support technical decision-making can have its output mis-used and can generate misunderstanding and unnecessary alarm. This potential is exacerbated when an emotionally-charged measure such as fatality possibility is being used as a measure of CoF. Given the conceptual difficulties in population-based estimates versus estimates for individual segment risks, the potential for misunderstanding is increased.

11.8.2.4 Value of statistical life and injury

Establishing a value of human life—a "statistical life," not an identified individual—is an emotional and controversial thing, as discussed in PRMM. Despite some on-going resistance, such valuations are becoming commonplace. Not only do they provide logical and necessary inputs to decision-makers, they are already ubiquitous, in the sense that any company's decision-making can be dissected to reveal a de facto valuation on human life, even if one is not explicitly stated.

Valuations in the US currently range from about $5 million up to about $15 million per statistical fatality avoided [1044]. To select a single estimate without researching the rationale behind the many valuations used for many different purposes, values used by government agencies in determining cost/benefit of proposed regulations can be used. This perhaps has the added benefit of de-personalizing the choice in value—it is not generated by the asset owner but rather by an unbiased agency representing the public interest.

A 2013 memorandum [1045] published by the US DoT provides guidance on values of statistical life (VSL), suggesting a value of $9.1 million be used and annually adjusted proportionally with changes in real income (estimated to increase by factor of 1.07 percent per year, for use in estimating future).

For future years, the formula for calculating future values of VSL is therefore:

$$VSL_{2012+N} = VSL_{2012} \times 1.0107^{N}$$

where VSL_{2012+N} is the VSL value N years after 2012

and VSL_{2012} is the VSL value in 2012 (i.e., $9.1 million).

Among its objectives in publishing this guidance, this ref [1044] states:

> Prevention of an expected fatality is assigned a single, nationwide value in each year, regardless of the age, income, or other distinct characteristics of the affected population, the mode of travel, or the nature of the risk. When Departmental actions have distinct impacts on infants, disabled passengers, or the elderly, no adjustment to VSL should be made, but analysts should call the attention of decision-makers to the special character of the beneficiaries.

This same ref [1044] offers guidance on economic valuations for injuries:

- Nonfatal injuries are far more common than fatalities and vary widely in severity, as well as probability.
- Each type of accidental injury is rated (in terms of severity and duration) on a scale of quality-adjusted life years (QALYs), in comparison with the alternative of perfect health. These scores are grouped, according to the Abbreviated Injury Scale (AIS), yielding coefficients that can be applied to VSL to assign each injury class a value corresponding to a fraction of a fatality.

The fractions shown Table 11.9 should be multiplied by the current VSL to obtain the values of preventing injuries of the types affected by the government action being analyzed.

Table 11.9
Relative Disutility Factors by Injury Severity Level (AIS)
For Use with 3% or 7% Discount Rate

AIS Level	Severity	Fraction of VSL
AIS 1	Minor	0.003
AIS 2	Moderate	0.047
AIS 3	Serious	0.105
AIS 4	Severe	0.266
AIS 5	Critical	0.593
AIS 6	Unsurvivable	1.000

Another reference states that, based on a willingness-to-pay study of road accidents, costs of serious and slight injuries are approximately 10% and 0.8% of the cost of a life, respectively.

The use of valuations for human suffering and fatality is a source of discomfort for some. Realistically, however, such valuations have always been implicitly employed, though often not documented. Failure to document does not prevent a company's VSL beliefs from being known. A company's implied VSL valuations, used in their decision-making, can be derived by their choices in design, operations, and maintenance practices, coupled with their incident history or some other representative history.

11.8.2.5 Historical Losses

It is useful to examine historical rates of population effects. In the US, the following rates have been observed, based on reporting of 'significant' and 'serious' pipeline incidents. For an approximate time period of 1992 to 2012, the following costs per incident were reported.

Table 11.10
Examples of human fatality/injury rates

	Hazardous Liq			Gas Transmission			Gas Distribution		
	fat/incid	inj/ incid	$prop/incid	fat/incid	inj/ incid	$prop/incid	fat/incid	inj/ incid	$prop/incid
max	0.026	0.179	2,704,031	0.197	0.545	3,649,280	0.427	2.027	2,952,663
avg	0.008	0.040	478,195	0.028	0.125	698,084	0.115	0.436	327,653
min	0.000	0.000	112,248	0.000	0.008	171,443	0.040	0.214	112,894

The maximum and minimum values are the highest annual per-incident rates in the time period. These values suggest the range of possibilities, at least for annual counts. Note that these are related to a certain type of incident, ie, 'significant' or 'serious'. Rates for all incidents would logically be much lower.

11.8.3 Property-related Losses

Property damage potential can be assessed through an examination of the following variables: population, property type (commercial, residential, industrial, etc.), property value, landscape value, roadway vulnerability, and highway vulnerability, and other considerations. Damage rates can be correlated to the threshold intensity effects tabulated in a previous section. Valuations are readily obtained from numerous published data on market values and construction/reconstruction costs for a region. Examples follow.

11.8.3.1 Damage Rates

One study used the PIR based on thermal intensities from natural gas jet fires (see previous discussion) to represent potential damage rates. With an observation that each combination of heat flux and duration associated with particular levels of damage falls at a specific normalized multiple of the PIR, the following damage distances emerge, expressed as a multiple of the PIR distance: ~1.6*PIR for severe damage, ~0.75*PIR for moderate damage, and ~0.5*PIR for minor damage.

Using these categories and various assumptions, a US government study [1015] assigned the following valuations:

- Severe indicates that a house is not safe to occupy and most likely needs to be demolished or completely renovated prior to occupancy. Valuations are set at 100% loss of $180K per building/house.
- Moderate indicates that a house has substantial damage and repairs are necessary prior to occupancy. Valuations for such damages are set at 50% of replacement value.
- Minor indicates that a house has the least amount of damage and could be legally occupied while repairs are made. Valuations for such damages are set at 20% of replacement value. [1015]

11.8.3.2 Costs

In the same study, density of dwellings was set at 12/acre or 6/building. For buildings with 4 or more stories, under a set of assumptions including a density of 0.5/acre, costs to repair minor damage from thermal effects of a pipeline release were set at $500K and moderate to severe damage set at $1,000K. Outside recreational facilities had valuations set at $250K for minor and $500K for moderate to severe damages. Parked vehicles, with an assumed densities ranging from 24 to 100 per acre, had damage valuations set at 0%, 30%, and 100%--corresponding to minor, moderate, severe, thermal radiation levels respectively—of a vehicle $17K retail value. Personal possessions that may be destroyed inside a building had damage valuations set at 5%, 15%, and 25% of building valuation--corresponding to minor, moderate, severe, thermal radiation levels respectively—yielding valuations of $9K, $27K, and $45K. [1015]

In another study, a sampling of 'above average incidents' costs is found in ref [1011]. Some of the oil spills listed in that reference are shown below.

Where incidents are 'reportable' per US regulations and costs are "Estimated cost of public and non-Operator private property damage paid/reimbursed by the Operator". Some of these incidents involved fatalities and injuries, but most represent property damage costs.

An examination of one set of US reportable incident data shows average property damage costs of about $700K per incident for natural gas transmission pipelines; $330K for natural gas distribution pipelines, and from $480K per incident for hazardous liquid pipelines. (See Table 11.11) Note that these types of pipeline operations have different criteria for 'reportable'. The hazardous liquid statistic involves many more incidents, normally very minor (for example, small leaks in facilities), accounting for the non-intuitive higher property damage costs for natural gas releases for which only relatively major incidents are reported. Note also that these 'per incident' costs are based on a subset of all incidents—costs per 'any' incident would logically be much lower.

Some key aspects of property damage potential will track population density. Therefore, property loss can also be estimated based on population density, in the absence of more definitive data.

Table 11.11
Sample incident costs

	gal/mcf	cost	$/gal or MCF
Crude Oil	843,444	725,000,000	$860
Crude Oil	158,928	4,194,715	$ 26
HVL (LPG/NGL)	137,886	1,811,756	$ 13
HVL (LPP/NGL)	130,368	524,275	$ 4
Gasoline	81,900	15,000,005	$ 183
Crude Oil	63,378	135,000,000	$ 2,130
Crude Oil	43,260	989,000	$ 23
Refined Products	38,640	13,184,000	$ 341
Refined Products	34,356	7,657,195	$ 223
Crude Oil	33,600	441,000	$ 13
Refined Products	29,988	831,750	$ 28
Natural Gas	83,487	734,698	$ 9
Natural Gas	79,000	1,883,770	$ 24
Natural Gas	61,700	2,310,000	$ 37
Natural Gas	50,555	6,700,000	$ 133
Natural Gas	47,600	375,363,000	$ 7,886
Natural Gas	41,176	406,699	$ 10
Natural Gas	34,455	117,000	$ 3
Natural Gas	14,980	116,000	$ 8

11.8.4 Environmental issues

Environmental damages are often very situation dependent given the wide array of possible biota that can be present and exposed for varying times under various scenarios. Environmental risk factors will overlap public safety risk factors to a large extent. See PRMM for a background discussion of environmental risk assessment.

11.8.4.1 Environmental sensitivity

Every potential spill site has some degree of sensitivity to a pipeline release. The environmental effects of a leak are partially recognized in the product hazard assessment. Liquid spills are generally more apt to be associated with chronic hazards. The modeling of liquid dispersions is a very complex undertaking as previously described.

In a risk assessment, there is usually an increased focus on more environmentally sensitive areas, with the implication that these locations carry a potential for greater or more lasting harm than most other locations. Areas more prone to damage and/or more difficult to re-mediate can be highlighted in the risk assessment. A strict definition of environmentally sensitive areas might not be absolutely necessary. A working definition by which most would recognize a sensitive area might suffice. Such a working definition would need to address rare plant and animal habitats, fragile ecosystems, impacts on biodiversity, and situations where conditions are predominantly in a natural state, undisturbed by man. To more fully distinguish sensitive areas, the definition should also address the ability of such areas to absorb or recover from contamination episodes.

The chronic aspect of a spill assesses the hazard potential of the product via characteristics such as aquatic toxicity, mammalian toxicity, chronic toxicity, potential carcinogenicity, and environmental persistence (volatility, hydrolysis, biodegradation, photolysis).

One method to quantify spill costs, specifically for oil spills, is available in ref [1030], with some excerpts below:

> To provide the EPA Oil Program Center with a simple, but sound methodology to estimate oil spill costs and damages, taking into account spill-specific factors for cost-benefit analyses and resource planning, the EPA Basic Oil Spill Cost Estimation Model (BOSCEM) was developed. EPA BOSCEM was developed as a custom modification to a proprietary cost modeling program, ERC BOSCEM, created by extensive analyses of oil spill response, socioeconomic, and environmental damage cost data from historical oil spill case studies and oil spill trajectory and impact analyses. In addition, elements of habitat equivalency analysis as applied in Natural Resource Damage Assessment (NRDA) and other environmental damage estimation methods, such as Washington State's Damage Compensation Schedule and Florida's Pollutant Discharge Natural Resource Damage Assessment Compensation Schedule were incorporated into the environmental damage estimation portion of ERC BOSCEM. Formulae, criteria, and cost modifier factors for estimating socioeconomic damages, including impacts to local and regional tourism, commercial fishing, lost-use of recreational facilities and parks, marinas, private property, and waterway and port closure, were derived from historical case studies of damage settlements and costs, as well as methods employed in other studies
>
> Input of spill criteria:
>
> 1. Specify amount of oil spilled (in gallons);
> 2. Specify basic oil type category;
> 3. Specify primary response methodology and effectiveness;
> 4. Specify medium type of spill location;
> 5. Specify socioeconomic and cultural value of spill location;
> 6. Specify freshwater vulnerability category of spill location;
> 7. Specify habitat and wildlife sensitivity category of spill location.
>
> Each oil spill is a ***unique event*** involving the spillage or discharge of a particular type of oil or combination of oils that may cause damage to the local and/or regional environment, wildlife, habitats, etc., as well as to third parties. No modeling method can ever exactly determine or predict costs of an oil spill.

Yet, there are patterns that emerge with respect to damages upon detailed analyses of oil spill case studies. For example, heavier oils are more persistent and present greater challenges – and thus costs – in oil removal operations than lighter oils, such as diesel fuel. Heavier oils, being more visible and persistent, have greater impacts on tourist beaches and private property. At the same time, lighter oils with their greater toxicity and solubility are more likely to cause impacts to groundwater and invertebrate populations. Greater effectiveness in oil removal tends to reduce environmental damages and socioeconomic impacts. Other factors, such as spill location, can also have significant impacts on spill costs and damages. A diesel fuel spill in an industrial area will likely have less impact and require a less expensive cleanup than one that occurs in or near a sensitive wetland. EPA BOSCEM incorporates these types of factors into a simple methodology for estimating the costs of "types of spills" that may be analyzed in a cost benefit analysis or for assessing which types of spills (oil type, location, etc.) that are causing the greatest impacts. The model allows for cost and damage estimation of different oil spill response methodologies, including different degrees of mechanical containment and recovery, as well as alternative response tools of dispersants and in situ burning that may have greater future applications in freshwater and inland settings. Response effectiveness can also be specified allowing for analysis of potential benefits of research and development into response improvements. [1030]

Methods such as this can be readily modified for other liquid spills. Insights from the ranges of adjustment factors—for example, what is the range of impacts from a socioeconomic perspective? or What habitat considerations are important?—can also be used to inform modeling of all releases, including gases and HVL's.

11.8.5 High-value areas

Beyond considerations of population, property, and sensitive environment, some areas near to a pipeline can be identified as "high-value" areas, independent of the typical population- or environmental sensitivity considerations. As used here, the term high-value area (HVA) is defined as a location whose harm in the event of a pipeline failure generate exceptional consequences. Examples are discussed in PRMM and include irreplaceable archaeological or cultural sites; science centers with rare specimens or equipment; and many others.

Higher receptor valuations, higher remediation costs, higher damage rates, and other inputs into the risk assessment can be used to reflect higher value areas.

Additional examples of the many areas or facilities that could warrant special attention as receptors—perhaps HVA's—include:

- School
- Church
- Hospital
- Limited mobility health centers

- Historic site
- Cemetery
- Busy harbor
- Airport
- University
- Industrial center
- Interstate highway / highway interchange
- Recreational area/parks
- Special agriculture
- Water treatment/source.

11.8.6 Combinations of receptors

The extremes of receptor damage potential will be intuitively obvious—the most environmentally sensitive area and the highest population class and the highest value areas co-located and all potentially seriously damaged by the same section would be the highest consequence section.

Non-extreme combinations of receptors are not always so obvious. There will normally be several types of receptors at a potential spill site, each with different vulnerabilities to a threat such as thermal radiation or contamination. The analysis difficulty can be addressed by assigning different damage rates to different receptors experiencing different hazard zone effects. Each damage rate corresponds to a certain receptor-damage state. Separate consequences values are generated for, as an example, fatalities, injuries, groundwater contamination, property damage values, etc, at several distances from the hazard origin.

11.8.7 Service Interruptions

While service interruption consequences are addressed as a different type of 'failure' and also as indirect costs, there is often an enormous direct cost of service interruption, even in risk assessment focus is purely on leak/rupture. The direct consequences of the interruption of a high volume delivery include loss of revenue, loss of product, and perhaps immediate contractual non-performance costs, even before indirect costs are considered. Examples include:

- high volume offshore gas pipelines as critical feeds to numerous and/or essential consumers
- deliveries or receipts tied to production (for example, processing plants, gathering systems, power generation, etc) whose re-start costs are enormous after relatively short interruption periods.

See related discussions under Chapter 12.4.2 Indirect Consequences.

11.8.7 Offshore CoF

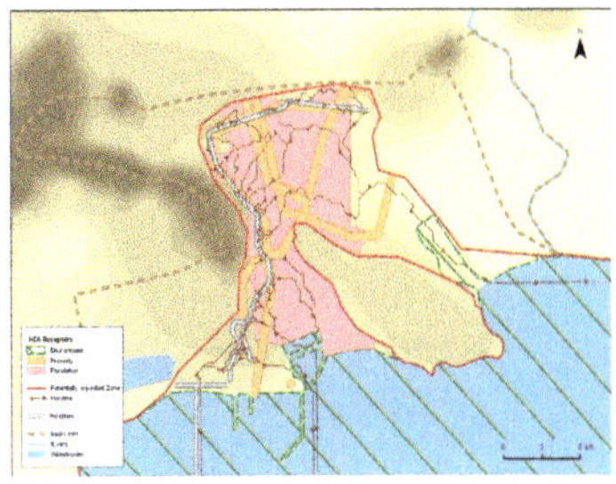

As with onshore spills, the type of product spilled, the distance to sensitive areas, and the ability to reduce spill damages will govern the consequence potential for offshore lines. Spills offshore can be assessed as they are in the onshore risk assessment model. This involves assessment of product hazard, spill size, dispersion potential, and vulnerable receptors within the hazard zone.

Offshore incidents are frequently more expensive due to increased costs of accessibility, repair, and return to service.

11.8.7.1 Receptors

Population density will not often be the dominant consequence for offshore pipeline failures. regulations in the US consider offshore pipelines to be in rural areas. Exceptions should be captured in the risk assessment, including proximity to recreational areas (beaches, fishing areas, etc.), harbors and docks, popular anchoring areas, ferry boat routes, commercial shipping lanes, commercial fishing and crabbing areas, etc.

11.8.7.2 Emergency response

Emergency response in offshore environments is usually more problematic than onshore due to the potential for liquid contaminant spread coupled with the remote, difficult-to-access locations of many offshore installations. The degree of dispersion of offshore liquid spills is a function of wind and current actions and product characteristics such as miscibility and environmental persistence. Conditions may change during a long event, further hampering response effectiveness.

11.8.8 Repair and Return-to-Service Costs

Repair and remediation costs can be a significant part of the cost of failure. The role of ancillary costs such as acquisition of necessary regulatory permits and permissions should not be underestimated. One source details an anomaly repair scenario where "This relatively common repair job, which could have been performed both safely and in an environmentally sound manner for under $90,000 within a few days, ended up costing in excess of $450,000 and requiring well over one-and-one-half years of preparation and planning time." [1005]

An assessment of repair costs can include the return-to-service costs associated with damaged system components. Damages to other nearby facilities are generally considered in receptor damages.

Factors impacting repair time and costs include:

- Type of repair
- Accessibility
- Need for and availability of special equipment and/or materials
- Need for and availability of special parts
- Component size (pipe sleeve cost, replacement pipe size, handling costs, etc)
- Need for and availability of special welders, welding materials, procedures, or qualifications.
- Need for and complexity in obtaining regulatory permits and/or landowner co-operation.

The ease-of-repair aspect, and hence costs, could be measured as a function of the variables such as:

- Topography/accessibility—Arctic, offshore, wetlands, unstable terrain, urban congestion, steep slopes, pavements, and numerous other environments are also associated with increased costs.
- Component size—damages to larger sized equipment often lead to more expensive repairs.
- Nearby facilities—stabilization, evacuation, and post-excavation repair of nearby facilities such as other utilities, buildings, roadways, and other structural features may add to repair costs.

11.8.8.1 Post Incident Investigations

> FOCUS POINT
> The higher the level of surprise at an incident, the higher the return to service costs.

Depending on type of damage causing the outage, extensive inspection along the entire pipeline might be warranted before it is prudent to return it to service. Costs of return to service inspection are appropriately included in a risk assessment. In some cases, the return-to-service inspection would be an accelerated schedule for already-planned inspection. In other cases, unplanned, widespread inspection would be prompted by an incident.

The PoF assessment can help to identify the extent of such post incident inspections. The number different locations contributing to similar PoF values is available in the risk assessment. Since the operator can often unilaterally choose how many locations to repair/reinforce/upgrade prior to resumption of service, the extent of damages is not a direct part of a failure consequence but would factor into risk management decisions. If an inspection is a mandatory requirement from a regulator, and would not otherwise have been performed by the owner, then it could be considered a cost of the failure.

These types of consequences vary based on failure type. This includes an assumption that incidents caused by time-dependent failure mechanisms such as corrosion prompt more extensive and expensive return-to-service actions compared to time-independent failures such as from vehicle impact. Incidents that lead to new or increased focus on failure mechanisms (for example, freeze, surge, or currently unknown mechanism) may also warrant treatment more costly return to service incidents.

Other return-to-service costs such as purging of pipeline and re-setting of instrumentation and equipment can similarly be included. Seasonal differences in accessibility and response efficiencies can also be recognized here.

Post-incident reaction, as a part of the return-to-service process, is a function of:

- Failure type—some will warrant more response than others
- Number of locations with 'similar' PoF values for failure type
- Number of locations with 'similar' PoF values for any failure type
- Extent of locations with 'similar' PoF values.

This mirrors the process by which an SME would gage the return-to-service efforts. If an incident is not unexpected—familiar failure mechanism that is reasonably foreseen by risk assessment—then it is extensive investigation may not be needed and resumption of operation may be easier. Similarly, when integrity knowledge is more complete—through recent and robust integrity verification, including DA-type assessments—then less post-incident integrity verification may be warranted. For example, a failure on uninspected system ABC may not prompt any actions on recently inspected system XYZ, even when the two are otherwise very similar. The calculated PoF values consider age and quality of integrity verification information, so using them directly to forecast post-incident reaction is consistent with the SME approach.

The level of surprise is also a factor. Lower PoF values will carry higher incident-reaction consequences. While this might at first appear counter intuitive, it is actually consistent with the SME decision-process. A failure at a low PoF location is a surprise. It challenges previously held beliefs about where and what types of potential failures warrant higher attention. This should prompt more investigation than a failure that is less surprising. The larger the surprise, the larger the reaction.

Regardless of how unexpected the event, it will usually suggest that more inspection on other, similar segments is prudent. A failure at any PoF often prompts an investigation of all pipe lengths with similar or worse PoF values (unless the failure is somehow uniquely possible only to the failed location). Lower PoF segments would prompt inspection of more length of pipe compared to higher PoF. Increased inspection will often generate the need for increased repairs, again adding to costs.

Outage periods which are extended by an incident-specific regulatory mandate following an incident are perhaps better captured in the indirect cost assessment.

11.8.9 Indirect costs

The consequence assessment is enhanced by recognizing that the direct costs of a pipeline failure are often not the only costs. In a certain public perception climate and with certain types of failures, total consequence of failure can be much higher than direct damages suggest. The factors impacting the level of indirect costs are numerous and frequently immeasurable and sometimes even inestimable with any degree of confidence. Even after an incident has occurred, obtaining an accurate assessment of indirect damages is often impossible. Implications regarding stock price, credit worthiness, lost opportunities, harm to current and future business negotiations, etc all make accurate assessments practically impossible.

Despite their challenges in quantification, some estimation is often warranted. Potential costs associated with a spill that may be considered indirect include:

- Fines and penalties
- Litigation
- Increased regulatory oversight
- Direct customer impacts
- Damage to corporate reputation
    - o decrease in stock value
    - o increased costs of financial dealings
    - o decrease in negotiating position
- Loss of company focus
    - o diversion of resources
        - mangement testimony
    - o hearings prep, action.

Given the difficulties in quantifying many of these indirect costs, use of a multiplying factor applied to estimates of direct cost is an appropriate approach. In this approach, the indirect costs are seen to be proportional to direct costs and therefore captured as an escalation factor.

11.8.9.1 Estimating Potential Damage to Corporate Reputation

Indirect consequences such as harm to corporate reputation, are often thought to closely parallel the risks to public and environment. Many of the same factors that suggest damage to corporate reputation would also precipitate other indirect costs. For instance, the potential for fines, litigation, and increased regulatory oversight following an incident would realistically be influenced by factors such as recent incident history and public perception climate—the same factors influencing corporate reputation.

A more robust analysis could include specific research into company's current or historical reputation. Financial ratings, stock analyst reports, consumer surveys, and similar assessments might provide a partial basis for such an evaluation. The extent of pipeline operations compared to the company's full suite of activities may be important.

Some investigation into the role of pipeline operations in a publicly traded company's overall business can be obtained taken from recent annual report. In a large hydrocarbon energy company, the financial activities related to exploration and production (E&P) may be a large part of the total business of a company. This fact is often relevant to the type of indirect damages predicted from an incident in that part of the business.

11.8.9.2 Example

Possible indirect consequence algorithms to assess this aspect of overall consequence are illustrated in the following example.

First, an estimate of the current condition is made. It is recognized that this 'current' conditions is highly variable—often a function of recent focus of news media. In this example, the corporate reputation for a large oil and gas company is currently judged as follows:

Pre-existing Reputation (scale can be viewed as "% mistrust" with 0% being neutral and -100% is most negative)

- Public perception of a company: % mistrust currently = 0%
    - o no damaging stories recently; generally neutral or favorable public impression of company
- Public perception of oil/gas industry: 50%
    - o Currently lower due to recent news headlines of legal actions in the industry; price of gasoline versus corporate profits is another relatively fresh issue that would probably be referenced by media; other pipeline incidents would likely be referenced in news stories.
- Public perception of large corporations: 20%
    - o Headlining corporate dishonesty episodes are fresh and likely to be referenced by media.
- Public perception of region: 80%
    - o nearby spill episodes still fresh

Total for Pre-existing reputation: 1-(1-0.5)x(1-0.2)x(1-0.8) = 92%. This is an OR gate combination, reflecting the fact that any single issue can overshadow the rest, as can an accumulation of lesser issues; for example, averaging is not appropriate, nor is choosing a maximum.

Pipeline headlines: (% worst case where 100% = 'incident on same pipeline within last year')

- Years since a media-covered incident similar to current:
    - o Anywhere: NA
    - o US: NA
    - o Neighbor/affiliate; pipeline incident in same region:

0.95 x (10-1yr old)/10 = 85.5% (a 95% similar incident occurred one year ago.)

o Company: 0
o same region: see above
o same asset class: 0
o same pipeline: 0
Total for Pipeline Headlines: 85.5% (others are insignificant)

Failure type (% perception where 100% is a failure type carrying stigma of 'negligence'):

- Corrosion (all forms): 100% (predictable, familiar failure mechanisms)
- Vehicle impact (company vehicle): 70%
- WIV: 50% (poorly understood mechanism)
- Geohazard meteorite strike: 0% (sympathy effect)
- Geohazard other: 50%
- Operational (slug, freeze, etc): 80%
- Sabotage: 0% (sympathy effect)

Offsets: (% of best possible offsets)

- Response reasonably fast and thorough, 50%
- Good content of early messages 80%
    - o minimal damage to environment;
    - o rapid public apology;
    - o rapid-immediate investigation and preliminary corrective actions;
- Follow-up actions are timely and well communicated 60%
- Media management is above average 70%

Corporate response effectiveness: 17% based on multiplying the above sub-factors

Competing news stories: arbitrary assumption of 20% (20% of damage that would otherwise occur is offset by coincident events deflecting media focus away from incident)

Total Offsets: 1-(1-0.17) x (1-0.2) = 33%

Scale = ([pre-exist rep] OR [pipeline headlines] OR [failure type])
AND [offsets]
[1-(1-0.92) x (1-0.855) x (1-[failure type])] x (1-0.33) = ~66% of scale

Scale limit currently set at 5: indirect costs can increase overall consequences by 5 times as a worst case. This magnitude reflects indirect costs that include damage to corporate reputation, litigation, fines/penalties, increased regulatory oversight and others.

Based on these variables and others, the indirect costs from damage to corporate reputation are judged to increase the direct costs by a factor of about 0.66 x 5 = 3.3. In the current climate, the failure mechanism is having only a slight impact on the indirect costs since other indicators are relatively high. The multiple reflects any type failure occurring in a climate already marked by suspicion or mistrust of regional operations, pipelines, oil and gas industry, and large businesses.

Since the multiplier is usually a constant, all failure scenarios of the same type and magnitude are equally affected. More discrimination is seen when corporate reactions or news worthiness are more variable (perhaps geographically sensitive) and in comparing various failure types.

11.8.10 Customer Impacts

See discussion in Chapter 12 Service Interruption Risk.

11.9 PROCESS OF ESTIMATING CONSEQUENCES

The key steps for the consequence assessment proposed here are:

1. Choose thresholds that determine hazard zone boundaries.
2. Identify consequence reduction measures.
3. Estimate hazard zone areas.
4. Characterize receptors within the hazard zone(s).
5. Include indirect consequence costs, if desired.
6. Calculate potential consequence per failure.

These ingredients are developed sequentially in the assessment process, with the 'per incident' expected loss values being the consequence measures that are combined with PoF estimates to obtain final risk estimates—in final units such as 'loss per year'.

11.10 EXAMPLE OF OVERALL EXPECTED LOSS CALCULATION

An example of the overall consequence estimation process is laid out in the following tables and discussion. Values shown are to illustrate the process only—they will ***not*** be realistic values for most pipelines and should not be used as a basis for any other estimates.

Table 11.12 shows how the hazard zone distances are estimated for this example. For the nine scenarios shown, maximum threshold distances range from 30' to 1500'. A distance of 1500' is considered to be the maximum impact distance for this location on the examined pipeline.

The analysis begins with estimates of hole size probabilities. Depending on the PoF analysis, the entry point can be either the relative hole size distribution or an 'absolute' hole size distribution. The former is illustrated here—the hole size distribution representing 100% of all possible failures; the relative chance of a certain size hole, given that some hole is present. The latter implies that several hole sizes have a specific probability of occurrence already estimated in the PoF assessment—there is a calculated probability of rupture, and a calculated probability of a pinhole, and so forth.

Table 11.12
Example Hazard Zone Distances and Probabilities

						Threshold Distances (ft)				
Product	Hole Size	Probability of Hole	Ignition Scenario	Probability of ignition scenario	Distance from source (ft)	Thermal impact	Overpress impact	Contamination Impact	Maximum Distance (ft)	Probability of Maximum Distance
propane	rupture	8%	immediate	60%	0	400	0	0	400	4.8%
			delayed	20%	300	400	800	0	1500	1.6%
			no ignition	20%	300	0	0	0	300	1.6%
	medium	12%	immediate	15%	0	300	0	0	300	1.8%
			delayed	15%	100	300	200	0	600	1.8%
			no ignition	70%	100	0	0	0	100	8.4%
	small	80%	immediate	10%	0	50	0	0	50	8.0%
			delayed	10%	30	50	0	0	80	8.0%
			no ignition	80%	30	0	0	0	30	64.0%
		100%								100.0%

These probabilities simulate a distribution of all possible hole sizes with their associated probabilities of occurrence. Such a distribution would be influenced by pipe material, stress level, and failure mechanism, as well as other considerations. In the table above, three relative hole size occurrence percentages are shown. They sum to 100%. Each will be multiplied by the PoF of all possible leak sizes—a very small number for most pipelines—to get absolute probabilities of occurrence. For instance, if the overall failure probability (all holes sizes) was estimated to be 1E-6 per mile-year, then the probability of a rupture is estimated to be 8% of that value or 0.08 x 1E-6 = 8E-8 = 0.000008% chance of rupture for each mile for each year. This also suggests 8E-8 ruptures per mile per year as an estimated frequency of occurrence.

Next, three ignition scenarios are modeled: 'immediate', 'delayed', and 'no' ignition. The probability of each scenario is estimated for each hole size scenario. In this sense, hole size is being used as a surrogate for leak size. Larger holes imply larger leaks and greater ignition potential. The three holes sizes and the three ignition possibilities will produce nine scenarios, thought to sufficiently represent the possibilities in this example.

The distance from source column represents the possible migration distance of spilled product from the leak source. It is based on dispersion modeling—vapor cloud drift—in the case of gaseous releases and overland flow modeling in the case of liquids. This distance is additive to thermal effects distances and contamination distances. The leaked product might travel some distance, ignite, and produce thermal damages from the ignition site, sometimes far from the leak site. In the contamination damage scenario, envision a pool of spilled liquid that accumulates some distance from the leak location and only then begins a more aggressive subsoil migration, causing a groundwater contamination plume spreading from the pool. Since propane—a highly volatile liquid—is the product in this example, no contamination impacts are foreseen.

Several thresholds are selected for production of hazard distance estimates. Shown are one thermal effects threshold, one overpressure threshold, and one contamination threshold. These must be defined in terms of some intensity level or some probable damage state before distances could be assigned. The evaluator will probably want to include multiple thermal and contamination thresholds to ensure that the full range of possibilities is portrayed. The distance for each threshold is estimated from appropriate models for the product released. A gaseous release might base the threshold on flame jet thermal radiation (as in ref [3], for example); an HVL release threshold might be based on overpressure distance as well as fireball or jet thermal radiation; and a liquid release is often based on pool fire thermal radiation or contamination level. In this example, the longest distance occurs with a delayed ignition scenario, allowing the vapor cloud to migrate before ignition initiates a thermal event, including overpressure, if the release is sufficiently large.

Figure 11.11 shows the resulting nine hazard zone distances.

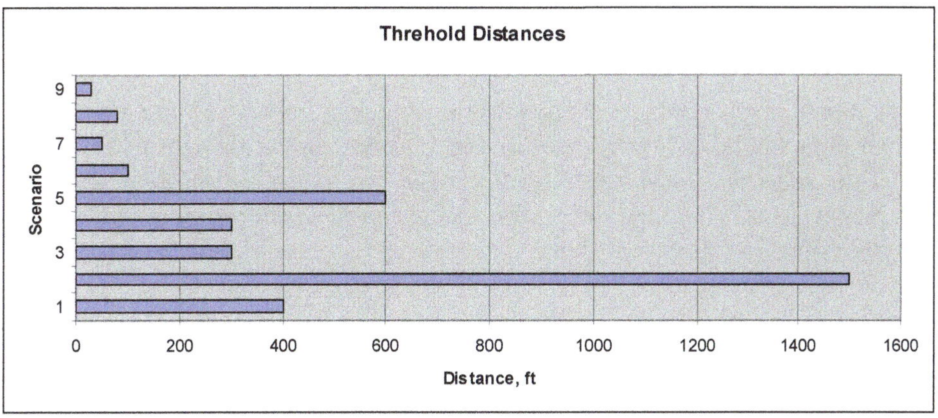

Figure 11.11 Visualizing Hazard Zone Distances

The probability of each scenario is calculated as the product of the hole size probability times the ignition scenario probability. These values can be multiplied by the overall PoF, to arrive at an absolute probability of each scenario. In the example tables, though, scenario probabilities assume that the pipeline failure has already occurred. Therefore, scenario probabilities sum to be 100%.

A simple plotting of distances such as shown below can be helpful. This grouping into zones is a modeling convenience that avoids having to perform receptor characterizations at too many distances.

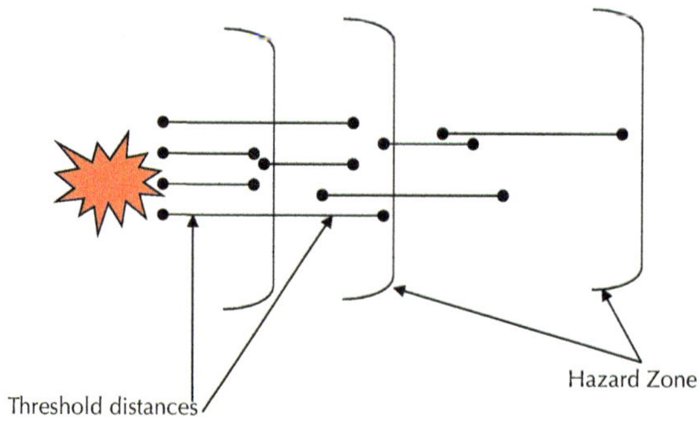

Figure 11.12 Visualizing Ranges of Thresholds and Grouping into Zones

In this example, the evaluator has grouped the threshold distances into three zones. This was done by setting some logical breakpoints. PIR is estimated to be 1500 ft and zones are defined as:

"less than 100 ft";

"from 100 ft to 50% of PIR (or 750 ft)"; and

"from 50% PIR to 100% PIR (or 750 ft to 1500 ft)".

The number of zones is up to the modeler. All events within a zone are treated as the same. This implies no differences in potential damages at the closest and farthest point of the zone. So, wider zones require more "averaging" of possibly widely-differing potentialities within the zone. More categories will result in more resolution but also more efforts in subsequent steps.

In this example, the modeler chose to use three zones. He also chose to make zones not equivalent in size—basing his groupings a non-linear reduction in impact intensity with increasing distance. Non-uniform zone sizes might also better represent the relative frequency of events. Perhaps scenarios leading to larger threshold distances are so rare, that a larger zone captures an equivalent number of scenarios as the smaller zones. Each grouping or zone will have a probability comprised of the probabilities of all the individual scenarios that can produce a threshold distance that falls in the zone.

Each zone represents a collection of numerous potential damage thresholds. There are no sharp demarcations between possible zones. For instance, 20% of the possible scenarios might produce hazard zones from 0 to 200 ft and 10% of the scenarios could produce distances of from 50 ft to 400 ft. These overlapping distances do not necessarily suggest break points for zones so any choice of break point is a compromise. A cumulative probability chart and graphical presentation of the various thresholds associated with various scenarios will help the modeler to establish zones and associated probabilities.

As is illustrated in Figure 11.12, there are some scenarios in the farthest zone that produce no impacts in the closest zone. For instance, a scenario where leaked product moves completely out of the closer zones (via sewer or puff cloud drifting, for example) before finding an ignition point. At the ignition point, the thermal effects are far from the release point and the receptors closer to the pipeline.

Each zone is assigned receptor damage rates based on the damages that would likely Each zone is assigned receptor damage rates based on the damages that would likely occur. For example, where very high heat radiation thresholds occur, higher fatality rates and higher property damage rates would be expected. The estimated damage rates are discussed in the next section.

Damage percentages are assumed to be 0% at distances beyond the PIR. The percentages will be used to calculate expected losses. They should be relatively conservative, reflect the modelers' experience and beliefs, and should be fully documented.

Again, this grouping of hazard distances is for modeling convenience. It is often easier to make the necessary receptor characterizations within a few zones rather than for each possible threshold distance. The trade-off is some measure of accuracy since compromises are made in setting the zones. All event scenarios occurring within a zone are treated equally, even though some occur at either extreme of the zone.

11.10.10.1 Step 4

Next, receptors are characterized within each hazard zone as is shown in Table 11.15. At three distances from the pipeline (maximum hazard distance divided into 3 zones), all receptors are characterized in terms of their number and types within each zone. In many cases, a circular hazard area is a fair representation. However, given certain topographies and/or meteorological phenomena, ellipses or other shapes might be more representative of true hazard areas.

The types of damages to each receptor that may occur in each zone should be considered. Characterization can be in terms of percentage of maximum damage or percentage chance of the maximum damage. For instance, in a zone close to the ignition point and following a very high consequence event, the damage state to humans might be 2% fatality and 100% injury. A more distant zone might be characterized as a damage state to humans of 0.1% fatality and 20% chance of injury. In the case of non-absolute damage states such as injuries or property damage, the percentage can be thought of as either x% chance of any damage, or a 100% chance of a damage that is x% of the maximum possible damage. Both conceptualizations are supported since the mathematical approach would be the same for each.

Recall that, as a modeling convenience, the probability of a certain hazard zone occurring is considered to also capture the diminished damage potential at the increasing distance.

Receptors at farther hazard zones produce lower expected losses since their probabilities of damage are lower. They are lower for two reasons: lower chance of that hazard distance happening, and lower intensities resulting in less damage to the receptor at farther distances.

Table 11.13
Damage State Estimates for Each Zone

Hazard Zone	Injury Rate	Fatality Rate	Environment Damage Rate	Service Interruption Rate
<100′	80%	8%	50%	100%
100′-50% PIR	50%	5%	30%	90%
50% -100% PIR	20%	2%	10%	80%

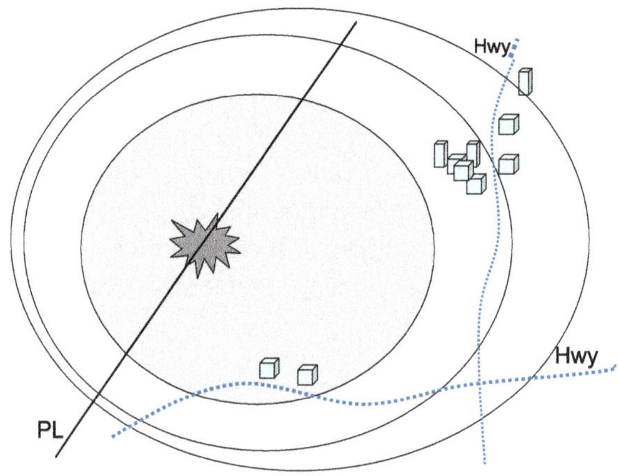

Figure 11.13 Multi-hazard zone analysis

Characterization of the receptors within each hazard zone includes count and type. Receptors can be efficiently quantified in terms of 'units' where each unit represents an individual or area (ft^2, m^2) of that type of receptor. The number of people impacts the injury and fatality potential. The area of environmental sensitivity impacts the clean up costs. The number of buildings impacts the property damage potential. The unitization can follow any logical means of quantification.

When consequences are monetized and risk expressed as EL, a unit is assigned a value, reflecting the cost of replacement, remediation, and other compensation. Environmental damages can be quantified in "environmental units", where the evaluator sets some equivalences among possible scenarios. For instance, an acre of 'old growth forest' may be set as 1 environmental unit, while a T&E species is set at 10 and an uncleanable aquifer at 15. In the absence of more definitive data, these are value judgments best established by knowledgeable environmental specialists along with company managers.

The receptor characterization will be determined by the scope of the assessment, with more robust assessments requiring more detailed characterization. For instance, some models will make distinctions among human populations—age, mobility, etc—for some thresholds. Consideration of shielding is another possible variable. Shielding of almost any kind is an effective reduction to radiant heat, minimizing damages or allowing more escape time. It can be incorporated into the receptor characterization or used as a stand-alone variable—a factor to reduce potential damages.

Steps 3 and 4 will have produced characterizations of possible receptor damages in each zone. Ideally, the risk evaluator will now have the ability to answer, at least generally, questions such as:

- How many people are typically in each zone?
- What is the potential rate of injuries, fatalities in each zone?
- What is the potential rate or % of other damages in each zone?
- How much property damage is likely in each zone?
- How much and what type environmental damage is possible in each zone?

He will also have gained the ability to answer these questions in somewhat quantitative terms, although many assumptions and uncertainties are usually embedded in such quantifications.

Three types of receptor-damages are recognized in this example: fatalities, injuries, and environmental damages. Other common receptors/damages include service interruption costs and property damages. Not shown in this table but used in the calculations, is a benefit from shielding. The evaluator estimates that, in this area, shielding from buildings, trees, etc; the amount of clothing normally worn; and the emissivity (heat movement through the atmosphere), a reduction factor of 30% should be applied to the injury and fatality rates. This assumption could also have been embedded in the overall damage rate estimates, but in this example, the modeler keeps this variable separate so that it can be a distinguishing factor when shielding conditions change. More detailed receptor characterizations are of course possible and supported by this approach. For instance, the population might be divided into groups based on increased susceptibility to injury or death, such as: "limited mobility"; "unshielded"; "weakened immune systems"; etc. Similarly, the environmental units could be categorized into many different subgroups. As with many aspects of modeling, the evaluator must make decisions involving tradeoffs between robustness and simplicity.

As another modeling convenience, receptors are measured in terms of units. A higher quantity or sensitivity of receptor type is captured in terms of more units. A dollar value is assigned to a unit of each type. In this example, an injury is valued at $100K, a fatality at $3.5M, and an environmental unit at $50K. Such valuations should be carefully set and fully documented.

Table 11.13 repeats some information from Table 11.12 and then shows how the scenarios are further developed using Table 11.14 & Table 11.15 and the valuations discussed.

Occurrence probabilities and valuations combine to arrive at expected losses for each receptor in each scenario. For instance, in the case of the first scenario, the human injury cost is estimated as the product of (scenario probability, over some time period) x (# of people) x (injury rate in zone "100' to 50% PIR") x (30% shielding benefit factor) x (cost of injury) = 4.8% x 5 x 50% x 30% x $100,000 = $3,600 per scenario. If the scenario frequency is estimated to be once every 10 years, then the expected human injury loss is $360 per year at this location.

Table 11.14
Characterization of Receptors Within Each Zone at a Particular Pipeline Location

Hazard Zone	No. of people	No. of Environ Units	No. of Service Interruption Units
<100′	1	0.5	1
100′-50% PIR	5	1	5
50% -100% PIR	10	1	10

Each scenario has an associated probability of occurrence, produces a certain hazard zone, and contains certain numbers and types of receptors with associated dollar values. Multiplying these values together and then summing the results for each hazard zone produces the expected loss for the pipeline segment.

The composite consequences per failure at this location on the pipeline is estimated to be ~$166K, as shown in Table 11.15. This is the expected loss from all pipeline failure scenarios. The annual expected loss is obtained by multiplying this value by the annual failure rate. If that value is 10^{-3} failures per mile-year and this "location" on the pipeline represents one mile, then the expected loss is ($166K per failure per year) x (10^{-3} failures per mile-year) = $55 per year. Therefore, over long periods of time, the cost of pipeline failures for this one mile of pipe is expected to average about $55 per year, as is shown in Table 11.16.

Table 11.15
Estimating Expected Loss from Hazard Zone Characteristics

		$100,000 $3,500,000 $ 50,000 Expected Loss				unit cost	unit cost		unit cost	
Hole Size	Ignition Scenario	Maximum Distance (ft)	Probability of Maximum Distance	Hazard Zone Group	# people	Human injury costs	Human fatality costs	# environ units	Environ Damage Costs	Probability weighted dollars per failure
	immediate	400	4.8%	100′-50% PIR	5	$ 3,600	$ 12,600	1	$ 720	$ 16,920
rupture	delayed	1500	1.6%	50% -100% PIR	10	$ 960	$ 3,360	1	$ 80	$ 4,400
	no ignition	300	1.6%	100′-50% PIR	5	$ 1,200	$ 4,200	1	$ 240	$ 5,640
	immediate	300	1.8%	100′-50% PIR	5	$ 1,350	$ 4,725	1	$ 270	$ 6,345
medium	delayed	600	1.8%	100′-50% PIR	5	$ 1,350	$ 4,725	1	$ 270	$ 6,345
	no ignition	100	8.4%	100′-50% PIR	5	$ 6,300	$ 22,050	1	$ 1,260	$ 29,610
	immediate	50	8.0%	<100′	1	$ 1,920	$ 6,720	0.5	$ 1,000	$ 9,640
small	delayed	80	8.0%	<100′	1	$ 1,920	$ 6,720	0.5	$ 1,000	$ 9,640
	no ignition	30	64.0%	<100′	1	$15,360	$ 53,760	0.5	$ 8,000	$ 77,120
			100.0%	Total expected loss per failure at this location						$165,660

Table Notes:

Not shown is a Shielding factor: estimated as a percentage, this adjusts the damage estimate by considering protective benefits of all shielding opportunities including clothing, buildings, etc. in each hazard group and for each receptor type. In this example, 30% shielding factor is used.

Table 11.16
Final Expected Loss Values

EXPECTED LOSS			
Failure Rate (failures per mile-year)	Probability of Hazard Zone[1,2]	Probability weighted dollars[2,3]	Probability weighted dollars per mile-year
0.001	4.80%	$16,920	$0.81
	1.60%	$4,400	$0.07
	1.60%	$5,640	$0.09
	1.80%	$6,345	$0.11
	1.80%	$6,345	$0.11
	8.40%	$29,610	$2.49
	8.00%	$9,640	$0.77
	8.00%	$9,640	$0.77
	64.00%	$77,120	$49.36
	100.00%	$165,660	$54.59

Table Notes:
[1]after a failure has occurred
[2]from Table 11.13
[3](damage rate) x (value of receptors in hazard zone)

The expected loss values can be viewed as part of the cost of operations. They can be used in decision-making regarding appropriate spending levels. The expected loss for this segment can be combined with all other segments' expected losses to arrive at an expected loss for an entire pipeline or pipeline system. So, while $55 per year appears very low, a 500 mile pipeline with the same estimates as this segment, suggests an expected loss from failures of over $27,000 per year.

This example illustrates the representation of risk as a frequency distribution of all possible damage scenarios, including their respective probabilities and consequence costs. The distribution is characterized by a representative number of point estimates produced by this evaluation. The point estimates show the range of risks and can themselves be compiled into a single estimate for the entire range of possibilities.

When risk aversion—disproportionate costs for higher consequences—is also considered, the overall expected loss value should not be used in isolation. The very rare, but very consequential scenarios, are obscured when all scenarios are compiled into a single point estimate. The more consequential events might warrant further consideration.

"Logical consequences are the scarecrows of fools and the beacons of wise men."

Thomas Henry Huxley

12 SERVICE INTERRUPTION RISK

Highlights

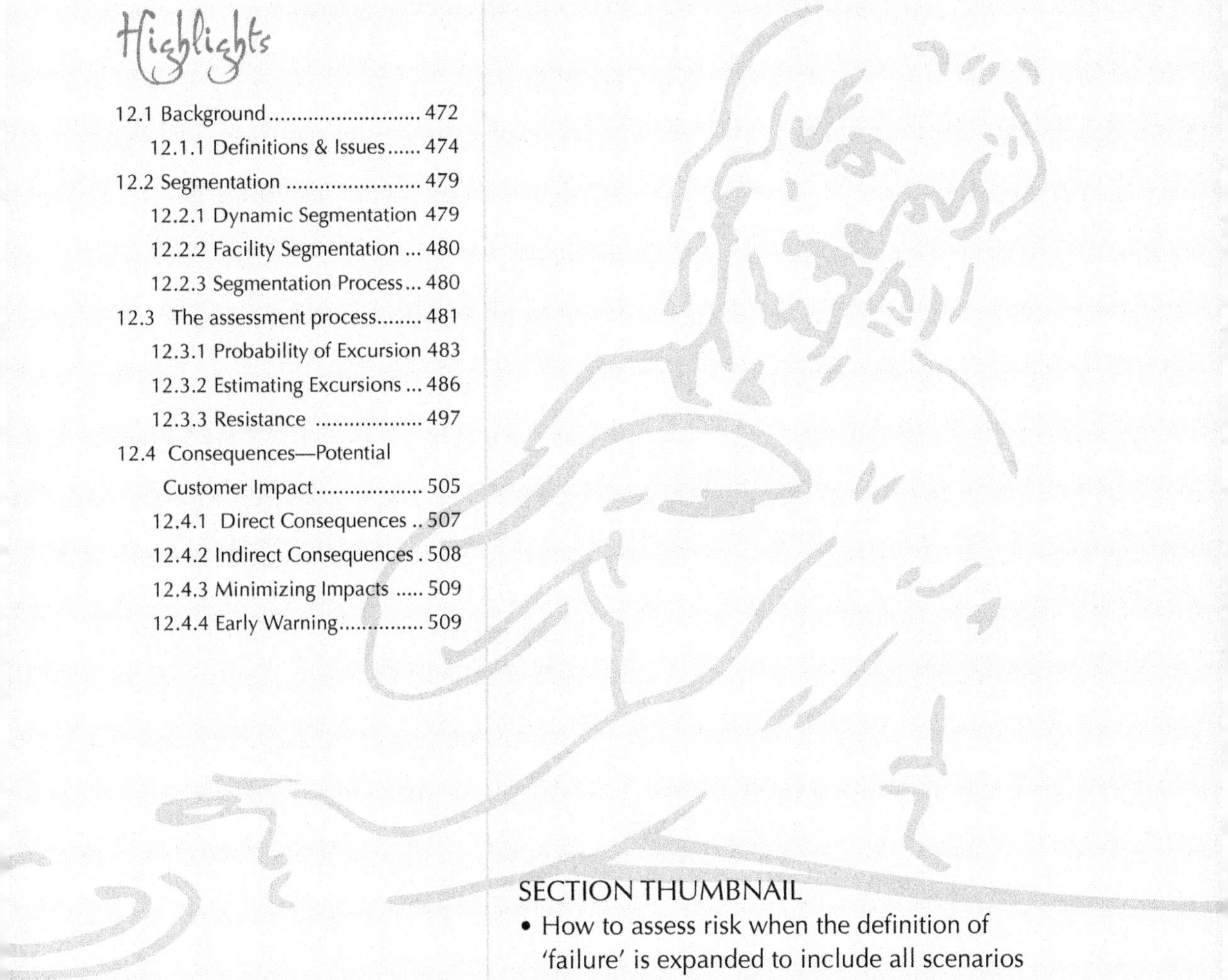

SECTION THUMBNAIL

- How to assess risk when the definition of 'failure' is expanded to include all scenarios that interrupt the desired use of the pipeline.
- The same risk assessment methodology can be used, but some analogous risk assessment elements warrant some discussion.

SECTION THUMBNAIL

- With an expanded definition of 'failure', service interruption risk becomes pertinent.
- Complexity is added since leak/rupture is only one of several ways 'failure' can occur.

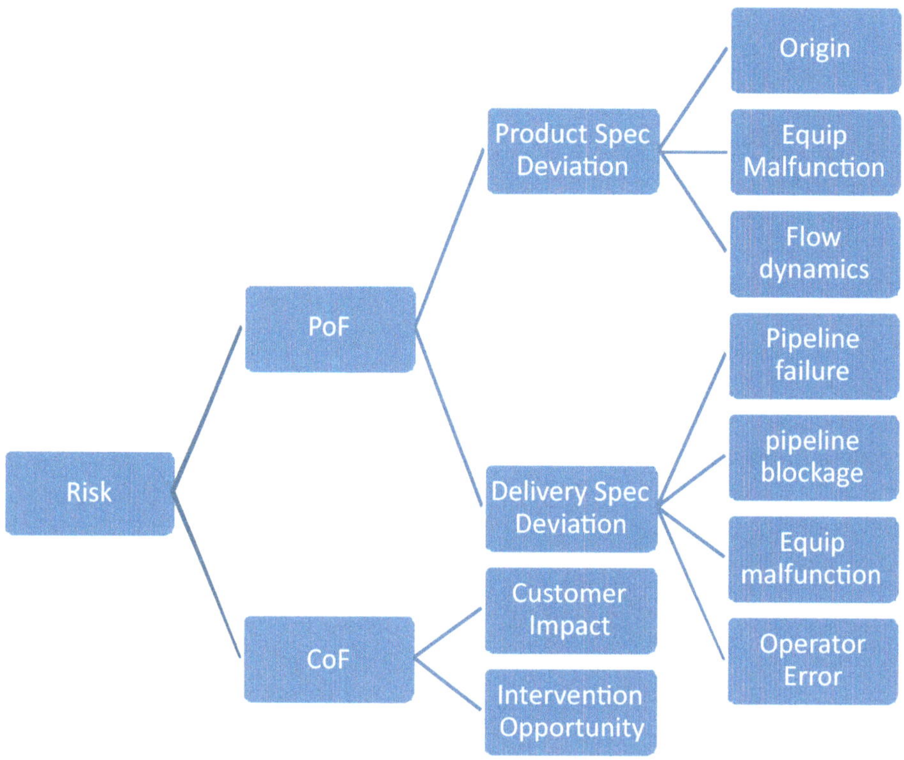

Figure 12.1 Risk of service interruption.

12.1 BACKGROUND

Up to now, the focus has been on assessing the risk of pipeline failure, with 'failure' defined as a leak or rupture. This is an integrity-focused risk assessment. Recall that a broader definition of failure for any engineered system is 'not meeting its intended purpose'. With a typical pipeline purpose of 'moving x volume of product y from point a to point b in time period z, within delivery parameters of a,b,c, etc., a pipeline has many ways to 'fail' that do not involve a leak or rupture. So, an expanded definition of 'failure' will often include 'service interruption'.

A service interruption can cause direct consequence to revenue generation, customer satisfaction, and other factors. In this chapter, the focus is on service interruption as a broader definition of failure, inclusive of all leak/rupture scenarios.

For this assessment, a service interruption is a deviation from product or delivery specifications that causes a negative impact to a customer. The definition implies the existence of a specification (an agreement stating the delivery parameters, including product quality), a time variable (duration of the deviation), a customer (an entity receiving service from the pipeline), and a consequence to that customer. These are discussed in this chapter. Additional terms and phrases such as excursions, upsets, 'off-spec', violations of delivery parameters, specification violations or non-compliances, will be used interchangeably in these discussions.

The quantification of service interruption risk will normally be meaningful only for the portion of the system directly connected to the customer. At all other locations, there is no customer to be harmed, so no potential consequences. It is only when the excursion manifests at a customer location that harm can occur. This is not to say that upstream portions do not contribute to service interruption potential—they certainly do. But since many systems have intervention opportunities, it is only after considering all interplays among excursion sources and remedies that the interruption potential at a given location can be known.

Note however, that the entire downstream portion of a pipeline system can be viewed as a customer of the segment being assessed.

The risk of service interruption is additive to the risk of pipeline leak/rupture. This makes the risk assessment more complicated because pipeline leak/rupture is only one of the often-numerous ways in which a service interruption can occur—leak/rupture is a subset of all possible service interruption scenarios. Service interruptions can be caused by contamination, blockages, under-performing equipment, and many others that in no way threaten system integrity. All must be assessed in order to fully measure service interruption risk. An event may or may not lead to a service interruption depending on how long the event lasts and the system's ability to respond to the event. So, the analyses must provide for the system's ability to absorb excursions without causing customer harm.

Ensuring an interruptible supply, ie, no service interruption, may conflict with ensuring minimum consequences to leak/rupture events. Scenarios such as erroneous valve closures or equipment failures normally cannot be tolerated from a service interruption viewpoint so steps are taken to limit the equipment and operational complexities that lead to unwanted interruptions. This may result in also limiting the necessary, desirable shutdowns for which the protective equipment is intended. This can present a design/philosophy challenge, especially when dealing with pipeline sections close to the customer where reaction times are minimal.

Including service interruption in the risk assessment is simply an expanded version of the failure = loss-of-integrity risk assessment methodology. The loss-of-integrity risk assessment is a part of the risk of service interruption assessment and is ready to be included into the expanded risk assessment.

Just as all causes of leaks and ruptures were itemized and evaluated, all causes of service interruptions must similarly be itemized and evaluated. Added to the probability of leak or rupture is the probabilities of all events that cause a service interruption but do not cause a leak or rupture. This involves identifying all possible excursions from delivery specifications, with no initial consideration for their ultimate potential for customer harm. For example, a blockage in a pipe segment should be treated as an excursion, even if that particular blockage does not directly impact any customer. A contaminant injection episode is an excursion even if it will be subsequently diluted to a level of insignificance. These would be excursions with no customer consequence, ie, no service interruption. Potential customer impacts, and how those translate to consequences for the service provider, are considered in the consequence of service interruption portion of the assessment.

Service interruption will normally include all of the leak/rupture failure mechanisms since all causes of leaks and ruptures usually cause service interruption. Some leak/rupture events may not, however, result in a service interruption. When an in-service repair such as a clamp can be implemented without interrupting the pipeline's operation, an excursion has occurred but the repair without halting flow has prevented a service interruption. The risk assessment should show both—the occurrence of the excursion and the lack of customer harm

The definition for service interruption contains reference to a time factor. Time is often a necessary consideration in a specification noncompliance. A customer's system might be able to tolerate certain excursions for some amount of time before losses are incurred. This is analogous to the measurement of 'resistance' in the leak/rupture assessment since some failure mechanisms can be resisted for longer times than others. When assessing customer sensitivity to specification deviations, the evaluator should consider tolerable excursion durations with probable durations. This is captured in the assessment through proper inclusion of excursion events and resistance estimates, as discussed in this chapter.

12.1.1 Definitions & Issues

Many issues are intertwined in a potential service interruption scenario. Again, a reductionist approach to the risk assessment—breaking the overall issue into smaller pieces—is efficient. This means that issues must be separated, measured independently, and those measurements must then be appropriately combined to reveal new knowledge. First, some definitions and issues will be presented to help ensure complete understanding of the assessment process.

Service

The service normally of interest here is the movement of products by pipeline under conditions agreed upon by the pipeline operator and a customer The focus here is on the service provider—normally the pipeline operator. The risk assessment produces estimates of frequencies and magnitudes of losses due to service interruptions, potentially suffered by the customer and for which the service provider is usually liable. Most of these loss scenarios arise because the customer does not receive the service that was promised. Beyond loss of revenue to the service provider, damages suffered by the customer due to the interruption will often also translate to losses to be borne by the service provider. So, the customer loss is linked to the service provider loss.

Service interruption

Defined by the definition of 'failure'. For purposes here, failure is defined as a deviation from product and/or delivery specifications that potentially causes an impact to a customer. A service interruption requires both a deviation from a service parameter and some impact to a customer.

Risk of service interruption

This is measured as follows:

Risk = Probability of Service Failure x Consequences

Exposure, mitigation, and resistance for each threat to service reliability must be measured to produce a PoF. Then, potential customer harm and related consequences are measured and combined with the PoF to yield the risk of service interruption. This is a separate branch in the risk assessment, additive to the leak/rupture risk assessment estimates.

Excursion

Any occurrence, along any point of a pipeline system, that potentially causes a service interruption. Any deviation from an intended product or transportation characteristic, for example product composition, flow rate, temperature, pressure, content, etc. is counted as an excursion, regardless of its ability to actually cause upset to a customer. For instance, even if a small amount of water carryover into a flowing pipeline will not result in a product spec violation by the time it reaches any customer, it is nonetheless an excursion. The probability of each excursion causing upset is considered separately from the identification of the excursion.

Voluntary Excursions

An operator may unilaterally decide to discontinue service for a variety reasons. It is usually a matter of choosing interruption as a less consequential course of action in light of other urgencies. Halting flow due to unacceptable product contamination, the need for emergency maintenance or repair on a segment, financial issues, non-performance of upstream supplier, weather events, are examples of many possible scenarios. If the operator must sacrifice service to one customer in order to continue service to another, that too obviously constitutes an excursion event of interest. See further discussion in later section.

Exposure, Exposure Event, Event

The frequency of unmitigated excursions. As in the integrity-focused risk assessment, the handling of some resistance issues can be via definitions of exposure events. When an excursion is defined as only those events 'large' enough to potentially cause a customer interruption (in the absence of mitigation), then the inherent ability to resist is captured in the assessment. Alternatively, all excursions, even very minor (as specified in the definition of 'excursion'), can be counted and then complete resistance to the more minor events is modeled.

Consequence

The amount of harm/damage/loss/upset potentially suffered by the pipeline owner/operator if the excursion reaches a customer facility and causes harm. Note that the implication is that the consequence of interest is the amount of customer harm that transfers to the owner/operator which might not be the entire amount of harm suffered by the customer. This helps to distinguish among various contracted pipeline services.

Offspec

A special type of excursion, this is an abbreviation for 'off specification' meaning failure to comply with an agreed upon specification that dictates the characteristics of the transportation or delivery service, including the characteristics of the delivered product.

Mitigation

Actions taken to reduce the frequency or magnitude of excursions. A mitigation prevents an excursion or reduces its severity and/or duration.

Resistance

Ability of the system to absorb excursions, preventing harm to customers. Resistance for these types of failure includes interventions (for example, engaging alternative supplies) and inherent system characteristics (for example, sufficient volume to dilute contaminants or sufficient pressure to temporarily withstand supply interruptions). Resistance does not prevent or reduce an excursion but prevents or reduces a service interruption. A resistance protects the customer from a service interruption even though an excursion has occurred.

Resistance: Intervention/Reactionary Type

Interventions, as used here, are a type of resistance to failure. They are actions taken. Examples of actions include blending (diluting) of contaminants until acceptable concentrations are achieved; turning off a contaminated supply and activating an alternate supply; increasing the flow contribution from another source to maintain pressures; etc.

Resistance: Inherent/System Characteristics Type

There are also inherent properties of the system that offer resistance to the excursion. Examples of resistance factors that are inherent include large volume segments that are able to absorb minor introductions of contaminants or abnormal flowrates (especially when compressible fluids are involved) with no impacts to customer. Some gas transmission pipeline segments, in effect being used for gas storage as they are intentionally 'packed' and 'unpacked' with gas, can absorb many excursions of inflows and outflows without interrupting a customer delivery.

Normalizing Exposures with Resistance and Consequences

Note that there may be, in some situations assessed, an overlap between exposure rate and system resistance customer impact. Resistance includes aspects like alternate supplies, ability to blend, etc. just as do some exposure rates from a source. Exposures below a certain threshold are insignificant to some customers; for example, the accidental introduction of a small amount of water into a large gas transmission pipeline. To keep the service interruption risk assessment efficient and organized, clarifying rules can distinguish when an aspect belongs to the exposure rate versus a resistance estimate.

The most robust analyses will pair excursion types with specific resistance capabilities and customer damages. This can be a complex, multi-dimensional analysis for each customer when all permutations of spec deviations and durations are judged against each customer's damage potential. Such rigor in an assessment is often unwarranted. A simple definition of excursion as 'failure to meet specifications' coupled with a customer damage rate, even when sometimes 'nearly zero damages' for certain excursions, is a simpler and often sufficiently accurate assessment approach. For example, a residential natural gas consumer is unaffected by slight deviations from natural gas specifications or delivery parameters so long as his appliances remain functional and undamaged.

To make the risk assessment more transparent, the agreed upon specifications for product quality and delivery parameters should define excursions. If a customer happens to be insensitive to certain spec deviations that should probably be captured in the consequence assessment. It should perhaps not be modeled as system resistance since the excursion has still occurred and has reached the customer.

Modeling choices should be made to ensure that exposure and resistance measurements employ a common definition. The most robust approach, counting exposure events by imagining absolutely no resistance, may not be warranted or practical in some assessments. The alternative—defining exposures as only events that can cause harm when 'standard' resistances are in place, may be a more desirable approach. See full discussion in Chapter 2 Definitions and Concepts.

Risk Overlaps

In addition to leak/rupture events being a subset of service interruption risk, there are other overlaps. For example, an offspec excursion such as introduction of water into a hydrocarbon stream is an event of interest to both leak/rupture assessment (internal corrosion) and service interruption. Service interruption, by definition, ie 'service', focuses only on potential customer impacts. This may include damages to non-owned (customer) facilities similar to damages experienced by the pipeline owner—internal corrosion, for example. This separation of consequences—those to the owner directly versus those incurred via a customer's consequence—is consistent with the reductionist approach of this recommended risk assessment methodology. Clarity is achieved by treating these consequences independently.

Reliability

Reliability issues overlap risk issues in many regards. This is especially true in stations where specialized and mission-critical equipment is often a part of the transportation, storage, and transfer operations. Those involved with station maintenance will often have long lists of variables that impact equipment reliability. Predictive-Preventive Maintenance (PPM) programs can be very data intensive—considering temperatures, vibrations, fuel consumption, filtering activity, etc. in very sophisticated statistical algorithms. When a risk assessment focuses solely on public safety, the emphasis is on failures that lead to loss of pipeline product. Since PPM variables measure all aspects of equipment availability, many are not pertinent to a risk assessment unless service interruption consequences are included in the assessment. Some PPM variables will of course apply to both and are appropriately included in any form of risk assessment.

12.2 SEGMENTATION

FOCUS POINT
The same segmentation strategy should be employed for service interruption risk as was used for leak/rupture risk.

Although segmentation occurs early in the risk assessment process, the ingredients needed for most efficient segmentation may not become apparent until service interruption scenarios are identified. The potential harm to each customer must be assessed at the customer's location along the pipeline, but the service interruption risk often involves all upstream portions and sometimes even from certain downstream locations. In most cases, all upstream segments connected to a customer-connected-segment, contribute to the service interruption risk for that customer—some by introducing excursion potential and some by providing intervention opportunities that may prevent excursions from causing a service interruption.

12.2.1 Dynamic Segmentation

As with integrity-focused risk assessment, dynamic segmentation is the best approach for modeling service interruption risk. Segment breaks are warranted only where there is significant—from a risk measurement viewpoint—change in any variable thought to impact service interruption probability or consequence. Putting aside leak/rupture segmentation for a moment, segments based on other service interruption factors can typically be longer than those generated in leak/rupture-focused risk assessments. An exception would be where customers or inflows are in close proximity to each other, for example in a distribution system or some gathering systems.

Sources of change for product or transportation characteristics typically include inflow locations, customer take-offs, pump stations, tank farms, pressure regulation points, and a few others. Relevant characteristics typically subject to gradual or abrupt changes along a pipeline system include pressure, volume, and flowrates. The first is common to most pipelines and, to some extent, drives change in the latter two. All may vary more with diameter changes along the route. These changing variables potentially impact dilution of contaminants and ability to meet delivery specifications and may therefore trigger new segments.

Since the service interruption consequence potential is assessed at the customer location, customer proximity along a pipeline will therefore also be a factor. The opportunity for reactionary interventions will often change with proximity to the customer—upstream/downstream volumes, pressures, flowrates, etc, partially determine the opportunity to intervene in an excursion scenario. A pipeline section very close to a customer, where early detection and intervention of an excursion is not possible, will show a greater risk than a section on the same line far enough away from the customer where detection and possibly avoidance of customer interruption are possible.

The frequency of segment-generation—for example, what change in pressure, flow, customer proximity, etc. warrants the creation of a new segment—depends on the desired rigor of the risk assessment.

The service interruption potential from one segment will usually transfer to the immediate downstream segment. Specifically, the excursion potential from upstream segments is normally relevant, since the upstream segment is essentially an inflow or source of product to the segment being assessed. Therefore, a segment near a customer will normally carry the excursion potential from many upstream segments. The risks—excursion and consequence potential—from all segments is of course relevant when aggregating the risk for the whole pipeline or any collection of segments.

12.2.2 Facility Segmentation

Segmentation within facilities is sometimes less intuitive. Each component or collection of components that potentially contributes to a service interruption should be assessed as a separate 'segment' for purposes of risk assessment. This contribution includes each component's role in leak/rupture potential as one possible scenario interruption scenario. Therefore, the same segmentation employed for leak/rupture assessment would normally be a starting point for service interruption assessment. See discussion under segmentation for integrity-focused risk assessment. Additional components may then be required to include components that play no role in leak/rupture potential but must be included as potential service interruption contributors.

For practical reasons including preliminary or very general assessments, an entire facility could be treated as a single source of potential excursions. The facility's collection of excursion scenarios would still need to be estimated with independent estimates of exposure, mitigation, and resistance. This requires at least a general consideration of the types and counts of potentially contributing components. Facilities with more numerous and/or more significant sources of excursion must be identified and their contribution to service interruption potential quantified in order to obtain an accurate risk assessment. Risk analyses tools such as HAZOPS are useful in collecting and assessing scenarios.

12.2.3 Segmentation Process

Most pipeline risk assessments will begin with an integrity-focused assessment—the risks from leak/rupture. These assessments will ideally be based upon a thorough dynamic segmentation process of all pipeline components, including station facilities such as tank farms, compressor stations, certain processing/treating locations, metering facilities, etc. Results from these assessments can be efficiently aggregated as detailed in Chapter 1 Risk Assessment at a Glance. Having a proper aggregation option, the numerous dynamic segments that went into the analyses do not necessarily have to be preserved for use in the service interruption risk assessment. Rather, the aggregated results—PoF (from leak/rupture) from point x to y—can be used as inputs to the service interruption risk assessment. This makes the service interruption risk assessment more intuitive.

Using this strategy, the following segmentation strategy can be efficient:

1. Identify non-leak/rupture factors contributing to excursion potential. Conceptually, this means working from customer locations upstream to any location with significant change or potential change in flow, pressure, product composition (treatment facilities, inflows, etc.), ability to change any of these (ie, available branch connections, pump/compressor stations, perhaps currently not used), or any other factors thought to be pertinent. In some cases, this will include special considerations for changes in potential for moving/entraining/sweeping of accumulated liquid/solid contaminants (for example, low spot accumulation points, critical angle exceedances, liquid drain traps, etc.), blockage formation likelihood; for example, hydrates, paraffins, etc.
2. Perform dynamic segmentation using these non-leak/rupture variables. This will normally result in fewer dynamic segments than produced from a complete leak/rupture assessment.
3. Aggregate PoF values from dynamic segments generated in the leak/rupture assessment. Apply these aggregated values to the service interruption segments, as appropriate.

12.3 THE ASSESSMENT PROCESS

> FOCUS POINT
> The same overall risk assessment process should be employed for service interruption risk as was used for leak/rupture risk.

As previously noted, service interruption risk assessment is a separate branch in the risk assessment, additive to the leak/rupture risk estimates when total risk is being measured. An excursion can occur in many different components (pipeline segments) so all portions of the pipeline contribute to the risk and must be included in the risk assessment. However, the potential consequences occur only at the customer, per the definition of service interruption.

The assessment of PoF, when failure is 'service interruption', follows the same format as for PoF when failure = loss of integrity (leaks/ruptures). Exposure, mitigation, and resistance for each threat to service reliability must be estimated.

Consistent with the definitions given previously. Risk is calculated as the product of the interruption likelihood and consequences:

Risk = Probability of Failure x Consequences

The PoF is the estimate of all pertinent likelihood elements—exposures, mitigations, resistance factors. Consequences represents the magnitude of potential damages arising from a service interruption. The PoF of each segment will usually contribute to the next downstream segment. The risk, however, remains with the customer location's segment since consequences are defined in terms of customer harm.

The overall process is generalized as follows:

1. Define all service interruption scenarios. What must happen and for how long? The transportation/delivery service contract may specify the parameters that constitute a failure in providing the service.
2. Identify all events that lead to service interruption. Each deviation parameter (for example: pressure, flow, quality, etc.) will normally have multiple causes—multiple underlying events. Techniques like HAZOPS are useful for this step. Assess the likelihood of each event.
3. Identify mitigating measures for each potential events. Multiple mitigation measures may be in place for each potential excursion event.
4. Identify all opportunities to intervene, once an excursion is underway. This is the estimate of resistance in terms of excursions that can be absorbed by the system, preventing customer harm. Note that sometimes a resistance measure can be taken far downstream of the excursion.
5. Define potential consequences to each customer for each type of service interruption. These consequences are normally expressed as monetary costs.
6. Using only non-leak/rupture variables, perform dynamic segmentation (see dynamic segmentation discussion at end of this chapter)
7. Using results from previous integrity-based assessments, calculate the aggregated leak/rupture PoF for each of the dynamic segments produced from previous step.
8. Determine the fraction of PoF leak/rupture events that could be addressed without interruption of service (for example: in service clamp repairs). Reduce the exposure from leak/rupture excursions by this fraction.
9. Perform risk assessment for all segments using exposure, mitigation, resistance, and consequence estimates as detailed in this chapter and elsewhere in this book.
10. To show overall service interruption risk for a pipeline (or portion of a pipeline) combine all pairs of PoF and customer CoF scenarios for each segment included in the summary.

The probability of excursion involves exposure and mitigation and is akin to the probability of damage PoD calculation for the leak/rupture assessment. The excursion probability of each segment will usually contribute to that of the next downstream segment.

The PoF uses this excursion probability and also captures the available resistances to interruption such as system redundancies, dilution volumes, and any intervention possibilities, where an excursion occurs along the pipeline, but resistance protects the customer from impact. Resistance to an excursion may not occur until some distance downstream of the location of the excursion. Consider a contamination excursion which eventually dilutes to insignificant levels, far from the origin of the excursion. A resistance will often transfer to the downstream segments. The risk, however, occurs at and remains with the customer location's segment since consequences are defined in terms of customer harm.

Excursion probability includes exposure and mitigation estimates. Service interruption probability includes excursion probability plus resistance. Service interruption risk includes service interruption probability plus potential customer impacts. The following sections are organized to follow this process flow:

1. Excursion probability
2. Service interruption probability (includes potential for customer impact)
3. Service interruption risk

12.3.1 Probability of Excursion

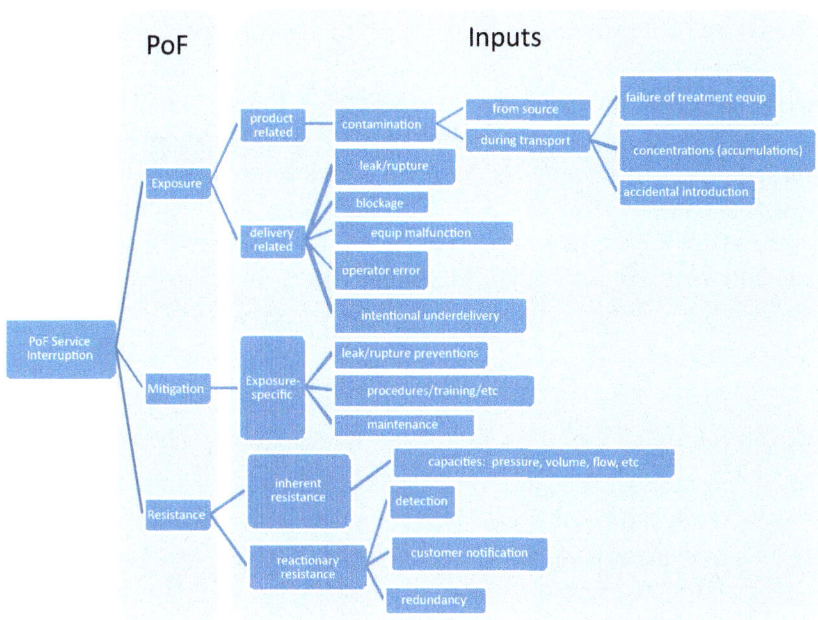

Figure 12.2 Exposure/Mitigation/Resistance Triad in PoF Service Interruption

Measuring the rate or probability of excursions combines exposure and mitigation estimates: [probability of excursion] = [exposure] x (1 – [mitigation]). An excursion source will often potentially affect long segments of the system and be insensitive to segment length. When lengths are relevant, such as for many leak/rupture, blockage, and pipeline dynamics excursions, event rates can be aggregated to include length effects. When event rates are sensitive to counts of components at the same location, for instance the number of independent shut down triggers at a facility, then event rates can again be aggregated to include component counts. Then, event rates in units of events/year rather than, say, events/mile-year can be efficiently used in service interruption risk estimates.

Probability of excursion should include all events that could potentially impact a customer in the absence of resistance. Extraction of resistance considerations—excluding them from the assessment—at this point in the analysis is important. For example, the fact that a contaminant introduced at point A will dilute to be inconsequential before the customer delivery at point B, does not negate the fact that the excursion has occurred. The customer impact—measured independently—can be zero, but the event is still counted in the probability of excursion estimation. While this may at first appear to be a complication, it actually adds clarity to the assessment. As with the integrity-focused risk assessment, failure to consider such factors independently weakens the analyses.

12.3.1.1 Excursion Exposure

Two general categories of excursions cover all possibilities: (1) deviations from product specifications and (2) deviations from specified delivery parameters. Each has its own set of exposures, mitigation measures, and resistance which will often overlap between the two types of upset.

We now look at the exposure, the excursion potential, in more detail. Using some of the factors first introduced in PRMM, the following overall equation is usually appropriate:

$$\text{Probability of Excursion} = (PSD + DPD)$$

Where

PSD = product specification deviation—the potential for the product transported to be off-spec—non compliant with a quality specification

DPD = delivery parameter deviation—the potential for some aspect of the delivery to be unacceptable—non compliant with the agreed upon terms of delivery

A breakdown of typical PSD and DPD exposure categories is as follows:

A. Product Specification Deviation (PSD)

A1. Product Origin

A2. Product Equipment Malfunctions
A3. Pipeline Dynamics
A4. Other

B. Delivery Parameter Deviation (DPD)
B1. Pipeline Failures
B2. Pipeline Blockages
B3. Equipment Failures
B4. Operator Error

An exposure estimate from each of these potential causes of excursion is part of the assessment. The exposure is the estimate of excursion frequency, in the absence of mitigation. The role of mitigation must be ignored when first generating exposure estimates. Discussion of exposure from each of these potential sources of excursion is in the following sections.

12.3.1.2 Excursion Mitigation

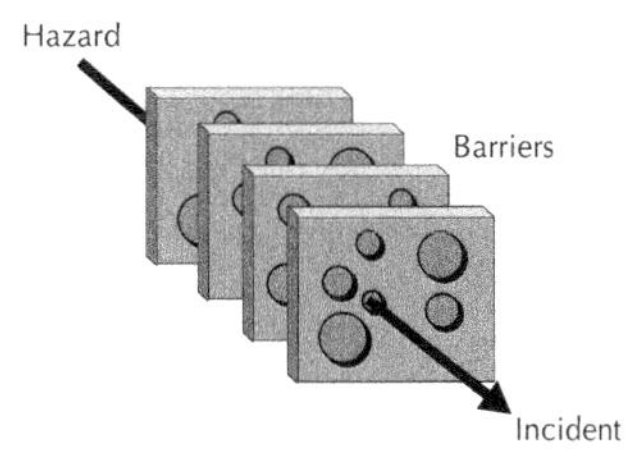

Once the exposure to excursions has been estimated, then mitigation measures can be identified and quantified. As with the integrity-focused assessment, the most robust assessment will pair specific exposures with specific mitigations. A more generalized assessment may take a short cut by assuming that some mitigations provide protection against all exposures, as long as excessive loss of accuracy does not accompany this short cut.

Mitigations are similar to those for leak/rupture prevention, especially those employed against human error, and include control and safety systems, procedures, training, SCADA, error preventors, etc. operator training and procedures often play a role in preventing or minimizing probabilities or consequences of service interruption episodes. These are important in calibration, maintenance, and servicing of detection and mitigation equipment as well as monitoring and taking action from a control room. The evaluator should look for active procedures and training programs that specifically address service interruption episodes. The availability of checklists, the use of procedures (especially when procedures are automatically computer displayed), and the knowledge of operators are all indicators of the strength of this mitigation.

Emergency/practice drills can play a role in preventing or minimizing service interruption excursions. While drilling can be seen as a part of operator training, it is a critical factor in optimizing response time and may be considered as a separate item in the assessment. Where regular drills indicate a highly reliable system, more effectiveness can be assumed. Especially when human intervention is required and especially where time is critical (as is usually the case), drilling should be regular enough that even unusual events will be handled with a minimum of reaction time.

See discussion in the integrity-focused assessment sections of this text for guidance on these and other general mitigation measures commonly employed to reduce both leak/rupture and service interruption events.

Additional mitigation to reduce service interruption excursions is available in the form of reliability programs such as PPM, real time monitoring, and others. Some exposure-specific mitigation measures are discussed in sections below. This list is not all-inclusive since mitigation opportunities are numerous and often customized to specific issues. All types of mitigation can and should be assessed for effectiveness, following the assessment guidance offered here and in the integrity-focused risk assessment discussions.

12.3.1.3 Excursion Resistance

Some resistance occurs at the point of the excursion; for example, immediate dilution, insignificant impact on pressure or flowrate, etc. while others provide resistance some distance from the excursion but prior to the customer location; for example, eventual dilution and recovery of pressure or flowrate, etc.

Excursion-specific resistance factors are discussed in the sections below while a general resistance discussion follows in an independent section.

12.3.2 Estimating Excursions

A. Product specification deviation (PSD)

The transportation of products by pipeline is a service normally governed by contracts that specify delivery parameters. These specifications will show the acceptable characteristics of the product moved as well as the acceptable delivery parameters such as temperature, pressure, and flowrate. Deviations from contract specifications can cause an interruption of service for customers. Even when formal contracts with such specifications do not exist, there is usually an implied agreement that the delivery will fit the customer's requirements.

In water pipelines, specifications vary depending on the type of water system. Potable water systems off-spec excursions include unacceptable levels of dissolved solids, metals, organic compounds, and others.

Off-spec episodes may involve product contamination. Some contaminants are also agents that promote internal corrosion in steel lines. Their potential introduction into a pipeline may have already been quantified in the integrity-focused risk assessment.

To assess the contamination potential, the evaluator should first define 'contamination'. A simple way to do this might be to define it as any product component that is outside the contract-specified limits of acceptability.

A list of all plausible scenarios that could produce contamination will be required in a robust risk assessment. For each potential offspec parameter, specific sources that generate or contribute to the excursion should be identified. This list will serve as a prompter for the assessments. At this point, no consideration for dilution, mitigation, or other contamination-reducing possibilities are included. Exposure estimates are independent of possible effects of mitigation and resistance—those considerations come later in the assessment.

A segment's exposure to excursions must include excursion potentials from all upstream sections. The general sources of offspec episodes or 'upsets' causing excursions are identified as:

- Product origin
- Product treatment equipment malfunctions
- Pipeline dynamics
- Other.

The assessment is to determine the frequency of future excursions from each specific source. To accomplish this, the evaluator should have a clear understanding of the possible excursion episodes. The historical perspective—details of previous incidents—will be important to the extent that previous experience is relevant to future performance. For example, conditions are similar.

Some specification parameters are put in place to control internal corrosion or other damages to the transportation equipment while others protect the customer's equipment and/or product quality. A list can be developed, based on customer specifications that show critical offspec parameters and intolerable concentrations. Additional columns for detectability, mitigation and customer sensitivity can be included to provide guidance for the next steps of the evaluation. This will also serve to better document the assessment.

A1. Product origin

The product's origin point, for example, delivery pipeline, storage facility, processing plant, ground well, etc, provides the first opportunity for excursion.

Changes of products in storage facilities and pipeline change-in-service situations, including batch deliveries, are also potential sources of deviation from product specifications. A composition change may also affect the density, viscosity, and dew point of a hydrocarbon stream. This can adversely impact processes that are intolerant to liquid formation or changes in those characteristics.

Even when a product originates directly from a single hydrocarbon processing plant, the composition may vary, depending on the processing variables and techniques. Temperature, pressure, or catalyst changes within the process will change the resulting stream to varying extents. Materials used to remove impurities from a product stream may themselves introduce a contamination. A carryover of glycol from a dehydration unit is one example; an over-injection of a corrosion inhibitor is another.

Inadequate processing of product or potential contaminant is another source of excursion. A CO_2 scrubber in an LPG processing plant, for example, might occasionally allow an unacceptably high level of CO_2 in the product stream to pass to the pipeline.

The use of drag reducing agents to enhance flowrates can also be a source of upset for sensitive customers.

The evaluator can seek evidence to assess the exposure—the unmitigated excursion potential--from changes at product origin, even when available evidence is based on the mitigated excursion potential.

Some qualitative examples of excursion estimation are shown in PRMM. These qualitative descriptors are reproduced as follows with possible quantitative estimates added.

High Rate; perhaps 0.5 to 500 events/year

Excursions are happening or have happened recently. Customer impacts occur routinely or are only narrowly avoided (near misses).

Medium Rate; perhaps 0.1 to 0.5 events/year

Excursions have happened in the past in essentially the same system, but not recently; or theoretically, a real possibility exists that a relatively simple (high-probability) event can precipitate an excursion.

Low Rate; perhaps 0.01 to 0.1 event/year

Rare excursions can theoretically occur under extreme conditions. Historical customer impacts are almost nonexistent.

No Exposure perhaps 0.00001 to 0.01 events/year

System configuration and/or customer insensitivity disallows upset possibility originating from source. A customer impact is virtually impossible in the present system configuration.

Prevention of offspec episodes and minimization of impacts is supported through close working relationships with customers and suppliers.

Mitigation of Exposures Arising from Source(s)

Because products often originate at facilities not under the control of the pipeline operator, there may be both foreign (owner of the origination point) mitigations and operator (of the segment being assessed) mitigations. Since it will often be difficult to assess and track changes in mitigations of non-owned facilities, it is often more efficient to include foreign mitigations in the exposure rate estimate assigned to the non-owned facility. Those mitigations are often still important to understand and perhaps quantify, but keeping them separate from mitigations applied by the owner of the assessed component is a modeling convenience.

Mitigation opportunities may be limited in some cases. However, common mitigation measures for non-owned/operated point-of-origin upset episodes include

- Real time or sampling-based monitoring of all pipeline entry points (and possibly even upstream of the pipeline—in the supplier facility itself—for early warning) to detect offspec episodes or their precursors at earliest opportunity

- Redundant decontamination/treatment/supply equipment for increased reliability on single source scenarios.
    - o Close working relationship with third-party suppliers
    - o Availability of multiple product stream sources at origin point (blending or partial shut in opportunities)
        - Arrangements of alternate supplies to shut off offending sources without disrupting pipeline supply
        - Provisions for rapid switches to alternate supplies
        - Plans and practiced procedures to switch to alternate supplies
    - o Automatic switching to alternate supplies l
- Operator training to ensure prompt and proper detection and reaction to excursions.

Any preventive actions should be factored into the assessment of excursion mitigation.

A2. Treatment equipment malfunctions

Pipeline equipment at, or downstream of, the product source, designed to control product specification parameters such as removal of impurities can malfunction and allow offspec episodes. This may overlap the previous assessment 'product origin' so care must be taken to count all events appropriately—neither over- nor under-counting.

Some on-line—during the transportation--equipment such as dehydrators help ensure product specification parameters including protecting the pipeline from possible corrosion agents. Hence, their reliability in preventing upsets will overlap previous analysis of their role in PoF from internal corrosion.

Injections of substances such as corrosion inhibitor liquids or flow-enhancing chemicals are examples of intentionally-introduced substances that may impact customers. Even when customers are unaffected by intended concentrations of such injected substances, equipment malfunction or flow regime changes may lead to higher concentrations of these products than what is tolerable by the customer.

Multi-phase pipelines, in which combined streams of hydrocarbon gas, liquids, and water are simultaneously transported, are often found in gathering systems and offshore production pipelines. Downstream receipts from such systems frequently rely on equipment to perform separation. When separation equipment fails, excursions occur.

When the equipment can potentially ***introduce*** a contaminant—for example, flow enhancer, glycol dehydration, corrosion inhibitor, etc.—an estimate of the unmitigated exposure, followed by the effectiveness of mitigation, is needed. When the equipment is ***preventing*** offspec excursions then its role as a mitigation measure against a continuous exposure needs to be estimated. In either case, estimation can be done in a very detailed, robust manner when critical consequence may emerge, or alternatively may be approximated by those knowledgeable of the system.

Unmitigated upset potential from on-line equipment malfunctions can range in event frequency from 'almost never' to 'continuous'. A detailed assessment may include formal equipment reliability modeling.

Mitigation

The following mitigation activities can be factored into the evaluation for excursions due to equipment malfunctions for both scenarios—equipment-generated exposures and equipment as excursion prevention:

- Strong equipment maintenance practices to prevent malfunctions
- Redundancy of systems (backups) to increase reliability of equipment or systems to reduce the probability of overall failures
- Early detection of malfunctions to allow action to be taken before a damaging excursion or a loss of function occurs.

A3. Pipeline dynamics

Another generator of excursion scenarios is liquids or solids concentrated in a product stream by a change in pipeline system dynamics. A possible source of solids could be foreign materials from original construction or subsequent repairs, materials originally introduced by within-spec streams, materials from offspec excursions, or materials generated within the pipeline during its operational history.

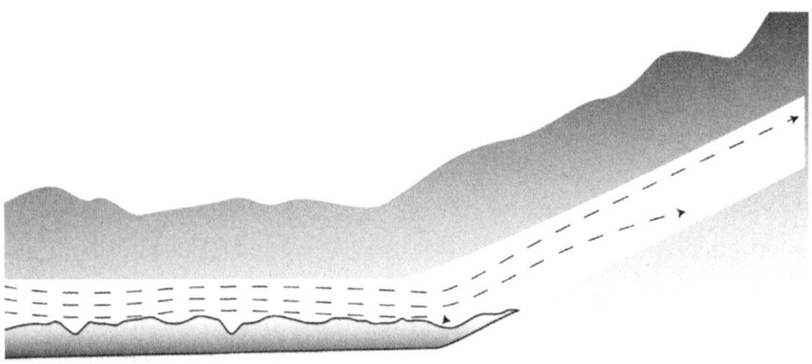

Figure 12.3 Critical Inclination Angle Exceeded, Resulting in Depositions

Free liquids, both water and heavier hydrocarbons, and solids may accumulate in low-lying areas of a pipeline transporting hydrocarbons.

Some pipelines also have potential for other types of accumulations. Hydrates, rust particles, debris from damaged pigs, or paraffin buildups displaced from the pipe wall are examples of materials generated during operations. (see also discussion of pipeline blockages) To cause this, the offending materials would have to be present initially, so an exposure estimate arises from that necessary condition. Added to this for the complete estimate of exposure, is the potential for an accompanying event causing a significant disturbance to the pipe displacing a large amount of the buildup at one time, leading to the customer impact.

Pipeline dynamics can also precipitate a service interruption by causing a delivery parameter to become offspec. Pressure surges or sudden changes in product flow may interrupt service as a control device engages or the customer equipment is exposed to unfavorable conditions. This halts flow, thereby interrupting the flowrate required by the specification.

Potential for upset from changes in pipeline dynamics is assessed in terms of exposure and mitigation, as are all types of service interruptions. Specific pairings of mitigations with affected exposures may be warranted since not all mitigations will affect all exposures. For instance, preventing excursions due to re-entrainment or sweeping of accumulations may have no benefit to the exposure of flow interruptions from inadvertent valve closures. Note also, that some mitigation measures will increase the potential for service interruptions. For instance, maintenance pigging carries a chance of flow interruption due to pig failure or formation of a blockage.

Mitigation

Prevention activities typically factored into the assessment for upset potential due to pipeline dynamics include:

- Performing pipeline pigging, cleaning, dehydration, etc., in manners that prevent later excursions.
- A protocol that requires experts to review any planned changes in pipeline dynamics. Such reviews are designed to detect hidden problems that might trigger an otherwise unexpected event.
- Close monitoring/control of flow parameters to avoid abrupt, unexpected shocks to the system.

Instrumentation calibration/maintenance to reduce unintentional activations. This is more appropriately included in the exposure estimate rather than as a mitigation measure, when the instrumentation is the initiator of the exposure.

A4. Other

As a special type of failure mechanism, the threat of sabotage may warrant special attention in service interruption risk, beyond its role in leak/rupture risk. Saboteur actions directed towards service interruption rather than leak/rupture can be included in this part of the assessment. With the change in definition of 'failure', this threat assessment will closely mirror the leak/rupture assessment. Different exposure types and frequencies must be identified, representing the product and delivery vulnerabilities rather than integrity vulnerabilities. Mitigations will be very similar for both types of 'failure'. The roll of resistance will need to be supplemented in the service interruption assessment since sabotage here may involve different types of excursions; for example, the introduction of an unexpected contaminant with different detectability and reaction opportunities.

Examples of additional upset scenarios that do not directly arise from a product in-coming source or from pipeline flowing dynamics include improper restoration to service after maintenance, change in service, infiltration of ground water into a low-pressure distribution system piping, incorrect handling of batched products, and others. When such scenarios are plausible, they should be included in the risk assessment with the same exposure-mitigation-resistance triad used in all PoF analyses.

B. Delivery parameters deviation (DPD)

General excursion scenarios that must be included in assessing the risk of service interruption are deviations from acceptable delivery parameters such as pressure, temperature, or flow. For example, when a city resident orders a connection to the municipal gas distribution system, the implied contract is that gas, appropriate in composition, will be supplied at sufficient flow and pressure to work satisfactorily in the customer's heating and cooking systems.

General causes of delivery parameter deviations include:

- Pipeline failures
- Pipeline blockages
- Equipment failures
- Operator error.

Since a customer impact is the consequence of interest, potential scenarios upstream of a customer normally generate the events of interest and are included in the evaluation. However, some downstream events may also generate upstream customer consequences. For instance, excessive flow entries or exits downstream may impact upstream pressure levels.

As with all exposures, a list of plausible scenarios should be developed. Critical delivery parameters, based on customer specifications, should be identified and linked to specific mechanisms that could upset those parameters.

Undersupply excursions—not meeting minimum pressure/flow specifications—are the most common types of excursions. These are judged to arise from two general types of exposure, each with specific contributors:

- insufficient delivery to customer
    - o intentional supply or inventory reductions
        - o reductions to accommodate seasonal-, business-, temporary maintenance-, other customer-needs, and other scenarios
    - o unintentional supply or inventory reductions
        - o leaks/ruptures in upstream segment(s)
        - o equipment failure
        - o operator error
        - o blockages

B1. Pipeline Leak/Rupture

A leak/rupture in a pipeline component will usually precipitate a delivery interruption. The possibility of this is assessed by performing the integrity-focused risk assessment (for leak/rupture) detailed in Chapter 2-11. The resulting estimate is a measure of this type of failure potential.

The excursion potential is equal to the PoF for leak/rupture estimated in the integrity-focused risk assessment, less the scenarios where no service interruption occurs despite there being a leak. When a leak can be repaired without interrupting flow, pressure, or other delivery parameter, for example, clamp installed, a service interruption has not occurred.

B2. Equipment failures

Equipment failures that can cause an unacceptable delivery parameter will normally need to be included in a service interruption assessment. Pumps, compressors, and valves are often critical since they directly control pressures and flowrates. These primary pieces of equipment are normally influenced by multiple secondary systems. Most modern pipeline control systems employ a complex network of manual and automatic monitoring, relief, and shut down instrumentation, as described in Chapter 8 Incorrect Operations. These same systems that reduce the probability of leak/rupture may increase the potential for service interruption. Erroneous equipment operations (inadvertent valve closure, pump stop, etc.), mis-calibration of instruments, or improper actions by operators or maintainers causing shut downs are examples.

Unintentional equipment activations—valves, rotating equipment, etc.—or equipment activations generated by abnormal conditions can cause flow restrictions. An "unwanted action" of such devices is normally not addressed in the basic risk assessment model because such malfunctions do not usually lead to pipeline leak/rupture. Therefore, this additional consideration must be added when service interruption is being evaluated.

Reliability improves when more than one line of defense exists in preventing excursions. For maximum benefits, there should be no single point of failure that would either create an excursion or disable the system's ability to prevent an excursion. Where redundant equipment or bypasses exist and can be activated in a timely manner, excursion probability is reduced.

Outages caused by weather or natural events such as hurricanes, earthquakes, fires, and floods are possible causes of leak/rupture and also considered in service interruption potential as possible sources of equipment failure excursion. A common example of a non-leak/rupture event of this type is an offshore pipeline system that is intentionally shut down whenever large storms threaten. Other examples include those typically covered under force majeure clauses in a legal contract.

The complexities and variabilities in pipelines and their associated control system designs prevents a detailed discussion of all possible interruption scenarios in this book. To generalize these scenarios, some categorizations of equipment potentially contributing to service interruption can be made. Here are some groupings and discussion to stimulate thinking on this topic.

Pressure and flow regulating equipment

Pumps and compressors used to maintain specified flows and pressures are more complex mechanical/electrical equipment that are more prone to service interruption. Relatively minor occurrences that will stop these devices in the interest of safety and prevention of serious equipment damage include those listed for leak/rupture prevention, such as pressure, flowrate, and tank levels. Additional parameters, associated with the prime movers and often threatening service interruption, but not immediate leak/rupture potential, include temperature, voltage, electrical current, vibration, sensor status, equipment position/status, and many more.

Valves

Flow stopping that halt flow through a pipeline are potential causes of specification violations. This includes shut-in devices from product origination points such as wells, and mainline block valves, including emergency shut-in, automatic, remote, check valves, and manual configurations are included here.

Safety/Control Systems

Instrumentation and devices intended to prevent damage to the system exist in virtually all pipeline delivery systems. Examples include regulator valves, relief valves, rupture disks, limit switches (which activate equipment upon certain pressure, temperature, tank level, electrical parameters, etc. limits or rate-of-change), and others that will normally impact ability to delivery when they activate.

Equipment controlling product properties during transportation can also be considered here. The number and nature of devices that could malfunction and cause a delivery parameter upset is normally important to a risk assessment. The phrase "single point of failure" is used to indicate that one component's failure is sufficient to precipitate a service interruption. This makes a system more vulnerable to excursion. Examples often include malfunction events associated with components such as instrument power supply, instrument supply lines, vent lines, valve seats, pressure sensors, relief valve springs, relief valve pilots, and SCADA signal processing.

Mitigation

Prevention (mitigation) activities for service interruptions caused by equipment malfunctions include:

- Measures to minimize potential for inadvertent equipment activations—fail safe logic, overrides, redundancies, etc.

- Measures to reduce rate of occurrence of abnormal conditions
- Equipment calibration and maintenance practices
- Inspections and calibrations including all monitoring and transmitting devices
- Redundancy preventing, for instance, one erroneous indication from unilaterally cause unnecessary device activations.

While these measures can be included in the assessment of exposure, it is often more useful to rather include them with mitigation. One benefit is the development of an argument, via cost/benefit analyses, for the increase or reduction in activities.

It will usually also be important to identify and include the presence of redundant systems that prevent customer impacts, even after component interruptions. Such systems were established for a reason and at a cost and therefore warrant consideration in the risk assessment.

Potential for delivery parameter deviation due to equipment failure is potentially high when excursions are happening or have happened recently--customer impacts occurring or are only narrowly avoided (near misses) by preventive actions. Frequent weather-related interruptions are additional indicators. Since such evidence is occurring with mitigation and resistance, exposure rates considered in the absence of mitigation and resistance may be especially high.

B3. Operator error

The potential for human errors and omissions is logically a part of service interruption potential. The risk analysis conducted for the leak/rupture risk assessment is normally a part of the service interruption assessment. Errors that lead to service interruption but not leak/ruptures, precipitate additional failure scenarios that are additive to the estimated error rates for leak/rupture.

Part of the service interruption assessment is the potential for an on-line operational error such as an inadvertent valve closure, unintentional halting of a pump or compressor, introduction of a contaminant or failure to remove a contaminant, or other errors that do not endanger the pipeline integrity but can temporarily interrupt pipeline operation. Note that the focus here is on accidental human activities. Willful actions are addressed as sabotage.

As with the potential for leak/rupture, the evaluator should begin the mitigation assessment with an examination of the training, testing, and procedures program to gauge the effectiveness of measures that are in place to generally avoid all errors. Error prevention activities also include visual/audible signs, signals, and alarms; the use of special checklists and procedures, and designs that allow excursions only under an unlikely sequence of errors.

B4. Pipeline blockages

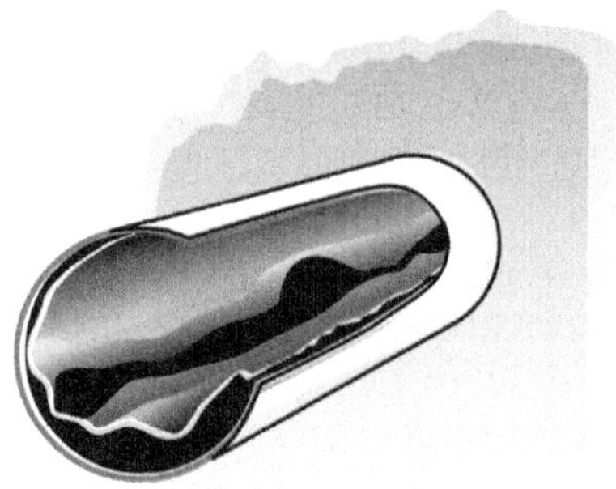

Figure 12.4 Interior wall build-ups, such as paraffin

Restricted or blocked flow in a pipeline may not lead to a leak/rupture but can generate a delivery parameter (such as pressure or flow) deviation.

The potential for unmitigated, unresisted blockage events may range from virtually zero events/yr, when potential is very low, to dozens of events/year when exposure is high.

Monitoring via pressure profile, internal inspection device, or others may provide early warning of impending blockages. Mitigative actions potentially taken include cleaning (mechanical, chemical, or thermochemical) at frequencies consistent with buildup rates; the introduction of chemical inhibitors to prevent or minimize buildup.

B5. Other

Examples of other delivery parameter excursions include voluntary deviations. When the operator chooses to create an excursion to avoid higher consequences, an excursion has nonetheless been created. Depending on issues such as contract provisions, the customer's impact and subsequent recovery of damages may differ from an accidental excursion. Voluntary or semi-voluntary excursion scenarios include:

- Weather events—operator chooses to interrupt service due to safety or system integrity issues; for example, halting operations during floods, hurricanes, ice storms, etc. These excursions differ from excursions generated by weather-related equipment failures in that no equipment failure has occurred and the operator is taking proactive measures.
- Financial events—these can range from choosing to supply one customer at the expense of another during a shortage to company bankruptcy. Intentional non-compliance with contracted terms of delivery could also be prompted by special financial issues.
- Other suppliers' non-performance—an example would be interruption of upstream supply causing downstream shortages.

- Urgent maintenance or repair—no failure has occurred but operator must respond to a failure precursor, perhaps identified during an inspection.

Exposure, mitigation, and resistance estimates can be assigned to these excursions and included in the assessment.

12.3.3 Resistance

In the integrity-focused risk assessment, the actions taken to prevent pipeline failures is included as mitigation in various threat assessments. A PoD estimate emerges from this. The system's ability to resist failure, given damage is occurring, is then measured as 'resistance'. PoF is calculated from PoD and resistance.

In the service interruption risk, actions to ***prevent*** events that lead to service interruptions are also assessed early in the assessment as mitigations and results in a 'probability of excursion'. Then, resistance to failure is added to produce a PoF, ie, probability of service interruption, since failure = service interruption here. Service interruption scenarios often have additional opportunities—beyond those available to leak/rupture prevention—for intervention after an excursion episode has occurred that would otherwise lead to a service interruption. System volumes, flow rates, pressures, redundancies, etc. all act to absorb the excursion, often by blending or diluting away the infraction, making it invisible to the customer.

Recall that the recommendation is to consider an excursion to be an event originating at the entry to the pipeline rather than at a customer. The consequence occurs at the customer. While both could be modeled as occurring only at the customer, excursions at other locations would still have to be assessed for their potential to reach the customer. Under this recommendation, that is the resistance estimate, to keep it independent from the initiating excursion event. This helps in diagnostics and risk management.

In the risk estimates, resistance shows when some segments are more capable of absorbing excursions and can at least partially recover from an episode before customer impact occurs. This exactly parallels the resistance estimate which distinguishes between PoD and PoF in an integrity-focused risk assessment. In both assessments, resistance is expressed as a fraction of failures avoided. A segment that is 90% resistive would experience a service interruption once out of every ten excursions (excursions that, despite mitigation, are occurring).

In some pipeline systems for which interruptible delivery is critical, extra provisions are usually made to prevent interruption. Timely reactions to events that would otherwise cause service interruptions are sometimes possible. Examples include halting the flow of an offending product stream and replacing it with an acceptable product stream, blending streams to reduce concentration levels, immediate treating of a contaminant, and immediate customer notifications when customers can prevent or minimize harm from an excursion.

Even a pipeline failure may not result in a service interruption. If the leak can be repaired without significant change to product flow (for example, a clamp repair) or an alternative supply is available to replace the lost supply to the customer.

Note that by considering interventions, a high-probability excursion that has a low probability of actually impacting the customer is recognized but shows lower risk than the same event that is more likely to impact the customer. This is important to the understanding and management of the risk.

Variable Resistance

Adding to the challenge of measuring resistance is the fact that some systems experience variable resistance. Seasonal changes in resistance are common, with supply-demand issues creating shortages and excesses. Some systems are highly variable, with inventories, system dynamics, and available options varying day-by-day or even hour-by-hour.

Inherent Resistance

System volumes, pressures, and dynamics play a role in resistance. Systems that are more able to absorb excursions in that they are slower to react to an upset or otherwise less sensitive to an excursion. For instance, a high-pressure, large-volume gas pipeline system in which outflows will only slowly depressure the system upon temporary loss of inflows. Contrast this with a small liquid system that is effectively “tight-lined” (inflows balance outflows with temporary imbalances resulting in immediate loss of pressure and flow). In this latter case, intervention opportunities will be limited and their effectiveness will be challenged.

Example: 12.1 Probability of Customer Impact from Delivery Excursion

A section of a high-pressure gas transmission system serves a customer with sensitivities to pressure and flowrate. Both must be kept within specified parameters.

Exposure and Mitigation Estimates

High side excursions—overpressure and excessive flowrate are both possible events in this segment and are deemed to be continuous exposures since the source holds pressure levels and generates flowrates that both exceed customer tolerance limits. The estimate of ‘continuous exposure’ carries an assignment of 5.3e5 events/yr (one event per minute) in this risk assessment’s protocol. Offsetting this exposure are mitigation measures, evaluated as shown below:

Protecting the customer from excessive pressures and flowrates are control devices and safety systems. Failure possibilities for mitigation equipment include mechanical or electrical failure of the systems, mis-calibration or failure of associated pressure/

flow sensors, loss of instrument power supply, incorrect signal from SCADA system, and others.

Mitigation of 'too much' pressure and flowrate excursions from equipment failure at the customer 'take-off' location are identified as:

- Pressure controller (pressure control valve) at customer gate—failure here can either interrupt service or allow too much pressure into customer facility. Controller failures leading to excessive pressure or flow are estimated to be 10-8 per year.
- Control valve at meter site Controller failures leading to excessive pressure or flow are estimated to be 10-7 per year.
- Additional mitigation offered by safety systems including high pressure and high flow automatic valves, are under the control of the customer and not included in this calculation. If they were to be included, they would modeled as generating redundant mitigation whereby both a controller and the safety system must simultaneously fail before the upset occurs.

In this scenario, either of the equipment failures would result in an event of interest, suggesting that they should be combined with an OR gate. This results in an estimate of 'probability of upset':

5.3e5 unmitigated exposure-events/yr x [(10e-8)+(10e-7)] upsets/exposure-events = 5.8e-2 upsets/year or an upset event about every 17 years.

Low-side excursions—not meeting minimum pressure/flow specifications, are judged (via HAZOPS) to arise from two general types of exposure, each with specific contributors:

- undersupply into segment = 0.2 events/year (an event every 5 years)
- pipeline leaks/ruptures on associated segments
- unintentional closures of any of three upstream automatic block valves unintentional halting of mainline compressor station where station bypass would not allow sufficient downstream pressure
- unplanned interruption of source flows
- improper planning of flows/inventories
- excessive outflows.

Emergency maintenance scenarios or = 0.01 events/year:

- pipeline leaks/ruptures on associated segments
- improper planning of flows/inventories
- others

SME's identify mitigation measures that are currently available to prevent the exposures that are not already fully analyzed (for example, the pipeline leak/rupture rates are already available). for a preliminary estimate, the SME team judges that mitigations are in place to offset approximately 8 out of 10 excursions of these types. This is a combined mitigation estimate that integrates each individual mitigation measures contribution as if all exposures are equally mitigated by each. This is a simplifying assumption, technically inaccurate (for example, pipeline failure rates are not mitigated by these mitigations) but deemed acceptable for current assessment needs. The team plans to update and improve upon these estimates with a detailed HAZOP later in the year.

Using these estimates, the probability of a low-side excursion is assessed to be:

(0.2 + 0.01) unmitigated events/year x (1 – 80%) upsets / event = 0.042 upsets/yr or an upset event about every 24 years.

Resistance estimates

Next, the SME team quantifies the ability of the system to resist the potential customer upset, given the occurrence of an upset event. To begin the analysis, all sources of resistance are identified and include:

- Line pack (inventory), normally of sufficient pressure/volume to allow several hours of undersupply into the segment, without impacting customer delivery parameters. This resistance effectively offsets the episodes that are of short duration. SME's assign a resistance benefit of 60% based on the fraction of shorter duration episodes possible.
- Redundancy: no redundancy of supply is available to this customer.
- Alternate supplies: the availability of contract provisions and relationships with product suppliers who would likely 'loan' product volume during a critical need, allows SME's to estimate that an additional 20% of the listed episodes would not lead to customer impact.

The combined resistance is therefore estimated to be: 60% OR 20% = 68% (68% of the episodes would not result in customer impacts).

The final probability of customer impact is estimated to be:

(0.06 + 0.04) upsets/yr x (1 – 68%) customer impacts/upset =
0.03 customer impacts per year
(or a customer impact about once every 31 years).

The fairly long history of 15 years with no excursions is used to partially validate this estimate.

Note that none of the equipment failures identified in the above example would cause a pipeline leak/rupture on the assessed segment, but rather serve to estimate a service interruption potential only.

A major delivery deviation would be consequential to this customer, requiring an emergency interruption of their processes and a multi-day resumption of service. Impacts to this customer are estimated to be $450,000 per delivery deviation. This, coupled with the previous estimate of impact probability, results in an EL = 0.03 x $450K = $14K per year.

Example: 12.2 Service interruption potential

Example 10.2 of PRMM can be improved by better quantifying the risk elements as follows: XYZ natural gas transmission pipeline has been sectioned and evaluated using a leak/rupture risk assessment model. This pipeline supplies the distribution systems of several municipalities, two industrial complexes, and one electric power generation plant. The most sensitive of the customers is usually the power generation plant. This is not always the case because some of the municipalities could only replace about 70% of the loss of gas on service interruption during a cold weather period. Therefore, there are periods when the municipalities might be critical customers. This is also the time when the supply to the power plant is most critical, so the scenarios are seen as equal.

Notification to customers minimizes the impact of the interruption because alternate supplies may be available at short notice. Early detection is possible for some excursion types, but for a block valve closure near the customer or for the sweeping of liquids into a customer service line, at most only a few minutes of advance warning can be assumed. There are no redundant supplies for this pipeline itself. The pipeline has been divided into sections for risk assessment. Section A is far enough away from the supplier so that early detection and notification of an excursion are always possible. Section B, however, includes an inflow metering station very close to the customer facilities. This station contains equipment that could malfunction and not allow any time for detection and remedy before the customer is impacted.

A preliminary and conservative, P90 risk of service interruption assessment is sought. Because each section includes common elements—conditions found in all sections--many input values will be the same for these two sections. The potential for excursions, considering all mitigations applied, for Section A and Section B is evaluated as follows:

Product specification deviation (PSD)

Product origin: 0.01 events/yr

Only one source, comprising approximately 20% of the gas stream, is suspect due to the gas arriving from offshore with entrained water. Onshore water removal facilities have occasionally failed to remove all liquids.

Equipment failure 0.2 events/yr

No gas treating equipment in this system. 0.0 events/yr

Pipeline dynamics 0.05 events/yr

Past episodes of sweeping of fluids have occurred when gas velocity increases appreciably. This is linked to the occasional introduction of water into the pipeline by the offshore supplier mentioned previously.

Other 0 events/yr

No other potential sources identified.

Delivery Parameter Deviations (DPD)

Pipeline failure:

0.0005 events/mile-year x 30 miles of pipeline = 0.0015 events/year

From previous integrity focused risk assessment.

Blockages: 0.000001 events/yr

No mechanisms to cause flow stream blockage, other than inadvertently closed valve, considered below.

Equipment: 0.06 events/yr

Automatic valves set to close on high rate of change in pressure have caused unintentional closures in the past. Installation of redundant instrumentation has theoretically minimized the potential for this event again. However, the evaluator feels that the potential still exists. Both sections have equivalent equipment failure potential.

Operator error (Section A)

Little chance for service interruption due to operator error. No automatic valves or rotating equipment. Manual block valves are locked shut. Control room interaction is always done. Mitigated error rate is estimated to be 0.05 events/year.

Operator error (Section B)

A higher chance for operator error due to the presence of automatic valves and other equipment in this section. Mitigated error rate from all plausible event scenarios is estimated, via a HAZOPS technique, to be 0.1 events/year.

Section A total = 0.01 + 0.2 + 0.05 + 0 + 0.0015 + 0.000001 + 0.06 + 0.05
= 0.37 excursions per year

Section B total = 0.01 + 0.2 + 0.05 + 0 + 0.0015 + 0.000001 + 0.06 + 0.1 + 0.37*
= 0.79 excursions per year

**Note that section A is an input to Section B. that is, all excursions originating and not eliminated in Section A, are excursions for Section B.*

The above values are analogous to the PoD values produced in the integrity-focused assessment. They reflect the frequency of events that could lead to failure, ie, customer harm.

Resistance

Next, resistance is estimated. Reactive and inherent interventions to excursion scenarios are available for both sections. For Section A, it is felt that system dynamics allow early detection and response to most of the excursions that have been identified. The volume and pressure of the pipeline downstream of Section A would dilute contaminants and allow an adequate response time to even a pipeline failure or valve closure in Section A. Fractions of events successfully resisted are assigned for blending/dilution (0.8), early detection and re-establishment of supply or establishment of alternative supply (0.3). These are thought to generally apply to all excursion types and, hence, establish the resistance via an OR gate. Therefore, Section A is 1 – (1 – 0.8) x (1- 0.3) = 86% resistive to the potential excursions. Section A is assessed to carry a service interruption potential of 0.37 excursions/year x (1 – 0.86) fraction resisted = 0.052 events/yr or a customer impact about once every 20 years.

Early notification is not able to provide enough warning for every excursion case in Section B, however. Therefore, reactive interventions will only apply to some excursions that can be detected and responded to, namely, those occurring upstream of Section B. For the types of excursions that can be detected in a timely manner, product origin and equipment problems, percentages are awarded for early detection (30), notification where the customer impact is reduced (10), and training (8). This analysis shows a much higher potential for service interruption for episodes occurring in Section B as opposed to episodes in Section A.

The customer consequence potential would be calculated next. A direct comparison between the two sections for the overall risk of service interruption can then be made.

12.3.3.1 Reactionary Resistance—Intervention Opportunities

To assess the availability and reliability of interventions, a compiled list of the effectivenesses of all the intervening actions that are plausible and available, is needed. Note that these actions may not apply to all identified episodes of product specification deviation or delivery parameter deviation and therefore may need to be paired with specific excursion types. If an action cannot reliably address excursions of all types, then intervention applies only to the benefiting excursion(s). For example, if an early detection system can react quickly to a pipeline failure but cannot detect a contamination episode, then the benefit applies only to leak scenario resistance.

Resistance percentages will be used in assessments for PSD and DPD independently. Again, reducing failure potential in this fashion does not indicate a reduced probability of the event, only the reduced probability of the event causing customer upset, ie, service interruption. This is an important distinction as it is in the integrity –focused risk assessment that discriminates between probability of damage and probability of failure.

Detection

When an excursion is not detectable, reactionary intervention action is not possible. When at least some of the possible excursions are detectable, additional intervention opportunities may be available. For resistance estimation, the ability to identify and provide some advance notice of an excursion plays a role only when it enables intervention.

The reliability and timeliness of detection should be assessed. Detection includes receiving, interpreting, and responding to the indications. Indirect indications, such as a pressure drop after an accidental valve closure, serve as detection mechanisms but often require diagnostic time.

A location on the pipeline near the customer may generate an excursion for which there would not be a possibility of early detection and timely reaction. When some excursion types can be detected and some may not be, or when detection/reaction is not reliable, effectiveness estimates should be accordingly adjusted and applied only to specific excursions.

12.3.3.2 Customer notification

In some cases, timely notification to a customer of an excursion can prevent an outage for that customer. In many cases, impacts can at least be reduced. This is discussed under consequences. Customer notification is generally not a resistance factor since it does not prevent the excursion from reaching the customer. Rather it is a part of consequence minimization.

12.3.3.3 Redundant equipment/supply

Resistance to excursion is available in system configurations that allow rerouting of product to blend a high contaminant concentration or otherwise keep the customer supplied with product that meets minimum quality and delivery specifications. The redundancy must be available in a time that will prevent customer harm. Factors impacting reliability may include the following:

- Degree of human intervention required
- Amount of automatic switching available
- Regular testing of switching to alternative sources
- Highly reliable switching equipment
- Knowledgeable personnel who are involved in
- Switching operations
- Contingency plans to handle possible problems during switching

12.4 CONSEQUENCES—POTENTIAL CUSTOMER IMPACT

As noted in the service interruption definitions, the consequence usually being measured in the risk assessment is the damages that occur ***to the pipeline owner/operator***, with the idea that it is the customer damages that are primarily driving the operator's damages. Both direct and indirect consequences should be recognized in the risk assessment.

A distinction between 'transportation event' vs 'delivery event' may be useful in some consequence assessments. Product ownership is often separated from transportation service in the pipeline industry. This separation has implications for costs of product loss during leak events and certain contract non-performance penalties. Whether the service interruption is interrupting the transportation or the delivery may be a subtle nuance that impacts costs.

A segment will often potentially impact multiple customers to varying degrees. Some sections of pipeline are therefore more capable than others of generating service interruption excursion. A transmission line excursion might impact several industrial users, other pipelines, or several entire distribution systems. In a distribution system, a failure on a 'main' will impact many end customers, whereas a service line failure will usually impact fewer. Number of customers is of course not the only metric of consequence. Some individual customers can be very high consumers of the pipeline service and/or have much higher consequences of service interruption, for example, an electrical power generation plant or a critical care health facility.

In distribution and gathering systems, meter counts and/or outflow volumes are normally available and can be linked to upstream portions of the pipeline system. A customer count or usage-adjusted count will normally be relevant to outage costs. It is often the most readily available metric of consequence and may appropriately serve as a surrogate for all consequences, pending further assessment.

Sometimes it is difficult to link customers with specific segments of a distribution or complex gathering system network, other than the piece that directly connects them to the system. Multiple, complex hydraulic modeling scenarios may be required to know possible impacts from portions of the system farther from a customer. Where it is not practical to link specific customers or even customer counts to all potential excursion-generating locations, approximations may be appropriate. The volume or pressure in any portion of the system or the count of customers downstream could be assumed to be directly proportional to the criticality of that supply at any location. Therefore, locations where higher flow rates, more downstream customers, etc. are potentially interrupted may be modeled to cause higher outage consequences.

A more robust risk assessment will include the specific sensitivity of the various customers. Both receipt- and delivery-customers should be included when either can be harmed by service interruption. A customer is not necessarily an outside party—internal customer harm is normally also of interest.

The customer tolerance to excursions is the key to consequences in service interruptions. The customer specifications should reflect the acceptable product and delivery parameters, which sets the definition of 'excursion'. However, when standardized specifications are used, there is often a difference between what can actually be tolerated versus what contract specifications allow. In some cases, customer sensitivity is fairly apparent. For instance, when the customer is a simple user of the product, such as a typical residential customer who uses natural gas for cooking and home heating, minor deviations from standard natural gas specifications are inconsequential. To determine at what point deviations do become consequential, the manufacturer of the customers' equipment (stove, heater, etc.) will be the more reliable information source. For more sophisticated consumers of transported products, interviews with the customer's process experts may be warranted. In many assessments, however, this is an unwarranted level of rigor. A simple definition of excursion as 'failure to meet specifications' coupled with an estimated customer damage rate, perhaps nearly 'zero damages' for certain excursions, is a simpler and often sufficiently accurate assessment approach.

There is often a time component to level of damage from an excursion. Some customers can incur large losses if interruption occurs for even short periods, as described in PRMM.

In a residential situation, if the pipeline provides heating fuel in cold weather conditions, loss of service can cause or cause or aggravate human health problems. Similarly, loss of power to critical operations such as hospitals, schools, and emergency service providers can have far-reaching repercussions. While electricity is the often most common need at such facilities, pipelines often provide the fuel for the primary generation of that electricity or the backup systems.

Some customers are only impacted if the interruption is for an extended period of time. Perhaps short time outages are tolerable and significant losses occur only with long term production interruption.

12.4.1 Direct Consequences

The most obvious cost of service interruption is the loss of pipeline revenue due to curtailment of product sales and/or transport fees during the excursion. This can be viewed as a direct cost of the interruption. Other direct costs are similar to those identified in the integrity-focused risk assessment and include:

- Fines, penalties
- Loss of product, if leak/rupture or if de-inventorying is necessary
- Clean up/remediation/restorations, if needed
- Repairs, if needed
- Return to service

The costs associated with a service interruption will usually be related to the duration of the outage.

12.4.1.1 Revenues

Revenues generated from the pipeline section being evaluated will often be a reasonable measure of the consequence potential of that section, from a provider-of-service (the pipeline owner/operator) view. A section's revenues should include revenues from all relevant up- and downstream sections whose ability to serve their customers may be simultaneously compromised by the outage. The entire downstream portion of a pipeline can be viewed as a customer of the segment being assessed. This captures the intuitive belief that a "header" or larger upstream section has higher consequence potential than a single-delivery downstream section.

12.4.1.2 Return to Service

Repair, outage, and other 'return to service' costs are an element of integrity-focused risk assessments, but since time is a critical aspect of many service interruptions, these processes must often also be included as an aspect of service interruption impacts. In addition to the direct costs associated with 'return to service', customer impacts related to outage periods are added here.

Consequences of distribution system failures can also be categorized as "outage related." These include damages arising from interruption of product delivery, including the relative time of the interruption. Some customers will be more damaged by loss of service than others.

The availability of make-up supply, can often require a complex network analysis with many assumptions and possible scenarios. As a modeling convenience, availability of replacement supply could be assumed to be inversely proportional to the normal flow rate under the premise that the greater the flow rate that is interrupted, the more difficult will be the replacement of that supply.

Other aspects of return to service costs include:

- Restoration priority (for example, the components of the system that would need to be repaired first, given that there are damages to or weaknesses within several portions)
- Extent of similar facilities
- Regulatory requirements related to return-to-service, if applicable (for example, inspection of similar facilities, if a leak/rupture has occurred)
- Spare parts inventories
- Reliability issues
- PPM programs

See Chapter 11 Consequence of Failure for further discussion of return-to-service costs. PRMM examples illustrate some rudimentary calculations of service interruption losses.

12.4.2 Indirect Consequences

Other costs, normally considered 'indirect costs', related to service interruption are also similar to leak/rupture indirect consequences and include those discussed in PRMM:

- Legal action directed against the pipeline operation
- Loss of contract negotiating power
- Loss of market share to competitors
- Loss of funding/support for future pipeline projects.
- Increased regulatory burdens

Legal implications can range from breach of contract actions to extra compensation for numerous types of customer indirect losses.

As discussed in PRMM, loss of credibility, loss of shareholder confidence, and imposition of new laws and regulations are all considered to be potential indirect costs of pipeline failure, whether that failure is a leak/rupture or a serious service interruption. The loss of service to more powerful political customers in certain socio-political environments, must sometimes be considered. A critical customer may have a degree of power or influence over the pipeline operation.

The CoF assessed in the integrity-focused risk assessment will overlap some aspects of the consequences of service interruption, where longer periods of interruption increase consequences (plant shut downs, lack of heating to homes and hospitals, etc.)

12.4.2.1 Indirect cost estimation

Indirect costs are difficult to calculate and are very situation specific, as also discussed in Chapter 11.8.9 Indirect costs. As with leak/rupture type failures, the indirect costs associated with service interruption may parallel the direct costs. That is, when no better information is available, a default percentage (or multiplying factor) of the direct costs can be used to represent the indirect costs. This is defensible since indirect costs are logically proportional to direct costs.

Of course, actual indirect costs can be dramatically higher in a specific situation, paralleling the situation-specific factors that determine when a leak/rupture scenario becomes more consequential. Until scenario-specific indirect costs can be more accurately estimated, the use of a simplification provides a convenient method to at least acknowledge the existence of indirect costs.

12.4.3 Minimizing Impacts

In this section, we examine actions taken that do not prevent the incident but lessen its impact ***after*** the excursion reaches the customer. This 'after reaching customer' distinction is important in discriminating between resistance and consequence minimization. Resistance measures the system's abilities to absorb the excursion and prevent it from reaching the customer. Here, we examine actions taken after customer impact is imminent.

Unlike spill consequence mitigation to reduce the consequences of pipeline leaks/ruptures, the service impact recognizes few opportunities for consequence mitigation. There are few analogous actions the pipeline operator can take to reduce customer impacts, once the excursion is being experienced by the customer. Note the distinction between mitigating the probability of an impact to a customer versus mitigating the impact once it has reached the customer. Recall that actions taken to either prevent excursions or prevent customer impact—blending, alternate supplies, etc.—are considered in the likelihood of service interruption. They act as mitigation or resistance measures to prevent customer impact.

Actions akin to emergency response as a consequence minimization for leak/rupture are not usually available under the assessment of service interruption (although they may be a part of the 'resistance' estimates, as part of the PoF assessment). This is chiefly due to the definition of service interruption.

Under our definition of 'service interruption', a consequence does not occur until/unless the event has reached the customer. Therefore, it is the customer who is able to take the most significant consequence mitigating actions, not the pipeline operator. So, unless the assessment evaluates the customer's internal abilities to mitigate an excursion, this aspect must be left largely unaddressable.

12.4.4 Early Warning

Early notification of an impending event is the chief ***consequence*** mitigation opportunity for service interruption risk. Especially when customer warning is sufficient to prevent an outage for that customer, consequences are minimized. This is a situation in which, by the action of notifying the customer of a pending specification violation, that customer can take action to prevent an outage. Coupled with a reliable early detection ability, this reduces the service interruption potential. An example would be an industrial consumer with alternative supplies where, on notification, the customer can easily switch to an alternate supply. Similarly, a delivering customer who has alternate delivery options to move his product may avoid harm when notified in sufficient time.

When a customer early warning is useful for minimizing impact but will not prevent an outage, the intervention affects consequences but not probability of upset. An example would be an industrial user who, on notification of a pending service interruption, can perform an orderly shutdown of an operation rather than an emergency shutdown with its inherent safety and equipment damage issues.

Even when intervention is not possible, early detection and timely notification is still valuable. Most customers will benefit from early warning. The customer's ability to react to the notification and adapt to the excursion can be estimated considering the range of possible detection/notification time periods. The value of the early detection and notification can be quantified by estimating the amount of consequence avoidance achieved.

13 **RISK MANAGEMENT**

Highlights

Situations in life often permit no delay; and when we cannot determine the action that is certainly the best, we must follow the action that is probably the best. If the action selected is indeed not good, at least the reasons for selecting it are excellent.

Descartes

SECTION THUMBNAIL

- Once risk assessment has been performed, how is risk management conducted?
- Cost/benefit analyses are important, but rarely the only consideration, given the often complex socio- economic ramifications of risk management decision-making.
- Purely objective, scientific, rational thinking may be insufficient in real-world risk communications and also in risk decision-making.
- Efficient risk management requires certain program elements, defining roles, responsibilities, processes, etc.

13.1 INTRODUCTION

Some may wonder why a book with *Pipeline Risk Management* in its title finally focuses on the 'management' aspect in the last chapter. Hopefully, it is apparent that in measuring risk—the risk assessment step—much of the management process becomes very apparent[1]. Full understanding of pipeline risk generates numerous opportunities to reduce that risk. So previous chapters have already identified risk mitigation opportunities. Reducing exposure, increasing mitigation or resistance, and minimizing consequences all serve to reduce risk.

Even if the risk quantification is imprecise, the exercise is important. The quantification puts a value on the depth of cover, patrol, ILI, pressure test, emergency response, leak detection, secondary containment, and the numerous other important determinants of risk, thereby providing the 'benefit' portion of cost/benefit analyses for these measures. Different mitigation measures will have different benefits (and costs) at various locations along a pipeline. The cost/benefit all along a pipeline guides decision-makers in risk management. Even when imprecise, the quantifications demonstrate a defensible, process-based approach to understanding and therefore managing risk.

However, even when the risk assessment is precise, there are still nuances and real challenges in risk management. For instance, knowing how and where risk reduction can/should be achieved still leaves not knowing ***when*** it should be done. Once a risk assessment has been completed and the results analyzed, the natural next step is risk management: "What, if anything, should be done about this risk picture that has now been painted?" This chapter can therefore focus on issues regarding the management

1 'apparent' but not always easy!

of pipeline risks and the strategies that will be required to balance the desire to reduce risk with limited available resources.

Reaction to risk should be appropriate—proportional.

13.2 RISK CONTEXT

Recall the earlier discussion in 'how to get answers quick!' To gain some sense of pipeline risks, an examination of historical incident statistics is useful. A sketch of 'typical' pipeline risk is readily available from statistics compiled by various sources, often governmental. Using these, a risk evaluator can keep some numbers handy to provide context when needed. For example, in the US, with about 300,000 miles of gas transmission pipeline and 175,000 miles of hazardous liquids pipeline (jurisdictional by US regulations), estimates of country-wide reportable failures per year is readily obtained. These data suggest that a value of 1 to 2 reportable accidents for every 2,000 mile-years of hydrocarbon transmission pipeline provides a rough US failure frequency value—0.0005 to 0.001 incidents per mile-year. So, a 100 mile pipeline could be expected[2] to experience a significant failure once every 10 to 20 years.

Similarly, one set of recent historical experience statistics suggests that losses of around \$500 to \$4,000 per mile-year can be associated with US regulated pipelines. So, an owner of a 100 mile pipeline may recognize that he is exposed to long term losses averaging from \$50,000 to \$400,000 per year.

This same exercise can be repeated to gain a general sense of fatality or injury potential from various types of pipeline systems, based on how large populations of pipeline segments have behaved over long periods of time.

Important cautions are in order when using any generic statistical data (a recurring cautionary statement in this text), especially such high-level summary data. There are many miles of pipeline in the US that will have no accidents nor losses of any kind in many decades of operation. There will also be some miles of pipeline with long term performance worse than suggested by the summary statistics.

Nonetheless, having numbers such as these can be the beginnings of comparisons with non-pipeline risks. See discussion later in this chapter.

13.3 APPLICATIONS

Once risk assessment has advanced to where the organization believes in produced results, the use of those results to support risk management can occur. Risk management

2 To the extent that it is represented by the population of pipeline segments from which the comparison statistic emerges.

plays numerous roles in decision support. PRMM discusses the following common and overlapping applications of a pipeline risk assessment/management program:

1. Identification of risks.
2. Reduction of risks.
3. Reduction of liability.
4. Resource allocations.
5. Project approvals.
6. Budget setting.
7. Due diligence.
8. Risk communications.

As well as the direct use of risk assessment results to support specific tasks in risk management, such as:

- Design an operating discipline
- Assist in route selection
- Optimize spending
- Strengthen project evaluation
- Determine project prioritization
- Determine resource allocation
- Ensure regulatory compliance

13.4 DESIGN PHASE RISK MANAGEMENT

From choices in routing and wall thickness to redundancy in control/safety systems, many risk impacting decisions are made in the design phase of a pipeline. Practitioners of ALARP recognize the need for risk assessment at the beginning of the design [1031]. The design process itself is an exercise in risk management, with specified 'reactions' tailored to changing risks along the pipeline route. Risk nonetheless varies along the length.

Risk management, practiced during the design process, establishes an initial safety margin. This may be mandated by regulations as previously discussed or chosen by the designer. As some modern design efforts move toward limit-state design approaches, the historical notions of safety margins and extra robustness in a design are being quantified and re-evaluated.

The safety margin is related to a target level of reliability, even if not explicitly stated. When stated, the use of event recurrence intervals is a common aspect. For instance, a structure could be designed to withstand a 100 year flood or alternatively, a 500 year flood; a 50 year recurrence interval seismic event or 100 year. There remains the potential, albeit remote, that a more severe event occurs in the structure's life. These considerations should be reflected in the risk assessment.

A pre-construction risk assessment is gaining in popularity since it helps owners understand another aspect of cost-of-ownership. These assessments will be based on the best available pre-construction information such as component design specifications, operational intent, maintenance plans, route surveys, soil investigations, geohazard threat assessments, and others. During construction and installation, new information pertinent to the risk assessment, will be available. This information usually deals with field-identified deviations from design intent and might include

- Minor deviations in intended route
- Unexpected subsurface conditions encountered
- Use of different pipe components (elbows versus field bends, etc.)
- Results of construction inspections and integrity tests
- Differences in actual vs minimum design requirements, such as depth of cover or need for protective caps.

While such changes are mostly covered by design and construction specifications, a certain amount of decision-making occurs informally on the job site. This is also the practice of risk management. As-built information will be very valuable for a detailed, initial risk assessment and future risk assessments.

An integrity verification, such as a pressure test and/or ILI, conducted immediately after installation, decreases the chance of failure from design-related issues and certain errors in manufacturing/construction. It also provides a baseline for comparisons to future integrity assessments, providing a means to determine the rate at which new damages are being introduced.

In some pipeline systems, such as gathering pipelines intended for finite service lives, some amount of degradation (corrosion) is accepted. This is normally an economical decision—given limited need for the asset, it is more cost effective to possibly need to repair/replace than protect. Most pipeline systems are designed to avoid all degradation mechanisms. This is in contrast to some engineered systems that have 'corrosion allowances' or other expectations of an amount of tolerable degradation or wear out. When a pipeline design document includes a 'design life' or similar metric, it is not usually intended as a measure of the structure's lifespan from a serviceability standpoint. It may indeed be a measure of some consumable aspect of the structure, such as an anode bed, designed to deplete over time. A design life may also indicate the period for which the asset is thought to be required, perhaps tied to the predicted life of a hydrocarbon production field. But, similar to a building, the life expectancy of a pipeline is indefinite when it is properly maintained. The use of design life to mean a period beyond which the pipeline structure becomes unserviceable, would be an extreme and unusual interpretation.

Specific risk elements can be better understood, and sometimes efficiently changed, in the design phase. Exposure can sometimes be changed by route selection; consequence can be changed by choices in route as well as product/pressure/volumes. Another interesting application of the recommended risk assessment approach is the ability to assess tradeoffs between increased mitigation and increased resistance during the design phase. Resistance options such as wall thickness often involve higher initial capital costs while many mitigation options involve either higher installation costs (for example, depth of cover) or on-going costs (for example, patrol, public education). Comparing the costs and risk reductions associated with such options strengthens the design and project economics.

See also the discussion of risk assessment and route selection in Chapter 13.9.4 Route alternatives.

13.5 MEASUREMENT TOOL

Formal risk assessment is a measuring tool. The measurements emerging from its application provide a consistent, defensible basis for risk management choices. As with any measuring process, uncertainty exists and should be acknowledged. Minor changes in risk results may not reflect actual changes but rather variations in an inherently 'noisy' set of data. Understanding of measurement variability begins with examination of the assessment results. Common sources of measurement variation—and possible contributors to measurement error—are listed in PRMM. The ability to distinguish real changes in risk level from changes due to measurement uncertainty will vary from assessment to assessment.

See Chapter 4.9 Data analysis and also PRMM for some simple statistical and graphical tools that can be used to further explore a risk assessment's capabilities as a measurement tool.

13.6 ACCEPTABLE RISK

Risk management will eventually require judgments regarding 'how safe is safe enough?" Value judgments associated with risk usually employ qualitative terms such as:

- Acceptable risk
- Tolerable risk
- Justifiable risk
- Negligible
- Trivial risks.

Choices in acceptable risk are complex, involving socio-economic and political considerations at a high level and human psychology on an individual level. To put a level of risk into perspective, it is instructive to look at the types of risks people are ordinarily exposed to during day-to-day life. There are voluntary activities (driving a car) and involuntary activities (being hit by lightning) that involve risks higher than those due to most pipeline components. But comparing voluntary to involuntary risks is not usually a sufficient argument for tolerability of risk.

13.6.1 Societal and individual risks

QRA has long-employed a distinction between individual risk and societal risk. Individual risk provides an estimate for the risk to an individual at a specific location for a specified period of time.

Societal risk usually represents the relationship between frequency of events and number of individuals that could suffer a specified harm from that event—for instance, the annual risk of death of a certain number of people in one pipeline incident. FN curves are commonly used to show the aggregation of many possible pairing scenarios—fatality count versus event frequency.

An important distinction between the two is the exposure created by a facility. Each pipeline component potentially generates a certain hazard radius. Only receptors within that radius are theoretically harmed by a leak/rupture from the component. A collection of components—for example, a long length of pipeline—will not expose the same receptors to potential harm. The maximum exposure occurs for an individual who is very near (perhaps directly over) the pipeline 24 hours of every day. The individual is exposed to pipeline failures immediately adjacent and for some distance along the pipeline to either side. Moving away from the line, risk decreases because he is exposed to less pipeline, based on simple geometry. Determining the length of pipe that can affect a single point is a consideration in individual risk estimates. Both individual and societal risk are important. The use of only societal risk to generate acceptable risk levels for instance, may result in areas with low receptor counts—for example, low population density—bearing a disproportionate amount of risk. The societal risk relationship may suggest that events with low counts of damage—for example, fatalities—are more tolerable and can therefore carry higher probabilities. Strict application of this may result in lower event probabilities only in areas where more receptors exist.

13.6.2 Reaction to Risk

A risk assessment will not accurately predict the next incident—what will happen, where, and when?—except in extreme cases. Our current understanding of real-world phenomena requires an allowance for randomness, expressed in our estimates as probability. So even if the risk assessment is as perfect as current understanding allows, it will still only be accurate for large populations of segments over long periods of time, again, not showing precisely where/when action should be taken. Therefore, it is common for decision-makers to take courses of action not fully supported by their risk assessment results. For example, a decision might be made to reduce certain high consequence, low probability events, even when such events carry a lower EL than other events (ie, lower consequence, higher probability events). This was noted in the discussion of matrix-style visualization tools. However, when this occurs, it must be recognized that one of two things are occurring:

1. The risk estimates are not trusted. There are several variations on this possibility. The most obvious is that the decision-maker feels something is omitted

or incorrectly assessed. Another possibility is that the decision-maker chooses a different confidence level. For instance, the risk assessment is conducted at a P90 level and the decision-maker, desiring a P99 level, overrides the assessment with what he believes accounts for the additional uncertainty.

2. The decision-maker is intentionally choosing an irrational path. A statement this absolute is possible since a risk assessment can include all available knowledge. If it does include all, then choosing a course contrary to that is supported by a trusted, complete, and logical assessment requires some valuation by the decision maker that is not supported by available knowledge. In the case of risk management, emotional decision-making is prevalent.

13.6.3 Risk Aversion

It is commonly accepted that our reactions to risk are not proportional. For instance, we are typically more outraged by—or more averse to—single events with larger consequences than multiple, smaller consequence events, even when the latter is ultimately more costly to society.

Visually, the slope of the common FN curve is said to display risk aversion. The shape of most FN curves shows increasingly lower chances of increasingly higher fatality count incidents. That is, the chances of a single event causing 100 fatalities should be much lower than 1/100 of the chance a single fatality. This reflects one aspect of risk aversion—the decreasing acceptability of single events that generate increasing consequences.

13.6.4 Decision points

Risk management requires that risk-altering decisions be made. Decision-making ultimately hinges on the concept of acceptable risk, even if not directly stated as such. Implicit in the notion of 'acceptable' risk is the determination of that risk level that will carry the designation. There must be a decision process to arrive at this believed-to-be-appropriate level of risk that will be called 'acceptable'.

Due to human risk perceptions, consequences often become more critical than probabilities in reactions to risk and, hence, in decision-making. An emphasis on dramatic but highly improbable scenarios is not always rational. In risk communications and regulatory decision making, this makes a formal study and quantification of incident event sequences more necessary. Many of the events in the sequences studied will be related to a particular damage state. The sequence begins with a failure probability but then follows paths that are ultimately measuring the likelihood of various consequence scenarios. Along the pathways to common consequences of interest are questions such as—is there immediate ignition or delayed ignition? How big a cloud may form? What are the likely temperature and wind conditions? What if an explosion occurs? How far are the vulnerable receptors?

The overall likelihood of failure of the pipeline—often the starting point for the event sequence—is a function of the PoF variables discussed in this book. Most risk management efforts should normally focus first on the probability of failure. This is not only because failure frequency reduction is usually the best way to reduce risks, but also because so many variables impact failure frequency that a formal structure is needed to properly consider all of the important factors.

While risk estimates produced with a modern risk assessment are expressed in absolute terms (for example, failures/km-year, $/mile-year), it is often their relative value that prompts action. Especially when absolute action-criteria are not triggered but when action is nonetheless prudent, risk management can employ ranking and scaling to prioritize and schedule management activities.

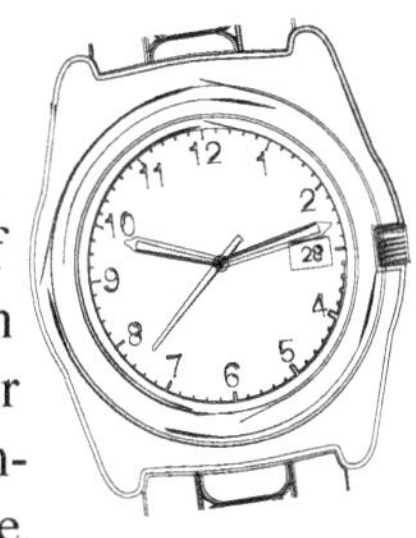

A complication in any decision process is the need for a time factor in setting a risk tolerance or an action trigger. A certain level of risk may be tolerable for some period of time, until the situation can be efficiently addressed. For instance, less-than-desired depth of cover may not require immediate attention and can be addressed in conjunction with other work planned in the area—perhaps months or years in the future. At some level, however, a risk is seen to be so unacceptable that immediate action, even the shutdown of the pipeline, may be warranted.

Recall that risk levels will generally rise over time, at least when uncertainty is modeled as increased risk. Any decision approach must acknowledge the potential increase over time. A certain portion of the risk management effort will often be going to offset natural increases in risk while the remainder advances the goal for risk reduction.

In many cases, the amount of available resources appears to set the de facto level of acceptable risk (beyond any compliance-based risk levels), since money usually runs out before the list of "things to do" is exhausted. Operators often generate/maintain an ongoing list of possible projects to manage the risk level on an asset but often fall short in establishing criteria for the criticality and timing of each potential project. Ideally, the budgets are themselves established by a consistent and defensible risk management strategy. A formal risk assessment is an essential element in the strategy.

With risk assessment results in hand, a risk management strategy can be developed to drive spending on all portions of all assets. A time horizon is an aspect of budget-setting; ie, how quickly are goals to be achieved? When the budgets are established with the aim to improve or maintain pre-established risk levels, then required actions are identified and appropriate levels of resources can be allocated.

Whether the exercise is to prioritize risk issues, rank projects, set annual spending budgets, or establish acceptable risk values, various risk management decision processes can be employed, as is discussed in the following section.

13.6.4.1 Comparative Criteria

Especially where quantitative acceptable risk criteria are not available, comparative risks are used to help judge acceptability. See examples and related discussions of risk comparisons and voluntary versus involuntary risks in PRMM.

Also relevant is the implied level of acceptable risk based on pipeline industry standards and regulations. As a comparison metric, these implied values can be used to suggest acceptability of risk. This is discussed in the next section of this chapter.

Changes in risk level also use comparisons—sometimes to emphasize a bias or position for or against some endeavor that generates the risk. For example, a change in risk from 5e-8 probability of fatality per year to 10e-8 probability of fatality per year can be described as either:

- A doubling of risk.
- A minor, insignificant increase in risk.

Both may be technically correct but sends dramatically different messages to an audience. Similar examples to suggest noteworthy or, alternatively, insignificant improvements in safety by the employment of new mitigation measures are common in debates over acceptable risk levels.

13.6.4.2 Numerical criteria

A numerical risk criterion is sometimes used at a decision point for risk management. Examples of specific criteria, usually used by regulatory agencies and expressed in terms of acceptable annual chances of fatality, are shown in PRMM. These values are sometimes used as actionable limits—"a risk above this line requires action; below the line is 'safe enough'."

For those wishing safety levels beyond regulatory minimum compliance levels that use such numerical criteria, it might be a starting point from which detailed risk management can begin.

Note that a numerical criteria for acceptable pipeline risk is often based on length, consistent with the definition of individual risk discussed earlier. This is logical since a long pipeline, while possibly exposing many receptors, does not increase the exposure to a given receptor due to its length. A criteria that does not consider this would make a criteria impossible to meet for a very long pipeline.

If criteria is based on unit length, then it must consider a very small unit length, eg inch, cm, mm, failure potentials. Otherwise, small but critical features can be masked by nearby very safe segments. Imagine an ILI-detected anomaly, only one mm in length but very deep, with failure imminent. If this is an isolated pit, the neighboring joints of pipe might be defect free for many meters and readily meet acceptable risk criteria. A per-km risk criteria could show acceptable risks despite the defect, due to its length contribution being so small, if an inappropriate risk aggregation strategy was used. A full and proper aggregation would ensure that the one mm feature results in an unacceptable per-km risk rate. See related discussions in Chapters 2 to 4.

13.6.4.3 Data-based criteria

Rather than an overall criteria for 'actionable' levels of risk, the analysis of values from a specific risk assessment can lead to the establishment of action triggers. This includes reactions to outliers (see later discussion) and continuous-improvement approaches, both of which react to results from specific assessments. PRMM discusses some data analyses techniques that might be useful in using risk assessment data to make risk management decisions.

A prudent philosophy to risk management may lie in continuous improvement but will also need to be supplemented by predetermined strategies that are at least loosely based on acceptability criteria. The operator can always be seeking risk reduction opportunities at all locations. However, for consistency and defensibility, the degree and speed with which risk reductions occur should be driven by pre-established trigger points (criteria), to ensure a predominantly 'continuous improvement' strategy is indeed reducing risks.

13.7 RISK CRITERIA

Establishment of risk criteria provides a way to confirm that acceptable or tolerable risk levels exist.

Both qualitative and quantitative risk criteria have been used. Numerical risk criteria can link quantitative risk estimates with subjective, qualitative decision criteria such as "insignificant risk" or "actionable risk."

13.7.1 ALARP

The concept of "as low as reasonably practical" (ALARP) is an example of such a linking and is widely recognized among risk assessment and risk management practitioners.

The ALARP principle generally requires facility owners to adopt all safety measures up to the point where the cost of the safety measure is "grossly disproportionate" to the risk reduction.

Even though quantitative criteria are used, the application of ALARP has a qualitative aspect to it. There are references that seek to quantify aspects such as 'grossly disproportionate' that are embedded in the ALARP definition.

Example: 13.1 This is illustrated in the following example:

Consider a catastrophic pipeline accident involving the death of two individuals and the loss of the pipeline with an estimated event frequency of 10-5 per mile-year. The threshold for disproportionate cost, using a disproportionality factor, is illustrated as follows:

The values and units in this example are:

10-5 accidents of this type per mile per year
58 miles length of pipeline
\$10M cost of fatality
2 person fatality per accident
6 is disproportionality factor, based on some guidance documents suggesting factors between 2 and 10
\$1.5M additional cost per accident for other losses

(10-5 × 58) accidents/year × (\$10,000,000 × 2 + \$1,500,000)/accident × 6
= \$75,000/year

In this example, \$21.5M is the cost of an accident of this type; \$12,500 is the annual risk from an accident of this type; and the \$75,000/year value is a theoretical maximum amount to be spent to reduce the chance of that accident. This is heavily influenced by the disproportionality factor.

This threshold for disproportionate cost is used in the following way: If it is possible to reduce the risk of the accident for less than \$75K/year then before the risk can be declared ALARP, it must be reduced. It may be possible to reduce the risk for much less. Alternatively, it may not be possible to significantly reduce the risk without spending vast amounts of money—in excess of the disproportionality-factor-adjusted avoided loss of \$75K/year. In this case, the risk would be determined to be ALARP and additional spending to reduce it is not warranted. Another example of when spending becomes 'grossly disproportionate' to the risk reduction benefits, is in the following section.

13.7.2 Examples of Established Quantitative Criteria:

Examples of numerical risk criteria can be found specifically for pipelines, more often for land-use planning, worker safety, and other industries such as chemicals processing and aerospace engineering. PRMM provides examples of risk criteria from around the world. Some additional examples follow.

13.7.2.1 Ireland

Ireland's Commission for Energy Regulation, in its ALARP recommendations [1031] recommends the following for 'petroleum undertakings':

- €2.4M as minimum value of 'implied cost of averting a fatality', based on work done by Ireland's National Roads Authority and comparable to UK HSE's 2003 valuation that equates to €2.25M in 2013.
- Grossly disproportionate is assumed to be more than 10X the benefit. Factors less than 10 will be considered but require 'a robust justification'. This factor also serves to better protect small populations exposed to the threat.
- Individual risk tolerability limits:<10-6 fatality per year is broadly acceptable, values of >10-4 for public or 10-3 for workers are unacceptable. This is reported to be comparable to criteria used in the Netherlands, Western Australia, and UK.
- Societal risk upper tolerances are established using 10-3 fatalities per year for 1 individual (y axis intersect) with a -1 slope on log-log plot of frequency versus number of fatalities (public only, not workers). The lower tolerability limit is two orders of magnitude below the upper.
- The use of a factor of at least 2 is seen in other disproportionality quantifications.

13.7.2.2 Latin America

A major pipeline operating country in Latin America used, for many years, a criteria of $5K/km as an unpublished criteria to determine actionable levels of risk. This was a maximum allowable risk level since it implicitly allowed segments with risk levels below this value to be unactionable.

13.7.3 Research

Recent work [777, 888] has suggested tolerable risk levels based on currently accepted standards of pipeline design, operation, and maintenance. These tolerable risk levels have been incorporated into Canadian pipeline standards[3] [9988] and were reportedly being considered for inclusion into US pipeline standards. Designed for onshore natural gas transmission pipelines, this assessment applies the concepts to the subject pipeline segments.

Reliability targets (excerpt from ref [888]):

> The goal of RBDA is to achieve tolerable and consistent risk levels for all pipelines. This is accomplished by setting a maximum permissible failure rate that is inversely proportional to the severity of the failure consequences for each limit state category. The reliability level corresponding to the maximum permissible failure rate is referred to as the target reliability level.

3 As a non-mandatory annex.

Tolerable SR levels were generated by calibration to current design codes and best North American industry practice as partly embodied in ASME B31.8, ASME B31.8S, and 94CFR192.327. Since new pipelines designed and maintained to the requirements of these standards are widely accepted as safe, the average level of SR associated with these pipelines was considered to be tolerable.

RBDA=Reliability based design and assessment

SR=societal risk

Limit state = a state beyond which the pipeline no longer satisfies a particular design or operating requirement. For this application, rupture and large leaks are the limit state of interest.

13.7.4 Offshore

Ref [999] recommends a risk based design standard for offshore pipelines based on safety classes. A safety class is determined by fluid transported, population density (location class), and consequence (safety class). Nominal target failure probabilities are set based on safety class. A reliability based design is an option under this design code and is summarized as follows:

Table 13.1
Nominal failure probabilities vs. safety classes

<table>
<tr><th rowspan="2">Limit States</th><th rowspan="2">Probability Bases</th><th colspan="4">Safety Classes</th></tr>
<tr><th>Low</th><th>Medium</th><th>High</th><th>Very High</th></tr>
<tr><td>SLS</td><td>Annual per Pipeline [1]</td><td>10^{-2}</td><td>10^{-3}</td><td>10^{-3}</td><td>10^{-4}</td></tr>
<tr><td>ULS[2]</td><td>Annual per Pipeline [1]</td><td rowspan="3">10^{-3}</td><td rowspan="3">10^{-4}</td><td rowspan="3">10^{-5}</td><td rowspan="3">10^{-6}</td></tr>
<tr><td>FLS</td><td>Annual per Pipeline [3]</td></tr>
<tr><td>ALS</td><td>Annual per Pipeline</td></tr>
<tr><td></td><td>Pressure containment</td><td>10^{-4}-10^{-5}</td><td>10^{-5}-10^{-6}</td><td>10^{-6}-10^{-7}</td><td>10^{-7}-10^{-8}</td></tr>
</table>

1) Or the time period of the temporary phase.
2) The failure probability for the bursting (pressure containment) shall be an order of magnitude lower than the general ULS criterion given in the table, in accordance with industry practice and reflected by the ISO requirements.
3) The failure probability will effectively be governed by the last year in operation or prior to inspection depending on the adopted inspection philosophy.

These nominal probabilities apply to an entire pipeline, according to the table shown.

Engineered structures placed in public areas, include not only pipelines, but also buildings, bridges, walls and numerous other structures. Therefore, building codes imply a level of acceptable risk which may be relevant to acceptable risks for a pipeline. PRMM lists examples of building reliability levels.

Table 13.2
Classification of safety classes

Safety class	Definition
Low	Where failure implies low risk of human injury and minor environmental and economic consequences. This is the usual classification for installation phase.
Medium	For temporary conditions where failure implies risk of human injury, significant environmental pollution or very high economic or political consequences. This is the usual classification for operation outside the platform area.
High	For operating conditions where failure implies high risk of human injury, significant environmental pollution or very high economic or political consequences. This is the usual classification during operation in location class 2.

Offshore Standard DNV-OS-F101, October 2007

13.8 RISK REDUCTION

Risk becomes zero when either the PoF or the CoF become zero. While zero risk is unrealistic for most industrial undertakings, it is useful to at least conceptually explore this scenario to confirm that the risk assessment appropriately captures such extreme scenarios. The probability of failure tends towards zero when any of three possible situations appear:

- No failure mechanisms exist—ie, exposure = 0
- Failure mechanisms are fully mitigated—ie, a threat exists but is prevented from acting on the system to the degree that a failure results. Mitigation = 100% results in no risk.
- The system is designed to fully withstand the threat—a failure mechanism acts on but cannot cause the system to fail. Resistance = 100%

CoF becomes zero when no damages can arise from the 'failure' being measured. For failure = leak/rupture, CoF becomes zero when any of the four subvariables is zero: product hazard, spill size, dispersion, or receptor damage potential.

13.8.1 Beginning Risk Management

Identifying when and where effectiveness risk reduction efforts should be applied can be a very complex process. In more extreme cases, the need and the urgency will be apparent. But, for most lengths of most pipelines, the seeking of incremental improvements rather than emergency reactions will guide risk management.

Pre-established decision criteria often provides the urgency of risk reduction—how fast should action be taken. Determination of 'outliers' versus 'systemic' type risk issues often provides the locations and extents where action is warranted. Finally, the risk assessment directs the identification and choices of risk reduction measures. The risk profile is the essential tool in managing pipeline risk.

13.8.2 Profiling

A risk profile—changes in risk along the pipeline route—is required to efficiently begin the process of pipeline risk management, whether the profile covers an entire pipeline system or a sub-section such as a HCA. The profile of changing risk along the length is the key to understanding and managing risk.

The profile instantly reveals the nature of the pipeline's risk. There may be extreme outliers, or stable but high risk, stable and low risk, rapid changes, and numerous other patterns. These patterns are critical in determining how to manage the risk.

The profile of any sub-part of risk may warrant examination. Certainly the interplay between PoF and CoF will influence risk management. But so too will changes in exposure, mitigation, and resistance inform decision-making, as will changes in hazard zone size and receptor populations/sensitivities.

Acceptable risk criteria and other pre-determined decision points (discussed previously) can be added to the profile. This clearly shows where action is warranted and not. Many applications of risk management will, however seek continuous improvement, where additional actions will be taken even where criteria are met. A comparative analyses is almost always a part of risk management that goes beyond meeting criteria. In all instances, the profile is the key tool.

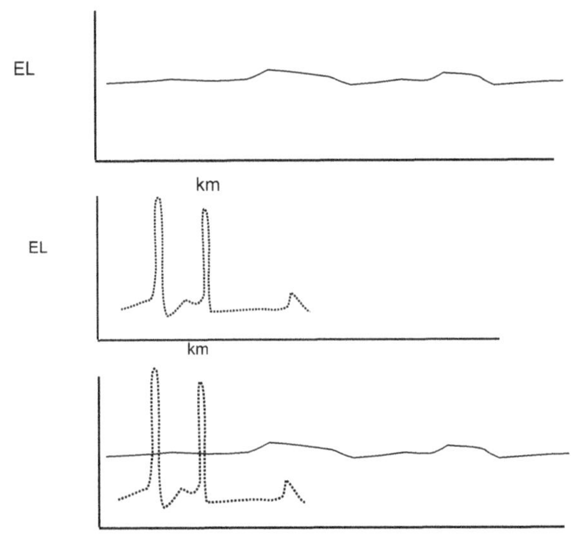

Figure 13.1 Use Profiles to Guide Risk Management

13.8.3 Outliers vs Systemic Issues

Pipelines or portions of pipelines may exhibit profiles such as the examples in Figure 13.1. Segment A in Figure 13.1 has some obvious outliers. These may also exceed acceptable criteria and hence warrant action—perhaps immediate action.

Segment B shows consistent risk—no obvious outliers. The entire length may meet criteria or it may alternatively be entirely out of compliance. This is the first determination to be made. If entirely failing to meet criteria, the risk issue is often systemic. That is, there is one or more risk-driving factors embedded along the entire length. Examples include a weak longitudinal weld seam, failing corrosion coating, sensitive and vulnerable receptors, etc. Knowing this, the risk management plan can be constructed accordingly.

A profile may show both A and B type behavior and alternate between the two or multiple variations of the two. This provides an opportunity to customize action plans to location-specific and issue-specific portions of the segment. With the initial determination of 'within/outside criteria' and then 'outliers' versus 'systemic' type issues, action planning can begin—tailoring possible actions to what is seen in the profile.

Candidate projects are identified based on the risk issue(s) needing to be addressed. A project may change exposure, mitigation, resistance, or consequence or it may impact more than one of these. But since at least one must be changed in order to change the risk, the exercise of identifying candidate projects is greatly facilitated by the risk assessment (which shows each of these components independently).

The location-specific or issue-specific portions of the segment will generally have remediation opportunities determined by what-if analyses. Potential projects are compared and chosen based on their cost/benefits.

13.8.4 Unit Length

Previously discussed 'unit risk' considerations will be important. A rank-ordering based on risk-per-foot will usually yield a different list than risk-per-segment, where a 'segment' is of varying lengths. Both lists are important—even a short stretch of disproportionately higher risk warrants attention, as does a segment whose cumulative risk is higher.

Segment length for risk management is often quite different than for risk assessment. Risk assessment is driven by the data. Aggregation approaches are used to subsequently collect multiple risk assessment segments into longer segments that will receive the same risk remediation.

13.8.5 Conservatism

As detailed in Chapter 2.16 Conservatism (PXX)—using an intentional bias towards overstating the actual risk—is a useful property of many risk assessments. Removal of such conservatism reduces apparent risk. Therefore, a legitimate form of risk management is often to remove uncertainty, thereby reducing the overstatements of risk and lowering the modeled risk.

As a subset of the conservatism role discussion, consider also the use of both measurements and estimates common in a modern risk assessment. Estimates must often be used when measurements are unavailable or carry too much uncertainty (see the Chapter 2.14 Measurements and Estimates discussion). A common risk issue identified for improvement will be any conservative estimates used. Replacing them with actual measurements is normally an uncertainty-reducing opportunity. Again, this reduction in uncertainty can be equated to reduction in risk, when using conservative inputs.

13.8.6 Mitigation options

The risk assessment focuses attention on risk reduction opportunities in several ways. Obviously, where risks are higher, more attention is probably warranted. Looking deeper, the risk assessment also shows the cause of the higher risk. Especially on a comparative basis, locations of higher exposure, less mitigation, and less resistance become apparent. This helps direct resources optimally. For instance, depth of cover or concrete slab protects a pipeline from third-party damage; cracks and corrosion flaws detected and removed while they are still of a size to have no impact on pipeline integrity ensures that TTF is sufficiently long to avoid failure. In practical terms, changing certain things are of course much more attractive than others.

Reducing risk by reducing the probability of failure—usually mitigating exposures identified in the PoF assessment—is normally the main risk management effort. Reducing potential consequences is usually more problematic due to the generally unchangeable nature, from a practical standpoint, of the consequence factors. It would require altering some aspect of the product stream and/or the pipeline's surroundings to effect the greatest change. Although some consequence elements such as emergency response and leak detection are very realistic opportunities to reduce consequences, their range of effectiveness and reliability do not often match the opportunities to impact the PoF.

Risk management may possibly even lead to the reduction or temporary elimination of certain mitigation activities in low-risk areas to allow more resources to go to higher risk segments. Intentionally permitting a risk increase in an area may be controversial and should only be done after careful and thoughtful analysis. Nonetheless, when additional resources are not available, redistribution of existing resources may be reasonable and prudent.

Table 13.1
Analyses of Changes

Change	Variables affected
Increase pipe wall thickness by 10%.	Resistance, all stress influenced factors, many associated changes if done on existing pipeline (new coating, depth cover, signs, etc.)
Reduce pipeline operating pressure by 10%.	Stress factors, leak size, hazard zone, MAOP potential, etc.
Improve leak detection on certain leak rate from 20min to 10min	Leak size, hazard zone (including reaction).
If population increases	Receptors, activity level for third party damages 22 per mile to 33 per mile (50% increase).
Increase air patrol frequency.	Third party damage, geohazards, sabotage, leak detection.
Improve depth-of-cover by 10%.	Third party damage (including impacts), geohazards, sabotage, corrosion.

13.8.7 Risks dominated by consequences

Since options for reducing potential consequences are normally fewer and more problematic, it is usually preferable to reduce risk by decreasing failure *potential*. Nonetheless, it is always useful and sometimes essential to examine consequence-reduction opportunities. The high level, simple multiplication of the 4 key leak/rupture consequence determinants introduced in Chapter 11 Consequence of Failure is useful here. The product of four variables essentially determines the magnitude of the potential consequences:

$$RI = PH \times RQ \times D \times R$$

Where

LI = Release impact (CoF)

PH = product hazard (toxicity, flammability, etc.)

RQ = release quantity (quantity of the liquid or vapor release)

D = dispersion (spread or range of the release)

R = receptors (all things that could be damaged by contact with the release).

Reducing any of the inputs results in CoF reduction.

For instance, changing the product type or pressure, installing secondary containment, relocating the pipeline or removing the nearby receptors, or reducing the size or flowrate are all risk reduction options, at least theoretically, but these are rarely realistic options due to economic considerations. Typically, the more practical opportunities for most pipelines involve improving leak detection and emergency response.

For service interruption risks, customer impact mitigations are similarly few compared to excursion avoidance opportunities. CoF reduction opportunities are detailed in Chapter 11 Consequence of Failure and Chapter 12 Service Interruption Risk.

Despite the more problematic nature of CoF reduction, occasionally, reducing failure probability is not enough to bring the risk to an acceptable level (by whatever acceptability criteria is being used). to explore additional leak/rupture risk reduction opportunities under this circumstance, one possible approach is as follows:

1. Determine to what level the PoF would need to be decreased in order for this risk to be brought in line with "normal" risk levels or some criteria of acceptability?
2. Is this level technically possible?
3. Is this level economically feasible?

If it is determined that acceptable risk levels cannot be achieved by lowering failure potential and the more practical CoF reductions are insufficient, then an examination of more extreme options is warranted.

- Can the product by modified to be less hazardous?
- Can alternative modes of transport result in lower risk?
- Can the pressure be reduced?
- Can the pipeline be relocated?
- Can the potential spill dispersion be reduced by secondary containment?

While these are a part of any risk management effort, they perhaps become especially critical when tolerable risk levels are most difficult to achieve.

13.8.8 Progress Tracking

Examining and tracking progress in risk reduction is efficiently accomplished via EL. Since EL values can be threat-specific, location-specific, consequence-specific, etc., various components of overall EL can be tracked as well as the total EL. For example, while an upgrade to a CP system should show improvement in overall EL, the impact on "external corrosion EL" will be driving that improvement and may warrant independent examination. An improvement in patrolling a pipeline may show significant improvements in "third party EL' and 'consequence reduction' (leak detection), again shown in the overall EL but perhaps also interesting to view independently.

13.9 SPENDING

Basing scheduling and resource allocation decisions on risk estimates should be a defensible, traceable process. The pipeline components with the highest and lowest risk estimates are obviously significant to risk management. A disproportionate amount of resources is justifiably spent on the higher risk segments. In a fully monetized risk assessment, appropriate amounts of spending are also suggested.

Underpinning the discussion of measuring risk avoidance costs should be the idea that analyses may ultimately prove that a venture is not worth pursuing. Once risk costs are added to capital and operating costs, there may be insufficient return on investment to justify the venture at all. A formal risk assessment provides the more objective means for such determinations. Experience-based judgment and perhaps even intuition will still be important in decision-making, but the structure and discipline of the risk assessment removes much of the subjectivity that would otherwise accompany such challenging determinations.

13.9.1 Cost of accidents

Risk reduction is intended to result in the avoided losses due to accidents. Avoided losses should include avoided indirect costs such as political and legal ramifications, contract violations, loss of customer confidence, and other considerations.

Chapter 11 Consequence of Failure discusses the estimation of potential loss, for example, the cost of accidents and shows some historical costs of incidents.

13.9.2 Cost of mitigation

Risk management seeks the most efficient attainment of acceptable risk. It is often appropriate to exhaust the lower cost risk reduction options before more expensive options are considered. A risk assessment 'values' mitigation activities based on their ability to reduce risk (specifically, reduce PoF), with no consideration given to the cost of the activity. Risk management adds mitigation cost considerations in order to optimize spending towards risk reduction. Some hypothetical projects, with example cost/benefit values, are shown in Table 13.2 (modified from original scoring examples shown in PRMM). Note that some actions have a very location-specific impact while others have a large system-wide impact. See the discussion on cumulative risk calculations earlier in this chapter.

Table 13.2
Sample mitigation project cost-benefit analysis

1 Action	2 Cost NPV ($K)	3 Failure mechanism impacted	4 Reduction in risk (%)
1000-ft pipe replacement	82	All	2,200
Increased training/procedures	25	Incorrect operations	20
Upgrade cathodic protection	46	Corrosion	54
Maps/records improvements	33	Third party; incorrect operations	8
Information management system improvements	19	All	17
Recoat 400ft	76	Corrosion	500

Note that some percentage changes represent orders of magnitude differences in 'before' and 'after' risks. This is consistent with real world experiences that demonstrate there are commonly multiple orders of magnitude differences between the higher and lower risk components.

Practitioners of ALARP are obliged to consider costs of mitigation as well as risks while conducting risk management. ALARP includes the generation of a cost/benefit analyses where the concept of potential mitigation that is grossly disproportionate to its benefit arises. Quantifying the point at which a potential mitigation becomes grossly disproportionate is debatable. One regulator states that a factor of 10 or more equates to disproportionality but provisions for lesser factors have been made. [1031] That regulator guidance document also addresses potential arguments, perhaps employed by some petitioners in the past that attempt to weaken the ALARP application:

The cost of the measure, against which the safety benefit is being compared, should be restricted to those costs that are solely required for the measure. Realistic costs should be used so that, for example, the measure is not over engineered to derive a large cost, distorting the comparison to conclude that it would be grossly disproportionate to implement.

If the cost of implementing a risk reduction measure is primarily lost or deferred production, the ALARP assessment should be undertaken for the two cases where lost or deferred production is and is not accounted for. If the decision is dependent on the additional cost of the lost or deferred production (i.e. the risk reduction measure would be installed without considering this cost), a highly robust and thorough argument as to why the measure could not be installed while losing less production (for example, at a shutdown) will be required if the measure is to be rejected.

If the lost production is actually deferred production (i.e. the life of the equipment is based on operating rather than calendar time), then the lost production should only take account of lost monetary interest on the lost production plus an allowance for operational costs during the implementation time, or potential increase in operational costs at the end of life.

If shortly after a design is frozen, or constructed, a risk reduction measure is identified that normally would have been implemented as part of a good design process, but has not been, it would normally be expected that the measure, or one that provides a similar safety benefit, is implemented. An argument of grossly disproportionate correction costs cannot be used to justify an incorrect design.

If the cost of a risk reduction measure is assessed to be in gross disproportion to the safety benefit it provides and it is not implemented because of a short remaining lifetime, it is expected that supporting analysis will be carried out for a number of different remaining lifetimes due to the inherent uncertainty in such a figure. The justification for a non-implementation decision that is dependent on a short lifetime assumption would have to be extremely robust. [1031]

An argument could be constructed that, for a reason such as the short remaining lifetime, the reinstatement cost of a previously functioning risk reduction measure is grossly disproportionate to the safety benefit that it achieves. This is commonly called *reverse ALARP*. In this case the test of Good Practice must still be met and, since the risk reduction measure was initially installed, it is Good Practice to reinstall or repair it. Reverse ALARP arguments will not be accepted in an ALARP demonstration. [1031]

Basic cost estimation practice is readily applied to risk management PRMM provides a more detailed discussion of estimating costs of risk mitigation

13.9.3 Consequences AND Probability

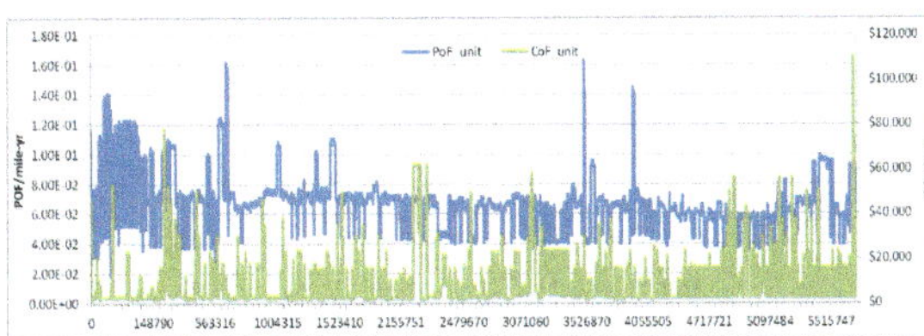

Risk management opportunities can be presented in a misleading way if both consequence and probability issues are not addressed. For instance, a government-sponsored study on the benefits of additional pipeline valve capabilities attempted to show a cost benefit conclusion. While it appropriately analyzed differences in consequence potential arising from increased shut in opportunities, it failed to provide the necessary context of how often would such 'savings' occur. The ref [1015] study on value of additional block valves [1050] discussion of cost/benefit concludes the following:

- "The study results further show that for natural gas release scenarios, block valve closure within 8 minutes after the break can result in a potential cost avoidance of at least $2,000,000 for 12-in nominal diameter natural gas pipelines and $8,000,000 for 42-in nominal diameter natural gas pipelines depending on the configuration of buildings within the Class 3 HCA."

- "The benefit in terms of cost avoidance for damage to buildings and personal property attributed to block valve closure swiftness increases as the duration of the block valve shutdown phase decreases. Risk analysis results for a hypothetical 30-in. nominal diameter hazardous liquid pipeline release of liquid propane show that the estimated avoided cost of moderate building and property damage resulting from block valve closure in 13 rather than 70 minutes is over $300,000,000."

Note that the above conclusions are not yet cost/benefit valuations. As presented, they do not consider the frequency of pertinent scenarios, a critical aspect in determining the risk reduction benefit, ie, how often the consequence avoidance is triggered.[4] Benefit realizations are also contingent upon outside factors, notably the ability of firefighters to be on scene within a specified time period.

At face value, these cost avoidance values may appear very attractive. However, the possibility of realizing such cost savings could be extremely remote. With a pertinent incident rate of, say 0.00001 per year, and cost of the additional capabilities being, perhaps $250,000 per installation, the attractiveness of the option is greatly reduced—ie, spending $250,000 to avoid $3,080/year of losses. ($308,000,000/incident x 0.00001 incidents/year = $3,080/year). On the other hand, if the incident rate is closer to 0.001, then the installation of the new capabilities is indeed very attractive.

Monitoring and linking costs to specific risk elements allows decision makers to more efficiently allocate resources. Safer practices may require extra operating costs but will ideally be offset by cost savings from the generally more efficient operation; for example, less downtime, employee absence, etc. Then, the focus can be on the value, from a risk perspective, of the activities.

13.9.4 Route alternatives

Much goes into the process of selecting a route for a pipeline or a site for an associated facility. An often overlooked aspect is that each potential route or site location carries a risk cost as well as an acquisition/installation cost and on-going operating/maintenance cost A less expensive installation route alternative may carry a "route penalty," as an offset to the cost savings once future risks are included in the analysis. This in effect assigns a cost to the condition(s) causing the increased risk. This is obvious in decisions such as avoiding populated areas when possible, but is less obvious for other elements of risk. Using a full and robust risk assessment ensures a complete understanding and improved decision-making.

For example, pipeline route A might be shorter than pipeline alternate route B. Suppose that the shorter distance would result in a savings of $665,000 in materials and installation costs. However, route A contains potential AC induced corrosion, more

4 They also, surprisingly, do not include any benefits from avoiding loss of life or injury.

corrosive soils, the presence of more buried foreign pipelines, and a higher potential incident rate of on-going excavation damage. Even after mitigating measures, these additional threats to pipeline integrity are estimated to cause the risk for route choice A to carry $135,000/year more risk (expected loss) than route B. Unless the facility is expected to only have a very short life span, the initial installation savings is quickly offset and the option is less attractive. Adding consequence considerations, a difference in pipeline routes involving, for example, differing population densities will often result in even more dramatic impacts on risk.

To support route selection, a robust risk assessment will assign a 'cost' to even an unchangeable condition along the pipeline. Examples include soil conditions, nearby population density, potential for earth movements, and nearby excavation activity levels. This 'cost' is the level of risk that is added by the condition. This is especially useful when alternate routes or site locations are considered in new pipeline or facility design.

13.10 RISK MANAGEMENT SUPPORT

As with any initiative, especially in larger organizations, the risk management program must itself be managed. This involves assignment of roles and responsibilities (ownerships) as well as written control documents guiding all aspects of the program. There are multiple examples of failed programs due to insufficient attention to the administrative aspects.

Similarly, an oft-overlooked aspect of pipeline risk management is risk communications. Risk can be a technically complex and emotionally charged topic. When competing interests and priorities are involved, as they often are among stakeholders of pipeline activities, communications should be done in a way that does not widen differences among those stakeholders. An audience can readily seize upon unfortunate 'sound bites' and an unintended messages and, depending on their bias, be too influenced or too dismissive of any risk assessment data. For instance, the priorities of neighbors to the pipeline are sometimes at odds with the owner/operators. The communications of risk 'facts' from the latter to the former has historically been problematic when not done with care and compassion.

PRMM details these concepts of program administration and risk communications. It also discusses the related issue of risk perception, an aspect that makes risk an emotional and more difficult topic to reach consensus. Comparative risks, including issues around voluntary versus involuntary risk, is a useful concept for enhancing risk understanding and in communications. This too is covered in the risk management chapter of PRMM with useful sample tables included.

A common observation emerging from mature pipeline risk management programs is that unforeseen benefits are numerous. From central repositories of information yielding new insights, to more consistent and defensible decision-making at many levels in the organization, new capabilities emerge and strengthen the corporation processes—if not the corporate culture itself.

A good plan, violently executed now, is better than a perfect plan next week.

George S. Patton

REFERENCES

1. "ALOHA (Areal Locations of Hazardous Atmospheres)," software for dispersions of contaminants in the atmosphere, developed by the National Oceanic and Atmospheric Administration and the Environmental Protection Agency, October 1997.
2. *AGA Plastic Pipe Manual for Gas Service*, Catalog No. XR 8902, Arlington, VA: American Gas Association, February 1989.
3. American Petroleum Institute, "Evaluation Methodology for Software Based Leak Detection Systems," API 1155, Washington, DC: API, February 1995.
4. American Petroleum Institute, "Pipeline Variable Uncertainties and Their Effect on Leak Detectability," API 1149, Washington, DC: API, November 1993.
5. "ARCHIE (Automated Resource for Chemical Hazard Incident Evaluation)," prepared for the Federal Emergency Management Agency, Department of Transportation, and Environmental Protection Agency, for Handbook of Chemical Hazard Analysis Procedures (approximate date 1989) and software for dispersion modeling, thermal, and overpressure impacts.
6. *ASME Code for Pressure Piping*, B31: "Gas Transmission and Distribution Piping Systems," ANSI/ASME B31.8,
7. ASTM, "Standard Test Methods for Notched Bar Impact Testing of Metallic Materials," E23–93a, American Society for Testing and Materialism, July 1993.
8. Baker, W. E., et al., *Explosion Hazards and Evaluation*, New York: Elsevier Scientific Publishing Co., 1986.
9. Battelle Columbus Division, "Guidelines for Hazard Evaluation Procedures," New York: American Institute of Chemical Engineers, 1985.
10. Bernstein, P. L., *Against the Gods: The Remarkable Story of Risk*, New York: John Wiley and Sons, 1998.
26. Dow Chemical, *Fire and Explosion Index Hazard Classification Guide*, 6th ed., Dow Chemical Co., May 1987.
42. Huges, D., *Assessing the Future: Water Utility Infrastructure Management*, American Water Works Association, 2002, Chap. 23.
47. Keyser, C. A., *Materials Science in Engineering, 3rd ed.*, Columbus, OH: Charles E. Merrill Publishing Co., 1980, pp. 75–101, 131–159.
48. Kiefner, J. F., "A Risk Management Tool for Establishing Budget Priorities," presented at the Risk Assessment/ Management of Regulated Pipelines Conference, a NACE TechEdge Series Program, Houston, TX, February 10–12, 1997.
51. Lockbaum, B. S., "Cast Iron Main Break Predictive Models Guide Maintenance Plans," *Pipe Line Industry*, April 1994.
52. Martinez, F. H., and Stafford, S. W. "EPNG Develops Model to Predict Potential Locations for SCC," *Pipeline Industry*, July 1994.

58. Morgan, B., "The Importance of Realistic Representation of Design Features in the Risk Assessment of High-Pressure Gas Pipelines," presented at Pipeline Reliability Conference, Houston, TX, September 1995.
59. Morgan, B., et al., "An Approach to the Risk Assessment of Gasoline Pipelines," presented at Pipeline Reliability Conference, Houston, TX, November 1996.
60. Moser, A. P., *Buried Pipe Design*, New York: McGraw-Hill, 1990.
67. Office of Gas Safety, "Guide to Quantitative Risk Assessment (QRA)," Standards Australia ME-038-01 (Committee on Pipelines: Gas and Liquid Petroleum), Risk and Reliability Associates Pty Ltd., April 2002.
69. Pipeline Industries Guild, *Pipelines: Design, Construction, and Operation*, London, New York: Construction Press, Inc., 1984.
71. *Proceedings of the International Workshop of Offshore Pipeline Safety* (D. V. Morris, Ed.), New Orleans, LA, December 4–6, 1991, College Station: Texas A&M University.
76. Rusin, M., and Savvides-Gellerson, E., *The Safety of Interstate Liquid Pipelines: An Evaluation of Present Levels and Proposals for Change*, Research Study 040, Washington, DC: American Petroleum Institute, July 1987.
78. Simiu, E., *Reliability of Offshore Operations: Proceedings of an International Workshop*, NIST Special Publication 833, Gaithersburg, MD: National Institute of Standards and Technology.
79. Smart, J. S., and Smith, G. L., "Pigging and Chemical Treatment Pipelines," presented at Pipeline Pigging and Inspection Technology Conference, Houston, TX, February 4–7, 1991.
83. Stephens, M. J., "A Model for Sizing High Consequence Areas Associated with Natural Gas Pipelines," C-FER Topical Report 99068, prepared for Gas Research Institute, Contract 8174, October 2000.
88. Vick, Reagan, et al., 1989.
89. Vick, S. G., *Degrees of Belief: Subjective Probability and Engineering Judgment*, Reston, VA: ASCE Press, 2002.
94. Wright, T., Colonial Pipeline Company, Atlanta, GA, personal communications.
95. Zimmerman, T., Chen, Q., and Pandey, M, "Target Reliability Levels for Pipeline Limit States Design," presented at ASME International Pipeline Conference, 1996.
222. PRMM: Muhlbauer, W. Kent. *Pipeline Risk Management Manual, 3rd ed.* Houston, Texas: Gulf Publishing Co, 2004.
333. Nessim et al. Target Reliability Levels for Design and Assessment of Onshore Natural Gas Pipelines. International Pipeline Conference, Calgary, Alberta, 2004
777. REVIEW OF RULE BASED DESIGN AND RELIABILITY BASED DESIGN FOR ONSHORE PIPELINES Joe Zhou and Brian Rothwell, TransCanada PipeLines Limited Calgary, Alberta, Canada; Maher Nessim and Wenxing Zhou, C-FER Technologies, Inc. Edmonton, Alberta, Canada

888. Proceedings of IPC 2004 International Pipeline Conference October 4 - 8, 2004 Calgary, Alberta, Canada IPC04-0321 TARGET RELIABILITY LEVELS FOR DESIGN AND ASSESSMENT OF ONSHORE NATURAL GAS PIPELINES Maher Nessim C-FER Technologies Wenxing Zhou C-FER Technologies Joe Zhou TransCanada Pipelines Limited Brian Rothwell TransCanada Pipelines Limited Martin McLamb BP Exploration Operating Company

9988 CSA Z662.1-07 Oil and gas pipeline systems, Annex O, Aug 2008

999 Recommended Practice DNV-RP-F105, Feb 2006

777 Response of Buried Pipelines Subject to Earthquake Effects; M.J. O'Rourke, X. Liu; Monograph Series; Multidisciplinary Center for Earthquake Engineering Research; Copyright © 1999 by the Research Foundation of the State University of New York and the Multidisciplinary Center for Earthquake Engineering Research.

1001 Sources for Hydrogen Gas Trapped in the Annuli of Pipeline Repair Sleeves SAER-6153 A. Lewis, H. Badairy, S. Duval, B. Isidro, Y. Al-Janabi, S. Mehta, H. Al-Mutairi, T. Newbound, W. Al-Obaid, S. Al-Rassam, A. Al-Shahrani, A. Sherik, I. Al-Thaiban Material Performance Group Research & Development Division Research & Development Center D. Catte, T. Lewis

1002 The Layer of Protection Analysis (LOPA) method Anton A. Frederickson, Mr., Dr. Independent Consultant – member of Safety Users Group Network; 01 April, 2002

1003 Comparison of PFD calculation, SIL requirements according to IEC/EN 61508 and ISA-TR84.0.02 (1998) Prof. Dr. Ing. Habil. Josef Borcsok, HIMA Paul Hildebrandt GmbH Co KG, Industrial Automation

1004 "Pipelines Prove Safer Than Road or Rail", D. Furchtgott-Roth, K.P. Green, *Pipeline & Gas Journal*, Dec 2013

1005 "Cost of Regulation Lessens With Coordination Among Agencies", M. Purpura, *Pipeline & Gas Journal*, Dec 2013

1006 June 2014; **http://opsweb.phmsa.dot.gov/pipeline_replacement/**

1007 "LDC's Continue to Upgrade the Nation's Gas Distribtuion Network", R. Tubb, *Pipeline & Gas Journal*, Dec 2013

1008 "Natural Gas Odorization monitoring for Safety and Consistency", D. Amirbekyan, N. Stylianos; *Pipeline & Gas Journal*, Dec 2013

1008 "Anchors and threats, do we know enough?", A. Hussain, S. Eldevik, L. Collberg, DNV GL, World Pipelines, May 2014

1009 "Cost effective application of the ALARP Principle", Dr Simon Hughes, Senior Safety Consutant, ERA Technology,

1010 "Solving the cybersecurity puzzle", D. Fox, URS Corporation; *Pipeline & Gas Journal*, Feb 2013.

1011 Leak Detection Study – DTPH56-11-D-000001, September 28, 2012, kiefner and associates Leak_Detection_Study__DTPH56-11-D-000001_R_Draft_final_10-04-2012.pdf Leak Detection Study – DTPH56-11-D-000001 FINAL 0339-1201 Kiefner and Associates, Inc. 3-22 October 2012

1012 Department of Housing and Urban Development. Safety Considerations in Siting Housing Projects, 1975. HUD Report 0050137

1013 K.S. Mudan and P.A. Croce. SFPE Handbook, chapter Fire Hazard Calculations for Large Open Hydrocarbon Fires. *National Fire Protection Association, Quincy*, Massachusetts, 2nd edition, 1995.4

1014 NISTIR 6546 Thermal Radiation from Large Pool Fires, Kevin B. McGrattan, Howard R. Baum, Anthony Hamins; *Fire Safety Engineering Division Building and Fire Research Laboratory*, November 2000, National Institute of Standards and Technology, U.S. Department of Commerce

1015 Studies for the Requirements of Automatic and Remotely Controlled Shutoff Valves on Hazardous Liquids and Natural Gas Pipelines with Respect to Public and Environmental Safety, for US DoT, PHMSA, Oak Ridge National Laboratory, ORNL/TM-2012/411, Oct 2012.

1016 **http://en.wikipedia.org/wiki/Debris_flow**

1017 PHMSA Advisory Bulletin on floods; ADB-2013-02

1018 Going by various names: NPHI, Natural Disaster Study, National Pipeline Risk Index **http://www.npms.rspa.dot.gov/data/data_natdis.htm**

1019 **http://www.nopsema.gov.au/resources/human-factors/human-error/**

1020 INGAA, Integrity Characteristics of Vintage Pipelines, Battelle Memorial Institute, F-2002-50435. prepared by E. F. Clark, B. N. Leis, R. J. Eiber, Oct 2004.

1021 API 579; Fitness-For-Service; API 579-1/ASME FFS-1, JUNE 5, 2007; (API 579 SECOND EDITION); American Petroleum Institute's Recommended Practice 579

1022 Keifner, John F. "Evaluating the Stability of Manufacturing and Construction Defects in Natural Gas Pipelines", April 26, 2007 Final Report No. 05-12R to US DoT, Office of Pipeline Safety.

1023 Proceedings of IPC2008, 7th International Pipeline Conference, IPC2008-64039; EVALUATING THE EFFECTS OF WRINKLE BENDS ON PIPELINE INTEGRITY Chris Alexander, Stress Engineering Services, Inc., Satish Kulkarni, El Paso Pipeline Group

1024 Mallaburn **http://pipelineandgasjournal.com/accurate-pipeline-inspection-data-requires-more-pig?page=4**; September 2014, Vol. 241, No. 9

1025 U.S. Coast Guard Hazard Assessment Handbook, Commandant Instruction Manual M 16465.13

1026 **http://www.corrosion-doctors.org/AtmCorros/CorrMaps.htm**

1027 **http://en.wikipedia.org/wiki/Metal_fatigue**

1028 Stewart, H.E. et al, "Pipeline Crossings of Railroads and Highways." American Gas Association Operating Section Proceedings. 91-DT-60, 1991, pp 443-468.

1032 report to congress; *Results of Hazardous Liquid Incidents at certain Inland Water Crossings Study*; Dec 2012 **http://www.phmsa.dot.gov/pv_obj_cache/pv_obj_id_F7EE2DB31D71255F6E1E3683FCDDC2A6635A1000/filename/Haz Liq Inci at Certain Inl Wat Cross Study - 12-27-12.pdf**

1033 Rikalla, "Methodologies of pipeline geohazard assessment" Moness Rizkalla, Visitless Integrity Assessment Ltd, Calgary, Alberta, Canada *Pipelines International* — March 2013
1034 FEDERAL EMERGENCY MANAGEMENT AGENCY FEMA-233/July 1992 Earthquake Resistant Construction of Gas and Liquid Fuel Pipeline Systems Serving, or Regulated by, the Federal Government Issued in Furtherance of the Decade for Natural Disaster Reduction Earthquake Hazard Reduction Series 67
1035 Oil Pipeline Characteristics and Risk Factors: Illustrations from the Decade of Construction A December 2001Report Prepared by John F. Kiefner, President, Kiefner & Associates, Inc.; Cheryl J. Trench, President, Allegro Energy Group; copyright API
1036 Mechanical Damage, Final Report, Integrity Management Program; Under Delivery Order DTRS56-02-D-70036 (Technical Task Order 16) submitted toU.S. Department of Transportation Pipeline and Hazardous Materials Safety Administration Office of Pipeline Safety; submitted by Michael Baker Jr., APRIL 2009
1037 Line Pipe Resistance to Outside Force Volume 2: Assessing Serviceability of Mechanical Damage PR-3-9305 Prepared for the Line Pipe Supervisory Committee Pipeline Research Committee of Pipeline Research Council International, Inc. Prepared by the following Research Agencies: Battelle Memorial Institute Authors: B. N. Leis, R. B. Francini Publication Date: 1999
1038 Aggregate Product Function extends SQL; Solution extends the capability of standard SQL by adding aggregate Product Function By Dr. Alexander Bell, 2006.
1039 BENDING MOMENT CAPACITY OF PIPES, Offshore Mechanical and Arctic Engineering, 1999, PL-99-5033, Søren Hauch and Yong Bai, American Bureau of Shipping, Offshore Technology Department, Houston, Texas
1040 Department of Transportation Research and Special Programs Administration Office of Pipeline Safety TTO Number 1 Integrity Management Program Delivery Order DTRS56-02-D-70036 Consequences of HVL Releases FINAL REPORT Submitted by: Michael Baker Jr., Inc. December 31, 2002
1041 Department of Transportation Research and Special Programs Administration Office of Pipeline Safety TTO Number 13 Integrity Management Program Delivery Order DTRS56-02-D-70036 Potential Impact Radius Formulae for Flammable Gases Other Than Natural Gas Subject to 49 CFR 192 FINAL REPORT Submitted by: Michael Baker Jr., Inc. June 2005
1042 Brooklyn QRA
1043 WTIA/APIA WELDED PIPELINE SYMPOSIUM, Sydney, Australia. 3 April 2009, PIPELINE RISK ASSESSMENT: NEW GUIDELINES, by Phil Hopkins1, Penspen, UK. Graham Goodfellow2, Penspen, UK. Roger Ellis 2, Shell, UK. Jane Haswell2, PIE, UK. Neil Jackson2, National Grid, UK.
1044 2013 MEMORANDUM TO: SECRETARIAL OFFICERS. MODAL ADMINISTRATORS,

From: Polly Trottenberg, Under Secretary for Policy; Robert S. Rivkin,General Counsel;
Subject: Guidance on Treatment of the Economic Value of a Statistical Life in U.S. Department of Transportation Analyses

1045 **http://www.questconsult.com/hazard.html**

1046 EPA 100-B-00-002 December 2000U.S. Environmental Protection Agency; RISK CHARACTERIZATION HANDBOOK; Prepared for the U.S. Environmental Protection Agency by members of the Risk Characterization Implementation Core Team, a group of EPA's Science Policy Council; Principal Authors: John R. Fowle III, Ph.D. Science Advisory Board Office of the Administrator; Kerry L. Dearfield, Ph.D. Office of Science Policy Office of Research and Development

1047 Risk Assessment – Recommended Practices for Municipalities and Industry; Canadian Society for Chemical Engineering; CSChE Risk Assessment – Recommended Practices

1048 Guidance Protocol for School Site Pipeline Risk Analysis; The California Department of Education (CDE), School Facilities Planning Division (SFPD) has established standards for use by Local Educational Agencies (LEAs) (i.e., school districts, county offices of education and charter school entities) in the selection of safe and educationally appropriate school sites (authority per Education Code section 17251). These standards have been adopted by the State Board of Education in the California Code of Regulations Title 5, Section 14010 – Standards for School Site Selection

1049 **Pipeline101.com**

1050 Reliability-based Prevention of Mechanical Damage to Pipelines. PR-244-9729; Prepared for the Offshore/Onshore Design Applications Supervisory Committee Pipeline Research Committee of Pipeline Research Council International, Inc

1051 Proceedings of IPC2004, International Pipeline Conference, IPC04-0541; A ROBUST APPROACH TO PIPELINE INTEGRITY MANAGEMENT USING DIRECT ASSESSMENT BASED ON STRUCTURAL RELIABILITY ANALYSIS Marcus McCallum, Advantica, UK, Andrew Francis, Andrew Francis & Associates, UK, Tim Illson Advantica, UK, Mark McQueen, Advantica, US, Mike Scott, Crosstex Energy, US, Steve Clarke, Crosstex Energy, US

1052 International Electrotechnical Commission: INTERNATIONAL STANDARD IEC/FDIS 31010 Risk management — Risk assessment techniques Reference number IEC/FDIS 31010:2009(E)

Index

D

N

O

P

Q

R

CPSIA information can be obtained
at www.ICGtesting.com
Printed in the USA
LVOW05*1907031016
507236LV00006B/6/P

9 780990 670001